AMBRIDGE

ghter Thinking

A Level Further Mathematics for AQA

Student Book 1 (AS/Year 1)

Stephen Ward, Paul Fannon, Vesna Kadelburg and Ben Woolley

CAMBRIDGE
UNIVERSITY PRESS

Shaftesbury Road, Cambridge CB2 8EA, United Kingdom
One Liberty Plaza, 20th Floor, New York, NY 10006, USA
477 Williamstown Road, Port Melbourne, VIC 3207, Australia
314–321, 3rd Floor, Plot 3, Splendor Forum, Jasola District Centre, New Delhi – 110025, India
103 Penang Road, #05-06/07, Visioncrest Commercial, Singapore 238467

Cambridge University Press is part of the University of Cambridge.
It furthers the University's mission by disseminating knowledge in the pursuit of education, learning and research at the highest international levels of excellence.

www.cambridge.org
Information on this title:
www.cambridge.org/9781316644430 (Paperback)
www.cambridge.org/9781316644294 (Paperback with Cambridge with Digital Access edition)

First published 2017
20 19 18 17 16 15 14 13 12 11 10 9

Printed in Great Britain by Ashford Colour Press Ltd.

A catalogue record for this publication is available from the British Library

ISBN 978-1-316-64443-0 Paperback
ISBN 978-1-316-64429-4 Paperback with Cambridge with Digital Access edition

Additional resources for this publication at www.cambridge.org/education

Message from AQA

This textbook has been approved by AQA for use with our qualification. This means that we have checked that it broadly covers the specification and we are satisfied with the overall quality. Full details of our approval process can be found on our website.

We approve textbooks because we know how important it is for teachers and students to have the right resources to support their teaching and learning. However, the publisher is ultimately responsible for the editorial control and quality of this book.

Please note that when teaching the A/AS Level Further Mathematics (7366, 7367) course, you must refer to AQA's specification as your definitive source of information. While this book has been written to match the specification, it cannot provide complete coverage of every aspect of the course.

A wide range of other useful resources can be found on the relevant subject pages of our website: www.aqa.org.uk

IMPORTANT NOTE
AQA has not approved any Cambridge with Digital Access content

Contents

Introduction

You have probably been told that mathematics is very useful, yet it can often seem like a lot of techniques that just have to be learnt to answer examination questions. You are now getting to the point where you will start to see where some of these techniques can be applied in solving real problems. However as well as seeing how maths can be useful we hope that anyone working through this book will realise that it can also be incredibly frustrating, surprising and ultimately beautiful.

The book is woven around three key themes from the new curriculum.

Proof

Maths is valued because it trains you to think logically and communicate precisely. At a high level maths is far less concerned about answers and more about the clear communication of ideas. It is not about being neat – although that might help! It is about creating a coherent argument which other people can easily follow but find difficult to refute. Have you ever tried looking at your own work? If you cannot follow it yourself it is unlikely anybody else will be able to understand it. In maths we communicate using a variety of means – feel free to use combinations of diagrams, words and algebra to aid your argument. And once you have attempted a proof, try presenting it to your peers. Look critically (but positively) at some other people's attempts. It is only through having your own attempts evaluated and trying to find flaws in other proofs that you will develop sophisticated mathematical thinking. This is why we have included lots of common errors in our Work it out boxes – just in case your friends don't make any mistakes!

Problem solving

Maths is valued because it trains you to look at situations in unusual, creative ways, to persevere and to evaluate solutions along the way. We have been heavily influenced by a great mathematician and maths educator George Polya who believed that students were not just born with problem-solving skills – they were developed by seeing problems being solved and reflecting on their solutions before trying similar problems. You may not realise it but good mathematicians spend most of their time being stuck. You need to spend some time on problems you can't do, trying out different possibilities. If after a while you have not cracked it then look at the solution and try a similar problem. Don't be disheartened if you cannot get it immediately – in fact, the longer you spend puzzling over a problem the more you will learn from the solution. You may never need to integrate a rational function in the future, but we firmly believe that the problem-solving skills you will develop by trying it can be applied to many other situations.

Modelling

Maths is valued because it helps us solve real-world problems. However, maths describes ideal situations and the real world is messy! Modelling is about deciding on the important features needed to describe the essence of a situation and turning that into a mathematical form, then using it to make predictions, compare to reality and possibly improve the model. In many situations the technical maths is actually the easy part – especially with modern technology. Deciding which features of reality to include or ignore and anticipating the consequences of these decisions is the hard part. Yet it is amazing how some fairly drastic assumptions – such as pretending a car is a single point or that people's votes are independent – can result in models which are surprisingly accurate.

More than anything else this book is about making links – links between the different chapters, the topics covered and the themes above, links to other subjects and links to the real world. We hope that you will grow to see maths as one great complex but beautiful web of interlinking ideas.

Maths is about so much more than examinations, but we hope that if you take on board these ideas (and do plenty of practice!) you will find maths examinations a much more approachable and possibly even enjoyable experience. However always remember that the results of what you write down in a few hours by yourself in silence under exam conditions are not the only measure you should consider when judging your mathematical ability – it is only one variable in a much more complicated mathematical model!

How to use this book

Throughout this book you will notice particular features that are designed to aid your learning. This section provides a brief overview of these features.

In this chapter you will learn how to:

- factorise polynomials and solve equations which may have complex roots
- link between the roots of a polynomial and its coefficients
- use substitutions to solve more complicated equations.

Learning objectives
A short summary of the content that you will learn in each chapter.

Before you start...

GCSE	You should be able to solve linear inequalities.	1 Solve $5-2x>3$.
AS Level Mathematics Student Book 1, Chapter 3	You should be able to solve quadratic inequalities.	2 Solve $x^2+6>5x$.

Before you start
Points you should know from your previous learning and questions to check that you're ready to start the chapter.

WORKED EXAMPLE
The left-hand side shows you how to set out your working. The right-hand side explains the more difficult steps and helps you understand why a particular method was chosen.

Key point
A summary of the most important methods, facts and formulae.

PROOF
Step-by-step walkthroughs of standard proofs and methods of proof.

Common error
Specific mistakes that are often made. These typically appear next to the point in the Worked example where the error could occur.

WORK IT OUT
Can you identify the correct solution and find the mistakes in the two incorrect solutions?

Tip
Useful guidance, including ways of calculating or checking answers and using technology.

Each chapter ends with a **Checklist of learning and understanding** and a **Mixed practice exercise**, which includes **past paper questions** marked with the icon.

In between chapters, you will find extra sections that bring together topics in a more synoptic way.

FOCUS ON...
Unique sections relating to the preceding chapters that develop your skills in proof, problem-solving and modelling.

CROSS-TOPIC REVIEW EXERCISE
Questions covering topics from across the preceding chapters, testing your ability to apply what you have learned.

You will find **practice papers** towards the end of the book, as well as a **glossary** of key terms (picked out in colour within the chapters), and **answers** to all questions. Full **worked solutions** can be found on the Cambridge Elevate digital platform, along with a **digital version** of this Student Book.

Maths is all about making links, which is why throughout this book you will find signposts emphasising connections between different topics, applications and suggestions for further research.

Rewind

Reminders of where to find useful information from earlier in your study.

Focus on...

Links to problem-solving, modelling or proof exercises that relate to the topic currently being studied.

Fast forward

Links to topics that you may cover in greater detail later in your study.

Did you know?

Interesting or historical information and links with other subjects to improve your awareness about how mathematics contributes to society.

Colour coding of exercises

The questions in the exercises are designed to provide careful progression, ranging from basic fluency to practice questions. They are uniquely colour-coded, as shown here.

1. A sequence is defined by $u_n = 2 \times 3^{n-1}$. Use the principle of mathematical induction to prove that $u_1 + u_2 + \ldots + u_n = 3^n - 1$.
2. Show that $1^2 + 2^2 + \ldots + n^2 = \frac{n(n+1)(2n+1)}{6}$
3. Show that $1^3 + 2^3 + \ldots + n^3 = \frac{n^2(n+1)^2}{4}$
4. Prove by induction that $\frac{1}{1\times 2} + \frac{1}{2\times 3} + \frac{1}{3\times 4} + \ldots + \frac{1}{n(n+1)} = \frac{n}{n+1}$
5. Prove by induction that $\frac{1}{1\times 3} + \frac{1}{3\times 5} + \frac{1}{5\times 7} + \ldots + \frac{1}{(2n-1)\times(2n+1)} = \frac{n}{2n+1}$
6. Prove that $1 \times 1! + 2 \times + 3 \times 3! \ldots + n \times n! = (n+1)! - 1$
7. Use the principle of mathematical induction to show that $1^2 - 2^2 + 3^2 - 4^2 + \ldots + (-1)^{n-1} n^2 = (-1)^{n-1} \frac{n(n+1)}{2}$.
8. Prove that $(n+1) + (n+2) + (n+3) + \ldots + (2n) = \frac{1}{2} n(3n+1)$
9. Prove using induction that $\sin\theta + \sin 3\theta + \ldots + \sin(2n-1)\theta = \frac{\sin^2 n\theta}{\sin\theta}$, $n \in \mathbb{Z}^+$
10. Prove that $\sum_{k=1}^{n} k\,2^k = (n-1)2^{n+1} + 2$

Black – practice questions which come in several parts, each with subparts **i** and **ii**. You only need attempt subpart **i** at first; subpart **ii** is essentially the same question, which you can use for further practice if you got part **i** wrong, for homework, or when you revisit the exercise during revision.

Green – practice questions at a basic level.

Blue – practice questions at an intermediate level.

Red – practice questions at an advanced level.

Purple – challenging questions that apply the concept of the current chapter across other areas of maths.

Yellow – designed to encourage reflection and discussion.

A – indicates content that is for A Level students only.

AS – indicates content that is for AS Level students only.

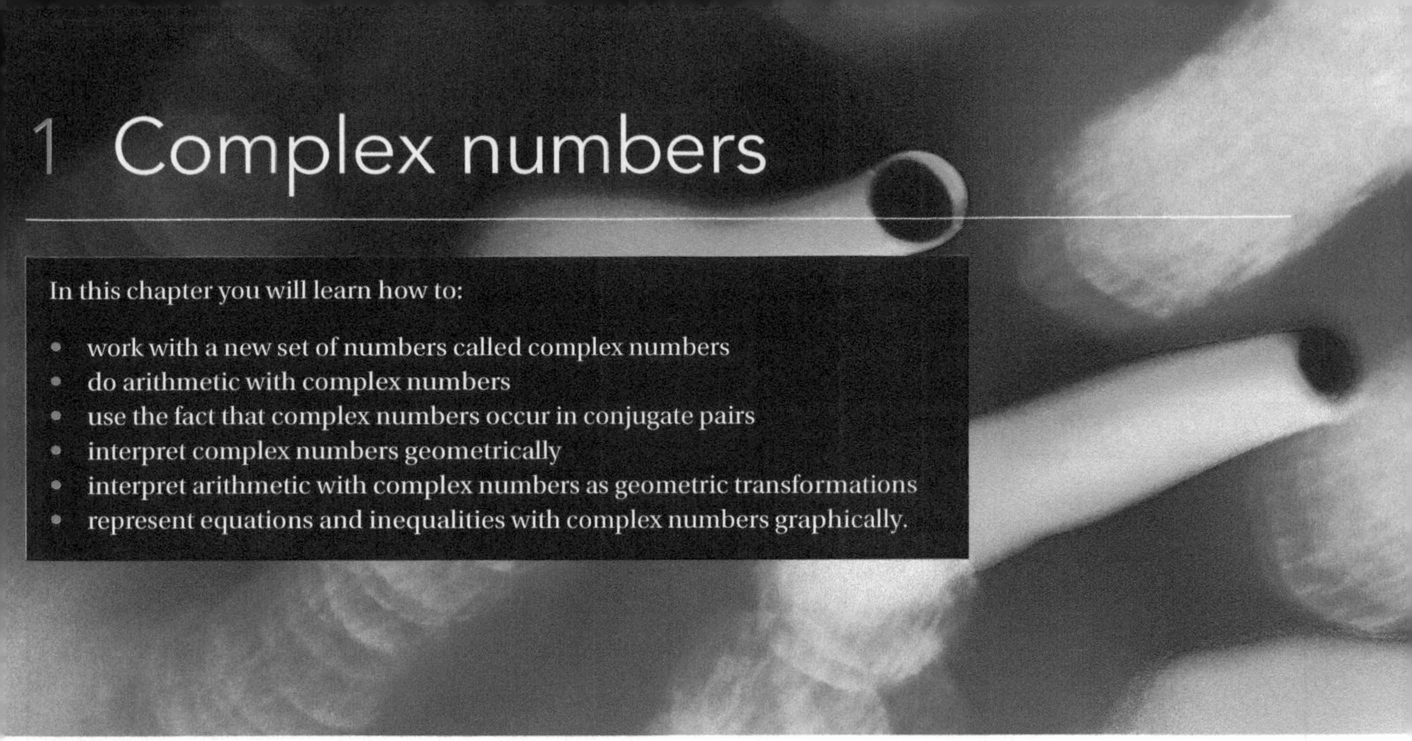

1 Complex numbers

In this chapter you will learn how to:

- work with a new set of numbers called complex numbers
- do arithmetic with complex numbers
- use the fact that complex numbers occur in conjugate pairs
- interpret complex numbers geometrically
- interpret arithmetic with complex numbers as geometric transformations
- represent equations and inequalities with complex numbers graphically.

Before you start...

GCSE	You should be able to use the quadratic formula.	1 Solve the equation $3x^2-6x+2=0$, giving your answers in simplified surd form.
GCSE	You should be able to represent a locus of points on a diagram.	2 Draw two points, P and Q, 5 cm apart. Shade the locus of all the points that are less than 3 cm from P and closer to Q than to P.
A Level Mathematics Student Book 1, Chapter 5	You should be able to solve simultaneous equations.	3 Solve these simultaneous equations. $\begin{cases} x+2y=5 \\ x^2+y^2=10 \end{cases}$

Finding the square roots of negative numbers

Until now, every number you have met has been a real number. You will have been told that it is not possible to find the square root of a negative number or equivalently, that no number squares to give a negative answer. However, in this chapter, you will learn that there is a number that squares to give −1, and so a whole new branch of numbers appears: complex numbers. Remarkably, these numbers have many useful applications in the real world, from electrical impedance in electronics to the Schrödinger equation in quantum mechanics!

Section 1: Definition and basic arithmetic of i

There is no real number that is a solution to the equation $x^2=-1$ but mathematicians have defined there to be an imaginary number that solves this equation. This imaginary number is given the symbol i.

Key point 1.1

$$i^2 = -1$$

or equivalently, $$i = \sqrt{-1}$$

In all other ways, i acts just like a normal constant.

WORKED EXAMPLE 1.1

Simplify these terms.

a i^3 **b** i^4 **c** i^5

a $i^3 = i^2 \times i$ — Use the normal laws of indices.

$= (-1)i$ — $i^2 = -1$.

$= -i$

b $i^4 = i^2 \times i^2 = (-1)^2$

$= 1$

c $i^5 = i^4 \times i = 1 \times i$

$= i$

You need to be familiar with some common terminology.

- A **complex number** is one that can be written in the form $x + iy$ where x and y are real. Commonly, z is used to denote an unknown complex number and $\mathbb{C}$ is used for the set of all complex numbers.
- In this definition, x is the **real part** of z and it is given the symbol $\text{Re}(z)$; y is the **imaginary part** of z and it is given the symbol $\text{Im}(z)$. So, for example, $\text{Re}(3 - i)$ is 3 and $\text{Im}(3 - i)$ is -1.

You can now do some arithmetic with complex numbers.

Tip

Your calculator may have a 'complex' mode, which you can use to check your calculations.

Common error

Remember that the imaginary part of the complex number $z = x + iy$ is the real number y, not iy.

WORKED EXAMPLE 1.2

Given that $z = 3 + i$ and $w = 5 - 2i$ find the value of each expression.

a $z + w$ **b** $z - w$ **c** zw **d** $w \div 2$

a $(3 + i) + (5 - 2i) = (3 + 5) + (i - 2i)$ — Group the real and the imaginary parts.

$= 8 - i$

b $(3 + i) - (5 - 2i) = (3 - 5) + (i + 2i)$

$= -2 + 3i$

Continues on next page

c $(3+i)\times(5-2i) = (3\times5)+(3\times-2i)+(i\times5)+(i\times-2i)$ Multiply out the brackets.

$= 15-6i+5i-2i^2$

$= 15-i-(2\times-1)$ Remember, $i^2 = -1$.

$= 17-i$

d $\frac{5-2i}{2} = \frac{5}{2} - \frac{2i}{2}$ Use the normal rules of fractions.

$= 2.5-i$

The original purpose of introducing complex numbers was to solve quadratic equations with negative **discriminants**.

Focus on...

See Focus on... Problem solving 1 to find out how some cubic equations can be solved by using a formula.

WORKED EXAMPLE 1.3

a Find the value of $\sqrt{-4}$.

b Hence solve the equation $x^2-4x+5=0$.

a $\sqrt{-4} = \sqrt{4}\sqrt{-1}$ Use the standard rules of square roots: $\sqrt{ab} = \sqrt{a}\sqrt{b}$.

$= 2i$ $i = \sqrt{-1}$.

b $x = \frac{-(-4)\pm\sqrt{16-4\times1\times5}}{2}$ Use the quadratic formula.

$= \frac{4\pm\sqrt{-4}}{2}$

$= \frac{4\pm2i}{2}$ Using part **a**.

$= 2\pm i$

Did you know

You may be sceptical about the idea of 'imagining' numbers. However, when negative numbers were introduced by the Indian mathematician Brahmagupta, in the 7th century, there was just as much scepticism. For example, in Europe, it took until the 17th century for negative numbers to be accepted. Mathematicians had worked successfully for thousands of years without using negative numbers, treating equations such as $x+3=0$ as having no solution. But once negative numbers were 'invented' it took only a hundred years or so to accept that equations such as $x^2+3=0$ also have solutions.

When you look at a new number system, one of the most important questions to ask is: 'When are two numbers equal?'

Key point 1.2

If two complex numbers are equal, then their real parts are the same and their imaginary parts are the same.

Although this may seem obvious, in mathematics it pays to be careful. For example, if two rational numbers $\frac{a}{b}$ and $\frac{c}{d}$ are equal, it does **not** mean that $a = c$ and $b = d$.

Despite its apparent simplicity, the result in Key point 1.2 has some remarkably powerful uses. One is to find square roots of complex numbers.

WORKED EXAMPLE 1.4

Solve the equation $z^2 = 8 - 6\text{i}$.

Let $z = x + \text{i}y$. — z is a complex number, so write $z = x + \text{i}y$, where x and y are real.

$$z^2 = 8 - 6\text{i}$$
$$(x+\text{i}y)^2 = 8 - 6\text{i}$$
$$x^2 + 2xy\text{i} + (\text{i}y)^2 = 8 - 6\text{i}$$
Expand the brackets.

$$x^2 + 2xy\text{i} - y^2 = 8 - 6\text{i}$$
$\text{i}^2 = -1$

Re: $x^2 - y^2 = 8$ (1)

Im: $2xy = -6$ (2)

Now equate the real parts and equate the imaginary parts.

From (2): $y = -\frac{3}{x}$ — Use substitution to solve these simultaneous equations.

Substituting into (1):

$$x^2 - \frac{9}{x^2} = 8$$
$$x^4 - 9 = 8x^2$$
$$x^4 - 8x^2 - 9 = 0$$
Multiplying through by x^2 leads to a disguised quadratic.

$$(x^2 - 9)(x^2 + 1) = 0$$
$$x^2 - 9 = 0 \quad \text{or} \quad x^2 + 1 = 0$$
$$x = \pm 3$$
There are two possible values for x^2. However, x is real so $x^2 + 1 = 0$ is impossible.

$$y = -\frac{3}{x}$$
$$= -1 \text{ or } 1$$
Find the value of y for each x value.

$\therefore z = 3 - \text{i}$ or $-3 + \text{i}$ — Write the answer in the form $z = x + \text{i}y$.

Notice that there are two numbers that have the same square; this is consistent with what you have already seen with real numbers.

EXERCISE 1A

1 State the imaginary part of each complex number.

a i $-3+5i$ ii $8-2i$ **b** i $6+i$ ii $19-i$

c i $2i-8$ ii $7i-2$ **d** i 15 ii $-3i$

e i i^2 ii $(1+i)-i$

f i $1+ai+b-i$ for $a, b \in \mathbb{R}$ ii $2-4i-(bi-a)$ for $a, b \in \mathbb{R}$

2 Simplify each expression, giving your answers in the form $x+iy$.

a i $2i+3i$ ii $i-9i$ **b** i $5i^2$ ii $-i^2$

c i $(-3i)^2$ ii $(4i)^2$ **d** i $(4i+3)-(6i-2)$ ii $2(2i-1)-3(4-2i)$

3 Simplify each expression, giving your answers in the form $x+iy$.

a i $i(1+i)$ ii $3i(2-5i)$ **b** i $(2+i)(1+2i)$ ii $(5+2i)(4+3i)$

c i $(2+3i)(1-2i)$ ii $(3+i)(5-i)$ **d** i $(3+i)^2$ ii $(4-3i)^2$

e i $\left(\frac{\sqrt{3}}{2}+\frac{1}{2}i\right)\left(\frac{\sqrt{3}}{2}-\frac{1}{2}i\right)$ ii $(3+2i)(3-2i)$

4 Simplify each expression, giving your answers in the form $x+iy$.

a i $\frac{6+8i}{2}$ ii $\frac{9-3i}{3}$ **b** i $\frac{5+2i}{10}$ ii $\frac{i-4}{8}$

c i $\frac{3+i}{2}+i$ ii $9i-\frac{6-4i}{2}$

5 Evaluate each expression, giving your answers in the form $x+iy$.

a i $\sqrt{-4}$ ii $\sqrt{-49}$ **b** i $\sqrt{-8}$ ii $\sqrt{-50}$

c i $\frac{4-\sqrt{-36}}{3}$ ii $\frac{-1+\sqrt{-25}}{3}$ **d** i $\frac{2+\sqrt{16-25}}{6}$ ii $\frac{5-2\sqrt{4-9}}{4}$

6 Solve the equations, simplifying your answers.

a i $x^2+9=0$ ii $x^2+36=0$ **b** i $x^2=-10$ ii $x^2=-13$

c i $x^2-2x+5=0$ ii $x^2-x+10=0$ **d** i $3x^2+20=6x$ ii $6x+5=-5x^2$

7 Evaluate, simplifying your answers.

a i i^3 ii i^4 **b** i $(-2i)^4$ ii $(-5i)^3$

c i $(1-\sqrt{3}i)^3$ ii $(\sqrt{3}+i)^3$ **d** i $\left(\frac{\sqrt{3}}{2}+\frac{1}{2}i\right)^3$ ii $\left(-\frac{\sqrt{3}}{2}+\frac{1}{2}i\right)^3$

8 Find real numbers a and b such that:

a i $(a+bi)(3-2i)=5i+1$ ii $(6+i)(a+bi)=2$

b i $(a+2i)(1+2i)=4-bi$ ii $(1+ai)(1+i)=b+2i$

c i $(a+bi)(2+i)=2a-(b-1)i$ ii $i(a+bi)=a-6i$

9 By writing $z=x+iy$, solve these equations.

a i $z^2=-4i$ ii $z^2=9i$ **b** i $z^2=2+2\sqrt{3}i$ ii $z^2=5+i$

10 Find the exact values of $a, b \in \mathbb{R}$ such that $(3+a\text{i})(b-\text{i}) = -4\text{i}$. Give your answers in the form $k\sqrt{3}$.

11 Find the exact values of $a, b \in \mathbb{R}$ such that $(1+a\text{i})(1+b\text{i}) = b+9\text{i}-a$.

12 By writing $z = x+\text{i}y$, solve the equation $\text{i}z+2 = \text{i}-3z$.

13 a Find values x and y such that $(x+\text{i}y)(2+\text{i}) = -\text{i}$.

b Hence express $-\frac{3\text{i}}{2+\text{i}}$ in the form $x+\text{i}y$.

14 By writing $z = x+\text{i}y$, solve the equation $z^2 = -3-4\text{i}$.

15 Solve the equation $z^2 = \text{i}$.

16 Use an algebraic method to solve the equation $z^2 = 12-5\text{i}$.

Section 2: Division and complex conjugates

In Section 1, you saw that quadratic equations may have two complex roots. In Worked example 1.3, the roots were $2+\text{i}$ and $2-\text{i}$; the imaginary part arises from the term after the $\pm$ sign in the quadratic formula. These two numbers form a **complex conjugate pair**. They differ only in the sign of the imaginary part.

Fast forward

In Chapter 2 you will use the fact that complex roots occur in conjugate pairs to factorise and solve cubic and quartic equations.

Key point 1.3

If $z = x+\text{i}y$, then the **complex conjugate** of z, $z^* = x-\text{i}y$

So for example, if $z = 2+\text{i}$ then $z^* = 2-\text{i}$, or if $w = 3-4\text{i}$ then $w^* = 3+4\text{i}$.

At first, the concept of conjugates may not appear particularly useful, but you will need them when you are dividing complex numbers.

WORKED EXAMPLE 1.5

Write $\frac{3+2\text{i}}{5-\text{i}}$ in the form $x+\text{i}y$.

$\frac{3+2\text{i}}{5-\text{i}} = \frac{(3+2\text{i})(5+\text{i})}{(5-\text{i})(5+\text{i})}$ — Multiply the numerator and the denominator by the complex conjugate of the denominator, $5+\text{i}$.

$= \frac{15+10\text{i}+3\text{i}+2\text{i}^2}{5^2-\text{i}^2}$ — Use the difference of two squares in the denominator.

$= \frac{13+13\text{i}}{25-(-1)}$ — $\text{i}^2 = -1$.

$= 0.5+0.5\text{i}$

Worked example 1.5 shows the general procedure for dividing by a complex number.

Key point 1.4

To divide by a complex number, write as a fraction and multiply the numerator and the denominator by the conjugate of the complex number in the denominator.

Tip

This procedure should remind you of rationalising the denominator when working with surds, which is also based on the difference of two squares.

You can prove that this procedure always results in a real number in the denominator.

PROOF 1

Prove that zz^* is always real.

Let $z = x + \text{i}y$, where x and y are real.

Then $z^* = x - \text{i}y$.

Writing z and z^* in terms of their real and imaginary parts is often the best way to prove results about complex conjugates.

So: $zz^* = (x + \text{i}y)(x - \text{i}y)$

$= x^2 - \text{i}xy + \text{i}yx - \text{i}^2y^2$

$= x^2 - (-y^2)$

$\text{i}^2 = -1$.

$= x^2 + y^2$

which is real.

x and y are real numbers.

Worked example 1.6 shows how to use the idea of equating real and imaginary parts to solve equations involving complex conjugates.

WORKED EXAMPLE 1.6

Find the complex number z such that $3z + 2z^* = 5 + 2\text{i}$.

Let $z = x + \text{i}y$.

As before, the best way to describe complex conjugates is in terms of their real and imaginary parts.

Then: $3z + 2z^* = 5 + 2\text{i}$

$3(x + \text{i}y) + 2(x - \text{i}y) = 5 + 2\text{i}$

$5x + \text{i}y = 5 + 2\text{i}$

$\text{Re}(z)$: $5x = 5$

$\text{Im}(z)$: $y = 2$

You can now equate real and imaginary parts.

$x = 1$, $y = 2$

So $z = 1 + 2\text{i}$

Being able to divide complex numbers means that you can solve more complicated equations.

WORKED EXAMPLE 1.7

Solve these simultaneous equations.

$$\begin{cases}(1+\mathrm{i})z+(2-\mathrm{i})w=3+4\mathrm{i}\\ \mathrm{i}z+(3+\mathrm{i})w=-1+5\mathrm{i}\end{cases}$$

$$\begin{cases}\mathrm{i}(1+\mathrm{i})z+\mathrm{i}(2-\mathrm{i})w=\mathrm{i}(3+4\mathrm{i})\\ (1+\mathrm{i})\mathrm{i}z+(1+\mathrm{i})(3+\mathrm{i})w=(1+\mathrm{i})(-1+5\mathrm{i})\end{cases}$$

The best method to use here is elimination: multiply the first equation by i and the second by $(1+\mathrm{i})$ ...

$$\begin{cases}(1+\mathrm{i})\mathrm{i}z+(2\mathrm{i}+1)w=-4+3\mathrm{i}\\ (1+\mathrm{i})\mathrm{i}z+(2+4\mathrm{i})w=-6+4\mathrm{i}\end{cases}$$

Subtracting:

$(-1-2\mathrm{i})w=2-\mathrm{i}$

... and then subtract them to eliminate z.

$$w=\frac{2-\mathrm{i}}{-1-2\mathrm{i}}\times\frac{-1+2\mathrm{i}}{-1+2\mathrm{i}}$$

$$=\frac{5\mathrm{i}}{1+4}=\mathrm{i}$$

Divide by $(-1-2\mathrm{i})$: multiply top and bottom by the complex conjugate.

Substituting into the second equation:

$\mathrm{i}z+(3+\mathrm{i})w=-1+5\mathrm{i}$

Substitute back to find z: use the second equation as it looks simpler.

$\mathrm{i}z+(3+\mathrm{i})(\mathrm{i})=-1+5\mathrm{i}$

$\mathrm{i}z+(-1+3\mathrm{i})=-1+5\mathrm{i}$

$\mathrm{i}z=2\mathrm{i}$

$z=2$

So $z=2, w=\mathrm{i}$.

EXERCISE 1B

1 Find the complex conjugate of each number.

a **i** $2-3\mathrm{i}$ **ii** $4+4\mathrm{i}$ **b** **i** $\mathrm{i}-3$ **ii** $3\mathrm{i}+2$

c **i** $3\mathrm{i}$ **ii** $-\mathrm{i}$ **d** **i** -45 **ii** 9

2 Write in the form $x+\mathrm{i}y$. Use a calculator to check your answers.

a **i** $\frac{3-2\mathrm{i}}{1+2\mathrm{i}}$ **ii** $\frac{4\mathrm{i}}{3-5\mathrm{i}}$ **b** **i** $\frac{4}{\mathrm{i}}$ **ii** $-\frac{1}{\mathrm{i}}$

c **i** $\frac{4+\mathrm{i}}{4-\mathrm{i}}$ **ii** $\frac{2\mathrm{i}+1}{2\mathrm{i}-1}$ **d** **i** $\frac{(1+\mathrm{i})^2}{1-\mathrm{i}}$ **ii** $\frac{(\mathrm{i}-2)^2}{\mathrm{i}+2}$

3 Solve these equations.

a $2z-3=4-3(\mathrm{i}+z)$ **b** $2\mathrm{i}z+1=4\mathrm{i}(z-3)$

4 Solve these simultaneous equations.

a $2z-3\mathrm{i}w=5$, $(1+\mathrm{i})z+3w=-4\mathrm{i}$

b $(1+\mathrm{i})z+(1-\mathrm{i})w=1$, $(1-\mathrm{i})z+2\mathrm{i}w=\mathrm{i}$

5 Find the complex number, z, if:

a $2z^* - 1 = 4i$ **b** $3z^* + 2 = 9i$

6 By writing $z = x + iy$, solve these equations.

a $z + 2z^* = 2 - 7i$ **b** $2z + iz^* = -3 - i$

7 If x and y are real numbers find the complex conjugate z^* when:

a **i** $z = 3 + (x + iy)$ **ii** $z = x - (2 - iy)$ **b** **i** $z = (x + 3iy) + (2 - i)$ **ii** $z = (3 + 3i) - (x - iy)$

c **i** $x + iy + \frac{1}{x + iy}$ **ii** $x + iy - \frac{1}{x + iy}$ **d** **i** $\frac{x}{x + iy} - \frac{x}{x - iy}$ **ii** $\frac{x}{x + iy} + \frac{x}{x - iy}$

8 Write $\frac{3 - 5i}{2i - 1}$ in the form $a + bi$. Show all your working.

9 **a** Let $z = x + iy$. Find, in terms of x and y, the real and imaginary parts of $3iz + 2z^*$.

b Find the complex number z such that $3iz + 2z^* = 4 - 4i$.

10 Find real numbers x and y such that $x + 3iy = z + 4iz^*$, where $z = 2 + i$.

11 By writing $z = x + iy$, prove that $(z^*)^2 = (z^2)^*$.

12 Solve $z + 3z^* = i$.

13 Solve $z + i = 1 - z^*$.

14 If $z = x + iy$, find the real and the imaginary parts of $\frac{z}{z + 1}$ in terms of x and y, simplifying your answers as far as possible.

Section 3: Geometric representation

You can represent real numbers on a number line.

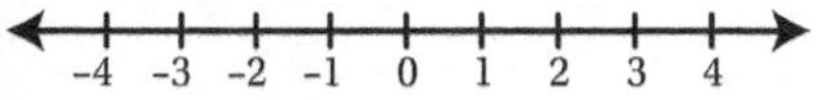

To represent complex numbers you can add another axis, perpendicular to the real number line, to show the imaginary part. This is called an **Argand diagram**.

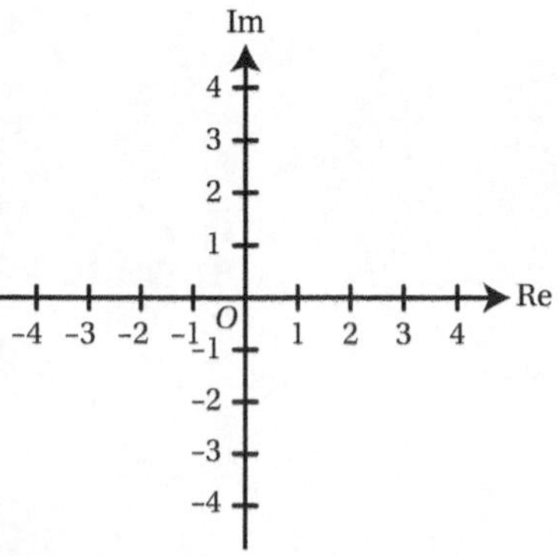

Did you know

Using real and imaginary axes was first suggested by the land surveyor Caspar Wessel. However, it is named after Jean-Robert Argand who popularised the idea.

WORKED EXAMPLE 1.8

Represent $3-2\text{i}$ on an Argand diagram.

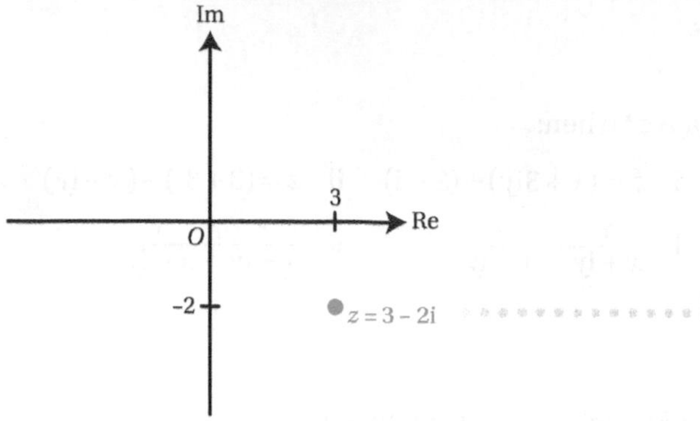

The point representing $3-2\text{i}$ has coordinates $(3, -2)$.

Operations with complex numbers have useful representations on Argand diagrams.

Consider two numbers, $z_1 = x_1 + \text{i}y_1$ and $z_2 = x_2 + \text{i}y_2$. On the diagram they are represented by points with coordinates (x_1, y_1) and (x_2, y_2). Their sum, $z_1 + z_2 = (x_1 + x_2) + \text{i}(y_1 + y_2)$, has coordinates $(x_1 + x_2, y_1 + y_2)$. But you can also think of coordinates as position vectors:

$$\begin{pmatrix} x_1 \\ y_1 \end{pmatrix} + \begin{pmatrix} x_2 \\ y_2 \end{pmatrix} = \begin{pmatrix} x_1 + x_2 \\ y_1 + y_2 \end{pmatrix}$$

So you can add complex numbers geometrically in the same way as you add vectors.

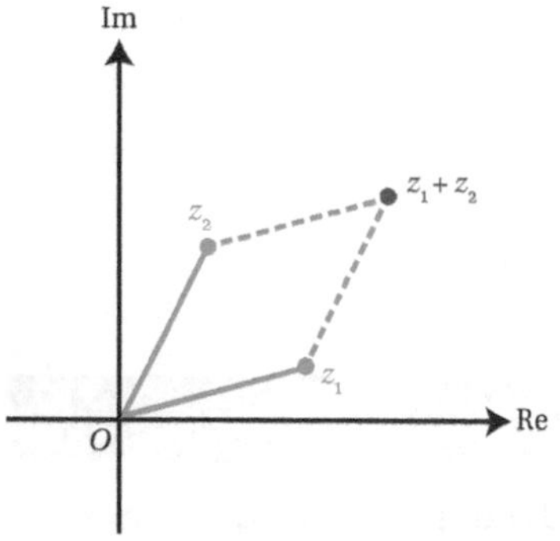

You can represent subtraction similarly, by adding the negative of the second number.

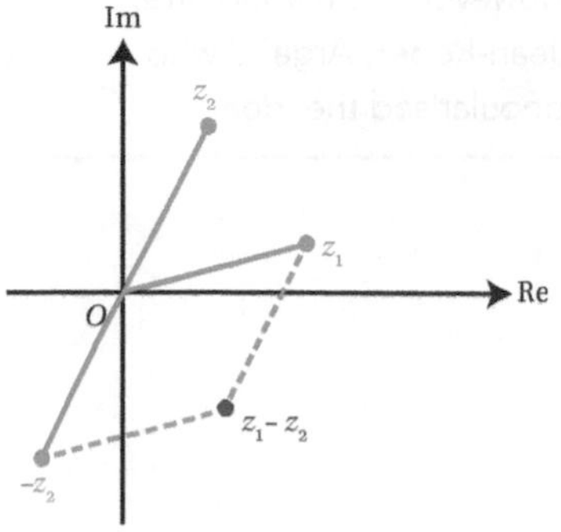

Taking a complex conjugate results in a reflection in the real axis.

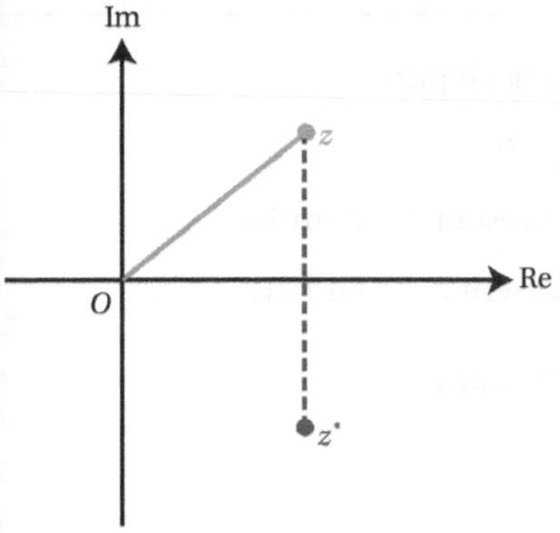

WORKED EXAMPLE 1.9

Two complex numbers, z and w, are shown on this Argand diagram.

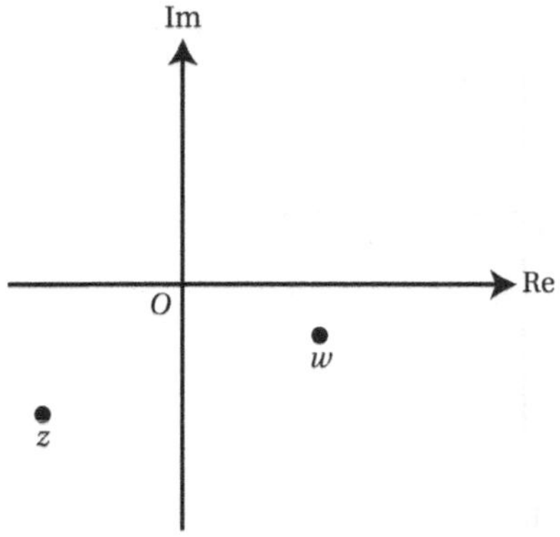

Represent these complex numbers on the same diagram.

a $w - z$ **b** z^*

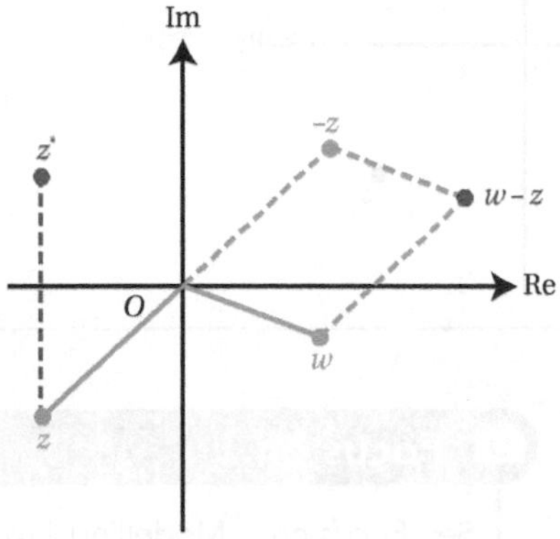

a To find $w - z$, first label $-z$ and then draw a parallelogram.

b z^* is the reflection of z in the real axis.

EXERCISE 1C

1 Represent each number on an Argand diagram. Use a separate diagram for each part.

a **i** $z = 4 + \mathrm{i}$ and $w = 2\mathrm{i}$ **ii** $z = -3 + \mathrm{i}$ and $w = -3\mathrm{i}$

b **i** Let $z = 4 + 3\mathrm{i}$. Represent z, $-z$ and $\mathrm{i}z$. **ii** Let $z = -2 + 5\mathrm{i}$. Represent z, $-z$ and $\mathrm{i}z$.

c **i** Let $z = 2 + \mathrm{i}$. Represent z, $3z$, $-2z$ and $-\mathrm{i}z$. **ii** Let $z = 1 - 3\mathrm{i}$. Represent z, $2z$ and $2\mathrm{i}z$.

2 Solve each quadratic equation and represent the solutions on an Argand diagram.

a **i** $z^2 - 4z + 13 = 0$ **ii** $z^2 - 10z + 26 = 0$

b **i** $z^2 + 2z + 17 = 0$ **ii** $z^2 + 6z + 13 = 0$

3 Each diagram shows two complex numbers, z and w. Copy the diagrams and add the points corresponding to z^*, $-w$ and $z + w$.

a **i**

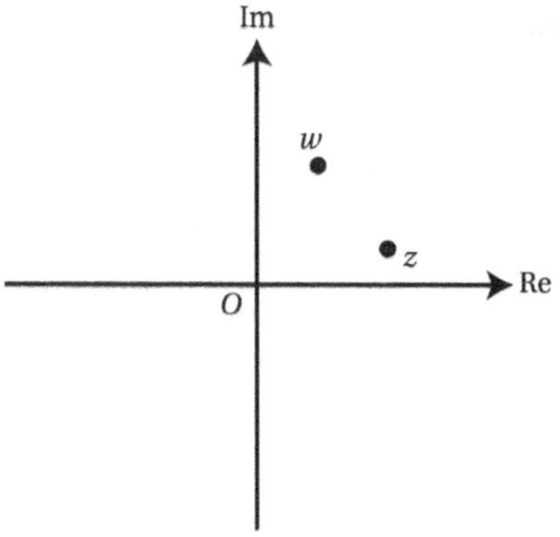

ii

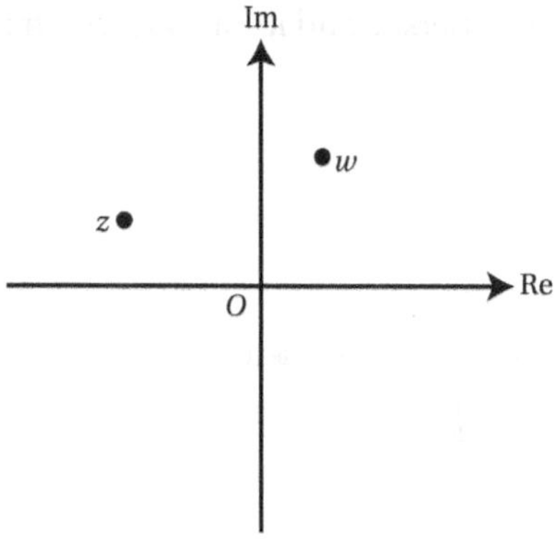

b **i**

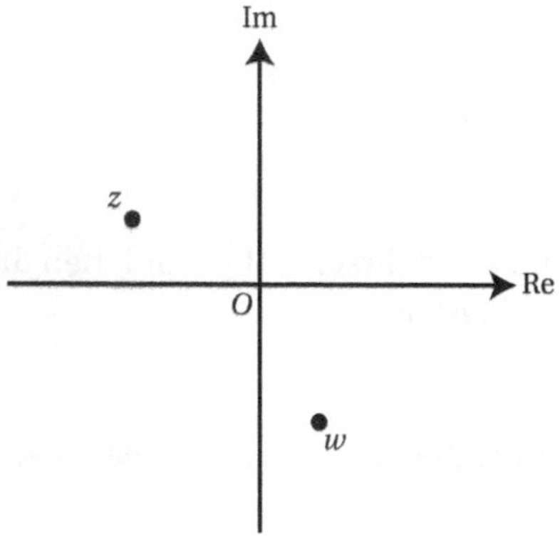

ii

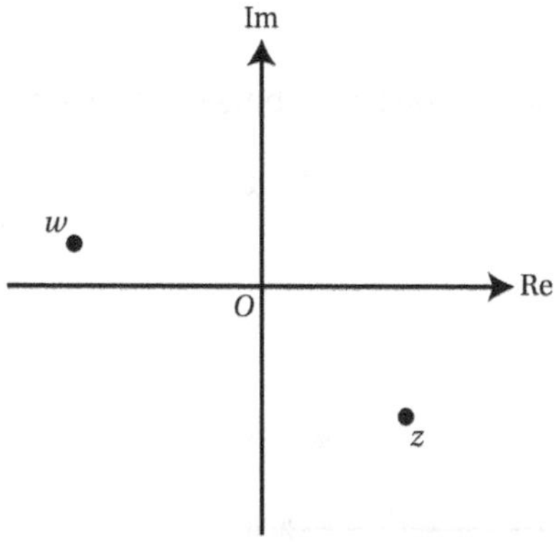

Modulus and argument

When a complex number is represented on an Argand diagram, its distance from the origin is called the **modulus**. There are lots of complex numbers with the same modulus; they form a circle centred at the origin. You can uniquely describe a particular complex number by giving its angle relative to the real axis; this is called the **argument**. The argument is conventionally measured anti-clockwise from the positive real axis.

Focus on...

See Focus on... Modelling 1 to find out how complex numbers can be used in electronics.

The symbols used for these are:

- $|z|$ or r for the modulus
- $\arg z$ or θ for the argument.

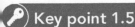

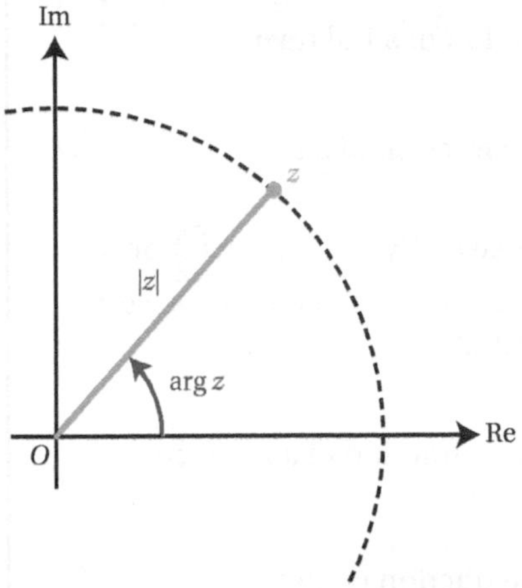

The notation $z = [r, \theta]$ is sometimes used for a complex number z with modulus r and argument θ.

This description of a complex number is called the **modulus–argument form.**

The description in terms of the real and imaginary parts, $z = x + \mathrm{i}y$, is called the **Cartesian form**.

Note here that the argument is not usually stated in degrees but in an alternative measure of angle called **radians**.

Radians

The **radian** is the most commonly used unit of angle in advanced mathematics.

Key point 1.5

$$360° = 2\pi \text{ radians}$$

You can deduce the sizes of other common angles: for example, a right angle is one quarter of a full turn, so it is $2\pi \div 4 = \frac{\pi}{2}$ radians.

Although sizes of common angles measured in radians are often expressed as fractions of π, you can also use decimal approximations: for example, a right angle measures approximately 1.57 radians.

Rewind

Radians are covered in more detail in A Level Mathematics Student Book 2, Chapter 7.

WORKED EXAMPLE 1.10

a Convert 75° to radians. **b** Convert 2.5 radians to degrees.

a $\frac{75}{360} = \frac{5}{24}$ — 75° is $\frac{75}{360}$ as a fraction of a full turn.

$\frac{5}{24} \times 2\pi = \frac{5\pi}{12}$ — Calculate the same fraction of 2π.

$\therefore 75° = \frac{5\pi}{12}$ radians — This is the exact answer. By evaluating $\frac{5\pi}{12}$ on a calculator you can also give a decimal answer: 75°= 1.31 radians (3 s.f.).

b $\frac{2.5}{2\pi}$ — 2.5 radians is $\frac{2.5}{2\pi}$ as a fraction of a full turn.

$\frac{2.5}{2\pi} \times 360 = 143.24\ldots$ — Calculate the same fraction of 360°.

$\therefore$ 2.5 radians = 143° (3 s.f.)

Tip

It is worth remembering some common angles in radians.

Degrees	0°	30°	45°	60°	90°	180°
Radians	0	$\frac{\pi}{6}$	$\frac{\pi}{4}$	$\frac{\pi}{3}$	$\frac{\pi}{2}$	π

Tip

When you need to evaluate trigonometric functions of angles in radians, make sure that your calculator is set to radian mode.

EXERCISE 1D

1 Express each angle in radians, giving your answer in terms of π.

a i 135° **ii** 45° **b i** 90° **ii** 270°

c i 120° **ii** 150° **d i** 50° **ii** 80°

2 Express each angle in radians, correct to 3 decimal places.

a i 320° **ii** 20° **b i** 270° **ii** 90°

c i 65° **ii** 145° **d i** 100° **ii** 83°

3 Express each angle in degrees.

a i $\frac{\pi}{3}$ **ii** $\frac{\pi}{4}$ **b i** $\frac{5\pi}{6}$ **ii** $\frac{2\pi}{3}$

c i $\frac{3\pi}{2}$ **ii** $\frac{5\pi}{3}$ **d i** 1.22 **ii** 4.63

Converting between the modulus–argument and Cartesian forms

To convert between the modulus-argument and Cartesian forms you can use a diagram and some trigonometry.

Key point 1.6

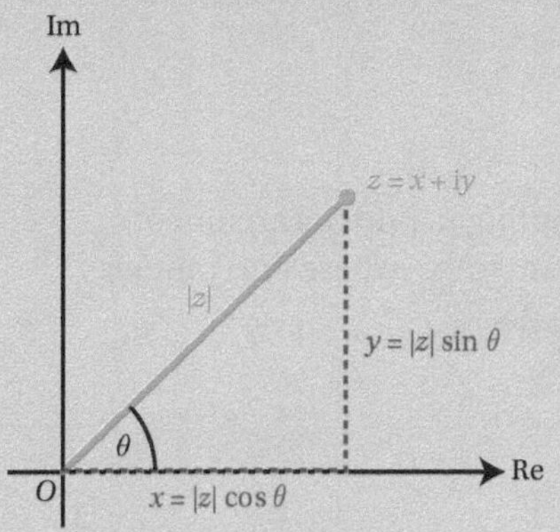

To convert from modulus-argument form to Cartesian form:

$$x = |z|\cos\theta$$

$$y = |z|\sin\theta$$

To convert from Cartesian to modulus-argument form:

$$|z| = \sqrt{x^2 + y^2}$$

$$\tan\theta = \frac{y}{x}$$

Tip

Always draw the complex number on an Argand diagram before finding the argument. It is important to know which angle you need to find.

The argument can be measured either between 0 and 2π or between $-\pi$ and π. It will be made clear in the question which is required.

WORKED EXAMPLE 1.11

A complex number z has modulus 3 and argument $\frac{\pi}{6}$. Write the number in Cartesian form.

$x = 3\cos\frac{\pi}{6} = \frac{3\sqrt{3}}{2}$

$y = 3\sin\frac{\pi}{6} = \frac{3}{2}$

Use $x = |z|\cos\theta$ and $y = |z|\sin\theta$.

$\therefore z = \frac{3\sqrt{3}}{2} + \frac{3}{2}\text{i}$

The Cartesian form is $z = x + \text{i}y$.

WORKED EXAMPLE 1.12

Write 2 – 3i in modulus-argument form, with the argument between 0 and 2π.

Give your answer correct to 3 s.f.

$|z| = \sqrt{2^2 + (-3)^2}$
$= \sqrt{13}$
$= 3.61$ (3 s.f.)

Use $|z| = \sqrt{x^2 + y^2}$.

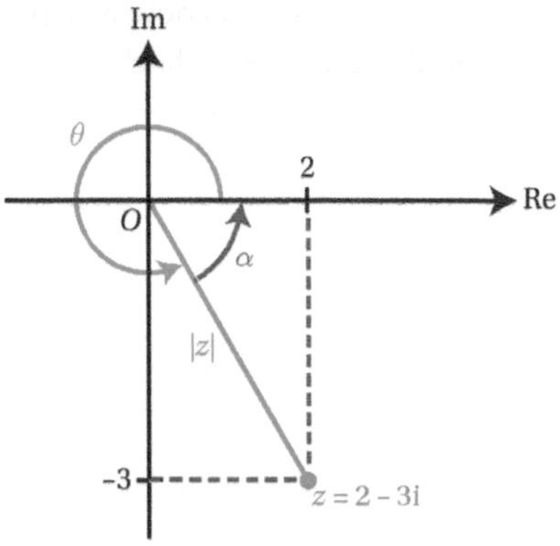

Before attempting to find the argument, draw a diagram to see where the complex number actually is.

$\tan \alpha = \frac{3}{2}$
$\therefore \alpha = 0.983$

Calculate the angle α. Remember to use radian mode on your calculator.

$\theta = 2\pi - 0.983$
$= 5.30$ (3 s.f.)

From the diagram, $\theta = 2\pi - \alpha$.

So $|z| = 3.61$ and $\arg z = 5.30$ (3 s.f.)

WORK IT OUT 1.1

Find the argument of $z = -5 - 3i$, where $-\pi < \arg z \leqslant \pi$.

Which is the correct solution? Identify the errors made in the incorrect solutions.

Solution 1	Solution 2	Solution 3
$\arctan\left(\frac{3}{5}\right) = 0.540$	$\arg z = -\arctan\left(\frac{3}{5}\right)$	$\arctan\left(\frac{-3}{-5}\right) = 0.540$
$\pi - 0.540 = 2.60$	$= -0.540$	$\arg z = 2\pi - 0.540$
$\therefore \arg z = -2.60$		$= 5.74$

There is another nice way of expressing complex numbers which, although strictly in Cartesian form, explicitly shows the modulus and argument.

$$\begin{aligned} z &= x + \mathrm{i}y \\ &= |z|\cos\theta + \mathrm{i}|z|\sin\theta \\ &= |z|(\cos\theta + \mathrm{i}\sin\theta) \end{aligned}$$

Key point 1.7

A complex number z with modulus r and argument θ can be written as

$$z = r(\cos\theta + \mathrm{i}\sin\theta)$$

Fast forward

In Further Mathematics Student Book 2 you'll see another way of writing complex numbers that involves the modulus and argument.

When writing a number in this form, it is important to notice that there must be a plus sign between the two terms and the modulus must be a positive number; otherwise you need to draw a diagram to find the argument.

WORKED EXAMPLE 1.13

Write $z = -2\left(\cos\frac{\pi}{3} - \mathrm{i}\sin\frac{\pi}{3}\right)$ in the form $r(\cos\theta + \mathrm{i}\sin\theta)$, with the argument between $-\pi$ and π.

$z = -2\cos\frac{\pi}{3} + 2\mathrm{i}\sin\frac{\pi}{3}$

In order to draw the diagram, you need to identify the signs of the real and imaginary parts.

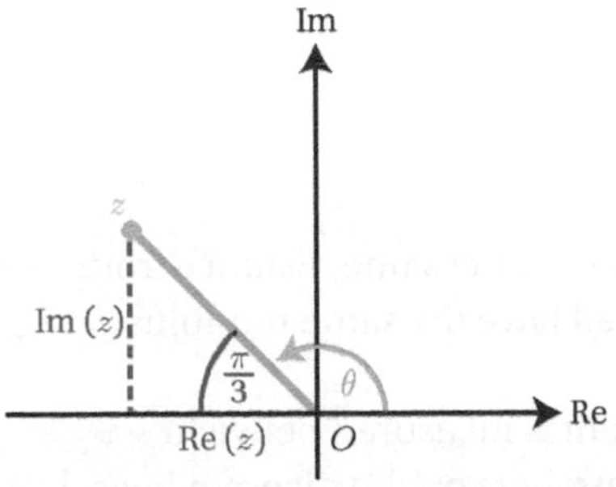

$\mathrm{Re}(z) = -2\cos\frac{\pi}{3} < 0$

$\mathrm{Im}(z) = 2\sin\frac{\pi}{3} > 0$

From the diagram:

$|z| = 2, \arg z = \theta = \frac{2\pi}{3}$

$\therefore z = 2\left(\cos\frac{2\pi}{3} + \mathrm{i}\sin\frac{2\pi}{3}\right)$

$\theta = \pi - \frac{\pi}{3} = \frac{2\pi}{3}$

An Argand diagram can help to identify the modulus and argument of a negative and complex conjugate of a number.

WORKED EXAMPLE 1.14

A complex number $z = r(\cos\theta + \mathrm{i}\sin\theta)$ is shown in the diagram. The argument is measured between $-\pi$ and π.

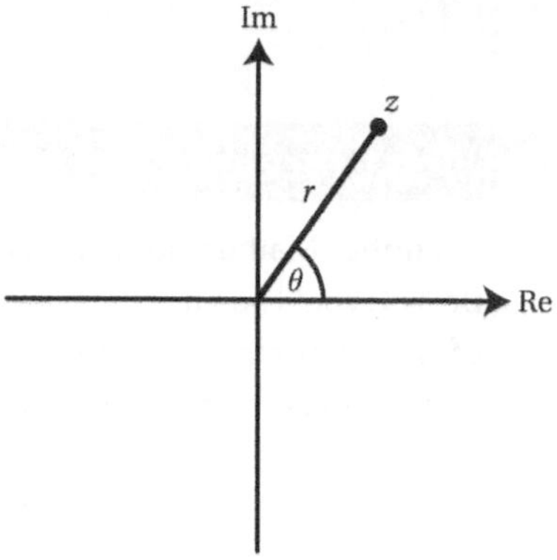

On the same diagram, mark the numbers $-z$ and z^*. Hence write $-z$ and z^* in modulus-argument form.

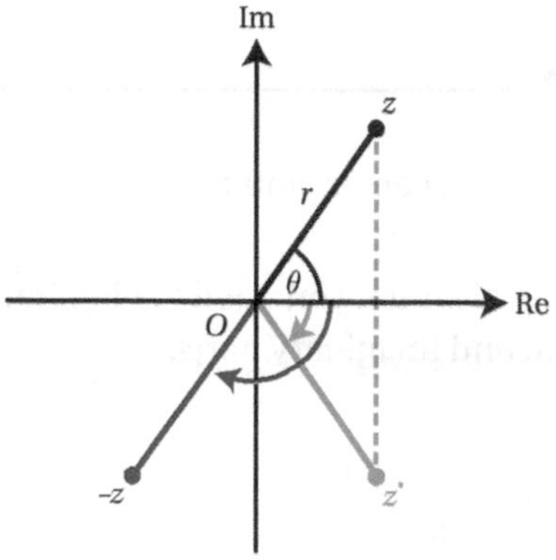

To find $-z$, make both real and imaginary parts negative.

z^* is the reflection of z in the real axis.

$|-z| = |z^*| = r$

All three points are the same distance from the origin, so all have the same modulus.

$$\arg(-z) = -(\pi - \theta)$$
$$= \theta - \pi$$
$$\arg(z^*) = -\theta$$

As the argument is measured between $-\pi$ and π, both numbers on the diagram have a negative argument.

$$\therefore -z = r(\cos(\theta - \pi) + \mathrm{i}\sin(\theta - \pi))$$
$$z^* = r(\cos(-\theta) + \mathrm{i}\sin(-\theta))$$

Write both numbers in modulus-argument form.

EXERCISE 1E

1. Find the modulus and the argument (measured between $-\pi$ and π) for each number.

a **i** 6	**ii** 13	**b** **i** -3	**ii** -1.6
c **i** $4\mathrm{i}$	**ii** $0.5\mathrm{i}$	**d** **i** $-2\mathrm{i}$	**ii** $-5\mathrm{i}$
e **i** $1+\mathrm{i}$	**ii** $2+\sqrt{3}\mathrm{i}$	**f** **i** $-1-\sqrt{3}\mathrm{i}$	**ii** $4-4\mathrm{i}$

2 Find the modulus and the argument (measured between 0 and 2π) for each number.

a **i** $4+2i$ **ii** $4-3i$

b **i** $i-\sqrt{3}$ **ii** $6i+\sqrt{2}$

c **i** $-3-i$ **ii** $-3+2i$

3 Given the modulus and the argument of z, write z in Cartesian form.

a **i** $|z|=4$, $\arg z=\frac{\pi}{3}$ **ii** $|z|=\sqrt{2}$, $\arg z=\frac{\pi}{4}$

b **i** $|z|=2$, $\arg z=\frac{3\pi}{4}$ **ii** $|z|=2$, $\arg z=\frac{2\pi}{3}$

c **i** $|z|=3$, $\arg z=-\frac{\pi}{2}$ **ii** $|z|=4$, $\arg z=-\pi$

4 Write each complex number in Cartesian form without using trigonometric functions. Display each one on an Argand diagram.

a **i** $\cos\frac{\pi}{3}+i\sin\frac{\pi}{3}$ **ii** $\cos\frac{3\pi}{4}+i\sin\frac{3\pi}{4}$

b **i** $3\left(\cos\frac{\pi}{2}+i\sin\frac{\pi}{2}\right)$ **ii** $5\left(\cos\left(-\frac{\pi}{2}\right)+i\sin\left(-\frac{\pi}{2}\right)\right)$

c **i** $4(\cos 0+i\sin 0)$ **ii** $\cos\pi+i\sin\pi$

5 Write each complex number in the form $|z|(\cos\theta+i\sin\theta)$.

a **i** $z=4i$ **ii** $z=-5$

b **i** $z=2-2\sqrt{3}i$ **ii** $z=\frac{\sqrt{3}+i}{3}$

6 Find the modulus and the argument of each number.

a **i** $4\left(\cos\frac{\pi}{3}+i\sin\frac{\pi}{3}\right)$ **ii** $\sqrt{7}\left(\cos\frac{3\pi}{7}+i\sin\frac{3\pi}{7}\right)$

b **i** $\cos\frac{\pi}{5}+i\sin\frac{\pi}{5}$ **ii** $\cos\left(-\frac{\pi}{4}\right)+i\sin\left(-\frac{\pi}{4}\right)$

c **i** $3\left(\cos\frac{\pi}{8}-i\sin\frac{\pi}{8}\right)$ **ii** $7\left(\cos\frac{4\pi}{5}-i\sin\frac{4\pi}{5}\right)$

d **i** $-10\left(\cos\frac{\pi}{3}+i\sin\frac{\pi}{3}\right)$ **ii** $-2\left(\cos\frac{\pi}{6}+i\sin\frac{\pi}{6}\right)$

e **i** $6\left(\cos\left(-\frac{\pi}{10}\right)+i\sin\left(-\frac{\pi}{10}\right)\right)$ **ii** $\frac{1}{2}\left(\cos\left(-\frac{\pi}{3}\right)-i\sin\left(-\frac{\pi}{3}\right)\right)$

7 **a** Write $3\left(\cos\frac{7\pi}{4}+i\sin\frac{7\pi}{4}\right)$ in Cartesian form in terms of surds only.

b Write $4i-4$ in the form $|z|(\cos\theta+i\sin\theta)$.

8 Let $z=1+\sqrt{3}i$ and $w=3\sqrt{3}-3i$.

a Find the modulus and the argument of z and w.

b Represent z and w on the same Argand diagram.

c Find the modulus and the argument of zw. Comment on your answer.

9 If $z = r(\cos\theta + \mathrm{i}\sin\theta)$, write these numbers in terms of r and θ, simplifying your answers as far as possible.

a $z + z^*$ **b** zz^* **c** $\frac{z}{z^*}$

10 If $z = \cos\theta + \mathrm{i}\sin\theta$, express the real and imaginary parts of $\frac{z-1}{z+1}$ in terms of θ, simplifying your answer as far as possible.

Rewind

Part **c** requires a double angle formula from A Level Mathematics Student Book 2, Chapter 8.

Section 4: Locus in the complex plane

You have met various examples of representing equations and inequalities graphically. You can represent an equation involving two variables, such as $x - 2y = 5$, on a graph, show an inequality such as $2 < x \leqslant 5$ on the number line and shade a region corresponding to an inequality in two variables, such as $y \geqslant 3x^2$, on a graph.

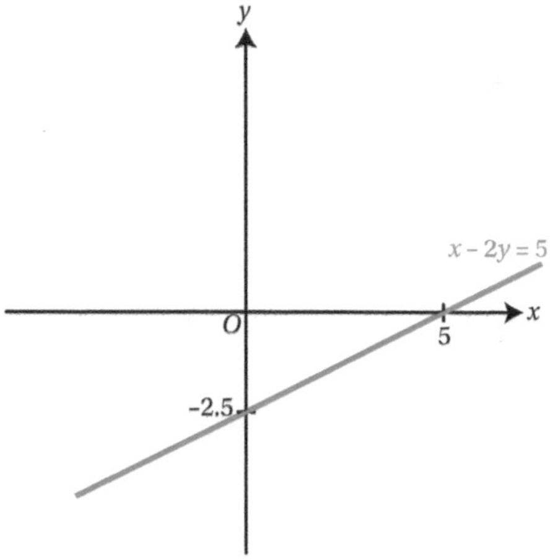

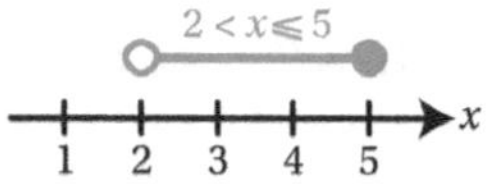

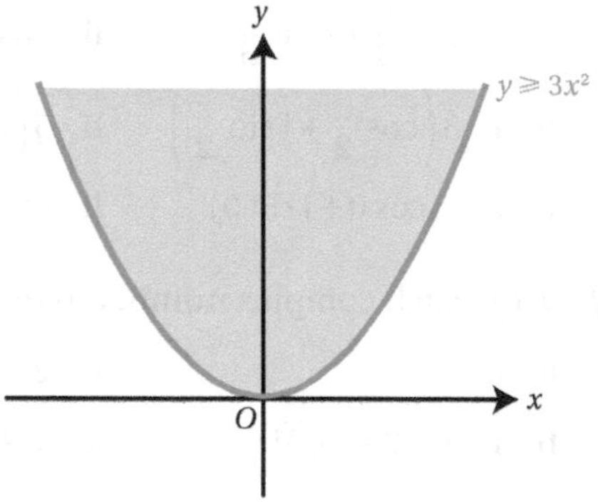

Now that you can represent complex numbers as points on an Argand diagram, there is a way to show equations and inequalities involving complex numbers graphically.

Locus involving the modulus

In Section 3, you defined the modulus of a complex number, z, as its distance from the origin, O, on an Argand diagram. This means that all complex numbers with the same modulus, r, form a circle around the origin, with radius r.

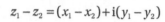

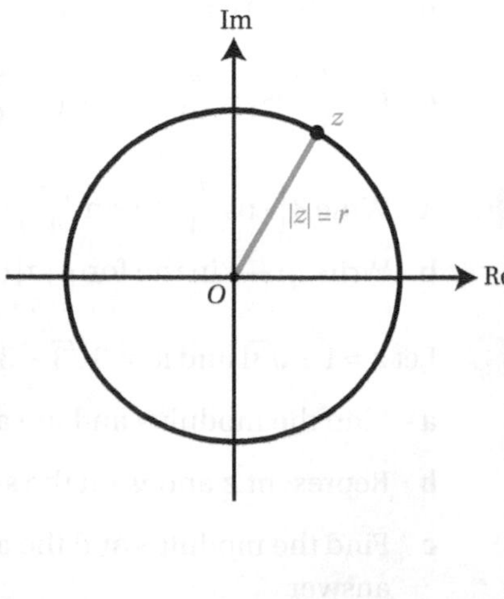

You can extend this idea to measure the distance between any two complex numbers on an Argand diagram.

If you write two numbers in Cartesian form, $z_1 = x_1 + \mathrm{i}y_1$ and $z_2 = x_2 + \mathrm{i}y_2$, then:

$$z_1 - z_2 = (x_1 - x_2) + \mathrm{i}(y_1 - y_2)$$

so:

$$|z_1 - z_2| = \sqrt{(x_1 - x_2)^2 + (y_1 - y_2)^2}$$

Notice that the last expression is the distance between the points with coordinates (x_1, y_1) and (x_2, y_2).

Key point 1.8

The distance between points representing complex numbers z_1 and z_2 on the Argand diagram is given by $|z_1 - z_2|$.

You can now find complex numbers that satisfy equations or inequalities involving the modulus by thinking about the geometric interpretation, rather than by doing calculations.

WORKED EXAMPLE 1.15

Show on an Argand diagram a set of points satisfying the equation $|z-2|=3$.

The points will lie on a circle, centre 2, radius 3.

The equation says 'the distance between z and 2 is 3.'

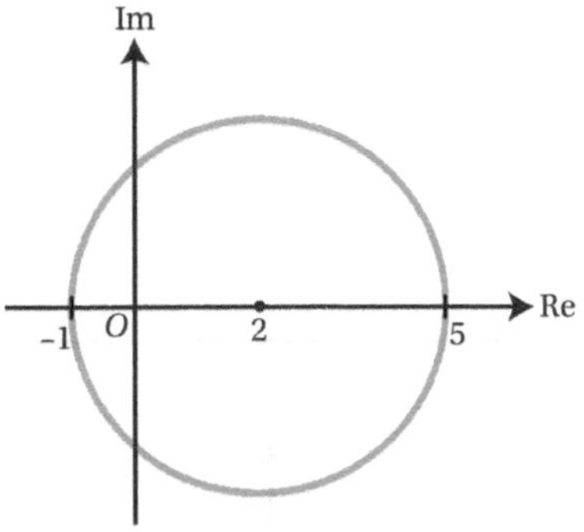

Key point 1.9

The equation $|z-a|=r$ represents a circle with radius r and centre at the point a.

WORKED EXAMPLE 1.16

Shade on an Argand diagram the set of points satisfying the inequality $|z-2+3\text{i}|<2$.

$|z-2+3\text{i}|<2$

$|z-(2-3\text{i})|<2$

You need to write the inequality in the form $|z-a|<2$.

The points will lie on a circle, centre $2-3\text{i}$, radius 2.

The inequality can be interpreted as saying, 'the distance between z and $2-3\text{i}$ is less than 2.'

Remember that the dashed line means that the circle itself is **not** included.

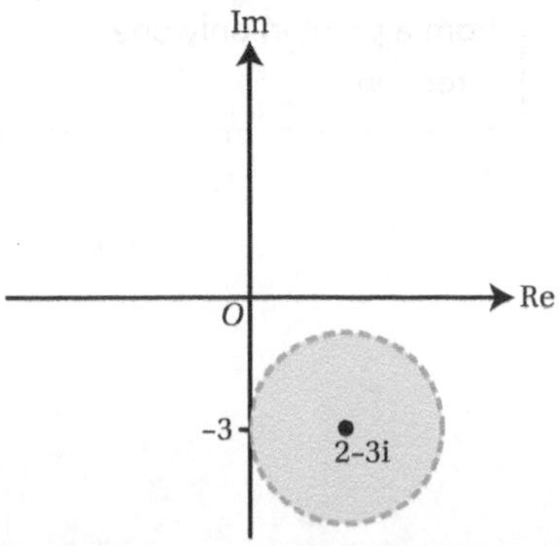

You can also describe the inequality in Worked example 1.16 by using (x, y) coordinates and set theory notation. The circle with centre (2, −3) and radius 2 has equation $(x-2)^2+(y+3)^2=4$, so the set of all points satisfying the inequality is $\{x+\text{i}y: (x-2)^2+(y+3)^2<4\}$.

Tip

A **locus** is a set of all the points that satisfy a given condition.

WORKED EXAMPLE 1.17

Sketch the locus of points in the Argand diagram that satisfy $|z-4|=|z-2\text{i}|$.

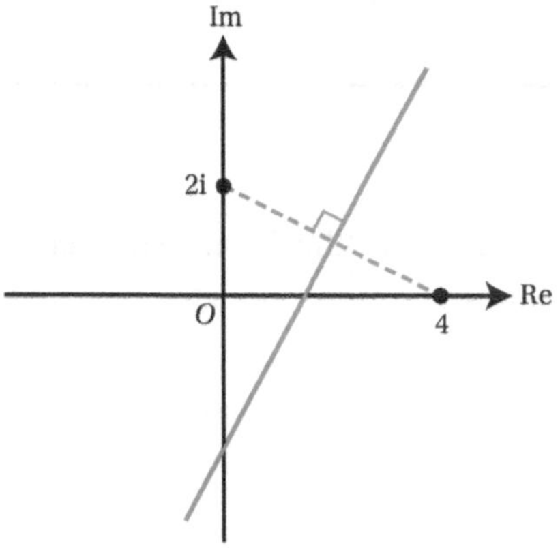

The equation says 'the distance from z to 4 is the same as the distance from z to 2i.'

This is the perpendicular bisector of the line segment joining the points 4 and 2i.

Locus involving the argument

The argument measures the angle that the line connecting z to the origin makes with the real axis. So the three points shown on the diagram all have the same argument, θ.

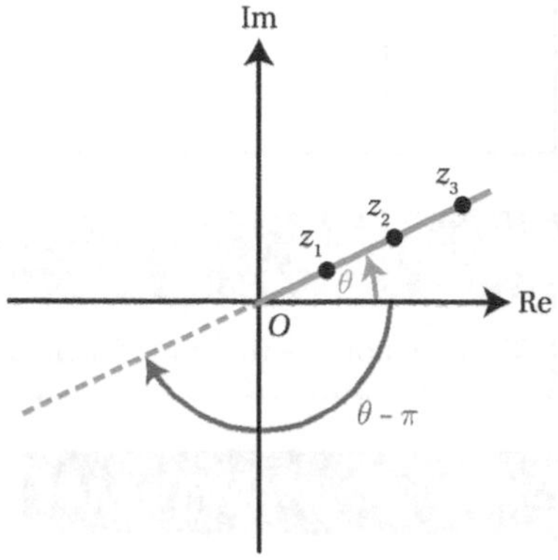

Since the argument is measured from the **positive** x-axis, any number on the dashed part of the line would have argument $\theta+\pi$ (or $\theta-\pi$ if the argument is negative).

If you replace z by $z-a$, the line shifts from the origin to the point a.

Key point 1.10

The locus of points satisfying $\arg(z-a)=\theta$ is the half-line starting from the point a and making angle θ with the positive x-axis.

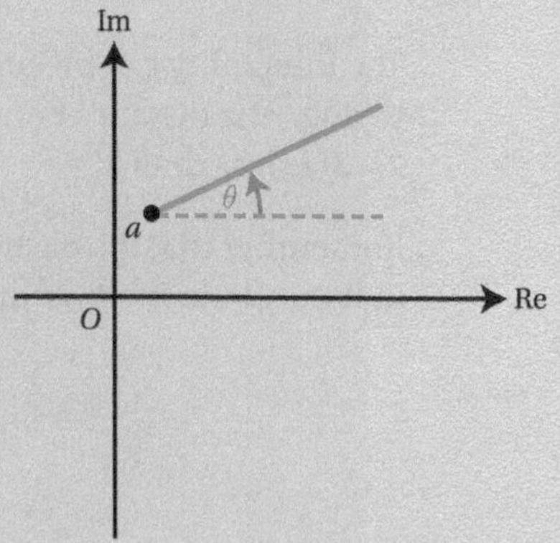

Tip

A half-line is a line extending from a point in only one direction.

WORKED EXAMPLE 1.18

On the Argand diagram, sketch the locus of points that satisfy $\arg(z-3+2i)=\frac{\pi}{4}$.

$\arg(z-3+2i)=\frac{\pi}{4}$

$\arg(z-(3-2i))=\frac{\pi}{4}$

You need to write the expression in brackets in the form $\arg(z-a)=\theta$.

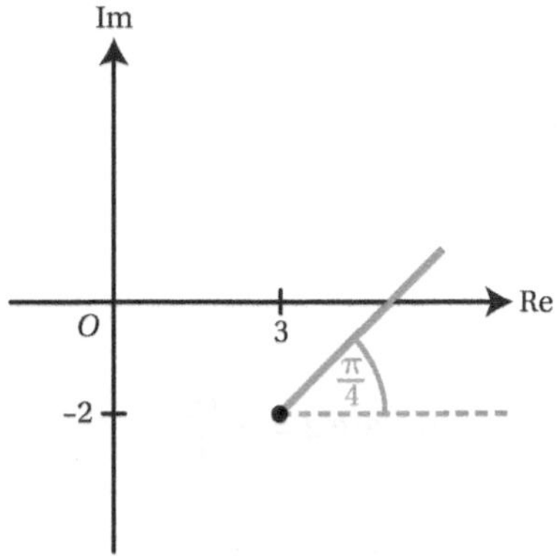

The half-line starts from the point $3-2i$ and makes a 45° angle with the horizontal.

WORKED EXAMPLE 1.19

On a single diagram, shade the locus of points that satisfy the inequalities $0<\arg(z+1-i)<\frac{2\pi}{3}$ and $|z|\leqslant 3$.

$0<\arg(z+1-i)<\frac{2\pi}{3}$

$0<\arg(z-(-1+i))<\frac{2\pi}{3}$

The first inequality represents the region between two half-lines, starting from the point $-1+i$ and making angles 0 and $\frac{2\pi}{3}$ with the vertical axis.

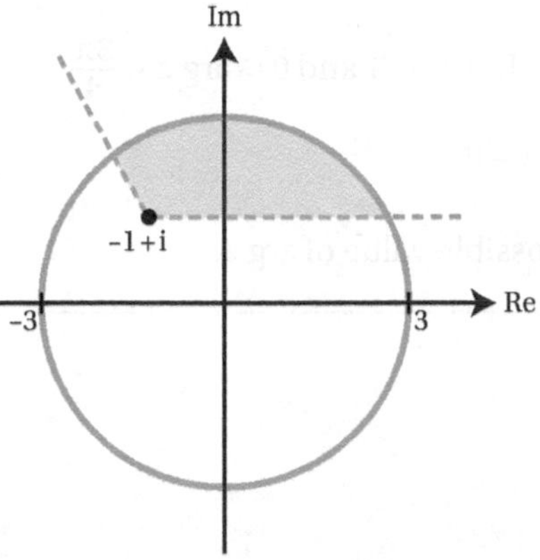

The second inequality represents the inside of the circle with radius 3 and centre at the origin.

The half-lines should be dashed and the circle solid.

You can indicate the required region by shading.

EXERCISE 1F

1 Sketch each locus on an Argand diagram. Shade the required regions where relevant.

a i $|z-2\mathrm{i}|=5$ ii $|z-3|=5$ b i $|z+4|=1$ ii $|z+3\mathrm{i}|=2$

c i $|z-\mathrm{i}|\leqslant 2$ ii $|z+\mathrm{i}|>3$ d i $|z-3+\mathrm{i}|>2$ ii $|z+1-2\mathrm{i}|\leqslant 1$

2 The set of points from Question 1 part **ai** can be described as $\{x+\mathrm{i}y: x^2+(y-2)^2=25\}$.

Write the sets of points for the rest of Question 1 using similar notation.

3 Sketch each locus on an Argand diagram. Shade the required regions where relevant.

a i $|z-2|=|z+2\mathrm{i}|$ ii $|z+1|=|z-\mathrm{i}|$

b i $|z-3\mathrm{i}|<|z+\mathrm{i}|$ ii $|z+2|>|z-3|$

4 Sketch each locus on an Argand diagram.

a i $\arg z=\frac{\pi}{3}$ ii $\arg z=\frac{\pi}{4}$ b i $\arg z=-\frac{\pi}{6}$ ii $\arg z=-\frac{3\pi}{4}$

c i $\arg z=\frac{\pi}{2}$ ii $\arg z=\pi$

5 On an Argand diagram, shade the region where $1<|z-3\mathrm{i}|\leqslant 3$.

6 On an Argand diagram, sketch the locus of points where $|z-3\mathrm{i}|=|z+6|$.

7 On an Argand diagram, shade the region where $\frac{\pi}{6}<\arg z<\frac{2\pi}{3}$.

8 a On the same diagram, sketch the loci of $|z+1|=|z-3|$ and $\arg z=\frac{\pi}{4}$.

b Hence find the complex number z that satisfies both equations.

9 On an Argand diagram, shade the region where $|z-3-\mathrm{i}|<3$ and $\arg z<\frac{\pi}{3}$.

10 On an Argand diagram, shade the locus of points that satisfy $\mathrm{Re}(z)\leqslant 1$, $\mathrm{Im}(z)<3$ and $0\leqslant\arg z<\frac{3\pi}{4}$.

11 Find the complex number z that satisfies $|z-3\mathrm{i}|=|z+\mathrm{i}|$ and $|z-1|=|z-\mathrm{i}|$.

12 z is a complex number satisfying $|z-2-2\mathrm{i}|=2$. Find the maximum possible value of $\arg z$.

Section 5: Operations in modulus–argument form

Addition and subtraction are quite easy in Cartesian form but multiplication and division are more difficult. Raising to a large power is even harder. These operations are much easier in modulus–argument form, thanks to the following result.

When you multiply two complex numbers you **multiply** their moduli and **add** their arguments.

Key point 1.11

For complex numbers z and w,

- $|zw| = |z||w|$
- $\arg(zw) = \arg z + \arg w$

The proof requires the use of compound angle formulae from the A Level Mathematics course:

$$\sin(A+B) = \sin A\cos B + \sin B\cos A$$

$$\cos(A+B) = \cos A\cos B - \sin A\sin B$$

Rewind

Compound angle formulae are covered in A Level Mathematics Student Book 2, Chapter 8.

PROOF 2

Let: $z_1 = r_1(\cos\theta_1 + \mathrm{i}\sin\theta_1)$ and
$z_2 = r_2(\cos\theta_2 + \mathrm{i}\sin\theta_2)$

Start by introducing variables.

$$z_1z_2 = r_1r_2(\cos\theta_1 + \mathrm{i}\sin\theta_1)(\cos\theta_2 + \mathrm{i}\sin\theta_2)$$
$$= r_1r_2(\cos\theta_1\cos\theta_2 - \sin\theta_1\sin\theta_2 + \mathrm{i}(\sin\theta_1\cos\theta_2 + \sin\theta_2\cos\theta_1))$$

Multiply them together and group real and imaginary parts.

$$= r_1r_2(\cos(\theta_1+\theta_2) + \mathrm{i}\sin(\theta_1+\theta_2))$$

Now use the compound angle formulae given above.

So: $|z_1z_2| = r_1r_2$

$\arg(z_1z_2) = \theta_1 + \theta_2$

This is in the form $|z|(\cos\theta + \mathrm{i}\sin\theta)$ so, by comparison, you can state the modulus and argument of z_1z_2.

i.e. $|z_1z_2| = |z_1||z_2|$
and $\arg(z_1z_2) = \arg z_1 + \arg z_2$

Finish the proof with a conclusion.

A similar proof gives the result for dividing complex numbers.

When you divide two complex numbers you **divide** their moduli and **subtract** their arguments.

Key point 1.12

For complex numbers z and w,

- $\left|\dfrac{z}{w}\right| = \dfrac{|z|}{|w|}$
- $\arg\left(\dfrac{z}{w}\right) = \arg z - \arg w$

WORKED EXAMPLE 1.20

a If $z = 3\left(\cos\frac{\pi}{3} + \text{i}\sin\frac{\pi}{3}\right)$ and $w = 5\left(\cos\frac{\pi}{4} + \text{i}\sin\frac{\pi}{4}\right)$, write zw in the form $r(\cos\theta + \text{i}\sin\theta)$.

b Write $\dfrac{6\left(\cos\frac{2\pi}{3} + \text{i}\sin\frac{2\pi}{3}\right)}{3\left(\cos\frac{\pi}{6} + \text{i}\sin\frac{\pi}{6}\right)}$ in the form $x + \text{i}y$.

a $|zw| = 3 \times 5 = 15$

Multiply the moduli and add the arguments.

$\arg(zw) = \frac{\pi}{3} + \frac{\pi}{4} = \frac{7\pi}{12}$

$\therefore zw = 15\left(\cos\frac{7\pi}{12} + \text{i}\sin\frac{7\pi}{12}\right)$

b $r = \frac{6}{3} = 2$

$\theta = \frac{2\pi}{3} - \frac{\pi}{6} = \frac{\pi}{2}$

$x = 2\cos\frac{\pi}{2} = 0$

Then convert to Cartesian form: $x = r\cos\theta$, $y = r\sin\theta$.

$y = 2\sin\frac{\pi}{2} = 2$

$$\therefore \frac{6\left(\cos\frac{2\pi}{3} + \text{i}\sin\frac{2\pi}{3}\right)}{3\left(\cos\frac{\pi}{6} + \text{i}\sin\frac{\pi}{6}\right)} = 2\text{i}$$

WORK IT OUT 1.2

Find the argument of $z = 3\left(\cos\frac{9\pi}{5} + \text{i}\sin\frac{9\pi}{5}\right) \times 5\left(\cos\frac{4\pi}{3} + \text{i}\sin\frac{4\pi}{3}\right)$.

Solution 1	Solution 2	Solution 3
$\arg z = \frac{9\pi}{5} \times \frac{4\pi}{3}$ $= \frac{12\pi^2}{5}$	$\frac{9\pi}{5} + \frac{4\pi}{3} = \frac{47\pi}{15}$ So: $\arg z = \frac{47\pi}{15} - 2\pi$ $= \frac{17\pi}{15}$	$\arg z = \frac{9\pi}{5} + \frac{4\pi}{3}$ $= \frac{47\pi}{15}$

EXERCISE 1G

1 Simplify each expression, giving your answers in the form $r(\cos\theta + \mathrm{i}\sin\theta)$.

a **i** $3\left(\cos\frac{\pi}{6}+\mathrm{i}\sin\frac{\pi}{6}\right)\times 7\left(\cos\frac{\pi}{5}+\mathrm{i}\sin\frac{\pi}{5}\right)$ **ii** $\left(\cos\left(-\frac{\pi}{9}\right)+\mathrm{i}\sin\left(-\frac{\pi}{9}\right)\right)\times 4\left(\cos\frac{\pi}{3}+\mathrm{i}\sin\frac{\pi}{3}\right)$

b **i** $\dfrac{8(\cos 6+\mathrm{i}\sin 6)}{2(\cos 2+\mathrm{i}\sin 2)}$ **ii** $\dfrac{15\left(\cos\frac{\pi}{7}+\mathrm{i}\sin\frac{\pi}{7}\right)}{5\left(\cos\frac{\pi}{2}+\mathrm{i}\sin\frac{\pi}{2}\right)}$

2 Write each of these expressions in the form $\cos\theta+\mathrm{i}\sin\theta$, where $-\pi<\arg z\leqslant\pi$.

a **i** $\left(\cos\frac{\pi}{3}+\mathrm{i}\sin\frac{\pi}{3}\right)\left(\cos\frac{3\pi}{4}+\mathrm{i}\sin\frac{3\pi}{4}\right)$ **ii** $\left(\cos\frac{2\pi}{5}+\mathrm{i}\sin\frac{2\pi}{5}\right)\left(\cos\frac{\pi}{4}+\mathrm{i}\sin\frac{\pi}{4}\right)$

b **i** $\left(\cos\frac{\pi}{3}+\mathrm{i}\sin\frac{\pi}{3}\right)\left(\cos\frac{\pi}{4}-\mathrm{i}\sin\frac{\pi}{4}\right)$ **ii** $\left(\cos\frac{2\pi}{3}+\mathrm{i}\sin\frac{2\pi}{3}\right)\left(\cos\frac{2\pi}{5}-\mathrm{i}\sin\frac{2\pi}{5}\right)$

c **i** $\dfrac{\cos\frac{3\pi}{5}+\mathrm{i}\sin\frac{3\pi}{5}}{\cos\frac{\pi}{4}+\mathrm{i}\sin\frac{\pi}{4}}$ **ii** $\dfrac{\cos\frac{\pi}{3}+\mathrm{i}\sin\frac{\pi}{3}}{\cos\frac{3\pi}{4}+\mathrm{i}\sin\frac{3\pi}{4}}$

d **i** $\dfrac{\cos\frac{\pi}{5}+\mathrm{i}\sin\frac{\pi}{5}}{\cos\frac{\pi}{4}-\mathrm{i}\sin\frac{\pi}{4}}$ **ii** $\dfrac{\cos\frac{\pi}{4}+\mathrm{i}\sin\frac{\pi}{4}}{\cos\frac{2\pi}{5}-\mathrm{i}\sin\frac{2\pi}{5}}$

3 Write $\dfrac{6\left(\cos\frac{2\pi}{3}+\mathrm{i}\sin\frac{2\pi}{3}\right)}{2\left(\cos\frac{\pi}{4}+\mathrm{i}\sin\frac{\pi}{4}\right)}$ in the form $r(\cos\theta+\mathrm{i}\sin\theta)$.

4 Write $\dfrac{8\left(\cos\frac{5\pi}{6}+\mathrm{i}\sin\frac{5\pi}{6}\right)}{4\left(\cos\frac{2\pi}{3}+\mathrm{i}\sin\frac{2\pi}{3}\right)}$ in the form $x+\mathrm{i}y$.

5 Let $z=\cos\frac{\pi}{4}+\mathrm{i}\sin\frac{\pi}{4}$ and $w=\cos\frac{\pi}{3}+\mathrm{i}\sin\frac{\pi}{3}$.

a Write zw in the form $x+\mathrm{i}y$, without using trigonometric functions.

b Hence find the exact value of $\tan\frac{7\pi}{12}$.

6 Use trigonometric identities to show that $\dfrac{1}{\cos\theta+\mathrm{i}\sin\theta}=\cos(-\theta)+\mathrm{i}\sin(-\theta)=\cos(2\pi-\theta)+\mathrm{i}\sin(2\pi-\theta)$.

Checklist of learning and understanding

- A complex number is of the form $z = x + \mathrm{i}y$, where x and y are real and $\mathrm{i} = \sqrt{-1}$.
  - Apart from the fact that $\mathrm{i}^2 = -1$, the arithmetic of complex numbers is the same as for real numbers.
  - To divide by a complex number, multiply top and bottom by its **complex conjugate** ($z^* = x - \mathrm{i}y$).
- It is useful to represent complex numbers geometrically using an **Argand diagram**. Addition can be represented on a diagram in the same way as adding vectors.
- There are two ways of describing numbers in the Argand diagram: **Cartesian** form $(x + \mathrm{i}y)$ and **modulus-argument** form $r(\cos\theta + \mathrm{i}\sin\theta)$. The two forms are linked by:
  - $r = \sqrt{x^2 + y^2}$
  - $\tan\theta = \frac{y}{x}$
  - $x = r\cos\theta$
  - $y = r\sin\theta$
- Multiplication and division are easier in modulus-argument form:
  - $\arg(zw) = \arg z + \arg w$
  - $\arg\left(\frac{z}{w}\right) = \arg z - \arg w$
  - $|zw| = |z||w|$
  - $\left|\frac{z}{w}\right| = \frac{|z|}{|w|}$
- Equations and inequalities can be represented on an Argand diagram:
  - $|z - a| = r$ represents a circle with centre a and radius r
  - $|z - a| = |z - b|$ represents the perpendicular bisector of the line segment connecting points a and b
  - $\arg(z - a) = \theta$ represents a half-line starting from a and making angle θ with the positive real axis
  - vertical and horizontal lines are represented by equations of the form $\mathrm{Re}(z) = c$ or $\mathrm{Im}(z) = c$, respectively.

Mixed practice 1

1 If $z=-2+3\text{i}$, find an expression for arg z. Choose from these options.

A $\tan^{-1}\left(\frac{3}{2}\right)$ **B** $\tan^{-1}\left(\frac{3}{-2}\right)$ **C** $\pi-\tan^{-1}\left(\frac{3}{2}\right)$ **D** $\pi+\tan^{-1}\left(\frac{3}{2}\right)$

2 What is $(zz^*)^*$ equivalent to? Choose from these options.

A z **B** z^*z **C** $(\text{Re}(z))^2$ **D** z^2

3 Let $z_1=1-\text{i}$ and $z_2=3+5\text{i}$.

Showing all your working clearly, find in the form $x+\text{i}y$:

a z_1z_2 **b** $\frac{z_1}{z_2}$

4 Express $z=3\text{i}-\frac{2}{\text{i}+\sqrt{3}}$ in the form $x+\text{i}y$.

5 **a** For the complex number $z=-2+5\text{i}$, find:

i $|z|$ **ii** arg z, where $-\pi<\arg z\leqslant\pi$.

b State the modulus and argument of z^*.

6 **a** Solve the equation $z^2+14z+53=0$.

b Represent the solutions on an Argand diagram.

7 Find the complex number z such that $3z-5z^*=4-3\text{i}$.

8 On an Argand diagram, illustrate the locus of points z that satisfy the inequality $3<|z-3+4\text{i}|\leqslant5$.

9 It is given that $z=x+\text{i}y$, where x and y are real numbers.

a Find, in terms of x and y, the real and imaginary parts of $(1-2\text{i})z-z^*$.

b Hence find the complex number z such that $(1-2\text{i})z-z^*=10(2+\text{i})$.

[©AQA 2010]

10 Find, in terms of w, the complex number that satisfies both $|z-w|=2\sqrt{2}$ and $\arg(z-w)=\frac{\pi}{4}$.

Choose from these options.

A $w+2\sqrt{2}$ **B** $w+1+\text{i}$ **C** $w+2+2\text{i}$ **D** $w+\sqrt{2}+\sqrt{2}\text{i}$

11 Given that $w=1+\sqrt{3}\text{i}$ and $z=1+\text{i}$ show that $\text{Re}\left(\frac{\sqrt{2}z+w}{\sqrt{2}z-w}\right)=0$.

12 Let z and w be complex numbers satisfying $\frac{w+\text{i}}{w-\text{i}}=\frac{z+1}{z-1}$.

a Express w in terms of z.

b Show that, if $\text{Im}(z)=0$, then $\text{Re}(w)=0$

13 Represent on an Argand diagram the region defined by the inequalities $0<\arg(z-2\text{i})\leqslant\frac{3\pi}{4}$ and $|z-2\text{i}|\geqslant2$.

14 Two complex numbers, a and b, are shown on the Argand diagram.

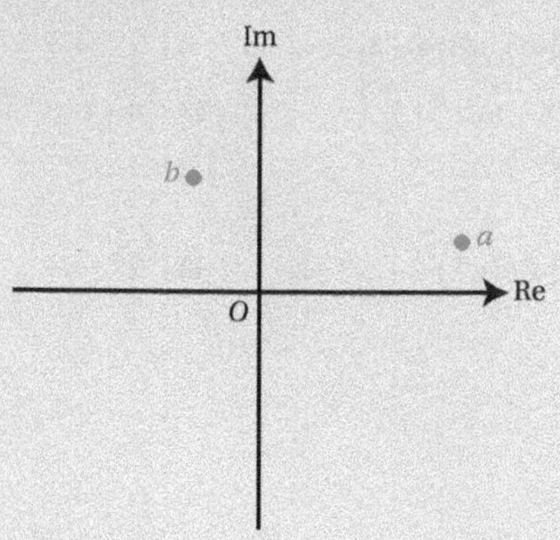

a Add the points representing the numbers $a+b$ and a^* to the diagram.

b Sketch the locus of points z that satisfy $|z-a|=|z-b|$.

15 Two loci, L_1 and L_2, on an Argand diagram are given by:

L_1: $|z+6-5\mathrm{i}|=4\sqrt{2}$

L_2: $\arg(z+\mathrm{i})=\frac{3\pi}{4}$

The point P represents the complex number $-2+\mathrm{i}$.

a Verify that the point P is a point of intersection of L_1 and L_2.

b Sketch L_1 and L_2 on one Argand diagram.

c The point Q is also a point of intersection of L_1 and L_2. Find the complex number that is represented by Q.

[©AQA 2013]

16 Let z and w be complex numbers such that $w=\frac{1}{1-z}$ and $|z|^2=1$. Find the real part of w.

17 If $z=\cos\theta+\mathrm{i}\sin\theta$ prove that $\frac{z^2-1}{z^2+1}=\mathrm{i}\tan\theta$.

18 If z and w are complex numbers, solve the simultaneous equations:

$$3z+w=9+11\mathrm{i}$$
$$\mathrm{i}w-z=-8-2\mathrm{i}$$

19 Let $z_1=\frac{\sqrt{6}+\mathrm{i}\sqrt{2}}{2}$ and $z_2=1-\mathrm{i}$.

a Show that $\frac{z_1}{z_2}=\cos\frac{5\pi}{12}+\mathrm{i}\sin\frac{5\pi}{12}$.

b Find the value of $\frac{z_1}{z_2}$ in the form $a+b\mathrm{i}$, where a and b are to be determined exactly in surd form. Hence find the exact values of $\cos\frac{5\pi}{12}$ and $\sin\frac{5\pi}{12}$.

20 By considering the product $(2+\mathrm{i})(3+\mathrm{i})$, show that $\arctan\frac{1}{2}+\arctan\frac{1}{3}=\frac{\pi}{4}$.

21 If $0<\theta<\frac{\pi}{2}$ and $z=(\sin\theta+\mathrm{i}(1-\cos\theta))^2$ find $\arg z$ in its simplest form.

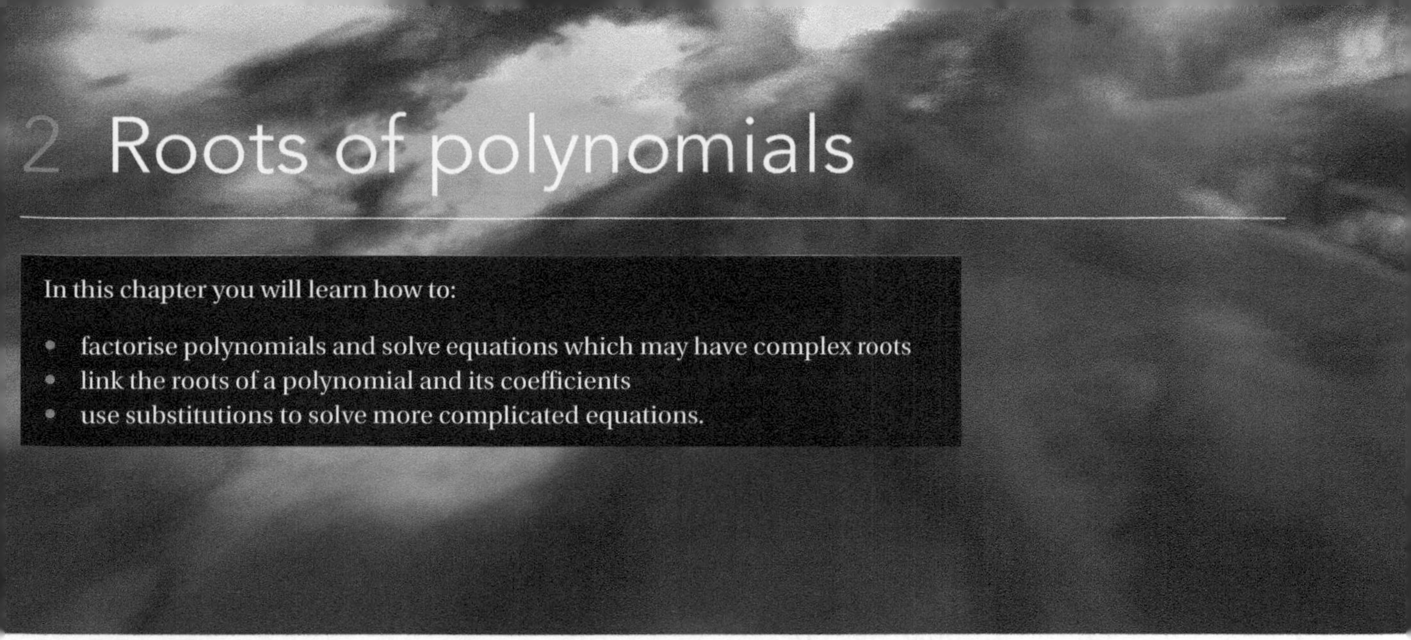

2 Roots of polynomials

In this chapter you will learn how to:

- factorise polynomials and solve equations which may have complex roots
- link the roots of a polynomial and its coefficients
- use substitutions to solve more complicated equations.

Before you start...

A Level Mathematics Student Book 1, Chapter 4	You should be able to use the factor theorem to factorise cubic polynomials.	1 a Show that $x=2$ is a root of $f(x)=2x^3-9x^2+7x+6$. b Hence factorise $f(x)$ completely.
Chapter 1	You should be able to perform arithmetic with complex numbers and solve quadratic equations with complex roots.	2 a Expand and simplify $(3-2i)(5+i)$. b Solve the equation $x^2-2x+5=0$.
Chapter 1	You should be able to work with complex conjugates.	3 Given that $z = 3 - 5i$, find: a $z+z^*$ b zz^*

Using complex numbers in factorising

This chapter draws together ideas from the AS Mathematics course about factorising cubics and quartics with the theory of complex numbers from Chapter 1 of this book to enable you to find complex roots of these polynomials. It also looks at the relationship between the coefficients of a polynomial and its roots.

Section 1: Factorising polynomials

You are already familiar with the link between factorising and real roots of a polynomial: for example, if $x=2$ is a root of the equation $p(x)=0$, then $(x-2)$ is a factor of the polynomial $p(x)$.

This is the **factor theorem** (which you met in A Level Mathematics Student Book 1, Chapter 4), and it also applies to polynomials with complex roots.

WORKED EXAMPLE 2.1

a Solve the equation $x^2 - 4x + 40 = 0$.
b Hence factorise $x^2 - 4x + 40$.

a $x = \frac{4 \pm \sqrt{(-4)^2 - 4 \times 40}}{2}$ ········ Use the quadratic formula.

$= \frac{4 \pm \sqrt{-144}}{2}$

$= \frac{4 \pm 12\mathrm{i}}{2}$

$= 2 \pm 6\mathrm{i}$

b Therefore: ········ Make the link between roots and factors.

$x^2 - 4x + 40 = (x - (2 + 6\mathrm{i}))(x - (2 - 6\mathrm{i}))$

$= (x - 2 - 6\mathrm{i})(x - 2 + 6\mathrm{i})$

You can use the same method to factorise polynomials of higher degree.

WORKED EXAMPLE 2.2

Let $\mathrm{f}(x) = x^3 - x^2 - x - 15$.

a Show that $\mathrm{f}(3) = 0$.
b Hence write $\mathrm{f}(x)$ as a product of a linear factor and a real quadratic factor.
c Solve the equation $\mathrm{f}(x) = 0$ and write $\mathrm{f}(x)$ as a product of three linear factors.

a $\mathrm{f}(3) = 3^3 - 3^2 - 3 - 15$ ········ Substitute $x = 3$ into $\mathrm{f}(x)$.

$= 27 - 9 - 3 - 15$

$= 0$

b $x^3 - x^2 - x + 15 = (x - 3)(x^2 + Bx + C)$ ········ As 3 is a root, $(x - 3)$ is one factor. You can find the quadratic factor by long division or comparing coefficients.

$\Rightarrow B = 2, C = 5$

$\therefore \mathrm{f}(x) = (x - 3)(x^2 + 2x + 5)$

c $\mathrm{f}(x) = 0$

$(x - 3)(x^2 + 2x + 5) = 0$

$x = 3$ or $x^2 + 2x + 5 = 0$

If $x^2 + 2x + 5 = 0$: ········ Use the formula to solve the quadratic.

$x = \frac{-2 \pm \sqrt{2^2 - 4 \times 5}}{2}$

$= \frac{-2 \pm \sqrt{-16}}{2}$

$= \frac{-2 \pm 4\mathrm{i}}{2}$

$= -1 \pm 2\mathrm{i}$

Continues on next page

$\therefore x = 3$ or $x = -1 \pm 2i$ — State all three roots of $f(x)$.

$\therefore f(x) = (x-3)(x-(-1+2i))(x-(-1-2i))$
$= (x-3)(x+1-2i)(x+1+2i)$ — Each root z has a corresponding factor $(x-z)$.

This means that you can now factorise some expressions that were impossible to factorise using just real numbers. A particularly useful case is this extension of the difference of two squares identity to the sum of two squares.

Key point 2.1

$$a^2 + b^2 = (a+bi)(a-bi)$$

WORKED EXAMPLE 2.3

Factorise $81z^4 - 16$.

$81z^4 - 16 = (9z^2 - 4)(9z^2 + 4)$ — This is a difference of two squares.

$= (3z-2)(3z+2)(3z-2i)(3z+2i)$ — The first factor is a difference of two squares again. The second factor is a sum of two squares.

EXERCISE 2A

1 Solve the equation $f(x) = 0$ and hence factorise $f(x)$.

a **i** $f(x) = x^2 - 2x + 2$ **ii** $f(x) = x^2 + 6x + 25$

b **i** $f(x) = x^2 + 3x + 4$ **ii** $f(x) = x^2 + 2x + 5$

c **i** $f(x) = 3x^2 - 2x + 10$ **ii** $f(x) = 5x^2 + 4x + 2$

2 Factorise each expression into linear factors.

a **i** $z^2 + 4$ **ii** $z^2 + 25$

b **i** $4z^2 + 49$ **ii** $9z^2 + 64$

c **i** $z^4 - 1$ **ii** $16z^4 - 81$

3 Given one root of the cubic polynomial, write it as a product of a linear factor and a real quadratic factor, and hence find the other two roots.

a **i** $x^3 + 2x^2 - x - 14$; root $x = 2$ **ii** $x^3 + 3x^2 + 7x + 5$; root $x = -1$

b **i** $2x^3 - 5x + 6$; root $x = -2$ **ii** $3x^3 - x^2 - 2$; root $x = 1$

4 $f(x) = x^3 - 13x^2 + 56x - 78$

Show that $(x-3)$ is a factor of $f(x)$ and find all the solutions of the equation $f(x) = 0$.

5 a Given that $(2x+1)$ is a factor of $2x^3+ax^2+16x+6$, show that $a=9$.

b Factorise $2x^3+9x^2+16x+6$ completely.

6 $f(x)=4x^4-4x^3-21x^2-9x$

a Show that $(x-3)$ is a factor of $f(x)$.

b Factorise $f(x)$ and hence solve the equation $f(x)=0$.

7 $f(x)=x^4+3x^3-x^2-13x-10$

a Show that $(x+1)$ and $(x-2)$ are factors of $f(x)$.

b Write $f(x)$ as a product of two linear factors and a quadratic factor. Hence find all solutions of the equation $f(x)=0$.

8 Find all solutions of the equation $x^4+21x^2-100=0$.

Section 2: Complex solutions to polynomial equations

You know from Chapter 1 that if a quadratic equation has two complex solutions, then they are always a conjugate pair: for example, the equation $x^2-6x+25=0$ has solutions $x=3+4\text{i}$ and $x=3-4\text{i}$. This happens because of the ± in the quadratic formula.

If you look at the cubic and quartic polynomials you factorised in Exercise 2A, you will see that the complex roots were either real or a complex conjugate pair. It can be proved that this result generalises to any **real polynomial** (a polynomial in which all of the coefficients are real numbers).

> **Did you know?**
>
> This result is **not** true if the polynomial has complex coefficients. For example, the equation $z^2-\text{i}z+2=0$ has solutions $-\text{i}$ and 2i.

> **Key point 2.2**
>
> Complex solutions of real polynomials come in conjugate pairs.
> If $f(z)=0$ then $f(z^*)=0$.

> **Focus on...**
>
> See Focus on... Proof 1 for a proof of this result.

This result can be very useful when you are factorising polynomials and solving equations. If you know one complex root, then you can immediately write down another one.

WORKED EXAMPLE 2.4

Given that $1-\text{i}\sqrt{6}$ is a root of the polynomial $f(x)=x^3+x^2+x+21$, find the other two roots.

$1-\text{i}\sqrt{6}$ is a root so $1+\text{i}\sqrt{6}$ is also a root. — Complex roots come in conjugate pairs.

So $(x-(1-\text{i}\sqrt{6}))$ and $(x-(1+\text{i}\sqrt{6}))$ are factors of $f(x)$. — Link roots to factors (using the factor theorem).

Therefore $(x-(1-\text{i}\sqrt{6}))(x-(1+\text{i}\sqrt{6}))$ is a factor.

Continues on next page

$(x-(1-i\sqrt{6}))(x-(1+i\sqrt{6}))$
$= x^2 - x(1+i\sqrt{6}) - x(1-i\sqrt{6}) + (1-i\sqrt{6})(1+i\sqrt{6})$
$= x^2 - x - xi\sqrt{6} - x + xi\sqrt{6} + 7$
$= x^2 - 2x + 7$

Expand and simplify.

$(x^3 + x^2 + x + 21) = (x+k)(x^2 - 2x + 7)$
By inspection, $k = 3$.
So the solutions are $x = 1 \pm i\sqrt{6}$ and $x = -3$.

Use polynomial division to find the third factor.

In Worked example 2.4, in order to find the quadratic factor of $f(x)$, you needed to expand $(x-(1-i\sqrt{6}))(x-(1+i\sqrt{6}))$, which is a product of the form $(x-z)(x-z^*)$. You get this type of expression whenever you are expanding two brackets corresponding to a pair of complex conjugate roots, and so it is useful to know a shortcut.

Key point 2.3

$$(x-z)(x-z^*) = x^2 - 2\text{Re}(z)x + |z|^2$$

In the case of $(x-(1-i\sqrt{6}))(x-(1+i\sqrt{6}))$, $z = 1 - i\sqrt{6}$ and so:

- $\text{Re}(z) = 1$
- $|z|^2 = 1^2 + (\sqrt{6})^2 = 7$

$\therefore (x-(1-i\sqrt{6}))(x-(1+i\sqrt{6})) = x^2 - 2\times 1x + 7 = x^2 - 2x + 7$

WORKED EXAMPLE 2.5

Given that one of the roots of the polynomial $f(x) = x^4 - 5x^3 + 26x^2 + 46x - 68$ is $3 - 5i$, find the remaining roots.

$3 - 5i$ is a root so $3 + 5i$ is also a root.

Complex roots come in conjugate pairs.

So $(x-(3-5i))$ and $(x-(3+5i))$ are factors of $f(x)$.

Link roots to factors (using the factor theorem).

Therefore:
$(x-(3-5i))(x-(3+5i)) = x^2 - 2\times\mathbf{3}x + \mathbf{34}$
$= x^2 - 6x + 34$
is a factor.

Multiply out the brackets (using the shortcut from Key point 2.3).

If $z = 3 + 5i$, then:
$\text{Re}(z) = 3$ and $|z|^2 = 3^2 + 5^2 = 34$.

Continues on next page

Factorising:
$x^4 - 5x^3 + 26x^2 + 46x - 68 = (x^2 - 6x + 34)(x^2 + x - 2)$

Divide f(x) by $x^2 - 6x + 34$ (using polynomial division or comparing coefficients).

For the roots of the second factor:
$x^2 + x - 2 = 0$
$(x+2)(x-1) = 0$
$x = -2$ or 1

You already know the roots of the first factor. Find the roots of the second factor by factorising.

So the roots of f(x) are: 1, −2 and $3 \pm 5\mathrm{i}$.

You can also find a polynomial with given roots.

WORKED EXAMPLE 2.6

Find a cubic polynomial with roots 3 and $4 + \mathrm{i}$.

Give your answer in the form $x^3 + bx^2 + cx + d$.

3 is a root of f(x), so $(x - 3)$ is a factor.

From the factor theorem, if z is a root then $(x - z)$ is a factor.

$4 + \mathrm{i}$ is a root, so $4 - \mathrm{i}$ is another root.

Complex roots occur in conjugate pairs.

So f(x) also has factors $(x - (4 + \mathrm{i}))$ and $(x - (4 - \mathrm{i}))$

Hence: $\mathrm{f}(x) = (x-3)(x-(4+\mathrm{i}))(x-(4-\mathrm{i}))$
$= (x-3)(x^2 - 2\times \mathbf{4}x + \mathbf{17})$
$= (x-3)(x^2 - 8x + 17)$
$= x^3 - 11x^2 + 41x - 51$

Use the result of Key point 2.3 to expand the brackets corresponding to the complex roots. If $z = 4 + \mathrm{i}$ then: $\mathrm{Re}(z) = 4$ and $|z|^2 = 4^2 + 1^2 = 17$

Expand and simplify.

WORKED EXAMPLE 2.7

The polynomial $\mathrm{f}(x) = x^4 + bx^3 + cx^2 + dx + e$ has roots $3\mathrm{i}$ and $5 - \mathrm{i}$. Find the values of the real numbers b, c, d and e.

The other two roots are $-3\mathrm{i}$ and $5 + \mathrm{i}$.

Write down the remaining roots and use them to find factors.

Hence the two quadratic factors are:
$(x + 3\mathrm{i})(x - 3\mathrm{i}) = x^2 + 9$
and:
$(x - (5 - \mathrm{i}))(x - (5 + \mathrm{i})) = x^2 - 10x + 26$

The complex conjugate pairs will give quadratic factors.

Continues on next page

Hence: $f(x) = (x^2 + 9)(x^2 - 10x + 26)$
$= x^4 - 10x^3 + 35x^2 - 90x + 234$

Multiply the two factors to get the polynomial.

So $b = -10, c = 35, d = -90, e = 234$.

WORK IT OUT 2.1

Find a cubic polynomial with real coefficients, given that two of its roots are 2 and $2\text{i} - 1$.

Which is the correct solution? Identify the errors made in the incorrect solutions.

Solution 1	Solution 2	Solution 3
The other complex root is $2\text{i} + 1$, so the polynomial is: $(x-2)(x-(2\text{i}-1))(x-(2\text{i}+1))$ $=(x-2)(x^2-4\text{i}x-5)$ $=x^3-(2+4\text{i})x^2+(8\text{i}-5)x+10$	The other complex root is $-2\text{i} - 1$, so the polynomial is: $(x+2)(x-(2\text{i}-1))(x-(-2\text{i}-1))$ $=(x+2)(x^2+2x+5)$ $=x^3+4x^2+9x+10$	The other complex root is $-1 - 2\text{i}$, so the polynomial is: $(x-2)(x-(-1+2\text{i}))(x-(-1-2\text{i}))$ $=(x-2)(x^2+2x+5)$ $=x^3+x-10$

EXERCISE 2B

1 Find the real values of a and b such that the quadratic equation $x^2 + ax + b = 0$ has the given roots.

a i 5i and –5i ii –3i and 3i

b i $3 - 4\text{i}$ and $3 + 4\text{i}$ ii $1 + 2\text{i}$ and $1 - 2\text{i}$

2 Given one complex root of a cubic polynomial, factorise the polynomial and write down all its roots.

a i $x^3 - 11x^2 + 43x - 65$; root $x = 3 + 2\text{i}$ ii $x^3 - x^2 - 7x + 15$; root $x = 2 + \text{i}$

b i $x^3 - 3x^2 + 7x - 5$; root $x = 1 - 2\text{i}$ ii $x^3 - 2x^2 - 14x + 40$; root $x = 3 - \text{i}$

3 Given one complex root of the quartic polynomial, find one real quadratic factor and hence find all four roots.

a i $x^4 - 2x^3 + 14x^2 - 8x + 40$; root $x = 1 + 3\text{i}$ ii $x^4 - 6x^3 + 11x^2 - 6x + 10$; root $x = 3 - \text{i}$

b i $x^4 - 2x^3 + 8x^2 + 14x + 39$; root $x = 2 - 3\text{i}$ ii $x^4 - 3x^3 + 27x^2 + 21x + 58$; root $x = 2 + 5\text{i}$

c i $x^4 + 2x^3 + 10x^2 + 8x + 24$; root -2i ii $x^4 - 4x^3 + 21x^2 - 64x + 80$; root 4i

4 Given that $5 - \text{i}$ is one root of the equation $x^3 - 8x^2 + 6x + 52 = 0$, find the remaining two roots.

5 $\text{p}(x) = x^3 + 3x^2 + 16x + 48$

a Show that $\text{p}(4\text{i}) = 0$. **b** Hence solve the equation $\text{p}(x) = 0$.

6 $2 + 5\text{i}$ is a root of the equation $x^4 - 4x^3 + 30x^2 - 4x + 29 = 0$.

a Write down another complex root of the equation.

b Find the remaining two roots.

7 Two roots of the equation $x^4 - 8x^3 + 21x^2 - 32x + 68 = 0$ are 2i and $4 - \text{i}$. Write down the other two roots and hence write $x^4 - 8x^3 + 21x^2 - 32x + 68$ as a product of two real quadratic factors.

8 $f(z) = z^4 + z^3 + 5z^2 + 4z + 4$

a Show that $f(2\text{i}) = 0$.

b Write $f(z)$ as a product of two real quadratic factors.

c Hence find the remaining solutions of the equation $f(z) = 0$.

9 Find a quartic equation with real coefficients and roots 4i and $3 - 2\text{i}$.

Section 3: Roots and coefficients

When you first learned to factorise quadratic equations you were probably told to look for two numbers that add up to the middle coefficient and multiply to give the constant term. For example, $x^2 - 10x + 21$ factorises as $(x-3)(x-7)$ because $(-3)+(-7) = -10$ and $(-3)\times(-7) = 21$. This means that the roots, 3 and 7, add up to give the negative of the middle coefficient, 10, and multiply to give the constant term, 21.

However, there are infinitely many other quadratic equations with the same roots, because you can multiply the whole equation by a constant. For example, another equation with roots 3 and 7 is $3x^2 - 30x + 63 = 0$. The two roots still add up to 10, which is $\frac{30}{3}$, and multiply to give 21, which is $\frac{63}{3}$. This is a particular example of a very useful general result.

Key point 2.4

If p and q are the roots of the quadratic $ax^2 + bx + c = 0$ then:

- $p + q = -\frac{b}{a}$
- $pq = \frac{c}{a}$

PROOF 3

If a quadratic equation has roots p and q then: — Write the equation in factorised form.

$$(x-p)(x-q) = 0$$

$$x^2 - (p+q)x + pq = 0$$

$$ax^2 - a(p+q)x + apq = 0$$

— You can multiply the whole equation by a number, and it will still have the same roots.

Hence: $b = -a(p+q)$ and $c = apq$ — Compare coefficients with $ax^2 + bx + c$.

So: $p + q = -\frac{b}{a}$ and $pq = \frac{c}{a}$

You can also find other functions of roots. This requires some algebraic manipulation.

WORKED EXAMPLE 2.8

The equation $5x^2 + 3x + 1 = 0$ has roots p and q. Find the value of:

a $\frac{1}{p} + \frac{1}{q}$ **b** $p^2 + q^2$.

$p + q = -\frac{b}{a} = -\frac{3}{5}$

$pq = \frac{c}{a} = \frac{1}{5}$

You could actually find the roots by using the quadratic formula, but Key point 2.4 provides a much quicker way to answer the question.

a $\frac{1}{p} + \frac{1}{q} = \frac{p+q}{pq}$

It is often helpful to combine fractions into a single fraction.

$= \frac{\left(-\frac{3}{5}\right)}{\left(\frac{1}{5}\right)}$

$= -3$

b $(p+q)^2 = p^2 + 2pq + q^2$

You can get $p^2 + q^2$ by squaring $p + q$.

$\Rightarrow p^2 + q^2 = (p+q)^2 - 2pq$

$= \left(-\frac{3}{5}\right)^2 - 2\left(\frac{1}{5}\right)$

$= -\frac{1}{25}$

$p^2 + q^2$ is a sum of two squares, so you may think that you made a mistake because the answer is negative. However, p and q can be complex numbers, so this is in fact possible!

The Greek letters α, and β are often used instead of p and q for the roots.

WORKED EXAMPLE 2.9

The equation $3x^2 + kx - (k+1) = 0$ has roots α and β.

a Write down expressions for $\alpha + \beta$ and $\alpha\beta$ in terms of k.
b Find $\alpha^3 + \beta^3$ in terms of k.

a $\alpha + \beta = \frac{-k}{3}$

$\alpha\beta = \frac{-(k+1)}{3}$

Apply Key point 2.4.

$\alpha + \beta = -\frac{b}{a}$

$\alpha\beta = \frac{c}{a}$

Continues on next page

b $(\alpha+\beta)^3 = \alpha^3+3\alpha^2\beta+3\alpha\beta^2+\beta^3$
$= \alpha^3+\beta^3+3\alpha\beta(\alpha+\beta)$

You can find $\alpha^3+\beta^3$ from the binomial expansion of $(\alpha+\beta)^3$.

$$\Rightarrow \alpha^3+\beta^3 = (\alpha+\beta)^3 - 3\alpha\beta(\alpha+\beta)$$
$$= -\frac{k^3}{27} - 3\left(-\frac{k+1}{3}\right)\left(-\frac{k}{3}\right)$$
$$= -\frac{k^3}{27} - \frac{k(k+1)}{3}$$
$$= -\frac{k^3+9k^2+9k}{27}$$

You need to try to express everything in terms of $\alpha+\beta$ and $\alpha\beta$.

EXERCISE 2C

1 The equation $x^2-kx+2k=0$ has roots α and β. Find the value of each expression in terms of k.

a $\alpha+\beta$ **b** $2\alpha\beta$ **c** $\alpha^2+\beta^2$

d $\frac{1}{\alpha}+\frac{1}{\beta}$ **e** $\alpha^3+\beta^3$ **f** $\frac{1}{\alpha^2}+\frac{1}{\beta^2}$

2 The equation $ax^2+3x-a^2=0$ has roots p and q. Find the value of each expression in terms of a.

a $3p+3q$ **b** p^2q^2 **c** p^2+q^2 **d** $(p-q)^2$

Cubic and quartic equations

For cubic equations there are three relationships between roots and coefficients. Just as in the proof of the relationships between coefficients and roots of a quadratic equation, if the roots of the equation $ax^3+bx^2+cx+d=0$ are p, q and r then you can write:

$$ax^3+bx^2+cx+d = a(x-p)(x-q)(x-r)$$

Expanding and comparing coefficients then gives the results.

Key point 2.5

If p, q and r are the roots of the cubic $ax^3+bx^2+cx+d=0$ then:

- $p+q+r = -\frac{b}{a}$
- $pq+qr+rp = \frac{c}{a}$
- $pqr = -\frac{d}{a}$

You can again combine these with algebraic identities to find other combinations of roots.

WORKED EXAMPLE 2.10

The equation $x^3 - 3x^2 + 4x + 7 = 0$ has roots α, β and γ.

Find the value of $\alpha^2 + \beta^2 + \gamma^2$.

$$\begin{aligned}(\alpha+\beta+\gamma)^2 &= (\alpha+(\beta+\gamma))^2 \\ &= \alpha^2 + 2\alpha(\beta+\gamma) + (\beta+\gamma)^2 \\ &= \alpha^2 + 2\alpha\beta + 2\alpha\gamma + \beta^2 + \gamma^2 + 2\beta\gamma \\ &= \alpha^2 + \beta^2 + \gamma^2 + 2(\alpha\beta + \beta\gamma + \gamma\alpha)\end{aligned}$$

You can find $\alpha^2 + \beta^2 + \gamma^2$ from the expansion of $(\alpha+\beta+\gamma)^2$.

$$\alpha + \beta + \gamma = -\frac{-3}{1} = 3$$

$$\alpha\beta + \beta\gamma + \gamma\alpha = \frac{4}{1} = 4$$

Now use Key point 2.5.

$$\alpha + \beta + \gamma = -\frac{b}{a}$$

$$\alpha\beta + \beta\gamma + \gamma\alpha = \frac{c}{a}$$

$$\begin{aligned}\therefore \alpha^2 + \beta^2 + \gamma^2 &= (3)^2 - 2(4) \\ &= 1\end{aligned}$$

Similar relationships between roots and coefficients can be found for polynomial equations of any degree. The expressions get increasingly complicated.

Key point 2.6

If p, q, r and s are the roots of the quartic $ax^4 + bx^3 + cx^2 + dx + e = 0$, then:

- $p + q + r + s = -\frac{b}{a}$
- $pq + pr + ps + qr + qs + rs = \frac{c}{a}$
- $pqr + pqs + prs + qrs = -\frac{d}{a}$
- $pqrs = \frac{e}{a}$

Tip

To help you to remember these equations, notice that the pattern is always the same.

- The sum of all the roots is related to $\frac{b}{a}$.
- The sum of all possible products of two roots is related to $\frac{c}{a}$.
- The sum of all possible products of three roots is related to $\frac{d}{a}$.

The signs alternate between – and +.

WORKED EXAMPLE 2.11

The equation $2x^4 - 3x^3 + x - 5 = 0$ has roots p, q, r and s.

Find the value of $\frac{1}{p} + \frac{1}{q} + \frac{1}{r} + \frac{1}{s}$.

$$\frac{1}{p} + \frac{1}{q} + \frac{1}{r} + \frac{1}{s} = \frac{pqr + pqs + prs + qrs}{pqrs}$$

Combine into a single fraction.

$$= \frac{\left(-\frac{1}{2}\right)}{\left(-\frac{5}{2}\right)}$$

$$= \frac{1}{5}$$

Use Key point 2.6.

$$pqr + pqs + prs + qrs = -\frac{d}{a}$$

$$pqrs = \frac{e}{a}$$

You can use information about the roots to prove facts about the coefficients of an equation.

Rewind

Remember that an arithmetic sequence is one in which each successive term changes by a constant number.

WORKED EXAMPLE 2.12

The roots p, q and r of the equation $x^3 - 3x^2 + cx + d = 0$ form an arithmetic sequence.

Show that $c + d = 2$.

$q - p = r - q$
$\Rightarrow p + r = 2q$

If p, q and r form an arithmetic sequence then $q - p = r - q$.

$p + q + r = -\frac{-3}{1} = 3$

You know three equations relating the roots to the coefficients. The most useful one seems to be $p + q + r = -\frac{b}{a}$.

$\therefore q + 2q = 3$
$q = 1$

Substitute from $p + r = 2q$.

$pq + qr + rp = c$
$p + r + rp = c$

To involve c and d you need to use the other two relationships as well. Start with $pq + qr + rp = \frac{c}{a}$ with $q = 1$.

$2 + rp = c$
$\Rightarrow rp = c - 2$

You know that $p + r = 2q = 2$.

$pqr = -d$
$pr = -d$

Finally use $pqr = -\frac{d}{a}$ with $q = 1$.

$\therefore c - 2 = -d$
$\Rightarrow c + d = 2$, as required.

Substitute from $rp = c - 2$.

EXERCISE 2D

1 The equation $3x^3 + 6x^2 + 12x - 4 = 0$ has roots p, q and r. Find the value of:

a $p + q + r$ **b** $\frac{1}{p} + \frac{1}{q} + \frac{1}{r}$ **c** $p^2qr + pq^2r + pqr^2$ **d** $p^2 + q^2 + r^2$.

2 α, β, γ and δ are the roots of the given quartic equation. In each case, find the value of the given expressions.

a $2x^4 - 4x^3 + 5x^2 + 3x - 1 = 0$; $\alpha\beta\gamma\delta$ and $\alpha\beta\gamma + \alpha\beta\delta + \alpha\gamma\delta + \beta\gamma\delta$.

b $3x^4 - 2x^2 + 1 = 0$; $\alpha + \beta + \gamma + \delta$ and $\alpha\beta + \alpha\gamma + \alpha\delta + \beta\gamma + \beta\delta + \gamma\delta$.

c $5x^4 - 3x - 8 = 0$; $\alpha + \beta + \gamma + \delta$ and $\alpha\beta\gamma\delta$.

3 The equation $4x^3 - 2x^2 + 4x + 1 = 0$ has roots p, q and r. Find the value of $p^2qr + pq^2r + pqr^2$.

4 The equation $3x^3 + 2x^2 - x + 5 = 0$ has roots a, b and c. Find the value of:

a $\frac{1}{a} + \frac{1}{b} + \frac{1}{c}$

b $\frac{1}{ab} + \frac{1}{bc} + \frac{1}{ca}$.

5 When two resistors of resistances R_1 and R_2 are connected in series in an electric circuit, the total resistance in the circuit is $R = R_1 + R_2$. When they are connected in parallel, the total resistance satisfies $\frac{1}{R} = \frac{1}{R_1} + \frac{1}{R_2}$.

Two resistors have resistances equal to the two roots of the quadratic equation $3R^2 - 12R + 4 = 0$. Find the total resistance in the circuit if the two resistors are connected:

a in series

b in parallel.

6 The cubic equation $3x^3 - 5x - 3 = 0$ has roots α, β and γ.

a Write down the value of $\alpha\beta\gamma$.

b Show that $\alpha + \beta = -\frac{1}{\alpha\beta}$.

7 A random-number generator can produce four possible values, all of which are equally likely. The four values satisfy the equation $x^4 - 9x^3 + 26x^2 - 29x + 10 = 0$.

If a large number of random values are generated, estimate their mean.

8 The equation $x^4 + bx^3 + cx^2 + dx + e = 0$ has roots p, $2p$, $3p$ and $4p$. Show that $125e = \frac{3}{10}b^4$.

9 The equation $x^3 + 2ax^2 + 3a^2x + 2 = 0$, where a is a real constant, has roots p, q and r.

a Find an expression for $p^2 + q^2 + r^2$ in terms of a.

b Explain why this implies that the roots are not all real.

Section 4: Finding an equation with given roots

You can use the relationships between roots of polynomials and their coefficients to find unknown coefficients in an equation with given roots.

Rewind

You can already do this by writing down the factors and expanding brackets (see Worked example 2.7), but this method is more direct.

WORKED EXAMPLE 2.13

The quadratic equation $5x^2 + bx + c = 0$ has real coefficients and one of its roots is $4 + 7i$. Find the values of b and c.

The other root is $4 - 7i$. — Complex roots occur in conjugate pairs.

Then: $(4 + 7i) + (4 - 7i) = -\frac{b}{5}$ — Use $p + q = -\frac{b}{a}$.

$8 = -\frac{b}{5}$

$b = -40$

Continues on next page

$(4+7\text{i})(4-7\text{i}) = \frac{c}{5}$ — Use $pq = \frac{c}{a}$.

$4^2 + 7^2 = \frac{c}{5}$ — Remember: $zz^* = |z|^2$.

$c = 325$

WORKED EXAMPLE 2.14

A quartic equation $x^4 + ax^3 + 14x^2 - 18x + b = 0$ has real coefficients and two of its roots are 3i and $1-2\text{i}$. Find the values of a and b.

The four roots are: $3\text{i}, -3\text{i}, 1+2\text{i}, 1-2\text{i}$ — As you know two of the complex roots, you can find the other two (their conjugates).

$(3\text{i}) + (-3\text{i}) + (1-2\text{i}) + (1+2\text{i}) = -\frac{a}{1}$ — Use $p+q+r+s = -\frac{b}{a}$, being careful to note how the coefficients have been labelled in this question (a is the coefficient of x^3 here).

$2 = -\frac{a}{1}$

$a = -2$

$(3\text{i})(-3\text{i})(1-2\text{i})(1+2\text{i}) = \frac{b}{1}$ — Use $pqrs = \frac{e}{a}$.

$9(1^2 + 2^2) = b$ — Use $zz^* = |z|^2$.

$b = 45$

You can also find a new equation with roots that are related to the roots of a given equation in some way, and you can do this without solving the equation. The strategy is to use the sum and product of roots of the first equation to find the sum and product of roots of the second equation.

WORKED EXAMPLE 2.15

The quadratic equation $3x^2 - 4x + 7 = 0$ has roots p and q. Find a quadratic equation with integer coefficients and roots p^2 and q^2.

$p + q = -\frac{-4}{3} = \frac{4}{3}$ — You don't need to find p and q - just their sum and product.

$pq = \frac{7}{3}$

Let the equation be $ax^2 + bx + c = 0$. — The coefficients of the new equation are related to the roots p^2 and q^2.

Then: $p^2 + q^2 = -\frac{b}{a}$ and $p^2q^2 = \frac{c}{a}$

Continues on next page

Set $a = 1$.

Then: $b = -(p^2 + q^2)$ and $c = p^2q^2$.

All equations with the required roots are multiples of each other, so you can set $a = 1$.

$$c = p^2q^2 = (pq)^2 = \left(\frac{7}{3}\right)^2 = \frac{49}{9}$$

You need to relate $p^2 + q^2$ and p^2q^2 to $p + q$ and pq. The second one is easier.

Substitute $pq = \frac{7}{3}$.

$$(p+q)^2 = p^2 + 2pq + q^2$$
$$\Rightarrow p^2 + q^2 = (p+q)^2 - 2pq$$

You can square $p + q$ to get $p^2 + q^2$.

$$b = -(p^2 + q^2) = -((p+q)^2 - 2pq) = -\left(\left(\frac{4}{3}\right)^2 - 2\left(\frac{7}{3}\right)\right) = \frac{26}{9}$$

Substitute $p + q = \frac{4}{3}$ and $pq = \frac{7}{3}$.

The equation is: $x^2 + \frac{26}{9}x + \frac{49}{9} = 0$

Substitute into $ax^2 + bx + c = 0$.

$$\Leftrightarrow 9x^2 + 26x + 49 = 0$$

You want the equation with integer coefficients, so multiply through by 9.

WORKED EXAMPLE 2.16

The equation $x^3 - 3x^2 + 2 = 0$ has roots p, q and r. Find a cubic equation with roots $3p$, $3q$ and $3r$.

For the equation $x^3 - 3x^2 + 2 = 0$:

$$p + q + r = 3$$
$$pq + qr + rp = 0$$
$$pqr = -2$$

Use Key point 2.5 for the original equation.

If the equation with roots $3p$, $3q$, and $3r$ is $x^3 + bx^2 + cx + d = 0$ then:

Now use the roots to find the coefficients. You can set the coefficient of x^3 to be 1.

$$3p + 3q + 3r = -b$$

Using the first part of Key point 2.5 for the new equation.

Continues on next page

$3(p+q+r)=-b$
$3(3)=-b$
$b=-9$

$p+q+r=3$

$(3p)(3q)+(3q)(3r)+(3r)(3p)=c$
$9(pq+qr+rp)=c$
$9(0)=c$
$c=0$

Using the second part of Key point 2.5 for the new equation.

$pq+qr+rp=0$

$(3p)(3q)(3r)=-d$
$27pqr=-d$
$27(-2)=-d$
$d=54$

Using the third part of Key point 2.5 for the new equation.

$pqr=-2$

The equation is $x^3-9x^2+54=0$.

Fast forward

You will see in Section 5 that you could also find the equation from Worked example 2.16 by using the substitution $u=3x$.

WORKED EXAMPLE 2.17

The equation $3x^3-5x^2+x-3=0$ has roots α, β and γ. Find a cubic equation with roots $\alpha\beta$, $\beta\gamma$ and $\gamma\alpha$.

For the equation $3x^3-5x^2+x-3=0$:

$\alpha+\beta+\gamma=\frac{5}{3}$

$\alpha\beta+\beta\gamma+\gamma\alpha=\frac{1}{3}$

$\alpha\beta\gamma=\frac{3}{3}$

Use Key point 2.5 for the original equation.

If the equation with roots $\alpha\beta$, $\beta\gamma$ and $\gamma\alpha$ is $x^3+bx^2+cx+d=0$ then:

You can take the coefficient of x^3 to be 1.

$\alpha\beta+\beta\gamma+\gamma\alpha=-b$

$\frac{1}{3}=-b$

$b=-\frac{1}{3}$

By Key point 2.5, the sum of the roots of the new equations is $-\frac{b}{1}$.

$(\alpha\beta)(\beta\gamma)+(\beta\gamma)(\gamma\alpha)+(\gamma\alpha)(\alpha\beta)=c$

Use the second part of Key point 2.5.

$\alpha\beta\gamma(\alpha+\beta+\gamma)=c$

Factorise.

$\left(\frac{3}{3}\right)\left(\frac{5}{3}\right)=c$

$c=\frac{5}{3}$

Continues on next page

$(\alpha\beta)(\beta\gamma)(\gamma\alpha) = -d$

$(\alpha\beta\gamma)^2 = -d$

$\left(\frac{3}{3}\right)^2 = -d$

$d = -1$

The product of the roots is $-\frac{d}{1}$ by Key point 2.5.

So the equation is

$x^3 - \frac{1}{3}x^2 + \frac{5}{3}x - 1 = 0.$

You can also write this equation as $3x^3 - x^2 + 5x - 3 = 0$.

EXERCISE 2E

1 Given the roots of the equations, find the missing coefficients.

a **i** $3x^3 - ax^2 - 3x + b = 0$; roots 1, −1, −2 **ii** $2x^3 + ax^2 + 10x + c = 0$; roots 1, 1, 2

b **i** $x^3 - ax^2 + 4x - b = 0$; roots 3, 2i, −2i **ii** $x^3 + bx^2 + 9x + d = 0$; roots 2 + i, 2 − i, 1

2 Find the polynomial of the lowest possible order with the given roots.

a **i** 5 + 2i, 5 − 2i **ii** 3 − i, 3 + i **b** **i** 1, 3i, −3i **ii** 5, i, −i

c **i** 2, −1, 1 + 2i, 1 − 2i **ii** −3, 1, 2 + i, 2 − i **d** **i** −2, 4 + 3i, 4 − 3i **ii** 1, −2 + 3i, −2 − 3i

3 A cubic equation has real coefficients and two of its roots are 2 and 4 − i.

a Write down the third root.

b Find the equation in the form $x^3 + bx^2 + cx + d = 0$.

4 The quartic equation $x^4 - ax^3 + bx^2 - cx + d = 0$ has real coefficients, and two of its roots are 3i and 3 − i.

a Write down the other two roots. **b** Hence find the values of a and d.

5 The equation $x^3 - 3x^2 + 4x + 1 = 0$ has roots p, q and r. Find a cubic equation with roots $3p$, $3q$ and $3r$.

6 The equation $4x^3 - 3x + 5 = 0$ has roots α, β and γ. Find a cubic equation with integer coefficients and roots $\alpha - 2$, $\beta - 2$ and $\gamma - 2$.

7 The quadratic equation $5x^2 - 3x + 2 = 0$ has roots p and q. Find a quadratic equation with roots $\frac{1}{p}$ and $\frac{1}{q}$.

8 The equation $x^4 - 3x^3 + x + 2 = 0$ has roots a, b, c and d. Find a quartic equation with roots $2a$, $2b$, $2c$ and $2d$.

9 Let p and q be the roots of the equation $5x^2 - 3x + 2 = 0$.

a Find the values of pq and $p^2 + q^2$.

b Hence find a quadratic equation with integer coefficients and roots p^2 and q^2.

10 The equation $x^3-3x^2+5=0$ has roots α, β and γ.

a Find the value of $\frac{1}{\alpha\beta}+\frac{1}{\beta\gamma}+\frac{1}{\gamma\alpha}$.

b Find a cubic equation with roots $\frac{1}{\alpha}$, $\frac{1}{\beta}$ and $\frac{1}{\gamma}$.

11 The equation $4x^3-3x+7=0$ has roots p, q and r. Find a cubic equation with integer coefficients and roots pq, qr and rp.

12 a Expand $(\alpha\beta+\beta\gamma+\gamma\alpha)^2$.

b If α, β and γ are the roots of the equation $x^3-x^2+2x+6=0$ find a cubic equation with roots α^2, β^2 and γ^2.

Section 5: Transforming equations

In Section 4, you saw how to use relationships between roots and coefficients of a polynomial equation to find another equation with roots that are related to the roots of the first one.

In some cases, however, there is an easier way to find this equation - by using a substitution.

Suppose the original equation, with roots p and q, is $5x^2-2x+1=0$.

If you set $x=\frac{u}{3}$ you get a new equation.

$$5\left(\frac{u}{3}\right)^2-2\left(\frac{u}{3}\right)+1=0$$
$$\Leftrightarrow 5u^2-6u+9=0$$

Since $u=3x$ this equation has roots $3p$ and $3q$.

Key point 2.7

Given an equation in x that has a root $x=p$, if you make a substitution $u=\mathrm{f}(x)$, then the resulting equation in u has a root $u=\mathrm{f}(p)$.

WORKED EXAMPLE 2.18

The equation $3x^3-x^2+2x+5=0$ has roots p, q and r.

a Find a cubic equation with roots $p-2$, $q-2$ and $r-2$.
b Hence find the value of $(p-2)(q-2)(r-2)$.

a $u=x-2$
$\Rightarrow x=u+2$

When $x=p$, $u=p-2$ so make the substitution $u=x-2$.

Continues on next page

$3(u+2)^3-(u+2)^2+2(u+2)+5=0$ — Substitute for x.

$3(u^3+6u^2+12u+8)-(u^2+4u+4)+2u+4+5=0$

$3u^3+17u^2+34u+29=0$

b $\therefore (p-2)(q-2)(r-2)=-\frac{29}{3}$ — This is the product of the roots of the equation in u.

Tip

If you were just asked to find $(p-2)(q-2)(r-2)$, you could expand $(p-2)(q-2)(r-2)$ and use the relationships between roots and coefficients of the original equation. You can decide for yourself whether this or the substitution method is simpler.

WORK IT OUT 2.2

The roots of the quadratic equation $3x^2-x+5=0$ are p and q. Find a quadratic equation with roots $\frac{p}{5}$ and $\frac{q}{5}$.

Which is the correct solution? Identify the errors made in the incorrect solutions.

Solution 1	Solution 2	Solution 3
From the first equation: $p+q=1$ and $pq=5$ So for the second equation: $\frac{p}{5}+\frac{q}{5}=\frac{1}{5}$ and $\frac{p}{5}\times\frac{q}{5}=\frac{5}{25}=\frac{1}{5}$ Hence the new equation is: $x^2-\frac{1}{5}x+\frac{1}{5}=0$ $\Leftrightarrow 5x^2-x+1=0$	Let $u=\frac{x}{5}$; then $\frac{p}{5}$ and $\frac{q}{5}$ are the roots for u. Make the substitution $x=5u$. $3(5u)^2-(5u)+5=0$ $\Leftrightarrow 15u^2-u+1=0$	Replace x by $\frac{x}{5}$: $3\left(\frac{x}{5}\right)^2-\left(\frac{x}{5}\right)+1=0$ $\Leftrightarrow 3x^2-5x+25=0$

A substitution is particularly useful if it transforms a difficult equation into one that you can solve easily.

WORKED EXAMPLE 2.19

Use the substitution $x = u - 1$ to solve the equation $x^4 + 4x^3 + 6x^2 + 4x - 80 = 0$.

$(u-1)^4 + 4(u-1)^3 + 6(u-1)^2 + 4(u-1) - 80 = 0$ — Make the given substitution.

$$\begin{aligned} &u^4 - 4u^3 + 6u^2 - 4u + 1 \\ &\quad + 4u^3 - 12u^2 + 12u - 4 \\ &\qquad + 6u^2 - 12u + 6 \\ &\qquad\quad + 4u - 4 - 80 = 0 \end{aligned}$$

It's a good idea to line up terms when expanding lots of brackets.

$u^4 - 81 = 0$ — You can solve the equation in u easily by factorising.

$(u^2 - 9)(u^2 + 9) = 0$

$u = \pm 3, u = \pm 3\text{i}$

$\therefore x = 2,\ x = -4,\ x = -1 + 3\text{i},\ x = -1 - 3\text{i}$ — Using $x = u - 1$.

EXERCISE 2F

1 Use the given substitution to transform the equation in x into a polynomial equation for u.

a **i** $x^3 - 3x + 1 = 0;\ x = 2u.$ **ii** $x^3 + 2x^2 + 5 = 0;\ x = 3u.$

b **i** $3x^3 - x + 4 = 0;\ x = u - 2.$ **ii** $2x^3 + x^2 + 1 = 0;\ x = u + 1.$

c **i** $x^3 - 3x^2 + 4x + 15 = 0;\ x = 2u + 1.$ **ii** $x^3 + x^2 - 6x + 10 = 0;\ x = 3u - 2.$

2 The equation $3x^3 - 4x + 2 = 0$ has roots p, q and r. Use a suitable substitution to find a cubic equation with roots $p - 1$, $q - 1$ and $r - 1$.

3 The equation $2x^4 + 2x + 5 = 0$ has roots a, b, c and d. Find a quartic equation with integer coefficients and roots $\frac{a}{2}, \frac{b}{2}, \frac{c}{2}$ and $\frac{d}{2}$.

4 **a** Show that the substitution $x = u - 2$ transforms the equation $x^3 + 6x^2 + 21x + 26 = 0$ into the equation $u^3 + 9u = 0$.

b Hence solve the equation $x^3 + 6x^2 + 21x + 26 = 0$.

5 **a** Find the value of c so that the substitution $x = u + c$ transforms the equation $x^3 - 12x^2 + 45x - 54 = 0$ into the equation $u^3 - 3u^2 = 0$.

b Hence find all the solutions of the equation $x^3 - 12x^2 + 45x - 54 = 0$.

6 The equation $3x^2 - 2x + 5 = 0$ has roots α and β. Using the substitution $x = 3u + 1$, or otherwise, find the value of $\left(\frac{\alpha - 1}{3}\right)\left(\frac{\beta - 1}{3}\right)$.

7 The equation $3x^2 - 9x + 1 = 0$ has roots α and β. Using a suitable substitution, or otherwise, find an equation with roots α^2 and β^2.

8 The substitution $x = u - k$ transforms the equation $x^4 + 4x^3 + 11x^2 + 14x + 10 = 0$ into the equation of the form $u^4 + bu^2 + c = 0$.

a Find the value of k.

b Hence solve the equation $x^4 + 4x^3 + 11x^2 + 14x + 10 = 0$.

Checklist of learning and understanding

- Complex roots of real polynomials occur in conjugate pairs: if $f(z) = 0$ then $f(z^*) = 0$. You can use this fact to factorise cubic and quartic polynomials.
- Coefficients of a polynomial can be expressed in terms of its roots.
  - For a quadratic equation $ax^2 + bx + c = 0$ with roots p and q:
    - $p + q = -\frac{b}{a}$
    - $pq = \frac{c}{a}$
  - For a cubic equation $ax^3 + bx^2 + cx + d = 0$ with roots p, q and r:
    - $p + q + r = -\frac{b}{a}$
    - $pq + qr + rp = \frac{c}{a}$
    - $pqr = -\frac{d}{a}$
  - For a quartic equation $ax^4 + bx^3 + cx^2 + dx + e = 0$ with roots p, q, r and s:
    - $p + q + r + s = -\frac{b}{a}$
    - $pq + pr + ps + qr + qs + rs = \frac{c}{a}$
    - $pqr + pqs + prs + qrs = -\frac{d}{a}$
    - $pqrs = \frac{e}{a}$
- You can use these relationships to find:
  - a polynomial with given roots
  - a polynomial with roots related to the roots of another polynomial.
- The second of these types of problem can sometimes also be solved by making a substitution.

Mixed practice 2

1. Find the sum of the roots of the equation $3x^4 + 4x^2 - x - 7 = 0$.

   Choose from these options.

   A $\frac{4}{3}$ B $-\frac{4}{3}$ C $-\frac{7}{3}$ D 0

2. The quadratic equation $x^2 - 2kx + k^2 = 0$ is transformed into an equation in u by the substitution $x = \frac{u-1}{2}$.

   Find the root of the transformed equation in terms of the constant k.

   Choose from these options.

   A $\frac{k-1}{2}$ B $\frac{-k-1}{2}$ C $2k+1$ D $-2k-1$

3. Given that $z = 3\text{i}$ is one root of the equation $z^3 - 2z^2 + 9z - 18 = 0$, find the other two roots.

4. $\text{f}(z) = z^3 + az^2 + bz + c$ where a, b, c are real constants. Two roots of $\text{f}(z) = 0$ are $z = 1$ and $z = 1 + 2\text{i}$. Find a, b and c.

5. One of the roots of the polynomial $\text{g}(x) = x^3 + 3x^2 - 7x + 15$ is $1 + \text{i}\sqrt{2}$.

   a Write down another complex root and hence find a real quadratic factor of $\text{g}(x)$.

   b Solve the equation $\text{g}(x) = 0$.

6. Find a quartic equation with real coefficients given that three of its roots are 2i, -1 and 5.

7. The quadratic equation $5x^2 - 7x + 1 = 0$ has roots α and β.

   a Write down the values of $\alpha + \beta$ and $\alpha\beta$.

   b Show that $\frac{\alpha}{\beta} + \frac{\beta}{\alpha} = \frac{39}{5}$.

   c Find a quadratic equation with integer coefficients, which has roots $\alpha + \frac{1}{\alpha}$ and $\beta + \frac{1}{\beta}$.

   [©AQA 2012]

8. Given that $z = 1 + 2\text{i}$ is one root of the equation $z^3 + z^2 - z + 15 = 0$, find the other two roots.

9. Two roots of the cubic equation $z^3 + bz^2 + cz + d = 0$ $(b, c, d \in \mathbb{R})$ are -2 and $2 - 3\text{i}$.

   a Write down the third root.

   b Find the values of b, c and d.

10. The polynomial $z^3 + az^2 + bz - 65$ has a factor of $(z - 2 - 3\text{i})$. Find the values of the real constants a and b.

11. a Show that $\alpha^3 + \beta^3 = (\alpha + \beta)^3 - 3\alpha\beta(\alpha + \beta)$.

    b Let α and β be the roots of the quadratic equation $x^2 + 7x + 2 = 0$. Find a quadratic equation with roots α^3 and β^3.

12 The equation $3x^3 - 4x^2 + 7x + 1 = 0$ has roots p, q and r. Use the substitution $u = x - 3$ to find a cubic equation with roots $p - 3$, $q - 3$ and $r - 3$.

13 The equation $5x^3 - 9x + 4 = 0$ has roots α, β and γ. Use a substitution of the form $u = kx$ to find a cubic equation with roots $\frac{\alpha}{2}$, $\frac{\beta}{2}$ and $\frac{\gamma}{2}$.

14 The cubic equation $z^3 + pz + q = 0$ has roots α, β and γ.

a i Write down the value of $\alpha + \beta + \gamma$. ii Express $\alpha\beta\gamma$ in terms of q.

b Show that $\alpha^3 + \beta^3 + \gamma^3 = 3\alpha\beta\gamma$.

c Given that $\alpha = 4 + 7\mathrm{i}$ and that p and q are real, find the values of:

i β and γ ii p and q.

d Find a cubic equation with integer coefficients which has roots $\frac{1}{\alpha}$, $\frac{1}{\beta}$ and $\frac{1}{\gamma}$.

[©AQA 2012]

A **15** **a** A cubic equation $ax^3 + bx^2 + cx + d = 0$ has roots x_1, x_2, x_3.

i Write down the values of $x_1 + x_2 + x_3$ and $x_1x_2x_3$ in terms of a, b, c and d.

ii Show that $x_1x_2 + x_2x_3 + x_3x_1 = \frac{c}{a}$.

b The roots α, β and γ of the equation $2x^3 + bx^2 + cx + 16 = 0$ form a geometric sequence.

i Show that $\beta = -2$.

ii Show that $c = 2b$.

16 **a** Show that:

i $p^2 + q^2 + r^2 = (p + q + r)^2 - 2(pq + qr + rp)$

ii $p^2q^2 + q^2r^2 + r^2p^2 = (pq + qr + rp)^2 - 2pqr(p + q + r)$.

b Given that the cubic equation $ax^3 + bx^2 + cx + d = 0$ has roots p, q and r:

i write down the values of $p + q + r$ and pqr in terms of a, b, c and d

ii show that $pq + qr + rp = \frac{c}{a}$.

c The equation $2x^3 - 5x + 2 = 0$ has roots x_1, x_2 and x_3.

i Show that $x_1^2 + x_2^2 + x_3^2 = 5$.

ii Find the values of $x_1^2x_2^2 + x_2^2x_3^2 + x_3^2x_1^2$ and $x_1^2x_2^2x_3^2$.

iii Hence find a cubic equation with integer coefficients and roots x_1^2, x_2^2 and x_3^2.

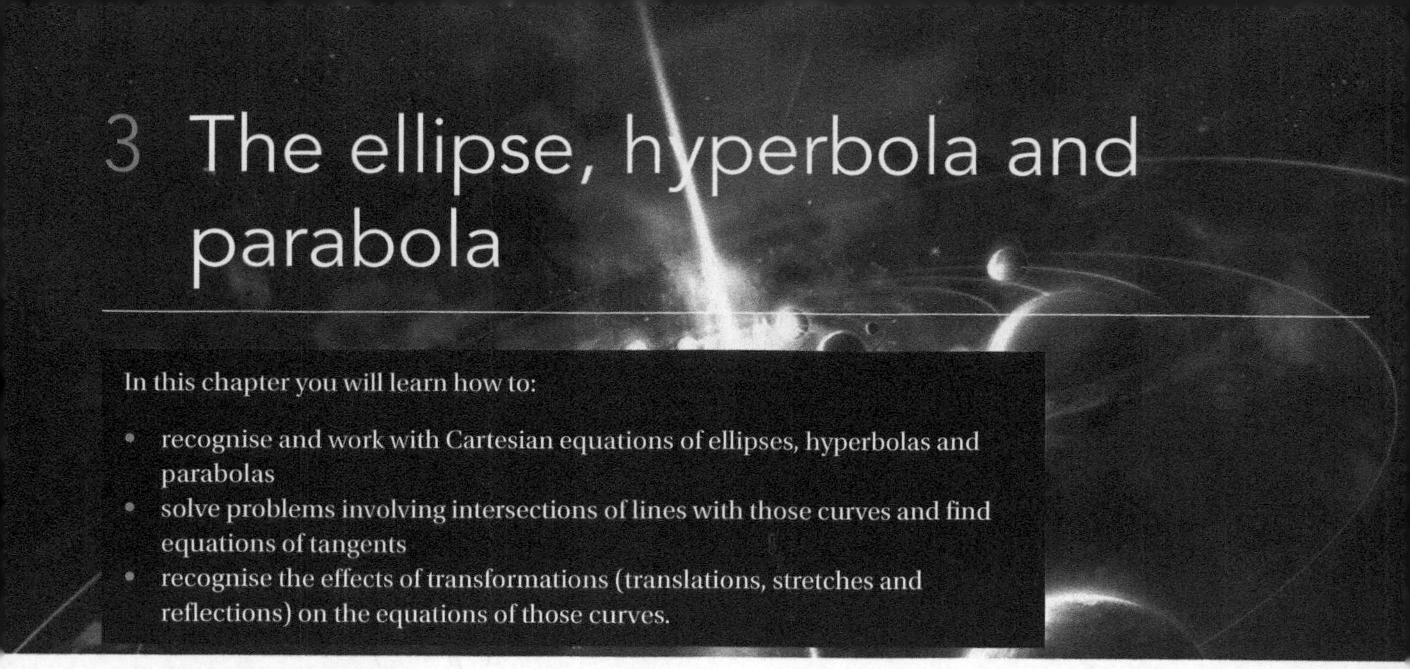

3 The ellipse, hyperbola and parabola

In this chapter you will learn how to:

- recognise and work with Cartesian equations of ellipses, hyperbolas and parabolas
- solve problems involving intersections of lines with those curves and find equations of tangents
- recognise the effects of transformations (translations, stretches and reflections) on the equations of those curves.

Before you start...

A Level Mathematics Student Book 1, Chapter 3	You should be able to use the discriminant to determine the number of solutions of a quadratic equation.	1 Find the set of values of k for which the equation $kx^2 - 2x + 5 = 0$ has exactly two real roots.
A Level Mathematics Student Book 1, Chapter 5	You should be able to interpret solutions of simultaneous equations as intersections of graphs.	2 Find the value of c for which the line $y = 3x + c$ is a tangent to the graph of $y = x^2 + 5$.
A Level Mathematics Student Book 1, Chapter 5	You should be able to recognise the graph of $y = \frac{k}{x}$, and know that it has asymptotes.	3 Sketch the graph of $y = \frac{3}{x}$ and state the equations of its asymptotes.
A Level Mathematics Student Book 1, Chapter 5	You should be able to recognise the relationship between transformations of graphs and their equations.	4 Find the equation of the resulting curve after the graph of $y = x^2 - 3x$ is: a translated 3 units in the positive x-direction b stretched vertically with scale factor 3.
A Level Mathematics Student Book 1, Chapter 6	You should be able to write down the equation of a circle with a given centre and radius, and to find the centre and radius from a given equation.	5 Find the centre and radius of the circle with equation $x^2 + 4x + y^2 - 10y = 7$.

Conic sections

In A Level Mathematics Student Book 1 you met two curves with equations that involve squared terms: the **parabola**, such as $y = 3x^2$, and the **circle**, such as $x^2 + y^2 = 5$. In this chapter you will meet some other curves with similar equations; for example: $y^2 = 3x$, $2x^2 + y^2 = 5$ and $x^2 - y^2 = 5$.

All such curves can be obtained as intersections of a plane with a cone; hence they are known as **conic sections**. As well as having many interesting mathematical properties, they have applications in modelling planetary orbits and the design of satellite dishes and radio telescopes.

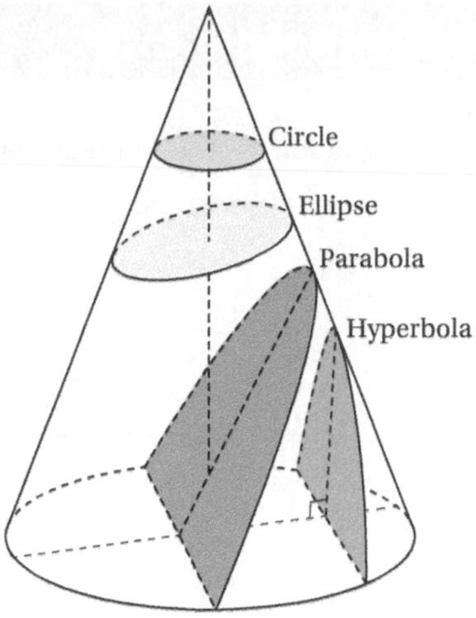

Section 1: Introducing the ellipse, hyperbola and parabola

The ellipse

You already know that the equation $x^2 + y^2 = 25$ represents a circle with radius 5, centred at the origin. Similarly, the equation $4x^2 + 4y^2 = 25$ can be rewritten as $x^2 + y^2 = \frac{25}{4}$, so it represents a circle with radius $\frac{5}{2}$. But what happens if the coefficients of x^2 and y^2 in this equation are not equal?

Use graphing software to investigate equations of this form. Here are some examples.

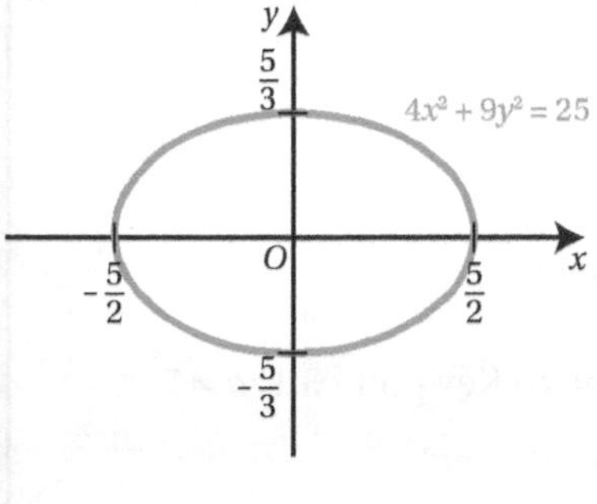

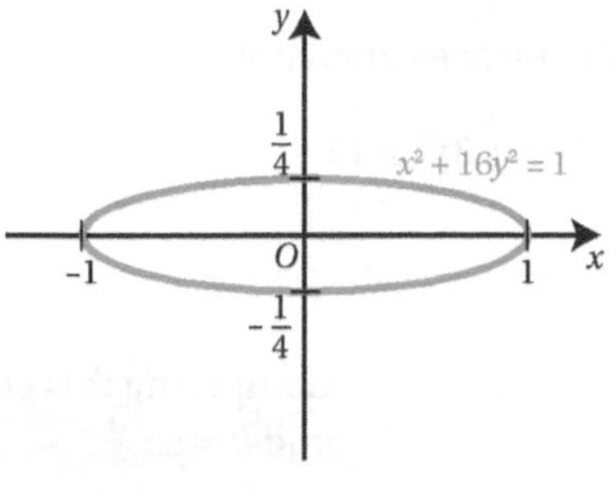

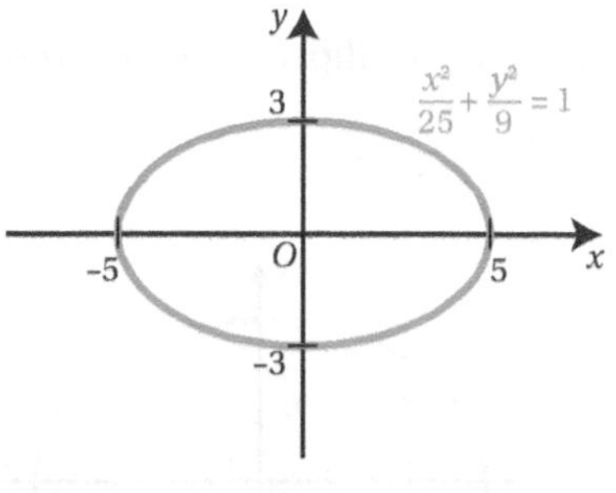

In each case there is a closed curve centred at the origin. The axes intercepts are most easily seen if the equation is written in the form $\frac{x^2}{a^2} + \frac{y^2}{b^2} = 1$.

Key point 3.1

The equation $\frac{x^2}{a^2}+\frac{y^2}{b^2}=1$ represents an **ellipse** centred at the origin, with x-intercepts $(\pm a, 0)$ and y-intercepts $(0, \pm b)$.

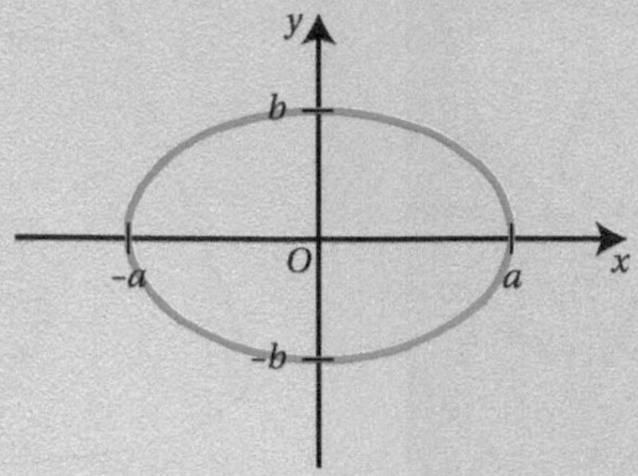

Did you know?

It can be shown, using the universal law of gravitation and Newton's laws of motion, that planets follow elliptical orbits.

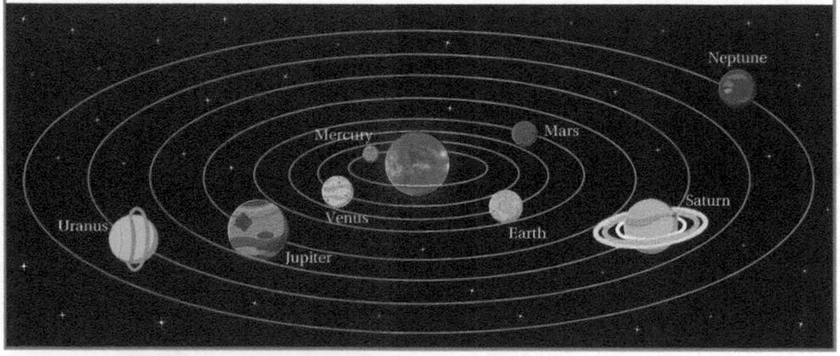

WORKED EXAMPLE 3.1

Sketch these ellipses, showing the coordinate axes intercepts.

a $\frac{x^2}{49}+\frac{y^2}{25}=1$ **b** $48x^2+3y^2=12$

a

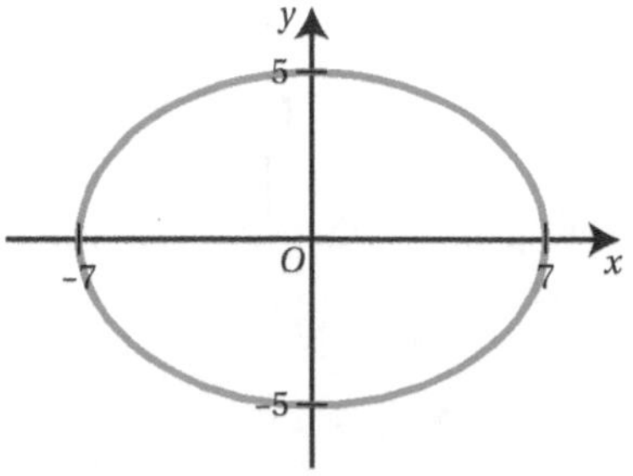

Comparing this equation to Key point 3.1, $a=7$ and $b=5$.

b $48x^2+3y^2=12$

$4x^2+\frac{1}{4}y^2=1$

To compare this to Key point 3.1, the right-hand side needs to equal 1. So divide the whole equation by 12.

Continues on next page

$\frac{x^2}{\frac{1}{4}} + \frac{y^2}{4} = 1$ — Multiplying by 4 is the same as dividing by $\frac{1}{4}$.

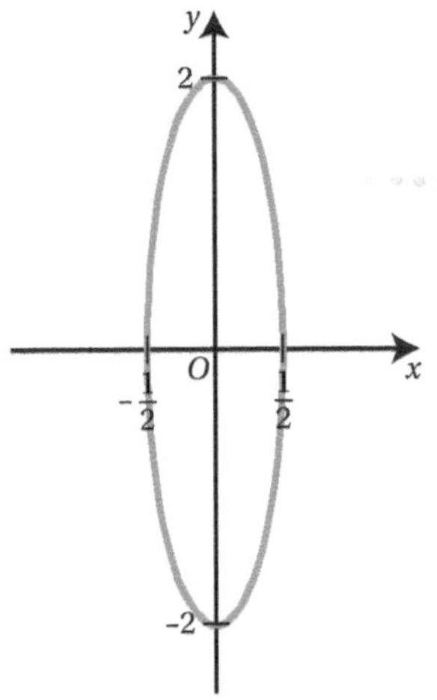

You can now see that this is an ellipse with $a = \frac{1}{2}$ and $b = 2$.

The hyperbola

What happens if, in the equation of an ellipse, the y^2 term is negative? Here are some examples.

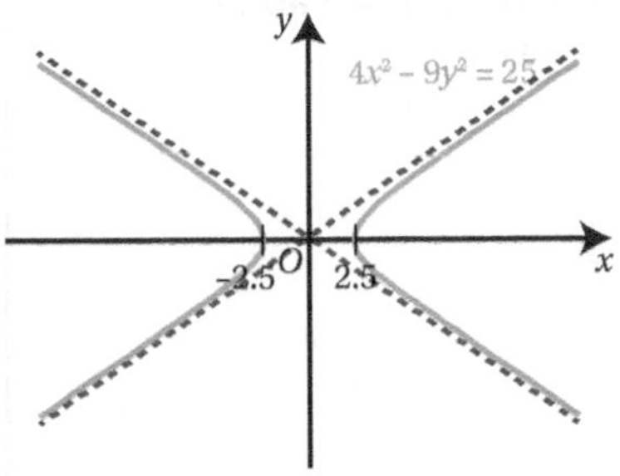

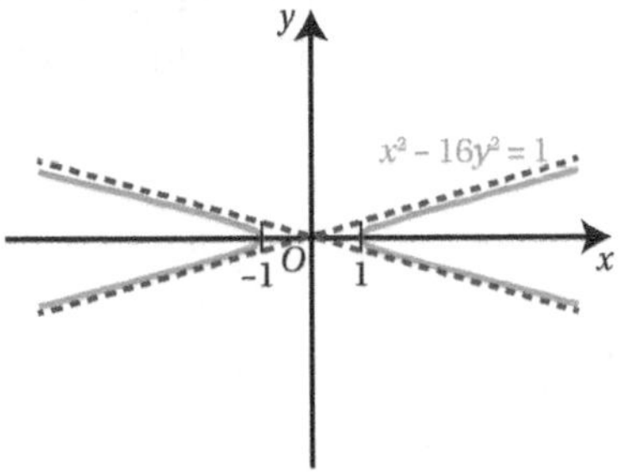

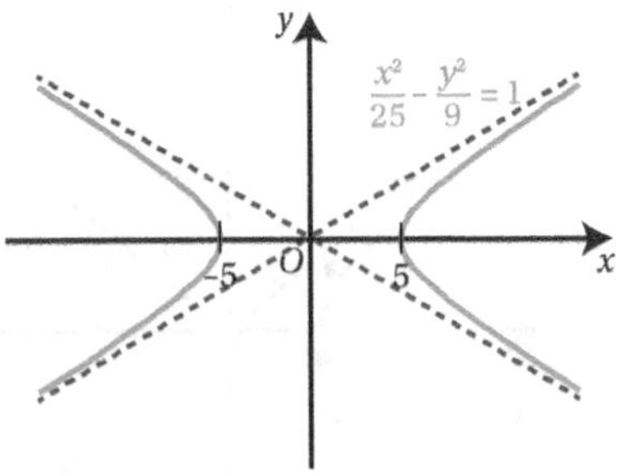

Each of these graphs is a curve called a **hyperbola**. As with the ellipse, it is convenient to write the equation in the form $\frac{x^2}{a^2} - \frac{y^2}{b^2} = 1$.

You can see several common features in all the graphs:

- The curve has the x and y-axes as lines of symmetry. This is because both x and y terms in the equation are squared.
- You can find the x-intercepts, which are also the vertices of the curve, by setting $y = 0$; so, for example, the curve with equation $\frac{x^2}{25} - \frac{y^2}{9} = 1$ crosses the x-axis at $(-5, 0)$ and $(5, 0)$.
- There is a range of values of x for which the curve is not defined. For example, the curve with equation $\frac{x^2}{25} - \frac{y^2}{9} = 1$ is not defined for $-5 < x < 5$.
- For large values of x and y, the curve seems to approach a straight line. Because of the symmetry of the curve, this line passes through the origin.

Tip

To see why $\frac{x^2}{25} - \frac{y^2}{9} = 1$ isn't defined for $-5 < x < 5$, rewrite the equation in the form $x^2 = 25\left(1 + \frac{y^2}{9}\right)$. Since y^2 is never negative, x^2 can't be smaller than $25(1 + 0) = 25$, so x can't be between -5 and 5.

Key point 3.2

The equation $\frac{x^2}{a^2} - \frac{y^2}{b^2} = 1$ represents a hyperbola with x-intercepts $(\pm a, 0)$ and asymptotes $y = \pm\frac{b}{a}x$.

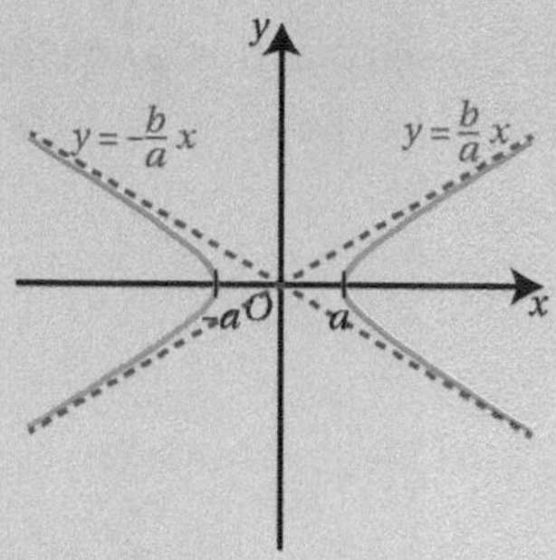

WORKED EXAMPLE 3.2

Sketch these hyperbolas, showing the axis intercepts and stating the equations of the asymptotes.

a $\frac{x^2}{36} - \frac{y^2}{25} = 1$ **b** $9x^2 - 25y^2 = 4$

a

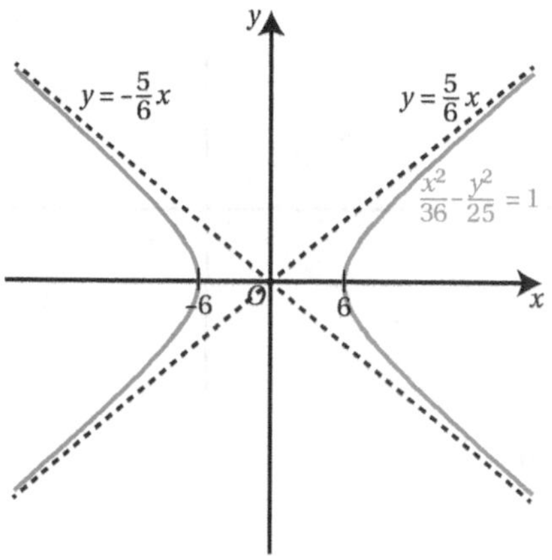

Comparing this to Key point 3.2, $a = 6$ and $b = 5$. The x-intercepts are $(\pm 6, 0)$.

The asymptotes are $y = \pm\frac{5}{6}x$.

The asymptotes are given by $y = \pm\frac{b}{a}x$.

b

$9x^2 - 25y^2 = 4$

First write the equation in the form $\frac{x^2}{a^2} - \frac{y^2}{b^2} = 1$.

$\frac{9x^2}{4} - \frac{25y^2}{4} = 1$

$\frac{x^2}{\frac{4}{9}} - \frac{y^2}{\frac{4}{25}} = 1$

Multiplying by $\frac{9}{4}$ is the same as dividing by $\frac{4}{9}$.

Continues on next page

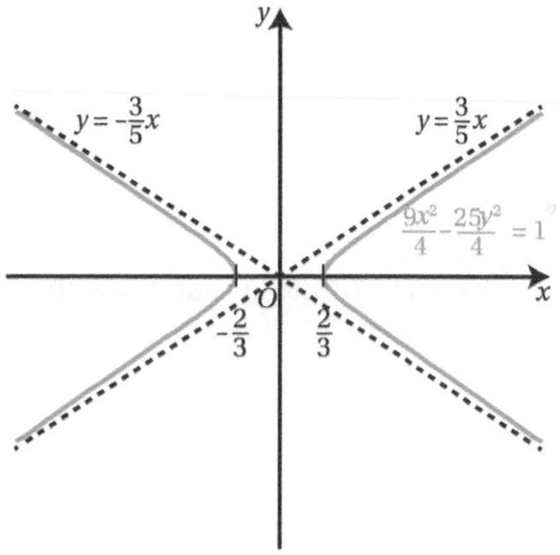

You can now see that this is a hyperbola with $a = \frac{2}{3}$ and $b = \frac{2}{5}$.

Hence the x-intercepts are $\left(\pm\frac{2}{3}, 0\right)$.

Asymptotes:

$y = \pm\frac{\frac{2}{5}}{\frac{2}{3}}x$

$y = \pm\frac{3}{5}x$

The asymptotes are given by $y = \pm\frac{b}{a}x$.

You can derive the equations of the asymptotes of a hyperbola (rather than just guessing them from the graph as you did in Worked example 3.2).

PROOF 4

Prove that the asymptotes of a hyperbola with equation $\frac{x^2}{a^2}-\frac{y^2}{b^2}=1$ are $y=\pm\frac{b}{a}x$.

The asymptotes pass through the origin, so their equations are $y = \pm mx$.

First, by symmetry of the curve, the asymptotes must pass through the origin so the y-intercept is zero. The two asymptotes must also be reflections of each other in the x-axis so their equations are of the form $y = mx$ and $y = -mx$.

Consider a line of the form $y = mx$. Whether or not this line crosses the hyperbola depends on the value of m.

The hyperbola never intersects its asymptotes, so consider the intersection of the curve with the lines of the form $y = \pm mx$.

$$\frac{x^2}{a^2}-\frac{(mx)^2}{b^2}=1$$
$$b^2x^2 - a^2m^2x^2 = a^2b^2$$
$$(b^2 - a^2m^2)x^2 = a^2b^2$$

To find any possible intersections, substitute $y = mx$ into the equation of the hyperbola.

Multiply both sides by the common denominator (a^2b^2) to clear fractions.

Solutions only exist if $b^2 - a^2m^2 > 0$

$$\Leftrightarrow a^2m^2 < b^2$$
$$\Leftrightarrow m^2 < \frac{b^2}{a^2}$$
$$\Leftrightarrow -\frac{b}{a} < m < \frac{b}{a}$$

The RHS is positive and x^2 can't be negative. This is only possible if the expression in the bracket on LHS is positive.

You can divide both sides by a^2 without affecting the direction of the inequality because a^2 is a positive number.

Continues on next page

Therefore for $m > 0$ if $m > \frac{b}{a}$ the line doesn't cross the hyperbola and if $m < \frac{b}{a}$ it does cross. The limiting case, when $m = \frac{b}{a}$, gives the gradient of the asymptote.

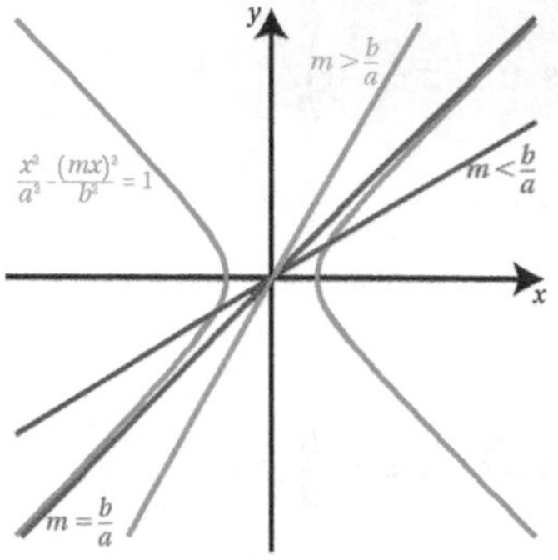

A graph shows how the limiting case gives the asymptotes.

Similarly for $m < 0$ to give the gradient of the other asymptote as $m = -\frac{b}{a}$. Hence the equations of the asymptotes are $y = \frac{b}{a}x$ and $y = -\frac{b}{a}x$.

You have actually met a special case of the hyperbola before: $y = \frac{1}{x}$ (or, more generally, $y = \frac{k}{x}$). It can be shown that this is indeed a hyperbola but rotated so that the asymptotes are the coordinate axes. If k is restricted to be a positive constant, then it can be written as $k = c^2$. To emphasise the symmetry between x and y the equation $y = \frac{c^2}{x}$ is usually written as $xy = c^2$.

Fast forward

You may be surprised that the equations $xy = c^2$ and $\frac{x^2}{a^2} - \frac{y^2}{b^2} = 1$ represent members of the same family of curves. In Further Mathematics Student Book 2 you will see how to prove that the curves are related by rotation.

Key point 3.3

The equation $xy = c^2$ represents a hyperbola with vertices at (c, c) and $(-c, -c)$ and coordinate axes as asymptotes.

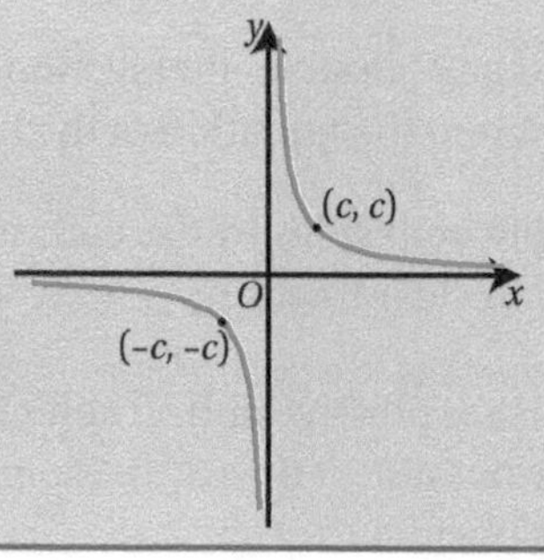

WORKED EXAMPLE 3.3

Sketch the curve with equation $4xy = 5$, labelling the coordinates of the vertices.

$4xy = 5$

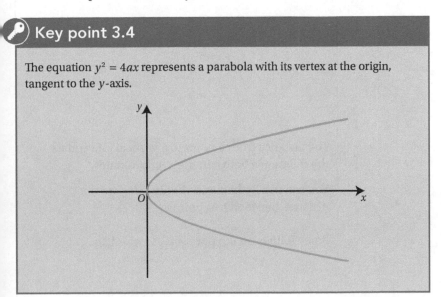

$xy = \frac{5}{4}$

To use Key point 3.3 the equation needs to be in the form $xy = c^2$.

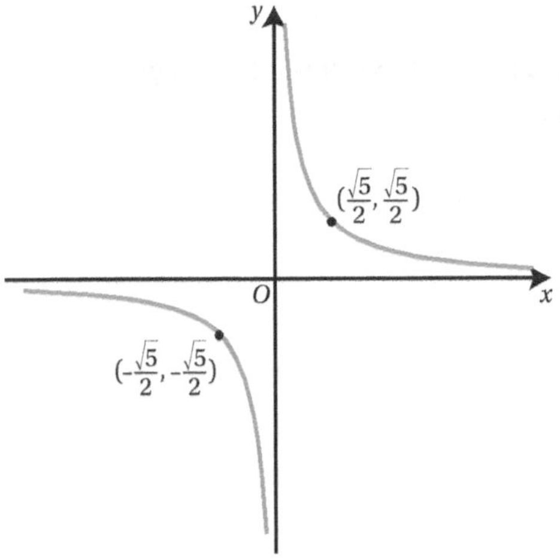

This is a hyperbola with the coordinate axes as asymptotes.

It has $c^2 = \frac{5}{4}$ so $c = \frac{\sqrt{5}}{2}$.

The parabola

In the equations of an ellipse and a hyperbola, both x and y terms are squared. You already know that $y = kx^2$ is an equation of a **parabola** with a vertex at the origin. If the y term is squared instead, the curve still has the same shape, but is 'sideways'.

Key point 3.4

The equation $y^2 = 4ax$ represents a parabola with its vertex at the origin, tangent to the y-axis.

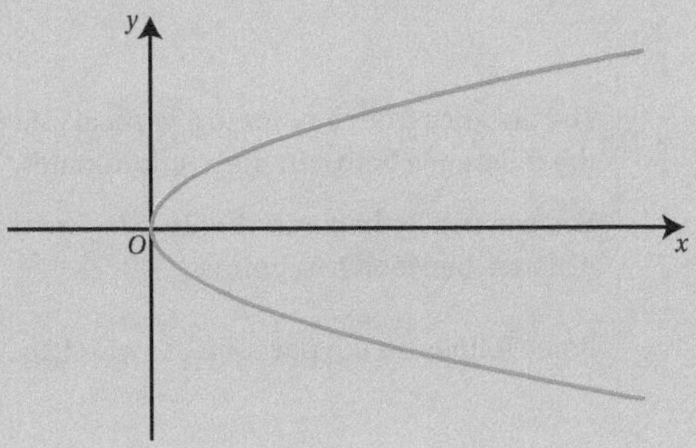

WORKED EXAMPLE 3.4

A curve has equation $y^2 = 12x$.

a Sketch the curve.

b Let P be a point on the curve with coordinates (p, q).
Show that the distance of P from the point $F(3, 0)$ equals the distance of P from the line $x = -3$.

a

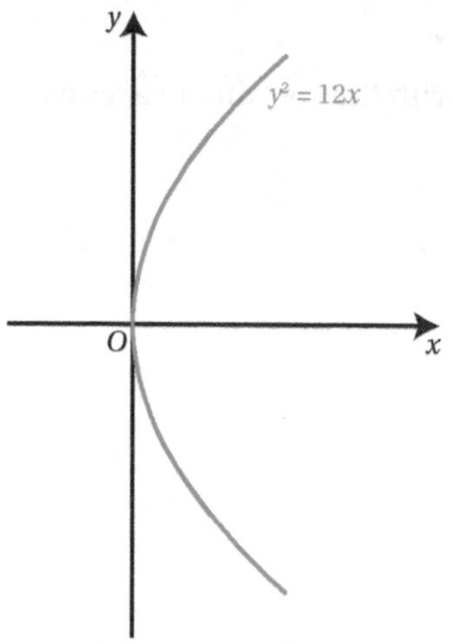

This is a parabola with vertex at the origin.

b

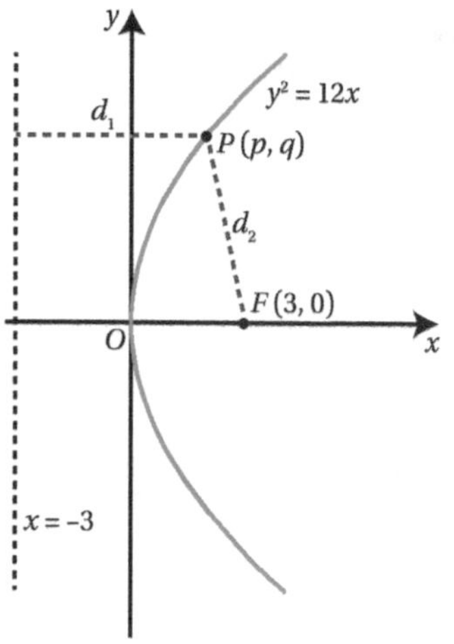

Sketch a graph showing the required distances.

$d_1 = p - (-3) = p + 3$

The distance from a point to a vertical line equals the difference between the x-coordinates.

$$
\begin{aligned}
d_2^2 &= (p-3)^2 + (q-0)^2 \\
&= p^2 - 6p + 9 + q^2 \\
&= p^2 - 6p + 9 + 12p \\
&= p^2 + 6p + 9
\end{aligned}
$$

You can use Pythagoras' theorem to find the distance between two points.

Point P lines on the parabola, so $q^2 = 12p$.

Hence $d_2 = p + 3 = d_1$, as required.

Did you know?

The property illustrated in Worked example 3.4 can also be taken as the defining feature of a parabola: it is the locus of the points that are equal distances from a given point (F) and a given line (l).

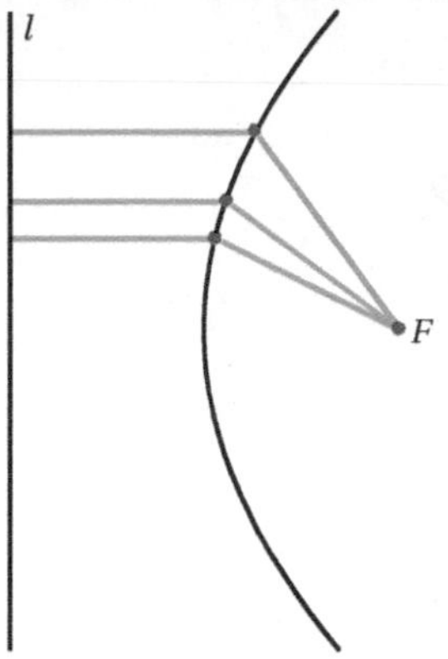

The point F in Worked example 3.4 is called the **focus** of the parabola, and has an important special property. Horizontal rays reflected off a parabola with equation $y^2 = 4ax$ all pass through the point F with coordinates $(a, 0)$.

Satellite dishes usually have a parabolic shape.

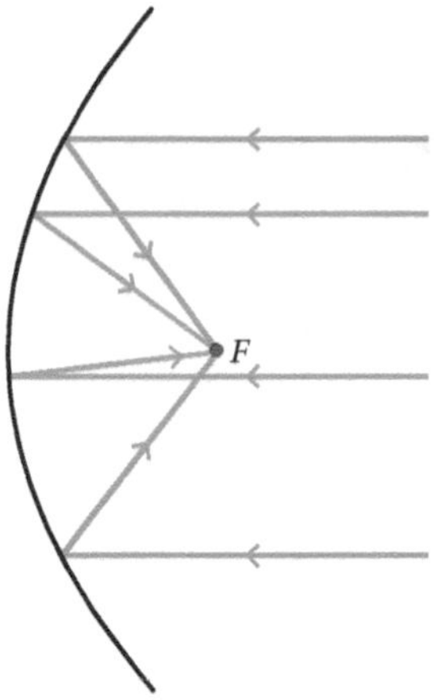

EXERCISE 3A

1 Sketch these ellipses, showing the coordinates of any axis intercepts.

a **i** $\frac{x^2}{4}+\frac{y^2}{9}=1$ **ii** $\frac{x^2}{25}+\frac{y^2}{4}=1$ **b** **i** $x^2+\frac{y^2}{9}=1$ **ii** $\frac{x^2}{16}+y^2=1$

c **i** $9x^2+16y^2=1$ **ii** $25x^2+9y^2=1$ **d** **i** $4x^2+25y^2=16$ **ii** $49x^2+16y^2=36$

2 Write down the equation of each ellipse.

a **i**

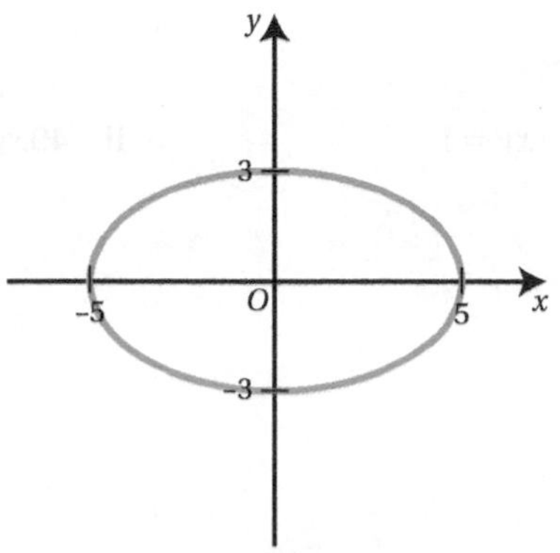

ii

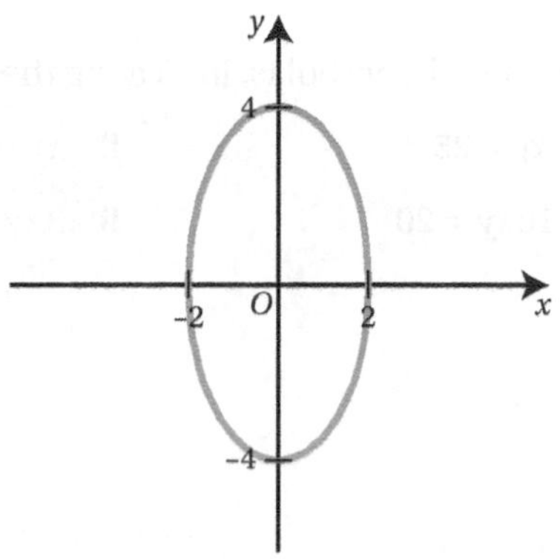

b **i**

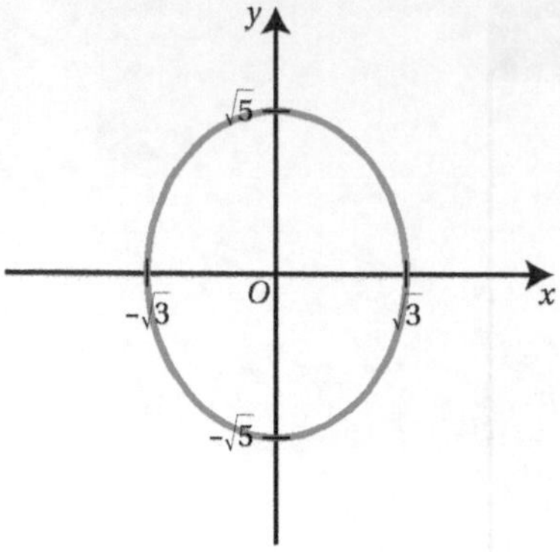

ii

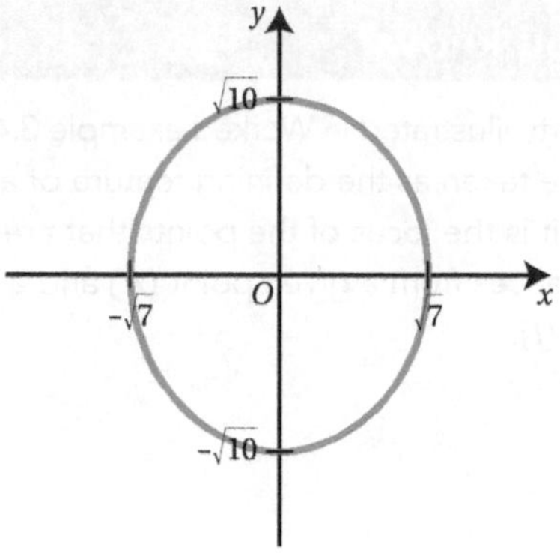

c **i**

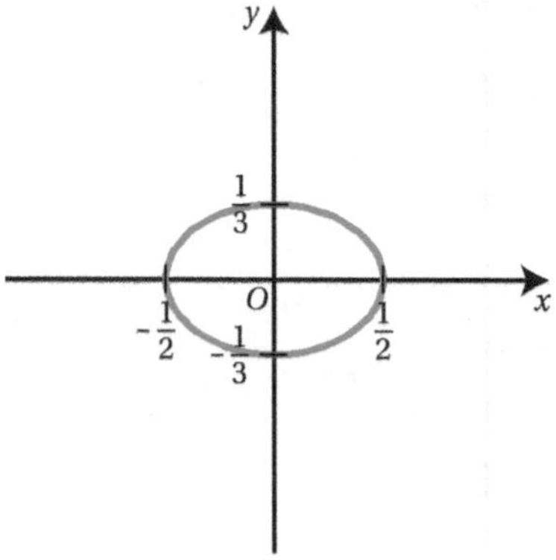

ii

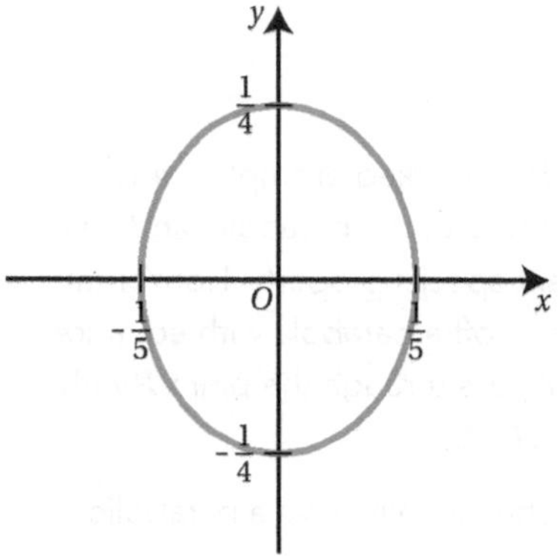

d **i**

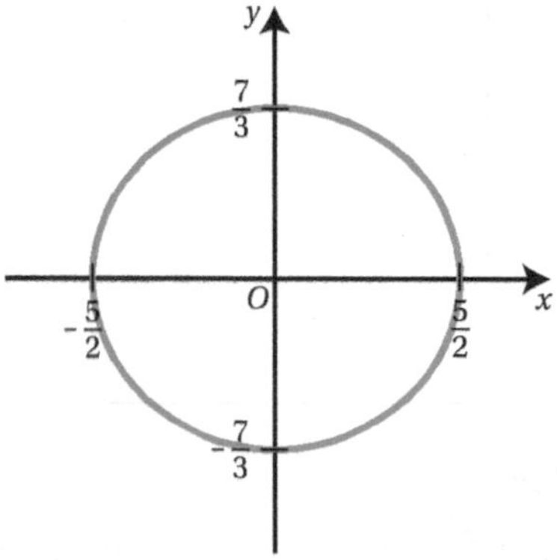

ii

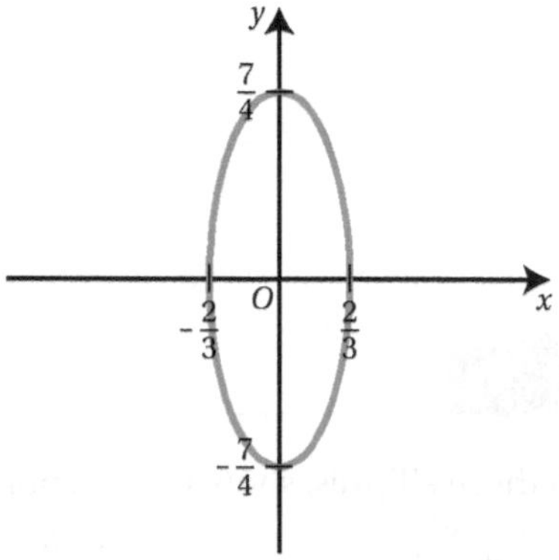

3 Sketch these hyperbolas, stating the coordinates of the vertices and the equations of the asymptotes.

a **i** $\frac{x^2}{4}-\frac{y^2}{9}=1$ **ii** $\frac{x^2}{25}-\frac{y^2}{4}=1$ **b** **i** $x^2-\frac{y^2}{9}=1$ **ii** $\frac{x^2}{16}-y^2=1$

c **i** $9x^2-16y^2=1$ **ii** $25x^2-9y^2=1$ **d** **i** $4x^2-25y^2=16$ **ii** $49x^2-16y^2=36$

4 Sketch these hyperbolas, indicating the coordinates of the vertices.

a **i** $xy=25$ **ii** $xy=16$ **b** **i** $4xy=1$ **ii** $49xy=1$

c **i** $16xy=20$ **ii** $3xy=25$

5 Find the equation of each hyperbola, giving your answers in the form $\frac{x^2}{a^2} - \frac{y^2}{b^2} = 1$.

a **i**

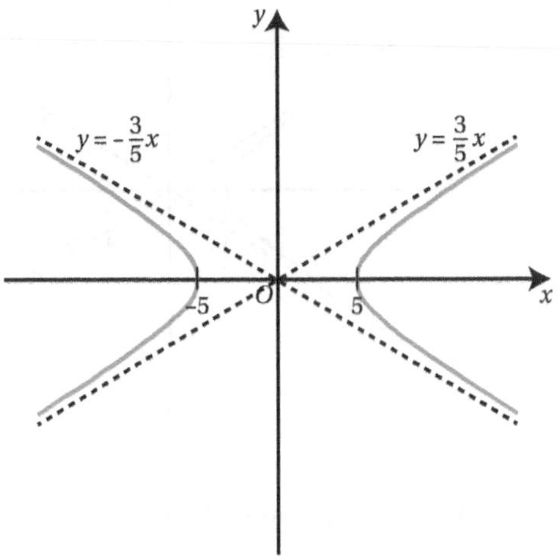

ii

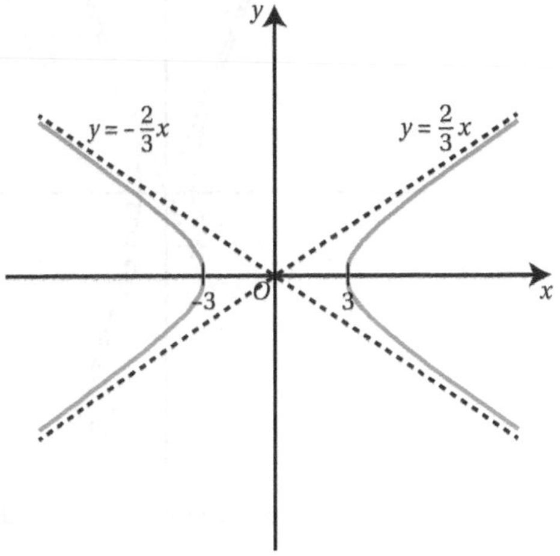

b **i**

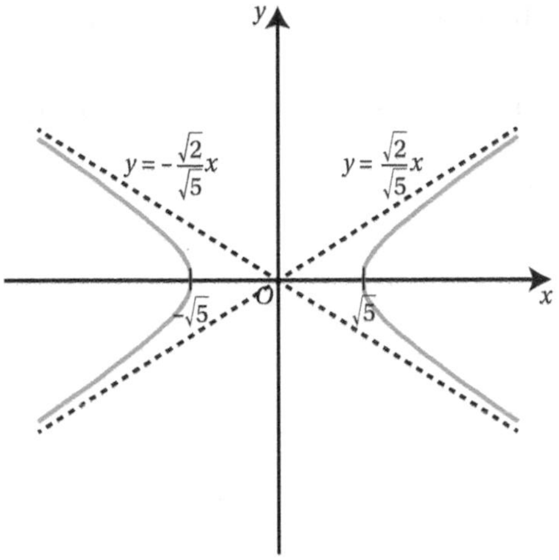

ii

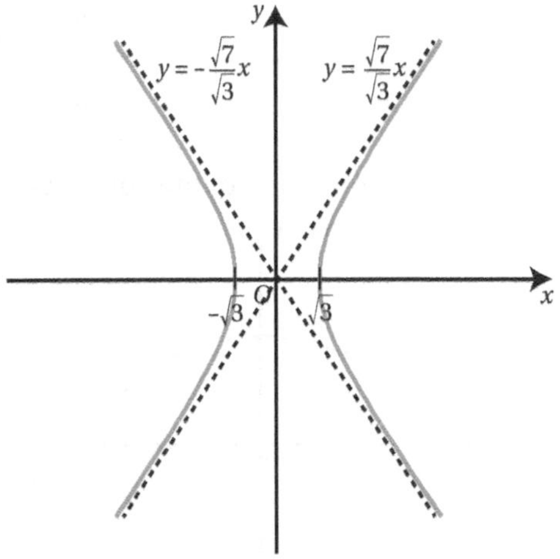

c **i**

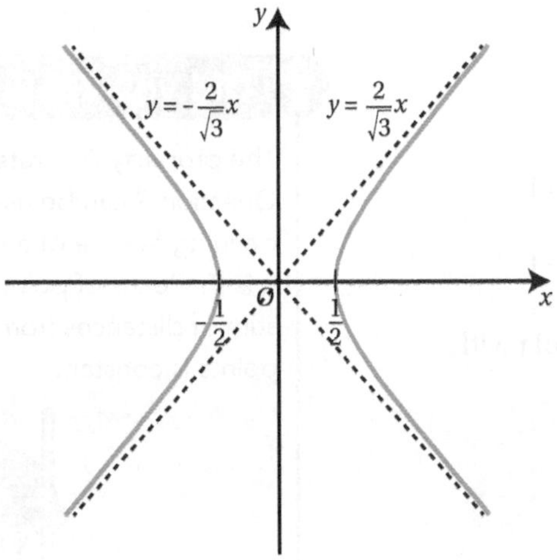

ii

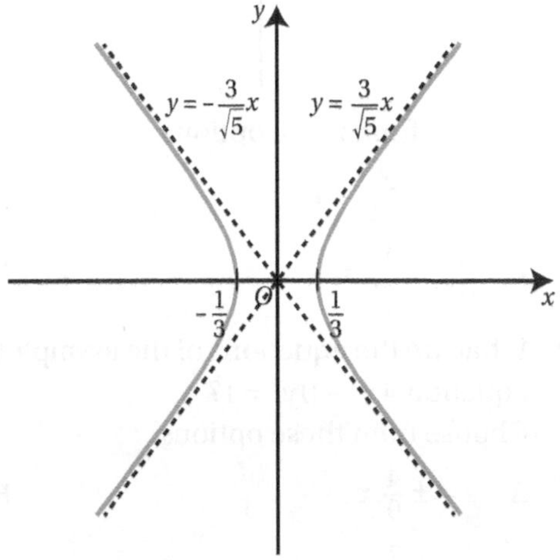

d i

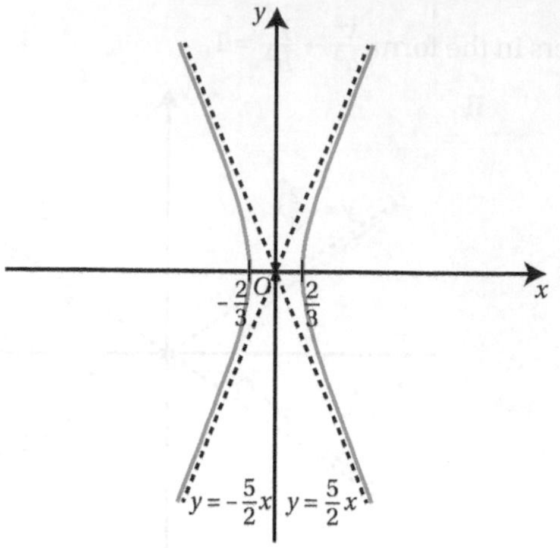

ii

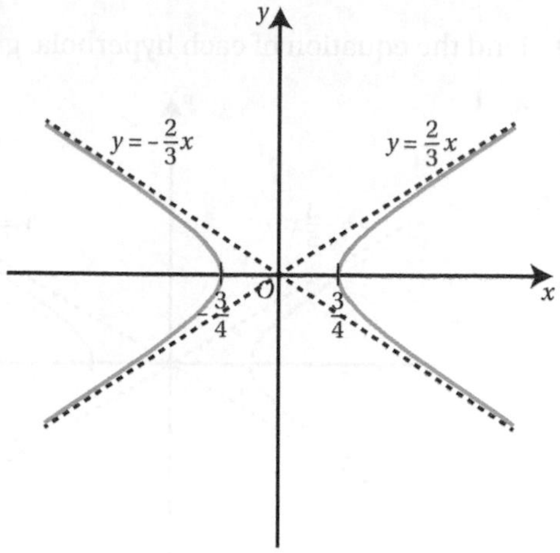

6 Sketch each of these curves, showing the coordinates of any axes intercepts and vertices, and equations of any asymptotes.

a i $\frac{x^2}{4}-\frac{y^2}{25}=1$ **ii** $\frac{x^2}{100}-\frac{y^2}{36}=1$ **b i** $y^2=8x$ **ii** $y^2=20x$

c i $\frac{x^2}{9}+4y^2=1$ **ii** $25x^2+\frac{y^2}{9}=1$ **d i** $xy=9$ **ii** $xy=144$

7 What is the equation of the curve in the diagram?

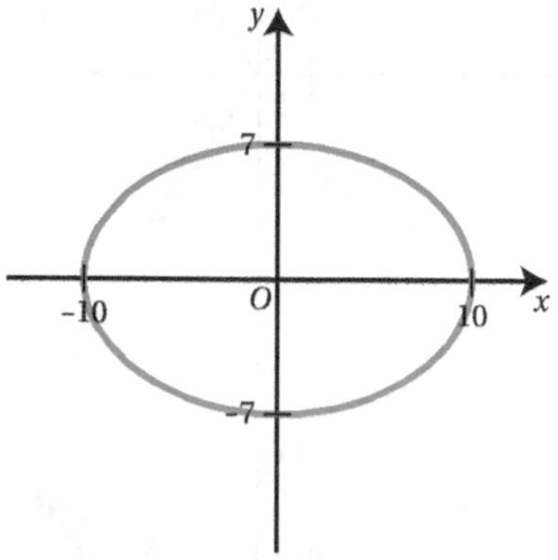

Choose from these options.

A $\frac{x^2}{100}-\frac{y^2}{49}=1$ **B** $100x^2+49y^2=1$

C $\frac{x^2}{100}+\frac{y^2}{49}=1$ **D** $49x^2-100y^2=1$

8 What are the equations of the asymptotes of the hyperbola with equation $4x^2-9y^2=1$?
Choose from these options.

A $y=\pm\frac{4}{9}x$ **B** $y=\pm\frac{9}{4}x$

C $y=\pm\frac{2}{3}x$ **D** $y=\pm\frac{3}{2}x$

9 An ellipse has equation $\frac{x^2}{25}+\frac{y^2}{9}=1$ and $P(p,q)$ is a point on the ellipse. Points F_1 and F_2 have coordinates $(-4,0)$ and $(4,0)$. Show that the sum of the distances PF_1+PF_2 does not depend on the value of p.

Did you know?

The property illustrated in Question 9 can be used as a defining feature of an ellipse: it is the locus of points whose sum of distances from two fixed points is constant.

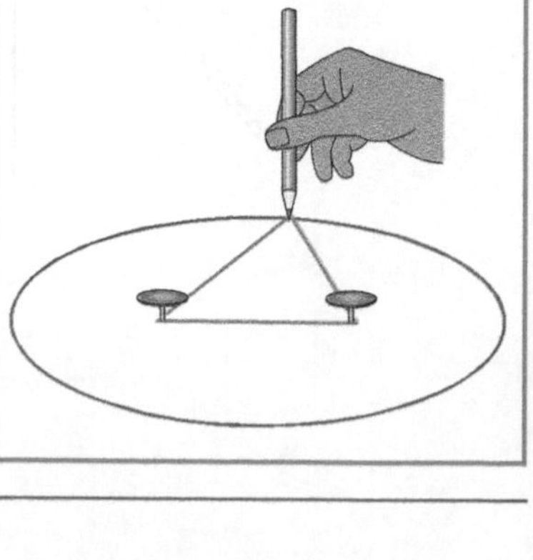

Section 2: Solving problems with ellipses, hyperbolas and parabolas

In A Level Mathematics Student Book 1 you learnt how to find intersections of lines with parabolas and circles. You also learnt how to use the **discriminant** to determine the number of intersections. You can use those methods to solve problems involving ellipses and hyperbolas.

Rewind

See A Level Mathematics Student Book 1, Chapter 3, for a reminder of quadratic equations and inequalities.

WORKED EXAMPLE 3.5

The line $y = 2x + c$ intersects the hyperbola with equation $\frac{x^2}{25} - \frac{y^2}{4} = 1$ at two points. Find the range of possible values of c.

$$\frac{x^2}{25} - \frac{(2x+c)^2}{4} = 1$$

To write an equation for the intersection points, substitute $y = 2x + c$ into the equation of the hyperbola.

$$4x^2 - 25(4x^2 + 4cx + c^2) = 100$$

Clear the fractions: multiply both sides by the common denominator.

$$96x^2 + 100cx + (100 + 25c^2) = 0$$

This is a quadratic equation in x. Make one side equal to zero.

There are two solutions:

$$(100c)^2 - 4(96)(100 + 25c^2) > 0$$

For two intersection points, the discriminant needs to be positive.

$$10000c^2 - 9600c^2 > 38400$$

$$c^2 > 96$$

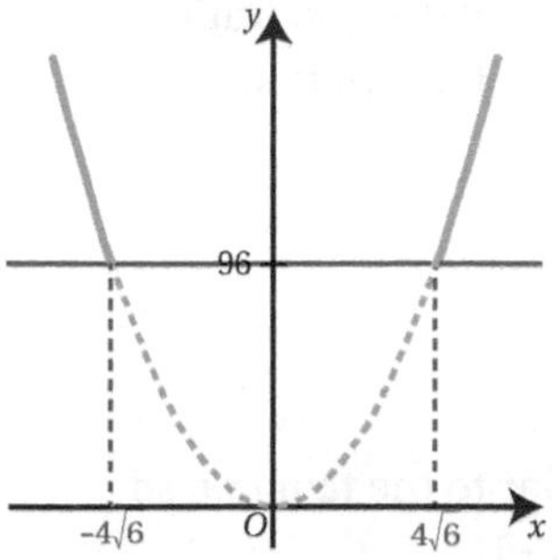

This is a quadratic inequality for c, so you can sketch the graph to see that there are two solution intervals.

$\therefore c < -4\sqrt{6}$ or $c > 4\sqrt{6}$

You can also use the discriminant to solve problems about tangents and normals.

Rewind

It is also possible to use implicit differentiation to find gradients of tangents and normals to these curves. This is covered in A Level Mathematics Student Book 2, Chapter 10.

WORKED EXAMPLE 3.6

The line $y = mx + 25$, with $m > 0$, is a tangent to the ellipse with equation $\frac{x^2}{400} + \frac{y^2}{225} = 1$.

a Find the value of m.

The line touches the ellipse at point Q.

b Find the equation of the normal to the ellipse at Q.

a

$$\frac{x^2}{400} + \frac{(mx+25)^2}{225} = 1$$

You need to find the value of m for which there is only one intersection point between the line and the ellipse. Start by writing an equation for the intersections: substitute $y = mx + 25$ into the equation of the ellipse.

$$225x^2 + 400(m^2x^2 + 50mx + 625) = 90\,000$$
$$9x^2 + 16(m^2x^2 + 50mx + 625) = 3600$$
$$(9 + 16m^2)x^2 + 800mx + 6400 = 0$$

This is a quadratic equation for x, so make one side equal to zero.

As there is one solution:

For the line to be a tangent, this equation should only have one solution. This happens when the discriminant is zero.

$$(800m)^2 - 4(9 + 16m^2)(6400) = 0$$
$$640\,000m^2 - (36 + 64m^2)(6400) = 0$$
$$100m^2 - 36 - 64m^2 = 0$$
$$36m^2 = 36$$
$$\therefore m = 1$$

The question says that m is positive.

b Coordinates of Q:

You first need to find the coordinates of Q. The x-coordinate is the solution of the quadratic equation with the value of m from part **a**.

$$25x^2 + 800x + 6400 = 0$$
$$x^2 + 32x + 256 = 0$$
$$(x + 16)^2 = 0$$
$$\therefore x = -16$$

$$y = mx + 25$$
$$= -16 + 25 = 9$$

Gradient of normal:

$$m_1 = -\frac{1}{m} = -1$$

The normal is perpendicular to the tangent, so their gradients multiply to –1.

Equation of normal:

$$y - 9 = -1(x + 16)$$
$$x + y + 7 = 0$$

Use $y - y_1 = m_1(x - x_1)$ for the equation of the normal.

Rewind

A normal to the curve is perpendicular to the tangent at the point of contact. See A Level Mathematics Student Book 1, Chapter 6, for a reminder of perpendicular lines.

WORKED EXAMPLE 3.7

A parabola has equation $y^2 = 20x$. A line passes through the point $(0, c)$, with $c \neq 0$, and has gradient m.

Given that the line is tangent to the parabola, express m in terms of c.

$y = mx + c \Rightarrow x = \frac{y-c}{m}$

Write the equation for the intersection of the line and the parabola: substitute x from the equation of the line into the equation of the parabola.

For the intersection with the parabola:

$y^2 = 20\left(\frac{y-c}{m}\right)$

$my^2 = 20y - 20c$

$my^2 - 20y + 20c = 0$

One solution:

$20^2 - 4m \times 20c = 0$

$80mc = 400$

If the line is a tangent to the parabola, this quadratic equation for y should have only one solution so the discriminant is zero.

$m = \frac{5}{c}$

Note that this gives a single value of m for every $c \neq 0$.

EXERCISE 3B

1 Find the value of m for which the line $y = mx + 2$ is tangent to the parabola with equation $y^2 = 5x$.

2 A hyperbola has equation $\frac{x^2}{16} - \frac{y^2}{25} = 1$.

a Sketch the hyperbola, stating the equations of the asymptotes.

b The line with equation $y = 2x + c$ is tangent to the hyperbola. Find the possible values of c.

3 The line $y = mx - 5$ is a tangent to the ellipse $\frac{x^2}{4} + \frac{y^2}{13} = 1$.
Find the possible values of m.

4 A parabola has equation $y^2 = 6x$.

a Show that the line with equation $y = x + \frac{3}{2}$ is tangent to the parabola and find the coordinates of the point of contact, T.

b The normal to the parabola at T intersects the parabola again at Q. Find the coordinates of Q.

5 The line $y = 3x + 5$ is tangent to the ellipse $\frac{x^2}{2} + \frac{y^2}{7} = 1$ at the point P.

a Find the coordinates of P.

The normal to the ellipse at P intersects the ellipse again at Q.

b Find, to 3 significant figures, the coordinates of Q.

6 A curve has equation $\frac{x^2}{16} + y^2 = 1$.

a Sketch the curve, showing the intercepts with the coordinate axes.

b A line with equation $y = mx + 2$ intersects the ellipse at two distinct points. Find the range of possible values of m.

7 A hyperbola has equation $\frac{x^2}{4} - \frac{y^2}{16} = 1$.

a Write down the equation of the asymptote with a positive gradient.

b Show that every other line parallel to this asymptote intersects the hyperbola exactly once.

8 A hyperbola has equation $x^2 - y^2 = 1$.

a The line $y = mx + 1$ is a tangent to the hyperbola. Find the possible values of m.

b Show that this hyperbola has no tangents which pass through the origin.

Section 3: Transformations of curves

In Sections 1 and 2 you only looked at ellipses and hyperbolas centred at the origin, and parabolas with vertex at the origin. You can use your knowledge of transformations of graphs to translate these curves.

Rewind

Transformations of graphs are covered in A Level Mathematics Student Book 1, Chapter 5.

Translations

You know how to apply a translation to a curve.

Key point 3.5

Replacing x by $(x - p)$ and y by $(y - q)$ results in a translation of the curve by the vector $\begin{pmatrix} p \\ q \end{pmatrix}$.

WORKED EXAMPLE 3.8

The hyperbola with equation $\frac{x^2}{10} - \frac{y^2}{6} = 1$ is translated by the vector $\begin{pmatrix} 2 \\ -7 \end{pmatrix}$.

Find the equation of the resulting curve and state the coordinates of its vertices.

$\frac{(x-2)^2}{10} - \frac{(y+7)^2}{6} = 1$ — Replace x by $(x-2)$ and y by $(y+7)$.

$3(x-2)^2 - 5(y+7)^2 = 30$ — Clear the fractions: multiply by the common denominator (30).

$(3x^2 - 12x + 12) - (5y^2 + 70y + 245) = 30$

$3x^2 - 12x - 5y^2 - 70y = 263$

The original hyperbola has vertices $(-\sqrt{10}, 0)$ and $(\sqrt{10}, 0)$, — The vertices of a hyperbola $\frac{x^2}{a^2} - \frac{y^2}{b^2} = 1$ are at $(\pm a, 0)$.

so the new hyperbola has vertices — The curve has been translated by $\begin{pmatrix} 2 \\ -7 \end{pmatrix}$.

$(2 - \sqrt{10}, -7)$ and $(2 + \sqrt{10}, -7)$

In A Level Mathematics Student Book 1 you learnt how to complete the square to find the centre and radius of a circle when the equation is written in an expanded form. You can apply the same technique to ellipses, hyperbolas and parabolas.

Rewind

See A Level Mathematics Student Book 1, Chapter 3, for a reminder of completing the square.

WORKED EXAMPLE 3.9

a Show that the equation $4x^2 + 16x - 9y^2 + 90y = 245$ represents a hyperbola, and find the coordinates of its vertices and the equations of its asymptotes.

b Hence sketch the curve.

a

$$\begin{aligned}4x^2+16x &= 4x(x^2+4x)\\ &= 4[(x+2)^2-4]\\ &= 4(x+2)^2-16\end{aligned}$$

You want to write the equation in the form $\frac{(x-p)^2}{a^2} - \frac{(y-q)^2}{b^2} = 1$, so start by completing the squares (separately for the x and y terms).

$$\begin{aligned}-9y^2+90y &= -9(y^2-10y)\\ &= -9[(y-5)^2-25]\\ &= -9(y-5)^2+225\end{aligned}$$

Remember to take out a factor of -9 from each term (rather than just 9).

Hence:

Now put these back into the original equation.

$$4x^2+16x-9y^2+90y=245$$

$$4(x+2)^2-16-9(y-5)^2+225=245$$

$$4(x+2)^2-9(y-5)^2=36$$

$$\frac{(x+2)^2}{9}-\frac{(y-5)^2}{4}=1$$

Divide both sides by 36 to get the equation in the required form.

This is a hyperbola with equation $\frac{x^2}{9}-\frac{y^2}{4}=1$ translated by the vector $\begin{pmatrix}-2\\5\end{pmatrix}$.

The vertices of the original hyperbola are at $(-3, 0)$ and $(3, 0)$, so the vertices of the new hyperbola are $(-5, 5)$ and $(1, 5)$.

The original hyperbola was translated 2 units to the left and 5 units up.

The asymptotes of the original hyperbola are $y = \frac{2}{3}x$ and $y = -\frac{2}{3}x$.

The asymptotes of a hyperbola are given by $y = \pm\frac{b}{a}x$.

The new asymptotes are:

The translation replaces x by $(x+2)$ and y by $(y-5)$.

$$(y-5)=\frac{2}{3}(x+2)$$

$$y=\frac{2}{3}x+\frac{19}{3}$$

and

$$(y-5)=-\frac{2}{3}(x+2)$$

$$y=-\frac{2}{3}x+\frac{11}{3}$$

Continues on next page

b

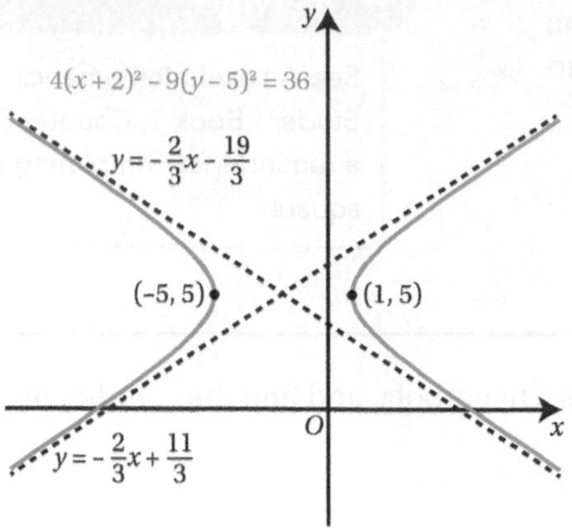

Stretches

You also know how to apply horizontal and vertical stretches to curves.

Key point 3.6

Replacing x by $\frac{x}{p}$ and y by $\frac{y}{q}$ results in a horizontal stretch with scale factor p and a vertical stretch with scale factor q.

WORKED EXAMPLE 3.10

Show that the ellipse with equation $4x^2+25y^2=9$ can be obtained from a circle with radius 1 by applying a horizontal stretch and a vertical stretch.

State the scale factor of each stretch.

$4x^2+25y^2=9$

$\frac{4x^2}{9}+\frac{25y^2}{9}=1$

You want to see the connection between the given ellipse and the circle $x^2+y^2=1$, so start by making the RHS of the equation equal to 1.

$\left(\frac{2x}{3}\right)^2+\left(\frac{5y}{3}\right)^2=1$

This is the equation of the circle, with x replaced by $\left(\frac{2x}{3}\right)$ and y replaced by $\left(\frac{5y}{3}\right)$.

Hence the ellipse is obtained from the circle $x^2+y^2=1$ by a horizontal stretch with scale factor $\frac{3}{2}$ and a vertical stretch with scale factor $\frac{3}{5}$.

Remember that a stretch with scale factor p replaces x by $\frac{x}{p}$.

Reflections

Key point 3.7

- Replacing x with $-x$ reflects a curve in the y-axis.
- Replacing y with $-y$ reflects a curve in the x-axis.
- Replacing x with y and y with x reflects a curve in the line $y = x$.
- Replacing x with $-y$ and y with $-x$ reflects a curve in the line $y = -x$.

WORKED EXAMPLE 3.11

A parabola has equation $y^2 = 12x$. Find the equation of the resulting curve when this parabola is:

a reflected in the y-axis

b reflected in the line $y = x$.

a $y^2 = 12(-x)$ — Reflection in the y-axis replaces x by $-x$.

$y^2 = -12x$

b $x^2 = 12y$ — Reflection in the line $y = x$ swaps x and y.

$y = \frac{1}{12}x^2$

WORKED EXAMPLE 3.12

Find the equation of the hyperbola shown in the diagram.

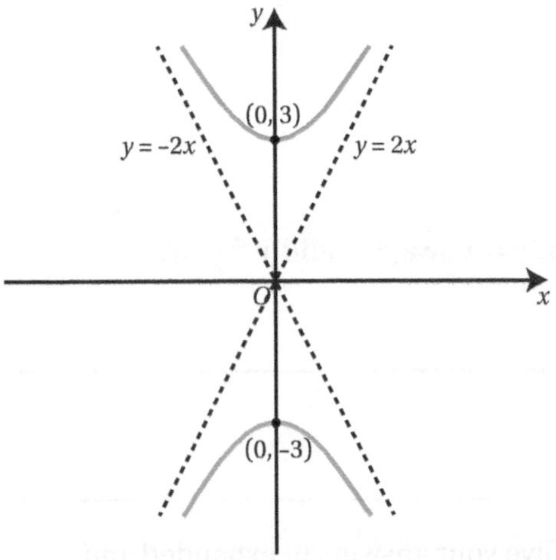

Continues on next page

Reflect the hyperbola in the line $y = x$.

You can obtain this hyperbola from a standard hyperbola by reflecting it in the line $y = x$.

First find the equation of the reflected hyperbola by using its vertex coordinates and equations of asymptotes.

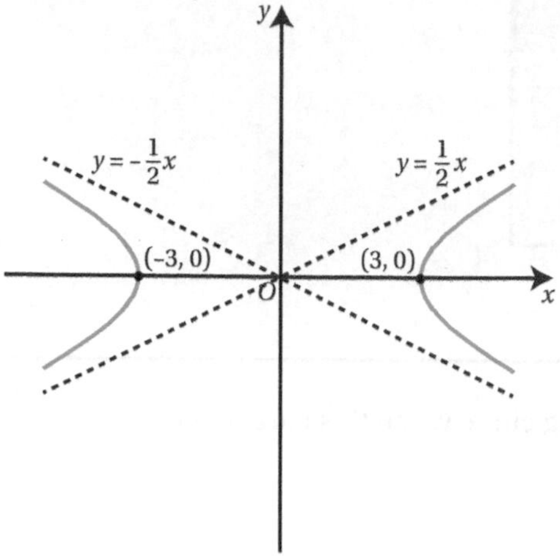

The asymptotes are:

$x = \pm 2y \Leftrightarrow y = \pm\frac{1}{2}x$

The reflection swaps x and y; you can use this to find the asymptotes of the reflected hyperbola.

Equation of the reflected hyperbola:

$$\frac{x^2}{a^2} - \frac{y^2}{b^2} = 1$$

From the vertices:

$$a = 3$$

The vertices are at $(\pm a, 0)$.

From the asymptotes:

$$\frac{b}{a} = \frac{1}{2} \Rightarrow b = \frac{3}{2}$$

The equations of the asymptotes are $y = \pm\frac{b}{a}x$.

The reflected hyperbola has equation

$$\frac{x^2}{3^2} - \frac{y^2}{\left(\frac{3}{2}\right)^2} = 1$$

$$\frac{x^2}{9} - \frac{4y^2}{9} = 1$$

The original hyperbola has equation

$$\frac{y^2}{9} - \frac{4x^2}{9} = 1$$

Reflection in $y = x$ swaps x and y.

EXERCISE 3C

1 Find the equation of each curve after the given transformation. Give your answers in expanded and simplified form.

a **i** $\frac{x^2}{6} + \frac{y^2}{2} = 1$; translation with vector $\begin{pmatrix} 3 \\ -2 \end{pmatrix}$

ii $\frac{x^2}{3} + \frac{y^2}{10} = 1$; translation with vector $\begin{pmatrix} -1 \\ 4 \end{pmatrix}$

b **i** $xy = 16$; translation with vector $\begin{pmatrix} -5 \\ 3 \end{pmatrix}$

ii $xy = 100$; translation with vector $\begin{pmatrix} 10 \\ 10 \end{pmatrix}$

c **i** $y^2 = 18x$; vertical stretch with scale factor 3

ii $y^2 = 12x$; horizontal stretch with scale factor 2

d **i** $\frac{x^2}{6}-\frac{y^2}{9}=1$; reflection in the line $y=x$ **ii** $\frac{x^2}{10}-\frac{y^2}{5}=1$; reflection in the line $y=-x$

e **i** $y^2=4x$; reflection in the y-axis **ii** $y^2=4x$; reflection in the x-axis

2 Each of these curves is a translation of a standard ellipse, hyperbola or parabola. Identify the translation vector, and hence sketch each curve.

a **i** $9x^2-18x+4y^2+8y-23=0$ **ii** $x^2+2x+25y^2-100y+76=0$

b **i** $9x^2-18x-4y^2+8y-31=0$ **ii** $4x^2-24x-y^2-2y+31=0$

c **i** $y^2-2y-8x+25=0$ **ii** $y^2+4y-12x+16=0$

d **i** $xy-3x-y-1=0$ **ii** $xy-x+2y-11=0$

3 An ellipse with equation $\frac{x^2}{5}+\frac{y^2}{3}=1$ is translated by the vector $\begin{pmatrix} p \\ q \end{pmatrix}$. The resulting curve has equation $3x^2-18x+5y^2+10y+17=0$.

Find the values of p and q.

4 A hyperbola with equation $3x^2-b^2y^2=1$ is translated by the vector $\begin{pmatrix} 1 \\ 4 \end{pmatrix}$. Given that the translated hyperbola passes through the origin, find the positive value of b.

5 The curve C_1 with equation $xy=16$, is translated by the vector $\begin{pmatrix} -3 \\ 8 \end{pmatrix}$. The resulting curve is C_2.

a Sketch the curve C_2, showing the coordinates of the vertices and equations of any asymptotes.

b Find the equation of C_2 in the form $y=\mathrm{f}(x)$.

6 A parabola with equation $y^2=4ax$ is stretched parallel to the x-axis. The equation of the resulting curve is $y^2=\frac{ax}{5}$. Find the scale factor of the stretch.

7 A parabola with equation $y^2=4ax$ is reflected in the line $y=x$. The two curves intersect at the origin and at one more point R. Find, in terms of a, the coordinates of R.

8 Sketch the curve with equation $2x^2-16x-9y^2+18y+12=0$, stating the coordinates of the vertices and equations of any asymptotes.

9 The curve C_1 has equation $\frac{x^2}{3}-\frac{y^2}{2}=1$.

a Sketch the curve C_1, showing the intercepts with the coordinate axes.

The curve C_2 is obtained from C_1 by a stretch with scale factor k parallel to the y-axis.

b Given that the point $(3, 5)$ lies on C_2, find the value of k.

10 The curve with equation $\frac{x^2}{a^2}+\frac{y^2}{b^2}=1$ is stretched parallel to the x-axis with scale factor $\frac{2}{3}$. The resulting curve is the circle with equation $x^2+y^2=4$. Find the positive values of a and b.

11 The line $y=mx$, where $m>0$, is tangent to the curve with equation $\frac{(x-4)^2}{2}+\frac{y^2}{5}=1$.

a Find, in exact form, the possible values of m.

b Hence find the equations of the tangents to the curve $\frac{x^2}{2}+\frac{y^2}{5}=1$ which pass through the point $(-4, 0)$.

12 The hyperbola with equation $\frac{x^2}{4}-\frac{y^2}{16}=1$ is translated by the vector $\begin{pmatrix} p \\ -3 \end{pmatrix}$. One asymptote of the new hyperbola is $y=2x-5$. Find the equation of the other asymptote.

13 The ellipse C_1 has equation $\frac{x^2}{48}+\frac{y^2}{36}=1$. C_1 is obtained from a circle C_2, with radius 6, by a horizontal stretch.

a State the scale factor of the stretch.

The point $P(6, 3)$ lies on C_1 and the point Q lies on C_2. P is the image of Q under the stretch.

b Find the coordinates of Q.

c Find the equation of the tangent to C_2 at Q.

d Hence find the equation of the **normal** to C_1 at P.

14 The diagram shows the hyperbola with equation $(x+3)^2-y^2=25$. The line $y=mx$ is a tangent to the hyperbola, and crosses the asymptotes at the points A and B.

a Find the value of m.

b Write down the equations of the asymptotes of the hyperbola.

c Find the exact distance AB.

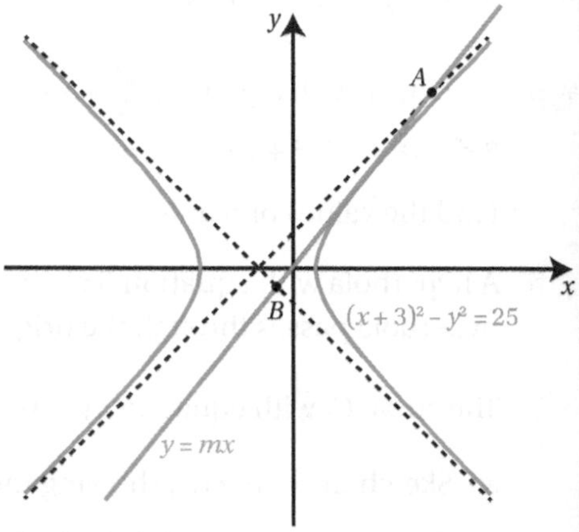

Checklist of learning and understanding

- The equation $\frac{x^2}{a^2}+\frac{y^2}{b^2}=1$ represents an ellipse with axis intercepts $(\pm a, 0)$ and $(0, \pm b)$.
- The equation $\frac{x^2}{a^2}-\frac{y^2}{b^2}=1$ represents a hyperbola with vertices at $(\pm a, 0)$ and asymptotes $y=\pm\frac{b}{a}x$.
- The equation $xy=c^2$ represents a hyperbola with vertices (c, c) and $(-c, -c)$ and the coordinate axes as asymptotes.
- The equation $y^2=4ax$ represents a parabola with the vertex at the origin and the x-axis as the line of symmetry.
- You can use the discriminant of a quadratic equation to solve problems about the number of intersections of a line with a parabola, hyperbola or an ellipse, including identifying tangents.
- You can apply transformations to curves by changing their equations.
  - Replacing x with $(x-p)$ and y with $(y-q)$ results in a translation with vector $\begin{pmatrix} p \\ q \end{pmatrix}$.
  - Replacing x with $\frac{x}{a}$ and y with $\frac{y}{b}$ results in a horizontal stretch with scale factor a and a vertical stretch with scale factor b.
  - Replacing x with $-x$ results in a reflection in the y-axis.
  - Replacing y with $-y$ results in a reflection in the x-axis.
  - Replacing x with y and y with x results in a reflection in the line $y=x$.
  - Replacing x with $-y$ and y with $-x$ results in a reflection in the line $y=-x$.

Mixed practice 3

1. What is a possible equation of the curve shown in the diagram?

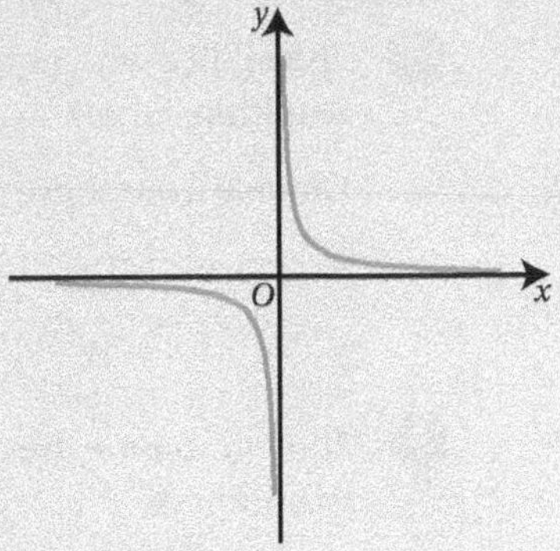

Choose from these options.

A $\frac{x^2}{3} - \frac{y^2}{4} = 1$

B $y^2 = 4x$

C $\frac{x^2}{3} + \frac{y^2}{4} = 1$

D $xy = 5$

2. What is the equation of the ellipse shown in the diagram?

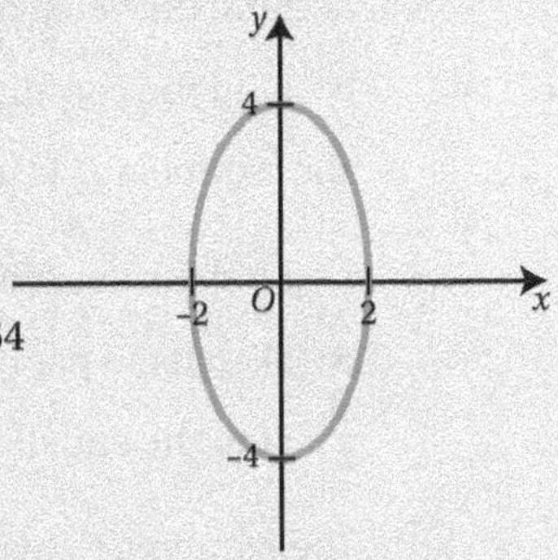

Choose from these options.

A $4x^2 + y^2 = 16$

B $\frac{x^2}{2} + \frac{y^2}{4} = 1$

C $4x^2 + 16y^2 = 1$

D $4x^2 + 16y^2 = 64$

3. A parabola with equation $y^2 = 3x$ is translated by the vector $\begin{pmatrix} -4 \\ 3 \end{pmatrix}$.

What is the equation of the resulting curve?

Choose from these options.

A $y^2 - 3 = 3x + 4$ B $(y-3)^2 = 3x + 4$ C $(y+3)^2 = 3(x-4)$ D $(y-3)^2 = 3x + 12$

4. The line $y = -4x + c$ is a tangent to the ellipse $\frac{x^2}{3} + \frac{y^2}{2} = 1$. Find the possible values of c.

5. The line $y = mx - 5$ is tangent to the hyperbola $\frac{x^2}{4} - \frac{y^2}{13} = 1$. Find the possible values of m.

6. The parabola C_1 has equation $y^2 = 12x$.

a Find the value of m for which the line $y = mx + 2$ is tangent to C_1, and find the coordinates of the point of contact.

The parabola C_1 is reflected in the y-axis to form a new curve, C_2.

b Write down the equation of C_2.

c Find the equation of the tangent to C_2 at the point $\left(-\frac{4}{3}, 4\right)$.

7. Point $P(3, 4)$ lies on the circle with equation $x^2 + y^2 = 25$.

a Show that the tangent to the circle at P has equation $3x + 4y = 25$.

The circle is stretched horizontally to form an ellipse with equation $x^2 + 4y^2 = 100$.

b Find the scale factor of the stretch.

c Find the equation of the tangent to the ellipse at the point $(6, 4)$.

8 The ellipse with equation $\frac{x^2}{3}+y^2=1$ is translated by the vector $\begin{pmatrix}3\\-2\end{pmatrix}$. What is the equation of the resulting curve?

Choose from these options.

A $x^2-6x+3y^2-12y+18=0$

B $x^2-6x+3y^2+12y+18=0$

C $x^2+6x+3y^2-12y+18=0$

D $x^2+6x+3y^2+12y+18=0$

9 The parabola P_1 has equation $y^2=8x$. The line l_1 with equation $y=mx+2$ is tangent to the parabola at the point A.

a Find the value of m.

b Find the coordinates of A.

The parabola P_1 is translated by the vector $\begin{pmatrix}-2\\3\end{pmatrix}$ to form a new parabola, P_2.

c Find the coordinates of the points where the parabola P_2 crosses the coordinate axes.

d Sketch the parabola P_2, showing the coordinates of its vertex.

e Find the equation of the tangent to P_2 at the point $(0, 7)$.

10 The parabola C_1, with equation $y^2=5x$, is reflected in the line $y=x$ to form a new parabola, C_2. The two parabolas intersect at the origin and another point, A.

a Write down the equation of C_2.

b Find the coordinates of A.

c Use differentiation to find the equation of the tangent to C_2 at A.

d Hence find the equation of the tangent to C_1 at A.

11 Point $P(p, q)$ lies on the parabola with equation $y^2=4ax$. For all values of p, the distance of the point P from the line $x=-5$ is the same as its distance from the point $(5, 0)$. Find the value of a.

12 A curve C_1 has equation $\frac{x^2}{9}-\frac{y^2}{16}=1$.

a Sketch the curve C_1, stating the values of its intercepts with the coordinate axes.

b The curve C_1 is translated by the vector $\begin{bmatrix}k\\0\end{bmatrix}$, where $k<0$, to give a curve C_2.

Given that C_2 passes through the origin $(0, 0)$, find the equations of the asymptotes of C_2.

[© AQA 2015]

13 An ellipse is shown on the right.

The ellipse intersects the x-axis at the points A and B. The equation of the ellipse is $\frac{(x-4)^2}{4}+y^2=1$.

a Find the x-coordinates of A and B.

b The line $y=mx\ (m>0)$ is a tangent to the ellipse, with point of contact P.

i Show that the x-coordinate of P satisfies the equation $(1+4m^2)x^2-8x+12=0$.

ii Hence find the exact value of m.

iii Find the coordinates of P.

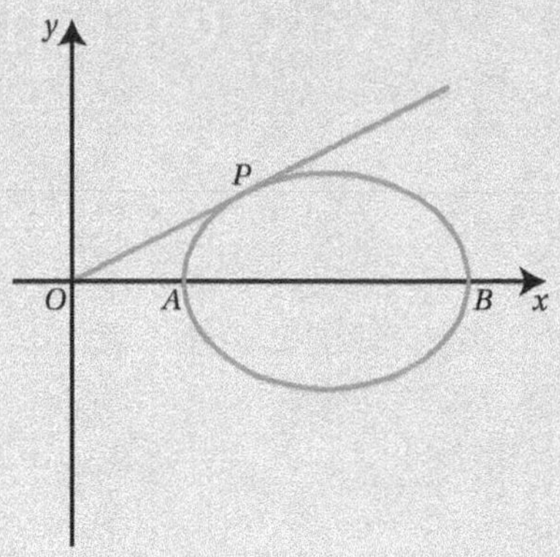

[©AQA 2013]

14 The hyperbola with equation $4x^2-9y^2=36$ is translated by the vector $\begin{pmatrix}5\\1\end{pmatrix}$. What are the asymptotes of the resulting curve?

Choose from these options.

A $4x-9y=1$ and $4x+9y=29$

B $2x-3y=7$ and $2x+3y=13$

C $4x-9y=-11$ and $4x+9y=-29$

D $2x-3y=13$ and $2x-3y=7$

15 A hyperbola with equation $x^2-\frac{y^2}{b^2}=1$, where b is a positive constant, is reflected in the line $y=x$. Given that the original hyperbola and its image intersect, find the range of possible values of b.

16 **a** Show that the line $y=mx+c$ is a tangent to the ellipse $\frac{x^2}{a^2}+\frac{y^2}{b^2}=1$ when $a^2m^2+b^2=c^2$.

b The line l is a tangent to both the ellipse $\frac{x^2}{25}+\frac{y^2}{21}=1$ and the ellipse $\frac{x^2}{16}+\frac{y^2}{57}=1$. Find the possible equations of l.

17 An ellipse E has equation

$$\frac{x^2}{16}+\frac{y^2}{9}=1$$

a Sketch the ellipse E, showing the values of the intercepts on the coordinate axes.

b Given that the line with equation $y=x+k$ intersects the ellipse E at two distinct points, show that $-5<k<5$.

c The ellipse E is translated by the vector $\begin{bmatrix}a\\b\end{bmatrix}$ to form another ellipse whose equation is $9x^2+16y^2+18x-64y=c$. Find the values of the constants a, b and c.

d Hence find an equation for each of the two tangents to the ellipse $9x^2+16y^2+18x-64y=c$ that are parallel to the line $y=x$.

[© AQA 2014]

4 Rational functions and inequalities

In this chapter you will learn how to:

- solve cubic and quartic inequalities
- sketch graphs of the form $y = \dfrac{ax+b}{cx+d}$
- solve inequalities of the form $\dfrac{ax+b}{cx+d} > ex+f$
- sketch graphs of the form $y = \dfrac{ax^2+bx+c}{dx^2+ex+f}$.

Before you start...

GCSE	You should be able to solve linear inequalities.	1 Solve $5-2x>3$.
A Level Mathematics Student Book 1, Chapter 3	You should be able to solve quadratic inequalities.	2 Solve $x^2+6>5x$.
A Level Mathematics Student Book 1, Chapter 3	You should be able to use the quadratic discriminant.	3 Find the values of k for which x^2+2x+k has **no** real solutions.
A Level Mathematics Student Book 1, Chapter 4	You should be able to sketch polynomials from their factorised form.	4 Sketch the graph of $y=(x+1)(x-3)^2$.
A Level Mathematics Student Book 1, Chapter 4	You should be able to use the factor theorem to factorise polynomials.	5 Factorise x^3-5x^2+8x-4.
A Level Mathematics Student Book 1, Chapter 5	You should be able to describe graphically simple transformations of functions.	6 What transformation turns $f(x)$ into $f(x-3)$?
A Level Mathematics Student Book 1, Chapter 5	You should be able to sketch graphs of reciprocal functions and know associated terminology.	7 What are the asymptotes of $y=\dfrac{2}{x}$?
A Level Mathematics Student Book 2, Chapter 5	You should be able to simplify rational functions.	8 Simplify $\dfrac{x^2+4}{x^2+1}$.

Why are inequalities important?

Although equations are much easier to deal with, in the real world inequalities are often more important. For example, you do not usually need a car with a fuel efficiency of exactly 50 miles per gallon – you want a fuel efficiency of at least 50 miles per gallon. You may want the temperature in a fridge to be at most 6 °C.

Unfortunately, not all of the theory you have learnt for solving equations will always work with inequalities. For example, you should know that if $x^2 > 4$ this does not mean that $x > \pm 2$. Instead you need to use graphs to help you solve inequalities.

In this chapter, you will explore the graphs of more complicated functions that are the ratio of two polynomials (called **rational functions**).

Section 1: Cubic and quartic inequalities

You already know that to solve quadratic inequalities you can sketch the graph. You can extend the same method to cubic and quartic inequalities.

Key point 4.1

To solve cubic or quartic inequalities, make one side zero and sketch the graph. Describe in terms of x the regions of the graph above or below the x-axis.

WORKED EXAMPLE 4.1

Solve $(x+2)(x-1)(x-3) > 0$.

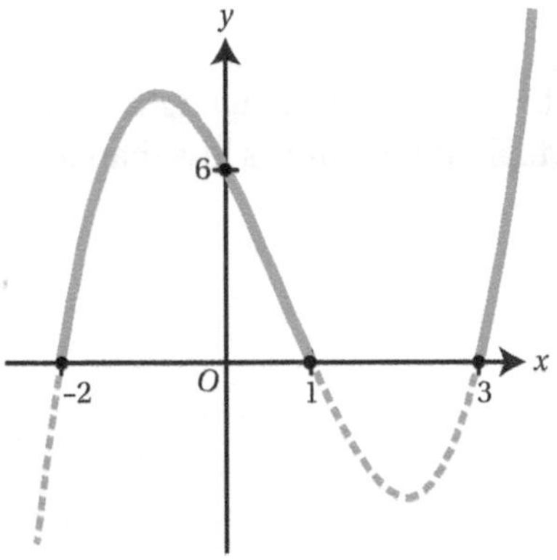

Sketch the curve of $y = (x+2)(x-1)(x-3)$. Highlight those parts of the graph for which the y-value is greater than 0.

So $-2 < x < 1$ or $x > 3$.

Describe, in terms of x, the highlighted sections. Notice that, since the inequality in the question is strict, the solution also involves strict inequalities.

Tip

You can also find this solution by examining the sign of each factor. For example, if $-2 < x < 1$, then the term in the first bracket is positive and the terms in the other two brackets are negative, so overall it will be positive.

If the inequality is not in a factorised form with one side zero, you may need to rearrange it and use the factor theorem.

WORKED EXAMPLE 4.2

Solve $x^4 + 14x^2 \leqslant 7x^3 + 8x$.

$x^4 - 7x^3 + 14x^2 - 8x \leqslant 0$
$x(x^3 - 7x^2 + 14x - 8) \leqslant 0$

If $f(x) = x^3 - 7x^2 + 14x - 8$, then:

To use the factor theorem try some small integer values. Once you have found one value you could then use **polynomial division** to find remaining factors, but it is often quicker just to find all three factors.

x	$f(x)$
−1	−30
0	−8
1	0
2	0
3	−2
4	0

So, from the factor theorem:
$f(x) = (x-1)(x-2)(x-4)$

Hence the inequality becomes:
$x(x-1)(x-2)(x-4) \leqslant 0$

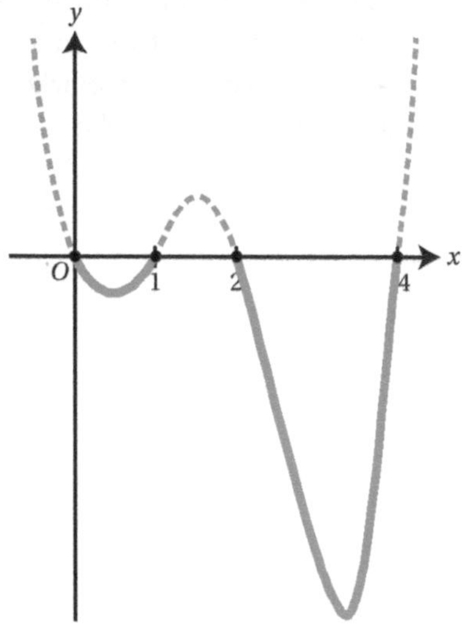

Now sketch the graph, and highlight those parts for which the y-value is less than 0.

$0 \leqslant x \leqslant 1$ or $2 \leqslant x \leqslant 4$

Describe the highlighted region on the graph in terms of x.

EXERCISE 4A

1 Solve each inequality.

a i $x(x-1)(x-2)>0$ ii $x(x+1)(x-4)>0$

b i $(x+3)(x-2)(x-5)\leqslant 0$ ii $(x-1)(x-3)(x-4)\leqslant 0$

c i $(x-1)(x-2)(x-3)(x-4)\leqslant 0$ ii $(x+2)(x+1)(x-3)(x-7)\leqslant 0$

d i $3(x+1)(x-2)^2\geqslant 0$ ii $5(x+3)^2(x-4)\geqslant 0$

e i $6(x+3)(x-1)(x^2+5)>0$ ii $2(x^2+3)(x-2)(x-4)>0$

f i $(3-x)(x-2)(x-5)<0$ ii $(4-x)(x-1)(x-8)<0$

2 Find all values of x that satisfy each inequality.

a i $x^3-4x<0$ ii $x^3-9x>0$

b i $x^4+2x^2\geqslant 3x^3$ ii $x^4+12x^2\geqslant 7x^3$

c i $x^3+2<2x^2+x$ ii $x^3+4<x^2+4x$

d i $x^3+3x^2-5x+1>0$ ii $x^3+3x^2+x-2<0$

3 Find all values of x such that $2x^3<7x^2+15x$.

4 Solve $a(x-b)(x-c)(x-d)>0$ if $0<a<b<c<d$.

5 a Show that $x=2$ is a root of $y=x^3+3x^2-11x+2$.

b Hence find the exact range of values of x such that $x^3+3x^2-7x+5>4x+3$.

6 Solve $(x^2-a^2)(x^2+b^2)\geqslant 0$ given that a and b are positive numbers.

7 A function f is given by $f(x)=3x^3+ax^2+bx+c$ where a, b, and c are integers.

Given that the solution to $f(x)>2$ is $-1<x<4$ or $x>5$, find a, b and c.

8 A function f is given by $f(x)=x^3+ax^2+bx+c$.

Given that the solution to $f(x)<2$ is $x<3$, $x\neq 1$, find a, b and c.

9 Find all values x such that $\sqrt{x}<x^2$.

10 Find a quartic inequality for which the solution set is:

a all real numbers

b exactly two numbers.

Section 2: Functions of the form $y = \dfrac{ax+b}{cx+d}$

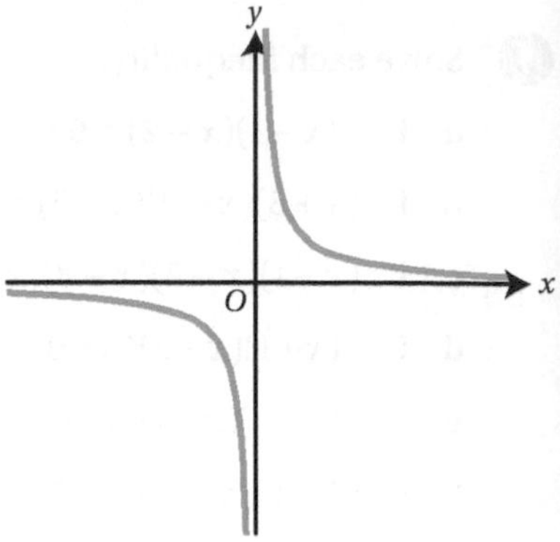

If $f(x) = \dfrac{1}{x}$, you know from Chapter 3 that the graph of $y = f(x)$ forms a **hyperbola**, with **asymptotes** at $x = 0$ and $y = 0$.

If you use your knowledge of transformations of functions, you can also sketch, say, $y = f(x-1)+2$, which is shifted one unit right and two units up. It therefore it has asymptotes at $x = 1$ and $y = 2$.

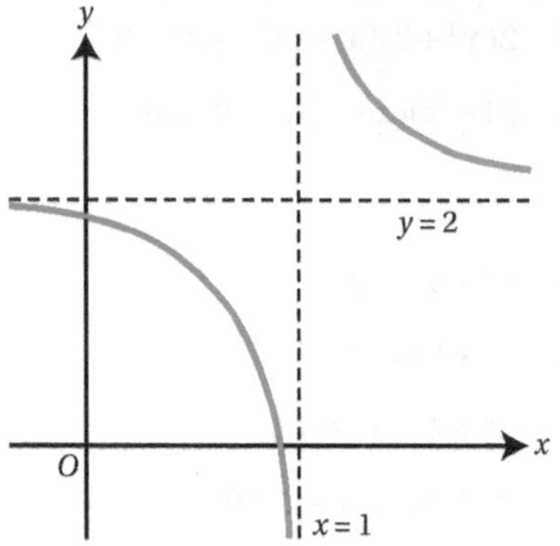

Algebraically:
$$\begin{aligned} f(x-1)+2 &= \frac{1}{x-1}+2 \\ &= \frac{1}{x-1}+\frac{2(x-1)}{x-1} \\ &= \frac{2x-1}{x-1} \end{aligned}$$

In fact, any function of the form $y = \dfrac{ax+b}{cx+d}$ will be a transformation of $y = \dfrac{1}{x}$, so it will also be a hyperbola. However, it can be difficult to determine which transformations have been applied. Instead you can analyse it by finding the axes intercepts and asymptotes directly.

- The y-intercept is when $x = 0$ so $y = \dfrac{b}{d}$.
- The x-intercept is when $y = 0$. This means the numerator is zero. This occurs when $ax+b = 0$ so $x = -\dfrac{b}{a}$.
- The vertical asymptote will occur when the denominator is zero. This occurs when $cx+d = 0$ so $x = -\dfrac{d}{c}$.
- The horizontal asymptote will occur at very large values of x. This means the denominator is approximately ax and the denominator is approximately cx so y tends towards $\dfrac{a}{c}$, which is the horizontal asymptote.

Key point 4.2

If $y = \dfrac{ax+b}{cx+d}$:

- the intercepts are $\left(0, \dfrac{b}{d}\right)$ and $\left(-\dfrac{b}{a}, 0\right)$
- the asymptotes are $x = -\dfrac{d}{c}$ and $y = \dfrac{a}{c}$.

WORKED EXAMPLE 4.3

a Sketch the function $y = \frac{x+2}{x+p}$ where $p > 2$. Label the asymptotes and intercepts.

b Solve $\frac{x+2}{x+p} \geqslant 2$.

a

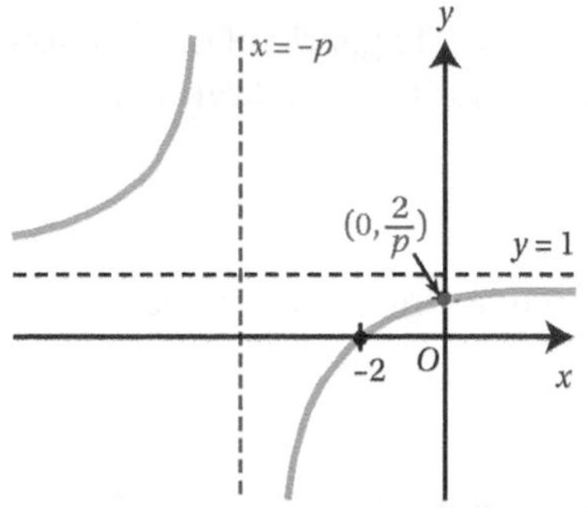

Using Key point 4.2 you can say that the intercepts are $(0, \frac{2}{p})$ and $(-2, 0)$.
The vertical asymptote is at $x = -p$ and the horizontal asymptote is at $y = 1$. This information, along with the fact that it is a hyperbola, is enough to sketch the curve.

b

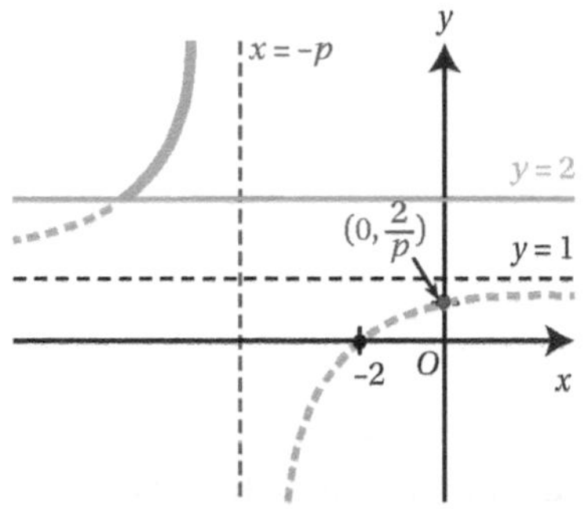

Add the line $y = 2$ to the graph. The only part of the hyperbola above this line is between the intersection and the vertical asymptote.

If $\frac{x+2}{x+p} = 2$ then:

$x + 2 = 2x + 2p$ so

$2 - 2p = x$

You can find the intersection point algebraically.

So the solution to the inequality is $2 - 2p \leqslant x < -p$.

Notice that the x can never equal $-p$ as that would result in division by zero.

In Worked example 4.3, you used the graph to help you solve the inequality. You can also do this algebraically, but you must take care. If you were just to multiply through by the denominator, you might be multiplying by a negative number, which would reverse the inequality. To avoid this you can use an algebraic trick.

Key point 4.3

When solving inequalities involving algebraic fractions, multiply both sides by the square of the denominator.

Tip

A graphical calculator does not allow you to sketch this function because it involves a parameter, p. However, if you have one you can check your answer for sense, by trying to sketch the function for particular values of p.

This works because square numbers are always positive, so you avoid ever having to reverse the inequality.

WORKED EXAMPLE 4.4

Use an algebraic method to solve $\frac{x-2}{x+p} \geqslant 2$.

$(x-2)(x+p) \geqslant 2(x+p)^2$ — Multiply through by $(x+p)^2$ to get rid of the fraction. Resist the temptation to multiply out brackets!

$0 \geqslant 2(x+p)^2 - (x-2)(x+p)$ — Make one side zero...

$0 \geqslant (x+p)(2x+2p-x+2)$ — ... then factorise by looking for a common factor - in this case $(x+p)$.

$0 \geqslant (x+p)(x+(2+2p))$

This is a quadratic inequality so sketch the graph.

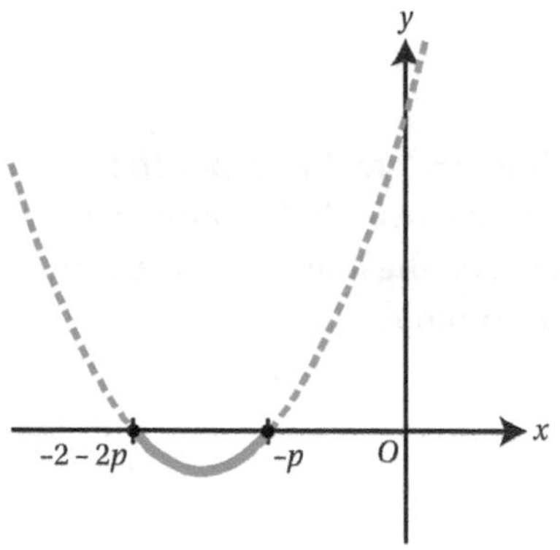

So $-2-2p \leqslant x \leqslant -p$ — Describe the part of the graph below the axis in terms of x.

However, $x \neq -p$ in the original equation, therefore: $-2-2p \leqslant x < -p$ — Remember to think about the solution in the context of the original problem.

EXERCISE 4B

1 Without using a calculator, determine exact values of any asymptotes and axis intersections and sketch the graph of each function $f(x)$.

a **i** $f(x) = \frac{2x-1}{x+4}$ **ii** $f(x) = \frac{3x-2}{x-1}$ **b** **i** $f(x) = 2\frac{x-4}{x+4}$ **ii** $f(x) = 1 - \frac{2x+3}{3-x}$

2 From the information given, write each function $f(x)$ in the form $f(x) = \frac{ax+b}{cx+d}$ by finding the values of a, b, c and d.

a **i** Asymptotes $y = 3$ and $x = 1$. Root at $x = 2$.

ii Asymptotes $y = 0.5$ and $x = -1$. y-intercept at $(0, 2)$.

b **i** Asymptote $y = 1$. Root at $x = 4$ and y-intercept $(0, -1)$.

ii Asymptote $x = 3$. Root at $x = 2$ and y-intercept $(0, 1)$.

3 Solve for x.

a **i** $\frac{x-3}{x+2}>1$ **ii** $\frac{x+4}{2x+3}<-2$ **b** **i** $\frac{2x-1}{x+5}\leqslant 3$ **ii** $\frac{2x-5}{x+1}\geqslant 1$

4 Solve for x, giving your answers as exact values.

a **i** $\frac{x-1}{x+2}>x-3$ **ii** $\frac{x+4}{2x+3}<x-2$ **b** **i** $\frac{2x-7}{x+1}<x-3$ **ii** $\frac{2x+5}{x+1}\geqslant x+1$

5 A function $f(x)$ is given by $f(x)=\frac{4x-1}{2x+3}$.

a Write down the vertical asymptote for $f(x)$.

The line l is given by equation $y=x-2$.

b **i** Sketch $y=f(x)$ and the line l on the same axes.

ii Find the values of x for which $f(x)>x-2$.

6 A curve has equation $y=\frac{2x-a}{x+b}$, where $a,b>0$.

a Sketch the curve, clearly giving the equations of all asymptotes and the coordinates of all axis intercepts.

b Hence solve the inequality

$$\frac{2x-a}{x+b}>3$$

7 The function $f(x)$ is given by $f(x)=\frac{5x}{x+4}$.

a The line l is given by the equation $y=mx+1$.

Given that l is a tangent to $y=f(x)$, calculate all possible values of m.

The function $g(x)$ is given by $g(x)=\frac{2-x}{x+1}$.

b **i** On the same axes, sketch $f(x)$ and $g(x)$.

ii Find the exact values of x for which $f(x)>g(x)$.

8 $f(x)=\frac{ax+b}{cx+d}$ for some values a,b,c and d with $a,c\neq 0$.

a Assuming $ad\neq bc$, state necessary and sufficient conditions in terms of a,b,c and d for $f(x)$ to have:

i a vertical asymptote at $x=k$

ii a horizontal asymptote at $y=p$.

b In part **a**, why was the condition $ad\neq bc$ required?

9 The solution set for $f(x)=\frac{16x+1}{px+1}>x+4$ is $x<q$ or $r<x<3$ for some values p,q and r.
Find p,q and r.

10 Curve C has equation $y=\frac{2x-5}{x-1}$.

a A line $y=mx+c$ is a tangent to the curve. Find a condition for c in terms of m.

Two parallel lines are each tangents to the curve, touching at points $P(p,q)$ and $R(r,s)$ with $p<1<r$.

b If the distance PR is d, find d^2 in terms of m, the gradient of each of the lines.

c By differentiating d^2 with respect to m, or otherwise, find the shortest distance between the two branches of the curve.

Section 3: Functions of the form $y = \dfrac{ax^2 + bx + c}{dx^2 + ex + f}$

If the top and bottom of a rational function are quadratic expressions, then the graph will no longer be a hyperbola, but you can use a similar analysis to that shown in Section 2.

Key point 4.4

If $y = \dfrac{ax^2 + bx + c}{dx^2 + ex + f}$ then:

- when $x = 0$, $y = \dfrac{c}{f}$
- when $y = 0$, x is a solution to $ax^2 + bx + c = 0$
- vertical asymptotes are solutions to $dx^2 + ex + f = 0$
- the horizontal asymptote is $y = \dfrac{a}{d}$.

WORKED EXAMPLE 4.5

a Sketch $y = \dfrac{x^2 + 4x + 3}{x^2 - 5x + 6}$.

b Hence solve the inequality $\dfrac{x^2 + 4x + 3}{x^2 - 5x + 6} \geqslant 4$.

a When $x = 0$, $y = 0.5$. — Find the y-intercept.

The horizontal asymptote is $y = 1$. — Consider what happens when x gets very large (or just use Key point 4.4).

When $y = 0$: — Find the x-intercepts by setting the numerator equal to zero.

$x^2 + 4x + 3 = 0$

$(x + 1)(x + 3) = 0$

$x = -1$ or $x = -3$

Vertical asymptotes occur when: — Find the vertical asymptotes by setting the denominator equal to zero.

$x^2 - 5x + 6 = 0$

$(x - 2)(x - 3) = 0$

$x = 2$ or $x = 3$

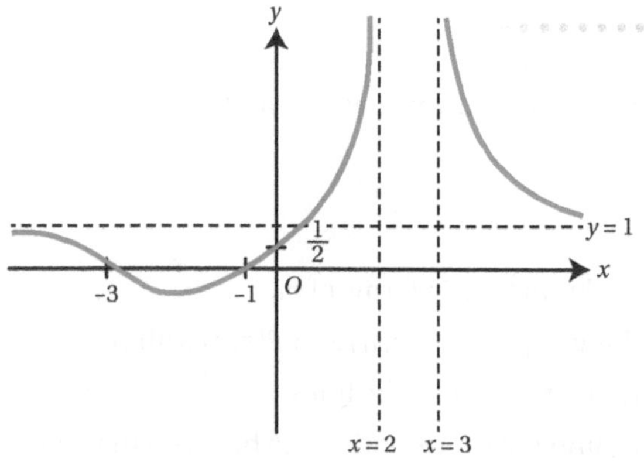

Put all the information together.

It is a little unclear whether the section between the asymptotes is positive or negative. To decide this, substitute $x = 2.5$ to find $y = -53$.

Continues on next page

b

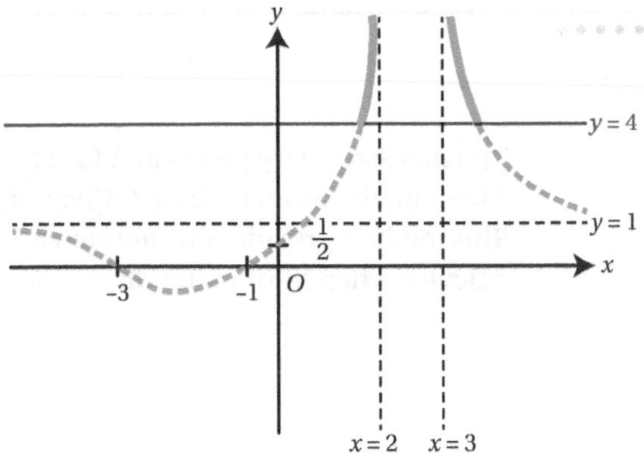

Add the line $y = 4$ to the diagram and see where the graph is above this line.

When $y = 4$:

$$4 = \frac{x^2 + 4x + 3}{x^2 - 5x + 6}$$

$$4x^2 - 20x + 24 = x^2 + 4x + 3$$

$$3x^2 - 24x + 21 = 0$$

$$x^2 - 8x + 7 = 0$$

$$(x - 1)(x - 7) = 0$$

$$x = 1 \text{ or } x = 7$$

Find the values of x at the intersection points.

So $1 \leqslant x < 2$ or $3 < x \leqslant 7$.

Describe the part of the graph above the line $y = 4$ in terms of x. Note that x can't be equal to 2 or 3 as there are asymptotes there.

One useful way of analysing these functions is to look at the possible y-values. If you intersect these functions with $y = k$ you get a quadratic equation. You are not particularly interested in the value of x that you get from these equations – you are more interested in whether or not a solution actually exists. To find out, use the **quadratic discriminant**.

Tip

If you have a graphical calculator you can try to check this graph – but you will find that it is very difficult to see all the relevant features in one window. Remember that a labelled diagram does not have to be to scale.

WORKED EXAMPLE 4.6

Find the stationary points on the graph $y = \frac{x^2 - 8x + 16}{x^2 - 4}$.

If $y = k$ then:

$$k = \frac{x^2 - 8x + 16}{x^2 - 4}$$

You met stationary points in A Level Mathematics Student Book 1, Chapter 14. This method provides an alternative to calculus when finding stationary points.

$kx^2 - 4k = x^2 - 8x + 16$

$(k-1)x^2 + 8x - (16 + 4k) = 0$

Discriminant $\Delta = 8^2 + 4(16 + 4k)(k-1)$

$= 16k^2 + 48k$

$= 16k(k+3)$

So there are solutions when $k > 0$ or $k < -3$.

You can interpret this as meaning that y-values do not occur between -3 and 0, so there must be a maximum at $y = -3$ and a minimum at $y = 0$.

A maximum occurs when $y = -3$ so $x = 1$.

Put $k = -3$ into the quadratic equation formed above to find the value of x.

A minimum occurs when $y = 0$ so $x = 4$.

EXERCISE 4C

1 Identify all axis intercepts and vertical and horizontal asymptotes for each function $y = f(x)$. Hence sketch the graph of each function.

a i $f(x) = \frac{(x-1)(x+2)}{(x+3)(x+5)}$ ii $f(x) = \frac{(x+1)(x+6)}{(x+3)(x+2)}$

b i $f(x) = \frac{5x^2 - x - 4}{2x^2 - 8}$ ii $f(x) = \frac{2x^2 - 5x - 3}{x^2 - 7x}$

c i $f(x) = 1 + \frac{8x - 9}{x^2 - 4x + 4}$ ii $f(x) = 2 - \frac{7x + 5}{1 - x^2}$

d i $f(x) = \frac{x^2 + 4}{x^2 + 1}$ ii $f(x) = \frac{x^2 - 16}{x^2 + 4}$

2 By finding a condition on k for the number of solutions to $f(x) = k$, identify the exact y-coordinates of each of the stationary points for each function.

a i $f(x) = \frac{x^2 - 4}{x^2 - 1}$ ii $f(x) = \frac{x^2 - 9}{x^2 - 3}$

b i $f(x) = \frac{x^2 - 1}{x^2 + 3x}$ ii $f(x) = \frac{x^2 - 4}{x^2 + 4x}$

c i $f(x) = \frac{x^2 - 4x + 3}{x^2 + 1}$ ii $f(x) = \frac{5x^2 - x - 4}{x^2 + 3x - 10}$

3 **a** Sketch the curve $y = \frac{7}{x^2 - 5}$, clearly labelling any asymptotes and axis intercepts.

b Hence solve the inequality $\frac{7}{x^2 - 5} > 4$.

4 One of the asymptotes of the graph of $y = \frac{x^2 + 2x + 1}{x^2 - kx + 3}$ is $x = 3$.

Find the value of k and give the equation of all other asymptotes.

5 Solve the inequality $\frac{3x^2 + 5x - 11}{2x^2 + x - 6} > 1$.

6 A curve has equation $y = \frac{x - 3}{x^2 + 16}$.

a **i** State the equations of any asymptotes.

ii State the coordinates of any axis intercepts.

b **i** If the line $y = k$ intersects the curve, show that:

$$64k^2 + 12k - 1 \leqslant 0$$

ii Hence find the coordinates of the turning points of the curve.

c Sketch the curve.

7 Sketch $y = \frac{(x-a)(x-c)}{(x-b)(x-d)}$ if $0 < a < b < c < d$.

8 The solution to $0 < \frac{3x^2 - 3x + 2}{x^2 + x - 2} < a$ is $b < x < 3$.

Find a and b.

9 f(x) is a function of the form $f(x) = \frac{ax^2 + bx + c}{x^2 + dx + e}$ where a is non-zero.

f(x) has no roots or vertical asymptotes, but has a maximum at 2, a minimum at 0.5, y-intercept at $(0, 1.5)$ and a horizontal asymptote $y = 1$.

a Show that $(b - 0.5d)^2 = (b - 2d)^2 = 2e$.

b Given the maximum and minimum points are at $x = 1$ and $x = -2$ respectively, find a, b, c, d and e.

10 Two particles are projected upwards at the same time.

Particle A begins 4.9 metres off the ground with an initial velocity of 98 m s^{-1}.

Particle B begins at ground level with an initial velocity 49 m s^{-1}.

$p(t)$ is the ratio of the heights above ground from $t = 0$ until the time the first particle hits the ground, and $p(0) = 0$.

a Taking acceleration due to gravity as 9.8 m s^{-2}, find $p(t)$.

b Find the maximum value of p, correct to three significant figures.

Section 4: Oblique asymptotes

Key point 4.4 states that the horizontal asymptote occurs at $y = \frac{a}{d}$. However, this does not work if $d = 0$. In this case, you need to simplify the rational function into a polynomial plus a rational function. This means that when x is very large, rather than tending towards a constant, the graph will tend towards a non-horizontal line.

WORKED EXAMPLE 4.7

Find the oblique asymptotes of $y = \frac{x^2+1}{x+1}$.

$$\frac{x^2+1}{x+1} \equiv \frac{x^2-1+2}{x+1}$$

You can also do this by polynomial long division, or by comparing coefficients.

$$\equiv \frac{(x-1)(x+1)}{x+1} + \frac{2}{x+1}$$

$$\equiv x - 1 + \frac{2}{x+1}$$

The oblique asymptote is $y = x - 1$.

You know this because when x is very large the $\frac{2}{x+1}$ term is very small.

EXERCISE 4D

1 Identify all axis intercepts and vertical and oblique asymptotes of $y = f(x)$ for each function. Hence sketch its graph.

a i $f(x) = \frac{x^2+4x-5}{x-2}$ ii $f(x) = \frac{4x^2+x}{x+1}$ b i $f(x) = x + \frac{2}{x+3}$ ii $f(x) = x - 3\frac{x+2}{x-2}$

2 Rewrite $f(x)$ in the form $f(x) = ax + b + \frac{cx+d}{rx^2+sx+t}$.

By solving $f'(x) = 0$, find the exact x-coordinates of the stationary points of each curve.

a i $f(x) = \frac{x^2-4}{x^2-1}$ ii $f(x) = \frac{x^2-1}{x^2+3x}$ b i $f(x) = \frac{x^2-4x+3}{x^2+1}$ ii $f(x) = \frac{5x^2-x-4}{x^2+3x-10}$

3 a Show that the function $f(x) = \frac{2x^2-x-3}{2x-5}$ can be written as $f(x) = Ax + B + \frac{C}{2x-5}$ where A, B and C are constants to be found.

b Hence write down the oblique linear asymptote for function $f(x)$.

c By finding a condition on k for the number of solutions to $f(x) = k$, find the y-coordinates of any stationary points of $f(x)$.

d Write down the axis intercepts and vertical asymptotes for $f(x)$.

e Use your answers to **b** to **d** to sketch a graph of $f(x)$.

4 A curve C is given by $y = \frac{x^2 + 2ax + a^2 - 1}{x + a}$, where $a > 0$.

a Find the equation of the oblique asymptote of C.

b By finding $\frac{dy}{dx}$ or otherwise, show that C has no stationary points.

c Sketch C, labelling all asymptotes and axis intercepts.

Checklist of learning and understanding

- To solve cubic or quartic inequalities, take all the non-zero terms to one side, so the other side is zero, and sketch the graph. Describe in terms of x the regions of the graphs above or below the x-axis.
- If $y = \frac{ax + b}{cx + d}$:
  - the intercepts are $(0, \frac{b}{d})$ and $(-\frac{b}{a}, 0)$
  - the asymptotes are $x = -\frac{d}{c}$ and $y = \frac{a}{c}$.
- When solving inequalities involving algebraic fractions, multiply both sides by the square of the denominator.
- If $y = \frac{ax^2 + bx + c}{dx^2 + ex + f}$ then:
  - when $x = 0$, $y = \frac{c}{f}$
  - when $y = 0$, x is a solution to $ax^2 + bx + c = 0$
  - vertical asymptotes are solutions to $dx^2 + ex + f = 0$
  - the horizontal asymptote is $y = \frac{a}{d}$.

Mixed practice 4

1 State the equations of the vertical asymptotes of the curve $y = \frac{x^2 - 4}{2x^2 - 3x}$.

Choose from these options.

A $x = 0$ and $x = \frac{3}{2}$

B $x = \frac{1}{2}$ and $x = \frac{4}{3}$

C $x = 2$ and $x = -2$

D $x = 2$ and $x = -3$

2 Solve the inequality

$$(1+x)(1-x)(2-x) < 0$$

Choose from these options.

A $x < -1$ or $1 < x < 2$

B $-2 < x < -1$ or $x > 1$

C $-1 < x < 1$ or $x > 2$

D $x < -2$ or $-1 < x < 1$

3 Solve the inequality

$$x^4 - 7x^3 + 10x^2 > 0$$

4 Curve C is given by the equation $y = \frac{6x^2 + 5x + 1}{x^2 - 3x - 4}$.

a Find all axis intercepts and asymptotes for C.

b Hence sketch C.

5 **a** **i** Write down the equations of the two asymptotes of the curve $y = \frac{1}{x-3}$.

ii Sketch the curve $y = \frac{1}{x-3}$, showing the coordinates of any points of intersection with the coordinate axes.

iii On the same axes, again showing the coordinates of any points of intersection with the coordinate axes, sketch the line $y = 2x - 5$.

b **i** Solve the equation

$$\frac{1}{x-3} = 2x - 5$$

ii Find the solution of the inequality

$$\frac{1}{x-3} < 2x - 5$$

[©AQA 2010]

6 **a** Show that $(x+2)$ is a factor of $x^3 + (b+2)x^2 + 2(b+2)x + 8$.

b $x < -2$ is the solution of the inequality

$$x^3 + (b+2)x^2 + 2(b+2)x + 8 < 0$$

Find the range of possible values of b.

7 The curve C has equation $y = \dfrac{2a-1}{x^2 - a^2}$, where $a > \dfrac{1}{2}$.

a State the equations of all asymptotes of C.

b Sketch C, labelling the coordinates of any axis intercepts

c Solve the inequality

$$\frac{2a-1}{x^2-a^2} < -1$$

8 The function $f(x)$ is given by $f(x) = \dfrac{3x^2 - 15x + 6}{x^2 - 5x + 6}$.

a State all axis intercepts and asymptotes for the graph of $y = f(x)$.

b Hence sketch the curve showing the asymptotes and axis intercepts.

9 Given that there are exactly two solutions to $\dfrac{x^2 + 2x + 1}{x^2 - 4} = k$, find the possible values of k.

10 Curve C has equation $y = \dfrac{ax^2 + bx + c}{x^2 + dx + e}$, and has asymptotes $y = k$, $x = k$ and $x = 1 - k$ where k is a real number. The x-axis intercepts of C are $(2k, 0)$ and $(-k, 0)$.

a Find values a, b, c, d and e in terms of k.

b Given the y-intercept of C is $(0, 9)$, find k.

11 Find a condition on c so that function $f(x) = \dfrac{x+c}{x^2 - 3x - c}$ has the whole of the real numbers as its range.

12 The diagram shows part of the curve with equation $y = \dfrac{x^2 - 3x}{x^2 - 2x + 1}$.

a Write down the coordinates of A.

b Write down the equation of any asymptotes for this curve.

c Determine the coordinates of the maximum point B and so find the least value k such that $\dfrac{x^2 - 3x}{x^2 - 2x + 1} \leqslant k$ for all x.

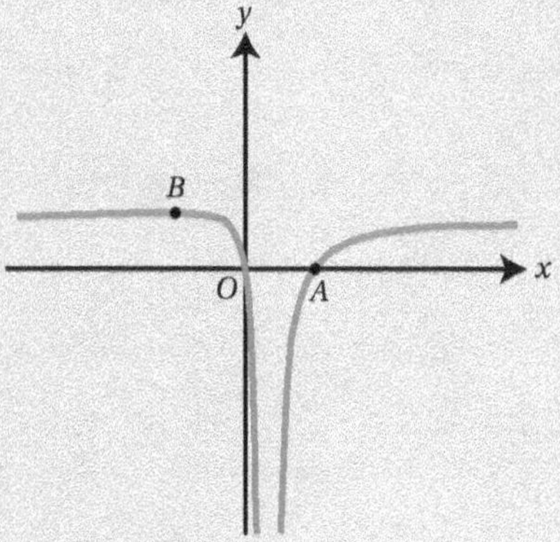

13 A curve C has equation $y = \dfrac{1}{x(x+2)}$.

a Write down the equations of all the asymptotes of C.

b The curve C has exactly one stationary point. The x-coordinate of the stationary point is -1.

i Find the y-coordinate of the stationary point.

ii Sketch the curve C.

c Solve the inequality $\dfrac{1}{x(x+2)} \leqslant \dfrac{1}{8}$.

[©AQA 2014]

14 A curve C has equation $y = \frac{x(x-3)}{x^2+3}$.

a State the equation of the asymptote of C.

b The line $y = k$ intersects the curve C. Show that $4k^2 - 4k - 3 \leqslant 0$.

c Hence find the coordinates of the stationary points of C.

(No credit will be given for solutions based on differentiation.)

[©AQA 2015]

A **15** $\mathrm{f}(x)$ is a rational function of the form $\mathrm{f}(x) = \frac{ax^2+bx+c}{dx+e}$ and the function $\mathrm{g}(x)$ is given by $\mathrm{g}(x) = \frac{4}{\mathrm{f}(x)}$.

It is known that $\mathrm{f}(x)$ has an oblique asymptote $y = x+1$ and $\mathrm{g}(x)$ has vertical asymptotes $x = \frac{1 \pm \sqrt{3}}{2}$.

Find all values of x for which $\mathrm{f}(x) = \mathrm{g}(x)$.

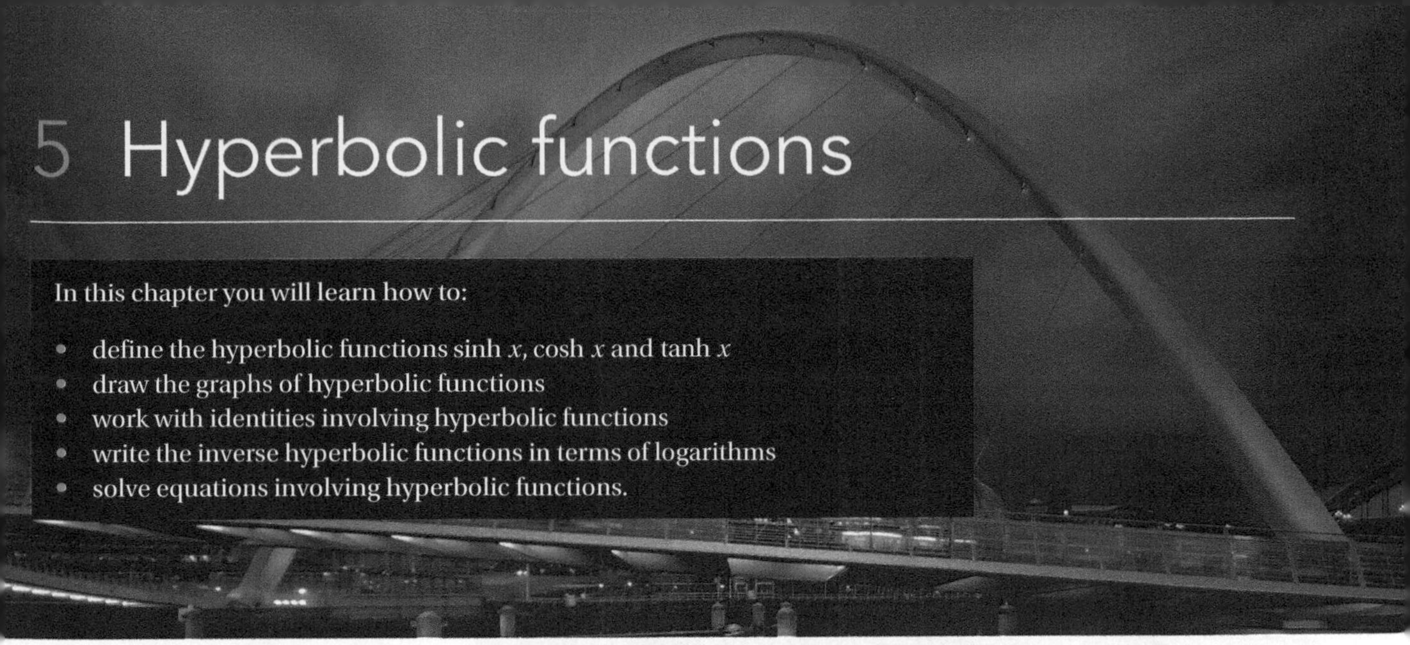

5 Hyperbolic functions

In this chapter you will learn how to:

- define the hyperbolic functions $\sinh x$, $\cosh x$ and $\tanh x$
- draw the graphs of hyperbolic functions
- work with identities involving hyperbolic functions
- write the inverse hyperbolic functions in terms of logarithms
- solve equations involving hyperbolic functions.

Before you start...

A Level Mathematics Student Book 1, Chapter 3	You should be able to solve quadratic equations.	1 Solve $x^2 + 3x = 1$.
A Level Mathematics Student Book 1, Chapter 5	You should be able to interpret transformations of graphs.	2 What transformation changes $f(x)$ to $3\,f(x+1)$?
A Level Mathematics Student Book 1, Chapter 7	You should be able to work with natural logarithms.	3 Given that $2^x = 5$, write x in the form $\frac{\ln a}{\ln b}$.
A Level Mathematics Student Book 1, Chapter 9	You should be able to complete binomial expansions of brackets with positive integer powers.	4 Expand $(2+x)^3$.
A Level Mathematics Student Book 1, Chapter 10	You should be able to work with the symmetries of trigonometric functions.	5 Given that $\sin 40° = a$, find $\sin 220°$.
A Level Mathematics Student Book 1, Chapter 10	You should be able to work with trigonometric identities.	6 Simplify $3\sin^2 2x + 3\cos^2 2x$.

What are hyperbolic functions?

Trigonometric functions are sometimes called **circular functions**. This is because of the definition that states: a point on the unit circle (with equation $x^2 + y^2 = 1$) at an angle θ to the positive x-axis, has coordinates $(\cos\theta, \sin\theta)$.

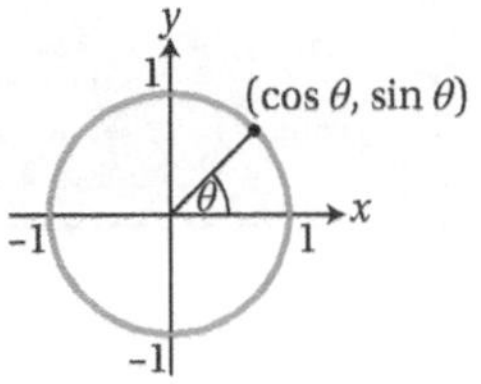

In Chapter 3 you met a related curve, with equation $x^2 - y^2 = 1$, called a **hyperbola**. Points on this hyperbola have coordinates $(\cosh\theta, \sinh\theta)$ although θ can no longer be interpreted as an angle.

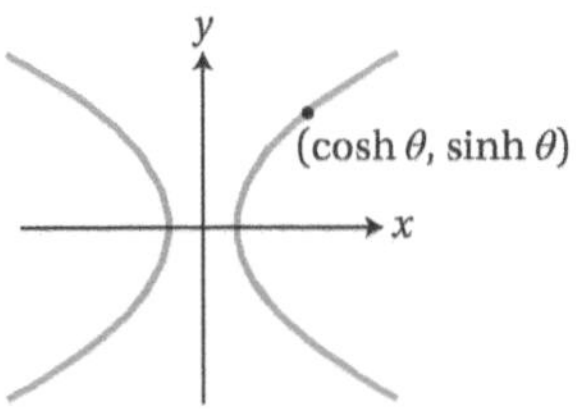

Section 1: Defining hyperbolic functions

Although the geometric definition of hyperbolic functions gives some helpful insight, a more useful definition is related to the number e.

Key point 5.1

- $\sinh x = \frac{e^x - e^{-x}}{2}$
- $\cosh x = \frac{e^x + e^{-x}}{2}$

You can define tanh x by analogy with the trigonometric definition (of tan x).

Key point 5.2

$$\tanh x \equiv \frac{\sinh x}{\cosh x} \equiv \frac{e^x - e^{-x}}{e^x + e^{-x}}$$

There are not many special values of these functions that you need to know, but from Key points 5.1 and 5.2 you should be able to see that $\cosh 0 = 1$, $\sinh 0 = 0$ and $\tanh 0 = 0$.

As for trigonometric functions, you need to know the graphs of hyperbolic functions.

Key point 5.3

The graph of $y = \sinh x$

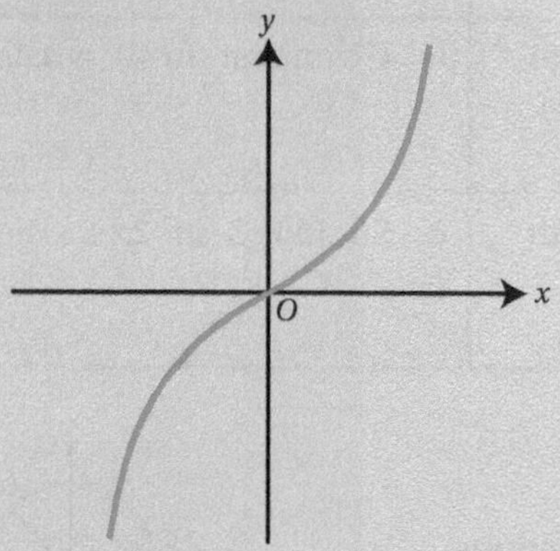

Fast forward

In Further Mathematics Student Book 2, you will see that the trigonometric functions can also be defined in a similar way in terms of complex numbers.

Tip

Cosh is pronounced as it reads, sinh is pronounced 'sinsh' or 'shine' and tanh is pronounced 'tansh'.

Did you know?

The tanh function is frequently used in physics, particularly in the context of special relativity and the study of entropy.

Key point 5.4

The graph of $y = \cosh x$

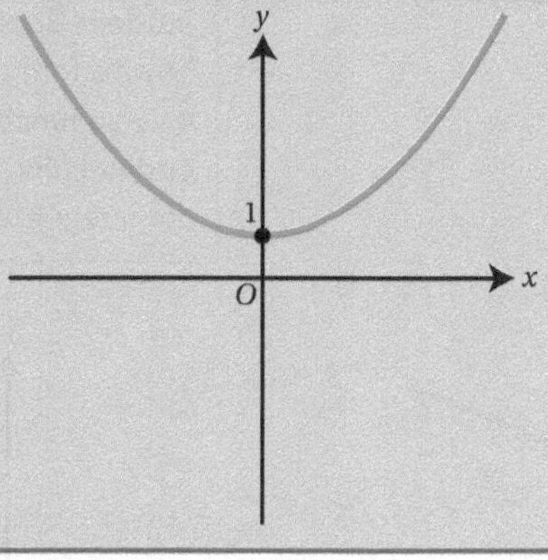

Did you know?

You may think that the graph of $\cosh x$ looks like a parabola, but it is slightly flatter. It is called a **catenary**, which is the shape formed by a hanging chain.

You can see that the minimum value of $\cosh x$ is 1, whereas $\sinh x$ has no minimum (or maximum).

Fast forward

In Further Mathematics Student Book 2 you will see the range of the sinh function is all real numbers and the range of the cosh function is $\cosh x \geq 1$.

Key point 5.5

The graph of $y = \tanh x$

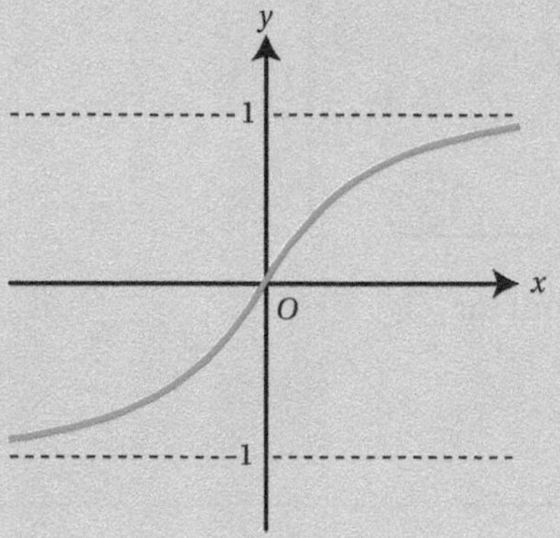

The tanh function has horizontal asymptotes at $y = 1$ and $y = -1$: $-1 < \tanh x < 1$.

Inverse hyperbolic functions

The **inverse functions** of the hyperbolic functions are called arsinh x, arcosh x and artanh x.

The graphs of these functions look like this.

> **Rewind**
>
> You saw in A Level Mathematics Student Book 2, Chapter 2, how to form the graphs of inverse functions from the original function by reflection in the line $y = x$.

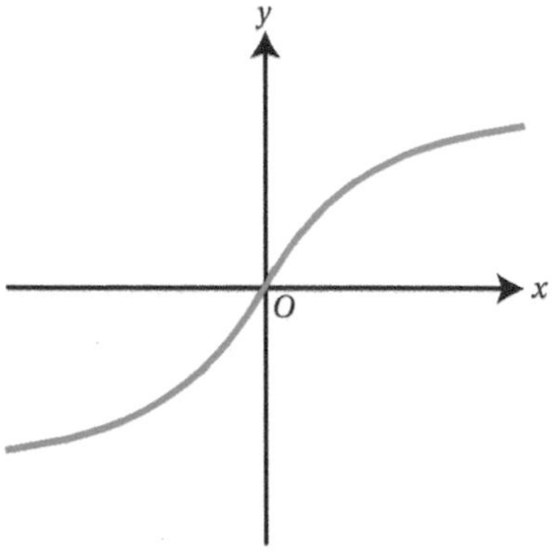

$y = \text{arsinh } x$

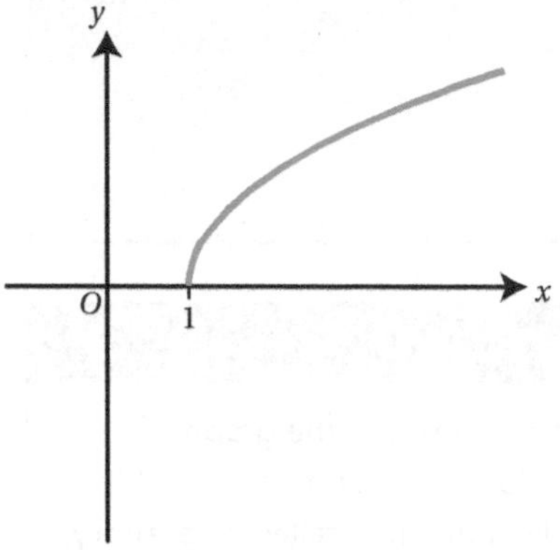

$y = \text{arcosh } x$

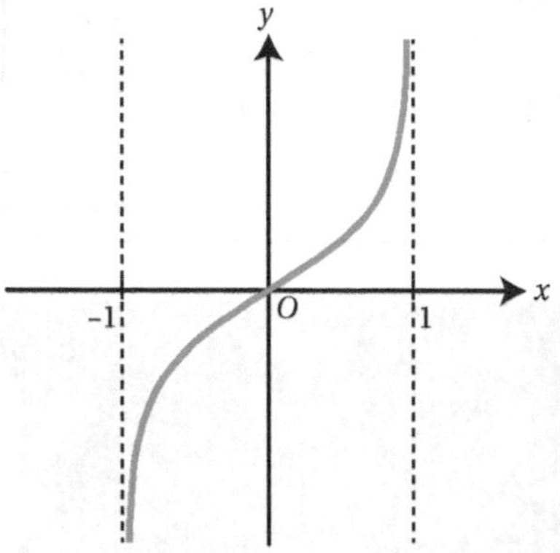

$y = \text{artanh } x$

You can use the inverse hyperbolic functions to solve simple equations involving hyperbolic functions. For example, if $\sinh x = 2$ then $x = \text{arsinh } 2$, which you can evaluate, on a calculator, as 1.4436....

> **Tip**
>
> On your calculator they might be called $\sinh^{-1} x$, $\cosh^{-1} x$ and $\tanh^{-1} x$.

However, you can use the definition of sinh x to derive a logarithmic form of this result. You can do this for all three inverse hyperbolic functions.

> **Key point 5.6**
>
> - $\text{arsinh } x = \ln\left(x+\sqrt{x^2+1}\right)$
> - $\text{arcosh } x = \ln\left(x+\sqrt{x^2-1}\right)$
> - $\text{artanh } x = \frac{1}{2}\ln\left(\frac{1+x}{1-x}\right)$
>
> **These will be given in your formula book.**

These results can all be proved in the same way. The proof for arcosh x is given here.

PROOF 5

Prove that $\text{arcosh } x = \ln\left(x+\sqrt{x^2-1}\right)$.

Let $y = \text{arcosh } x$ — Let $y = \text{arcosh } x$ and then look to find an expression for y.

Then $\cosh y = x$ — Take cosh of both sides.

Continues on next page

$\frac{e^y+e^{-y}}{2}=x$ — Use the definition of cosh y.

$e^y+e^{-y}=2x$ — Rearrange into a disguised quadratic in e^y.

$e^y+\frac{1}{e^y}=2x$

$(e^y)^2+1=2xe^y$

$(e^y)^2-2xe^y+1=0$

So — Use the quadratic formula.

$$e^y=\frac{2x\pm\sqrt{(2x)^2-4}}{2}$$

$$=\frac{2x\pm\sqrt{4x^2-4}}{2}$$

$$=\frac{2x\pm\sqrt{4(x^2-1)}}{2}$$

$$=\frac{2x\pm\sqrt{4}\sqrt{x^2-1}}{2}$$ — Use the algebra of surds to simplify the expression.

$$=x\pm\sqrt{x^2-1}$$

But arcosh x is a function so it can only take one value. — Conventionally, you take the positive root, so this makes $e^y>1$ and $y>0$.

$\therefore e^y=x+\sqrt{x^2-1}$

$y=\ln\left(x+\sqrt{x^2-1}\right)$

But $y=\text{arcosh } x$

So $\text{arcosh } x=\ln\left(x+\sqrt{x^2-1}\right)$

WORKED EXAMPLE 5.1

Solve $\sinh x = 7$, giving your answer in the form $\ln\left(a+\sqrt{b}\right)$ for integers a and b.

$\sinh x=7$ — Apply the inverse sinh function to find x.

$x=\text{arsinh } 7$

$=\ln\left(7+\sqrt{7^2+1}\right)$ — Use the logarithmic form $\text{arsinh } x=\ln\left(x+\sqrt{x^2+1}\right)$.

$=\ln\left(7+5\sqrt{2}\right)$

When you are trying to solve equations involving hyperbolic cosines, using the inverse function – unfortunately – does not give all the solutions: it just gives the positive one. Just as when taking the square root of both sides, you need to use a 'plus or minus' sign to get both possible solutions. This will be clear if you consider the graph of cosh x.

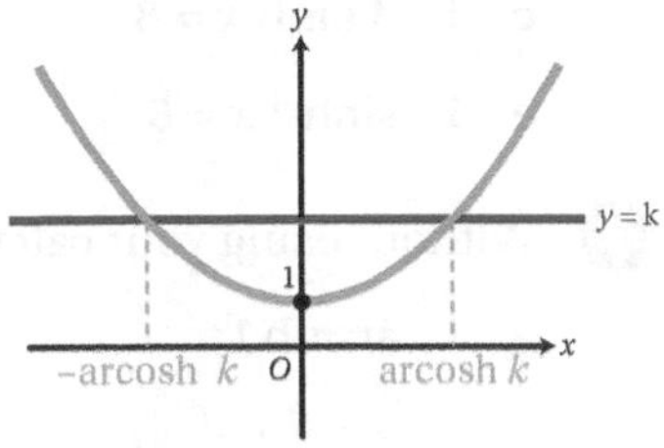

WORKED EXAMPLE 5.2

Given that $\cosh^2 x - \cosh x = 2$, express x in the form $\ln\left(a+\sqrt{b}\right)$ or $\ln\left(a-\sqrt{b}\right)$.

$\cosh^2 x - \cosh x - 2 = 0$

$(\cosh x - 2)(\cosh x + 1) = 0$

$\cosh x = 2$ or $\cosh x = -1$

The expression is a disguised quadratic, so rearrange it to make one side zero and then factorise it. You could also have used the quadratic formula.

But $\cosh x \geqslant 1$ so $\cosh x = -1$ has no solutions.

$x = \pm \text{arcosh}\, 2$

Use the inverse cosh function to find x. Remember that you need a plus or minus ($\pm$).

$= \pm \ln\left(2+\sqrt{2^2-1}\right)$

$= \pm \ln\left(2+\sqrt{3}\right)$

Use the logarithmic form $\text{arcosh}\, x = \ln\left(x+\sqrt{x^2-1}\right)$.

So

$x = \ln\left(2+\sqrt{3}\right)$ or $\ln\left(\frac{1}{2+\sqrt{3}}\right)$

Use the fact that $-\ln x \equiv \ln\left(x^{-1}\right) \equiv \ln\left(\frac{1}{x}\right)$.

$\ln\left(\frac{1}{2+\sqrt{3}}\right) = \ln\left(\frac{2-\sqrt{3}}{(2+\sqrt{3})(2-\sqrt{3})}\right)$

$= \ln\left(\frac{2-\sqrt{3}}{1}\right)$

Simplify the second solution by rationalising the denominator to produce the required form.

So $x = \ln\left(2+\sqrt{3}\right)$ or $x = \ln\left(2-\sqrt{3}\right)$

You can apply the method used at the end of Worked example 5.2 to write $-\ln\left(2+\sqrt{3}\right)$ as $\ln\left(2-\sqrt{3}\right)$ in general, so the two solutions to $\cosh x = k$ for $k > 1$ can always be written as $x = \ln\left(k \pm \sqrt{k^2-1}\right)$.

EXERCISE 5A

1 Use your calculator to evaluate each expression where possible.

a i $\cosh(-1)$ ii $\sinh 3$ **b** i $\tanh^2 1$ ii $\cosh^2 3$

c i $3\sinh(0.2)+1$ ii $5\tanh\left(\frac{1}{2}\right)+8$ **d** i $\sinh^{-1} 0.5$ ii $\cosh^{-1} 2$

e i $\tanh^{-1} 2$ ii $\cosh^{-1} 0$

2 Solve each equation, giving your answers to 3 significant figures.

a i $\sinh x = -2$ ii $\sinh x = 0.1$ **b** i $2\cosh x = 5$ ii $\cosh x = 4$

c i $4\tanh x = 3$ ii $\tanh x = 0.4$ **d** i $3\tanh x = 4$ ii $3\cosh x = 1$

e i $\sinh^{-1} x = 5$ ii $\cosh^{-1} x = 4$

3 Without using your calculator, find the exact value of each expression.

a i $\text{arsinh}\, 1$ ii $\text{arsinh}\, 2$ **b** i $\cosh^{-1} 2$ ii $\cosh^{-1} 3$

c i $\text{artanh}\,\frac{1}{3}$ ii $\text{artanh}\,\frac{1}{4}$

d i $\text{arcosh}\,(-3)$ ii $\text{artanh}\,(-2)$

e i $\sinh^{-1}\sqrt{2}$ ii $\tanh^{-1}\frac{1}{\sqrt{3}}$

4 Solve the equation $3\cosh(x-1)=5$.

5 Find and simplify the exact value of $\cosh(\ln 2)$.

6 Find and simplify a rational expression for $\tanh(\ln 3)$.

7 Solve the equation $\sinh^2 3x=5$.

8 Solve the equation $2\tanh^2 x+2=5\tanh x$.

9 Find and simplify an expression for $\cosh(\sinh^{-1} x)$.

10 Prove that $\sinh^{-1} x=\ln(x+\sqrt{x^2+1})$.

11 Prove that $\tanh^{-1} x=\frac{1}{2}\ln\left(\frac{1+x}{1-x}\right)$.

12 In the derivation of $\cosh^{-1} x$ you found that two possible expressions were $\ln(x+\sqrt{x^2-1})$ and $\ln(x-\sqrt{x^2-1})$. Show that their sum is zero and hence explain why the chosen expression is non-negative.

Section 2: Hyperbolic identities

Just as there is the identity $\sin^2 x+\cos^2 x\equiv 1$ linking the trigonometric functions $\sin^2 x$ and $\cos^2 x$, there is also an identity linking the hyperbolic functions $\sinh^2 x$ and $\cosh^2 x$.

Rewind

You met the identity $\sin^2 x+\cos^2 x\equiv 1$ in A Level Mathematics Student Book 1, Chapter 10.

Key point 5.7

$$\cosh^2 x-\sinh^2 x\equiv 1$$

This will be given in your formula book.

Fast forward

In Section 3 you will use this identity to solve equations.

You can prove this identity by using the definitions of sinh and cosh given in Key point 5.1.

WORKED EXAMPLE 5.3

Prove that $\cosh^2 x-\sinh^2 x\equiv 1$.

$$\cosh^2 x=\left(\frac{e^x+e^{-x}}{2}\right)^2$$

$$=\frac{e^{2x}+2e^x e^{-x}+e^{-2x}}{4}$$

Start from the definition of one of the hyperbolic functions. It doesn't matter which one. It is squared in the expression so square it and simplify.

Continues on next page

$$= \frac{e^{2x} + 2 + e^{-2x}}{4}$$

Since $e^x e^{-x} = 1$.

$$\sinh^2 x = \left(\frac{e^x - e^{-x}}{2}\right)^2$$

Repeat with the $\sinh^2 x$ term.

$$= \frac{e^{2x} - 2e^x e^{-x} + e^{-2x}}{4}$$

$$= \frac{e^{2x} - 2 + e^{-2x}}{4}$$

$e^x e^{-x} = 1$ again.

$$\cosh^2 x - \sinh^2 x \equiv \frac{e^{2x} + 2 + e^{-2x}}{4} - \frac{e^{2x} - 2 + e^{-2x}}{4}$$

Combine the two terms and simplify.

$$\equiv \frac{e^{2x} + 2 + e^{-2x} - (e^{2x} - 2 + e^{-2x})}{4}$$

$$\equiv \frac{e^{2x} + 2 + e^{-2x} - e^{2x} + 2 - e^{-2x}}{4}$$

$$\equiv \frac{4}{4}$$

$$\equiv 1$$

Although the only two hyperbolic identities you are expected to know for the AS Further Mathematics course are $\tanh x \equiv \frac{\sinh x}{\cosh x}$ (Key point 5.2) and $\cosh^2 x - \sinh^2 x \equiv 1$ (Key point 5.7), you may be asked to prove other unfamiliar hyperbolic identities. To do this, always return to the definitions of the functions and follow a process similar to that in Worked examples 5.3 and 5.4.

Rewind

The result in Worked example 5.3 proves that $(\cosh x, \sinh x)$ lies on the hyperbola $x^2 - y^2 = 1$, as described in the introduction to this chapter.

WORKED EXAMPLE 5.4

Prove that $\sinh 2x = 2 \sinh x \cosh x$.

$$\text{LHS} \equiv \frac{e^{2x} - e^{-2x}}{2}$$

On the LHS use the definition of $\sinh x$ and replace each x with $2x$.

$$\text{RHS} \equiv 2 \times \frac{e^x - e^{-x}}{2} \times \frac{e^x + e^{-x}}{2}$$

Then work from the RHS. Substitute the definitions of $\sinh x$ and $\cosh x$.

$$\equiv \frac{(e^x - e^{-x}) \times (e^x + e^{-x})}{2}$$

$$\equiv \frac{(e^x)^2 - (e^{-x})^2}{2}$$

Multiply out the brackets, using the difference of two squares.

$$\equiv \frac{e^{2x} - e^{-2x}}{2}$$

Using the rules of indices.

$$\equiv \text{LHS}$$

EXERCISE 5B

1 Prove that $\cosh x - \sinh x \equiv e^{-x}$.

2 Simplify $\sqrt{1+\sinh^2 x}$.

3 Prove that $\cosh 2x \equiv \cosh^2 x + \sinh^2 x$.

4 Prove that $1-\tanh^2 x \equiv \dfrac{1}{\cosh^2 x}$.

5 Prove that $\tanh 2x \equiv \dfrac{2\tanh x}{1+\tanh^2 x}$.

6 Prove that $\cosh x - 1 \equiv \frac{1}{2}\left(e^{0.5x}-e^{-0.5x}\right)^2$. Hence prove that $\cosh x \geqslant 1$.

7 Prove that $\cosh A + \cosh B \equiv 2\cosh\left(\dfrac{A+B}{2}\right)\cosh\left(\dfrac{A-B}{2}\right)$.

8 Use the binomial theorem to show that $\sinh^3 x \equiv \frac{1}{4}\sinh 3x - \frac{3}{4}\sinh x$.

9 **a** Explain why $(\cosh x + \sinh x)^3 \equiv \cosh 3x + \sinh 3x$ and $(\cosh x - \sinh x)^3 \equiv \cosh 3x - \sinh 3x$.
Hence show that $\cosh 3x \equiv \cosh^3 x + 3\cosh x \sinh^2 x$.

b Write $\cosh 3x$ in terms of $\cosh x$.

10 Given that $\tan y = \sinh x$ show that $\sin y = \pm\tanh x$.

Section 3: Solving harder hyperbolic equations

When you are solving equations involving hyperbolic functions you have several options:

- Rearrange to get a hyperbolic function that is equal to a constant and use inverse hyperbolic functions.
- Use the definition of hyperbolic functions to get an exponential function that is equal to a constant and use logarithms.
- Use an identity for hyperbolic functions to simplify the situation to one of the two preceding options.

It is only with experience that you will develop an instinct about which method will be most efficient.

WORKED EXAMPLE 5.5

Solve $\sinh x + \cosh x = 4$.

$$\sinh x + \cosh x = 4$$

$$\frac{e^x - e^{-x}}{2} + \frac{e^x + e^{-x}}{2} = 4$$

Use the definitions of $\sinh x$ and $\cosh x$.

$$\frac{2e^x}{2} = 4$$

$$e^x = 4$$

$$x = \ln 4$$

Tip

When you are dealing with the sum or difference of two hyperbolic functions, it is often useful to use the exponential form.

WORKED EXAMPLE 5.6

Solve $5\sinh x - 4\cosh x = 0$, giving your answer in the form $\ln a$.

$5\sinh x = 4\cosh x$

You could use the definitions of sinh and cosh, but it is easier to use the identity $\tanh x \equiv \frac{\sinh x}{\cosh x}$.

$\frac{\sinh x}{\cosh x} = \frac{4}{5}$

$\tanh x = \frac{4}{5}$

$x = \text{artanh}\,\frac{4}{5}$

Then use the logarithmic form of artanh.

$= \frac{1}{2}\ln\left(\frac{1+\frac{4}{5}}{1-\frac{4}{5}}\right)$

$= \frac{1}{2}\ln 9$

$= \ln 3$

$\frac{1}{2}\ln 9 = \ln 9^{\frac{1}{2}} = \ln 3$

WORKED EXAMPLE 5.7

Solve $\cosh^2 x + 1 = 3\sinh x$, giving your answer in logarithmic form.

$\cosh^2 x + 1 = 3\sinh x$

$(1+\sinh^2 x) + 1 = 3\sinh x$

The equation involves two types of function. You can use the identity $\cosh^2 x - \sinh^2 x \equiv 1$ to replace the $\cosh^2$ term.

$\sinh^2 x - 3\sinh x + 2 = 0$

Solve the resulting quadratic.

$(\sinh x - 1)(\sinh x - 2) = 0$

$\sinh x = 1$

$x = \text{arsinh}\,1$

$x = \ln(1+\sqrt{2})$

Use the logarithmic form of arsinh.

or

$\sinh x = 2$

$x = \text{arsinh}\,2$

$x = \ln(2+\sqrt{5})$

WORK IT OUT 5.1

Solve $\sinh 2x \cosh 2x = 6 \sinh 2x$.

Which is the correct solution? Identify the errors made in the incorrect solutions.

Solution 1	Solution 2	Solution 3
Dividing by 2: $\sinh x \cosh x = 3 \sinh x$ $\sinh x \cosh x - 3 \sinh x = 0$ $\sinh x(\cosh x - 3) = 0$ $\sinh x = 0$ or $\cosh x = 3$ $x = \sinh^{-1} 0$ or $x = \cosh^{-1} 3$ $x = 0$ or $x = 1.76$	Dividing by $\sinh 2x$: $\cosh 2x = 6$ $2x = \cosh^{-1} 6$ $= \ln(6+\sqrt{35})$ $x = \frac{1}{2}\ln(6+\sqrt{35})$	$\sinh 2x \cosh 2x = 6 \sinh 2x$ $\sinh 2x \cosh 2x - 6 \sinh 2x = 0$ $\sinh 2x(\cosh 2x - 6) = 0$ $\sinh 2x = 0$ or $\cosh 2x = 6$ $2x = \sinh^{-1} 0$ or $2x = \pm\cosh^{-1} 6$ $x = 0$ or $x = \pm 1.24$

EXERCISE 5C

1 Find the exact solution to $\cosh x = 5 - \sinh x$.

2 Solve $\cosh x - \sinh x = 2$, giving your answer in the form $\ln k$.

3 Solve $3(2\sinh x - 1)(\cosh x - 4) = 0$, giving your answers correct to 3 significant figures.

4 Solve $5\sinh x + 3\cosh x = 0$, giving your answer in the form $\ln k$, where k is a rational number.

5 Find the exact solution to $2\sinh x = 1 + \cosh x$.

6 Solve $6\sinh x - 2\cosh x = 7$, giving your answers in logarithmic form.

7 Solve $2\cosh^2 x - 5\sinh x = 5$, giving your answers in exact form.

8 Solve $\sinh^2 x = \cosh x + 1$, giving your answer in logarithmic form.

9 Solve $\tanh x = \dfrac{1}{\cosh x}$, giving your answer in logarithmic form.

10 Solve $\sinh x = \dfrac{1}{\cosh x}$, giving your answer in logarithmic form.

11 $$\sinh x + \sinh y = \frac{21}{8}$$
$$\cosh x + \cosh y = \frac{27}{8}$$

a Show that $e^x = 6 - e^y$ and $e^{-x} = 0.75 - e^{-y}$.

b Hence find the exact solutions to the simultaneous equations.

12 Find a sufficient condition on p, q and r for $p^2 \cosh x + q^2 \sinh x = r^2$ to have at least one solution.

Checklist of learning and understanding

- **Definitions of hyperbolic functions:**
  - $\sinh x = \frac{e^x - e^{-x}}{2}$
  - $\cosh x = \frac{e^x + e^{-x}}{2}$
  - $\tanh x \equiv \frac{\sinh x}{\cosh x} \equiv \frac{e^x - e^{-x}}{e^x + e^{-x}}$
- **Graphs of hyperbolic functions:**
  - $y = \sinh x$

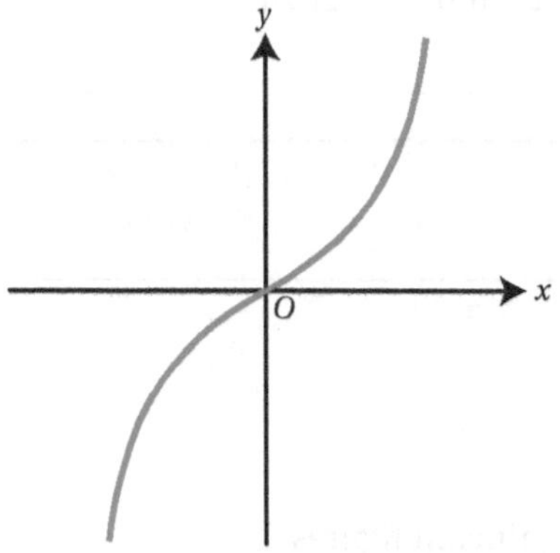

  - $y = \cosh x$

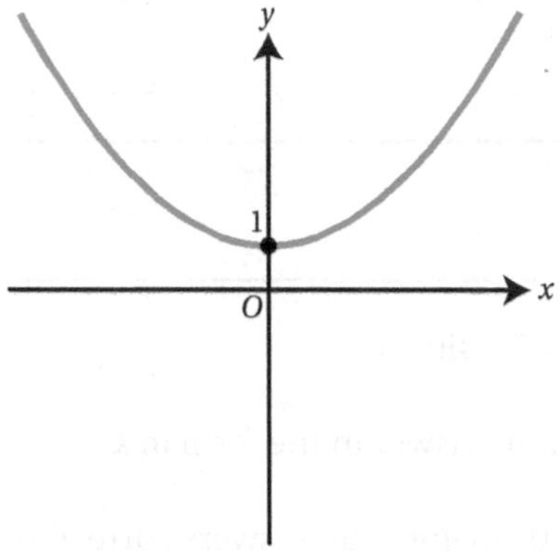

  - $y = \tanh x$

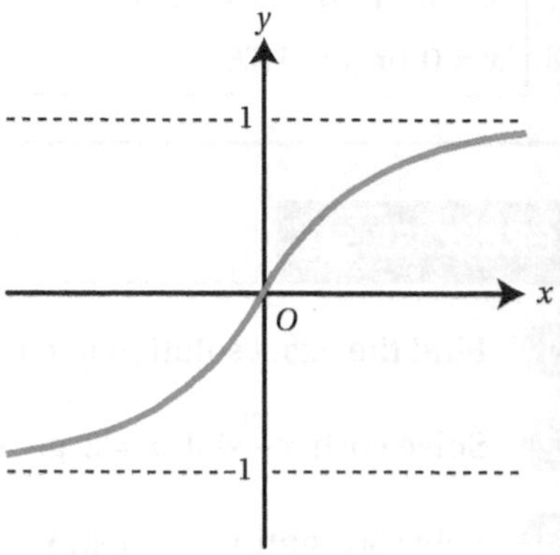

- **Logarithmic form of inverse hyperbolic functions:**
  - $\text{arsinh}\, x = \ln\left(x + \sqrt{x^2 + 1}\right)$
  - $\text{arcosh}\, x = \ln\left(x + \sqrt{x^2 - 1}\right)$
  - $\text{artanh}\, x = \frac{1}{2}\ln\left(\frac{1+x}{1-x}\right)$
- **The identity $\cosh^2 x - \sinh^2 x \equiv 1$**

Mixed practice 5

1 Evaluate $\tanh^{-1}\left(-\frac{1}{2}\right)$.

Choose from these options.

A Doesn't exist. **B** $\frac{1}{2}\ln 3$ **C** $\ln\frac{\sqrt{3}}{3}$ **D** 0

2 Solve $\sinh x + \cosh x = k$, giving your answer in terms of k.

Choose from these options.

A $\text{arsinh}\left(\frac{k}{2}\right)$ **B** $\text{arcosh}\,(2k)$ **C** e^k **D** $\ln k$

3 Simplify $\tanh(1 + \ln p)$.

4 Solve $\cosh(x+1) = 3$, giving your answer in terms of logarithms.

5 Solve $4\sinh 2x = \cosh 2x$, giving your answer correct to 3 significant figures.

6 Find the exact solutions to $16\cosh^2 x + 8\cosh x = 35$.

7 **a** Express $5\sinh x + \cosh x$ in the form $Ae^x + Be^{-x}$, where A and B are integers.

b Solve the equation $5\sinh x + \cosh x + 5 = 0$, giving your answer in the form $\ln a$, where a is a rational number.

[©AQA 2008]

8 Find the exact solutions to $2\cosh x + \sinh x = 2$.

9 Solve the equation $\sinh 3x\cosh^2 3x = 5\sinh 3x$, giving your answers in exact form.

10 Solve $3\sinh^2 x - 13\cosh x + 7 = 0$, giving your answers in terms of natural logarithms.

11 Prove that $\cosh x > \sinh x$ for all x.

12 Find and simplify an expression for $\tanh(\text{arsinh}\, x)$.

13 Use the binomial theorem to show that $\cosh^4 x \equiv \frac{1}{8}\cosh 4x + \frac{1}{2}\cosh 2x + \frac{3}{8}$.

14 **a** Sketch the graph of $y = \tanh x$.

b Given that $u = \tanh x$, use the definitions of $\sinh x$ and $\cosh x$ in terms of e^x and e^{-x} to show that $x = \frac{1}{2}\ln\left(\frac{1+u}{1-u}\right)$.

c **i** Show that the equation $\frac{3}{\cosh^2 x} + 7\tanh x = 5$ can be written as $3\tanh^2 x - 7\tanh x + 2 = 0$.

ii Show that the equation $3\tanh^2 x - 7\tanh x + 2 = 0$ has only one solution for x.

Find this solution in the form $\frac{1}{2}\ln a$ where a is an integer.

15 **a** Use the definitions $\sinh\theta = \frac{1}{2}(e^\theta - e^{-\theta})$ and $\cosh\theta = \frac{1}{2}(e^\theta + e^{-\theta})$ to show that $1 + 2\sinh^2\theta = \cosh 2\theta$.

b Solve the equation $3\cosh 2\theta = 2\sinh\theta + 11$, giving each of your answers in the form $\ln p$.

[©AQA 2009]

16 **a** Prove the identity $\cosh 3x = 4\cosh^3 x - 3\cosh x$.

b If $48u^3 - 36u - 13 = 0$ and $u = \cosh x$ find the value of x.

c Hence find the exact real solution to $48u^3 - 36u - 13 = 0$, giving your answer in a form without logarithms.

17 Solve these simultaneous equations, giving your answers in exact logarithmic form.

$\sinh x + \sinh y = 3.15$

$\cosh x + \cosh y = 3.85$

18 Using the logarithmic definition, prove that $\text{arsinh}(-x) = -\text{arsinh}\, x$.

19 Given that $\text{artanh}\, x + \text{artanh}\, y = \ln\sqrt{5}$, show that $y = \dfrac{3x-2}{2x-3}$.

20 **a** Using the definition $\sinh\theta = \frac{1}{2}(e^{\theta} - e^{-\theta})$, prove the identity $4\sinh^3\theta + 3\sinh\theta = \sinh 3\theta$.

b Given that $x = \sinh\theta$ and $16x^3 + 12x - 3 = 0$, find the value of θ in terms of a natural logarithm.

c Hence find the real root of the equation $16x^3 + 12x - 3 = 0$, giving your answer in the form $2^p - 2^q$, where p and q are rational numbers.

[©AQA 2014]

6 Polar coordinates

In this chapter you will learn how to:

- use polar coordinates to represent curves
- establish various properties of those curves
- convert between polar and Cartesian equations of a curve.

Before you start...

A Level Mathematics Student Book 2, Chapter 7	You should be able to use radians.	1 a Express $\frac{7\pi}{6}$ radians in degrees. b State the exact value of $\sin\frac{4\pi}{3}$.
A Level Mathematics Student Book 1, Chapter 10	You should be familiar with graphs of trigonometric functions.	2 a Find the set of values of θ, between 0 and 2π, for which $\cos\theta < 0$. b State the greatest possible value of $5 - 2\cos\theta$.

What are polar coordinates?

You are familiar with describing positions of points in the plane by using Cartesian coordinates, which represent the distance of a point from the x- and y-axes. But you are also familiar with bearings, which determine a direction in terms of an angle from a fixed line. If you know that a point lies on a certain bearing, you can describe its exact position by also specifying the distance from the origin. For example, this diagram shows that P is 4 cm from O on a bearing of 250°.

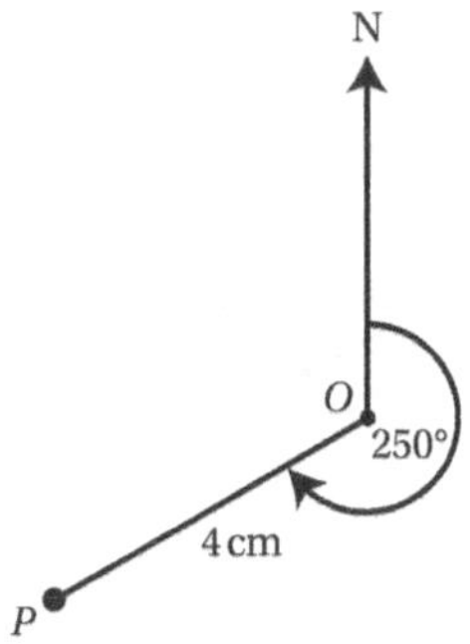

Polar coordinates use a similar idea: positions of points are described in terms of a direction and distance from the origin. They can be used to describe curves that cannot easily be represented in Cartesian coordinates. In this chapter, you will learn about equations of curves such as these.

Because the distance from the origin explicitly features as a variable, polar coordinates are often used to describe quantities that vary with distance, such as the strength of the gravitational field.

Section 1: Curves in polar coordinates

Polar coordinates describe the position of a point by specifying its distance from the origin (also called the **pole**) and the angle relative to a fixed line (called the **initial line**). You take the initial line to be the positive x-axis, and measure the angle anti-clockwise.

You write polar coordinates as (r, θ), where r is the distance and θ the angle. Angle θ is usually taken as lying between 0 and 2π. r is interpreted as being the distance from the pole in the direction of the angle, so that $r = 0$ gives the pole (irrespective of the value of θ), while negative r would produce a point 'behind' the pole. For example, point A in the diagram has polar coordinates $\left(3, \frac{\pi}{6}\right)$, point B has polar coordinates $(5, \pi)$ and point C has polar coordinates $\left(3, \frac{7\pi}{4}\right)$. Note that polar coordinates $\left(-3, \frac{7\pi}{6}\right)$, $(-5, 0)$ and $\left(-3, \frac{3\pi}{4}\right)$ would also describe points A, B and C respectively.

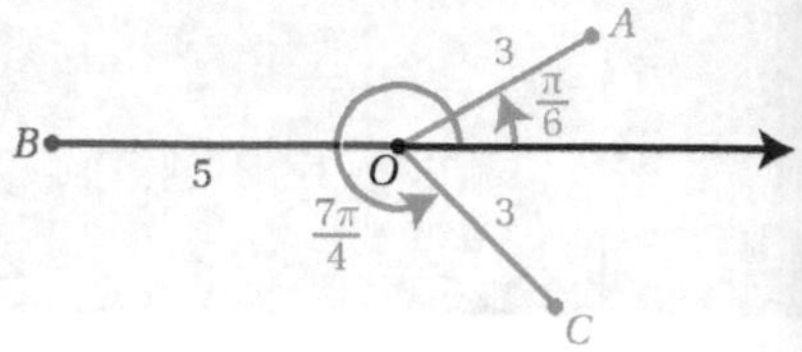

WORKED EXAMPLE 6.1

Points M and N have polar coordinates $\left(12, \frac{\pi}{4}\right)$ and $\left(9, \frac{5\pi}{6}\right)$. Find:

a the length MN

b the area of the triangle MON, where O is the pole.

a

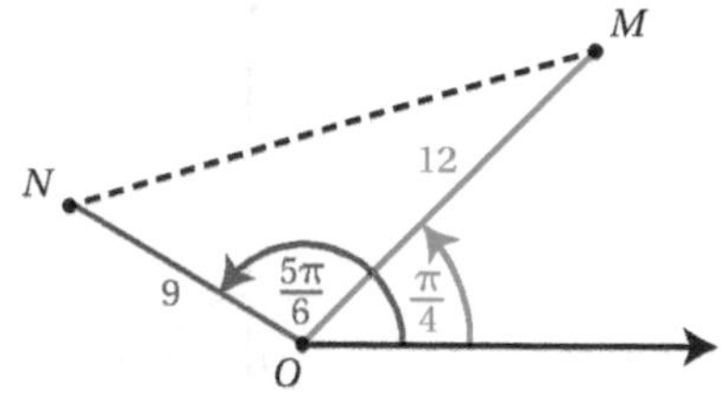

Polar coordinates give information about lengths and angles.

In triangle MON, $OM = 12$, $ON = 9$.

You can use the cosine rule in triangle MON.

$\angle MON = \frac{5\pi}{6} - \frac{\pi}{4} = \frac{7\pi}{12}$

So:

$MN^2 = 12^2 + 9^2 - 2 \times 12 \times 9 \cos \frac{7\pi}{12}$

$MN = \sqrt{280.9} = 16.8$

b Area $= \frac{1}{2}(12)(9)\sin\frac{7\pi}{12} = 52.2$

Use area of triangle $= \frac{1}{2}ab\sin C$.

In Cartesian coordinates, an equation of a curve gives a relationship between the x and y coordinates. Similarly, a polar equation of a curve is a relationship between r and θ that holds for any point on the curve.

WORKED EXAMPLE 6.2

a Make a table of values for the curve with polar equation $r = \sqrt{\theta}$ for $0 \leqslant \theta \leqslant 2\pi$.

b Hence sketch the curve.

a

θ	0	$\frac{\pi}{4}$	$\frac{\pi}{2}$	$\frac{3\pi}{4}$	π	$\frac{5\pi}{4}$	$\frac{3\pi}{2}$	$\frac{7\pi}{4}$	2π
r	0	0.89	1.25	1.53	1.77	1.98	2.17	2.34	2.51

Use $r = \sqrt{\theta}$ to calculate r for various values of θ.

b

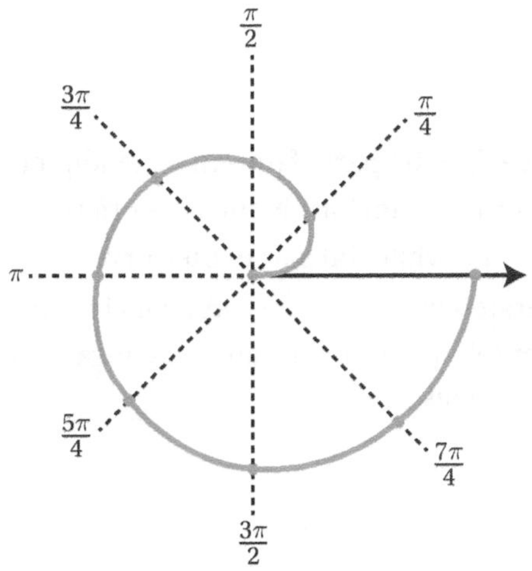

Plot the points and join them up.

Tip

If you have a graphical calculator you may be able to sketch a curve with a given polar equation.

WORKED EXAMPLE 6.3

Sketch the curve with equation $r = \sin 2\theta$ for $0 \leqslant \theta \leqslant 2\pi$.

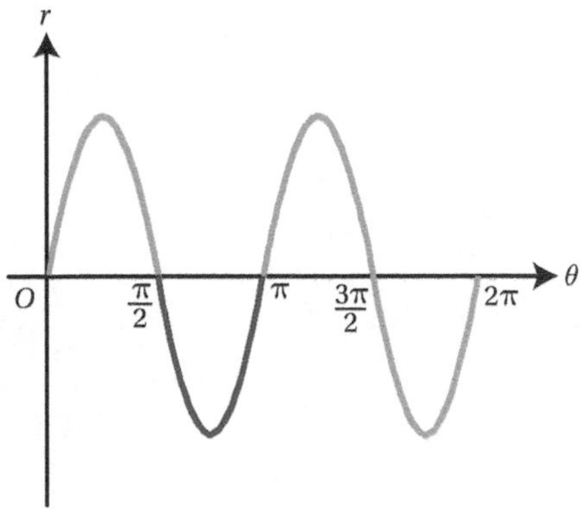

Sketch the graph of $r = \sin 2\theta$.

r is negative when:

$\frac{\pi}{2} < \theta < \pi$ or $\frac{3\pi}{2} < \theta < 2\pi$

Identify the values of θ for which r is negative.

Continues on next page

θ	0	$\frac{\pi}{4}$	$\frac{\pi}{2}$	$\frac{3\pi}{4}$	π	$\frac{5\pi}{4}$	$\frac{3\pi}{2}$	$\frac{7\pi}{4}$
r	0	1	0	−1	0	1	0	−1

You can also see that r increases for $0 < \theta < \frac{\pi}{4}$, then decreases for $\frac{\pi}{4} < \theta < \frac{\pi}{2}$, then repeats the same values between π and $\frac{3\pi}{2}$.

You can see a similar pattern but with negative values of r for $\frac{\pi}{4} < \theta < \frac{\pi}{2}$ and $\frac{3\pi}{2} < \theta < 2\pi$.

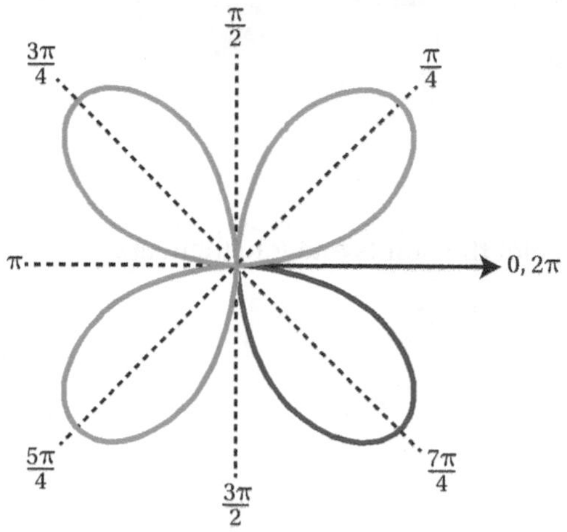

Notice that the part of the curve generated by $\frac{\pi}{2} < \theta < \pi$ appears in the lower right quadrant, while the part of the curve generated by $\frac{3\pi}{2} < \theta < 2\pi$ appears in the upper left quadrant, because r is negative for these angles.

Sometimes a polar curve will be defined with r restricted to $r \geqslant 0$. In these cases, the curve will not exist in regions where $r < 0$.

Just as in Worked example 6.3, you start by identifying values of θ for which $r < 0$. This time, though, you simply exclude these regions as the curve isn't defined there.

So, for example, if you had been asked to sketch $r = \sin 2\theta$, $r \geqslant 0$ for $0 \leqslant \theta \leqslant 2\pi$, you would get the same curve as in Worked example 6.3 but without the red and green parts, because $r < 0$ in the regions $\frac{\pi}{4} < \theta < \frac{\pi}{2}$ and $\frac{3\pi}{2} < \theta < 2\pi$.

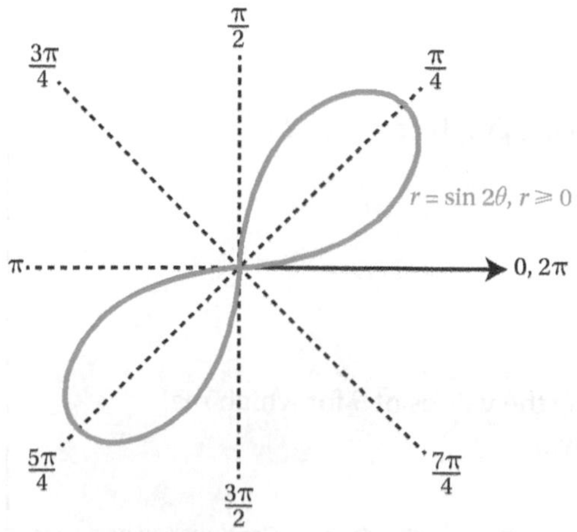

EXERCISE 6A

1 Plot the points with the given polar coordinates.

a i $\left(3, \frac{\pi}{4}\right)$ ii $\left(4, \frac{2\pi}{3}\right)$ b i $\left(5, \frac{7\pi}{6}\right)$ ii $\left(3, \frac{5\pi}{4}\right)$

c i $(2, \pi)$ ii $\left(1, \frac{3\pi}{2}\right)$ d i $\left(-2, \frac{7\pi}{4}\right)$ ii $\left(-2, \frac{5\pi}{6}\right)$

2 For points A and B with given polar coordinates, find the distance AB and the area of the triangle AOB.

a i $A\left(5, \frac{\pi}{6}\right), B\left(7, \frac{\pi}{4}\right)$ ii $A\left(2, \frac{\pi}{2}\right), B\left(5, \frac{2\pi}{3}\right)$

b i $A\left(10, \frac{\pi}{4}\right), B\left(8, \frac{7\pi}{6}\right)$ ii $A\left(4, \frac{3\pi}{4}\right), B\left(5, \frac{5\pi}{3}\right)$

c i $A\left(1, \frac{2\pi}{3}\right), B\left(2, \frac{11\pi}{6}\right)$ ii $A\left(6, \frac{7\pi}{4}\right), B\left(4, \frac{\pi}{4}\right)$

3 For each equation, make a table of values (for $0 \leqslant \theta \leqslant 2\pi$) and sketch the curve.

a i $r = 2\theta$ ii $r = \theta^2$

b i $r = \theta^2 - 5\theta + 6$ ii $r = \theta^2 - 2\theta$

c i $r = 2\cos 2\theta$ ii $r = \sin 3\theta$

> **Tip**
>
> If part of a curve occurs once for certain values of θ when r is positive and is then repeated for different values of θ when r is negative, you **do not** need to indicate the repetition by drawing over the curve again to make it bolder.

4 Shade the region described by each inequality. They are given in polar coordinates.

a $r \leqslant 2$ b $1 \leqslant r \leqslant 3$ c $\frac{\pi}{4} < \theta \leqslant \pi$

5 A curve has polar equation $r = \cos 2\theta$, $r \geqslant 0$ for $0 \leqslant \theta \leqslant 2\pi$.

a State the values of θ for which the curve is not defined.

b Hence sketch the curve.

6 A curve has polar equation $r = 4\sin\theta$, $0 \leqslant \theta \leqslant 2\pi$.

a Show that the points A and B, with polar coordinates $\left(2, \frac{\pi}{6}\right)$ and $\left(4, \frac{\pi}{2}\right)$, lie on the curve.

b Sketch the curve.

c Find the exact length of AB.

Section 2: Some features of polar curves

When sketching curves in Cartesian coordinates you normally mark the axis intercepts, maximum and minimum points. For polar curves, there are similar features that you can deduce from the equation.

Minimum and maximum values of r

Since r is a function of θ, you can use differentiation to find its minimum and maximum values.

Key point 6.1

The minimum and maximum values of r occur when $\frac{dr}{d\theta} = 0$.

WORKED EXAMPLE 6.4

A curve has polar equation $r = 150 + 9\theta^2 - 2\theta^3$ for $0 \leqslant \theta \leqslant 2\pi$.

a Find the minimum and maximum values of r, and the values of θ for which they occur.
b Sketch the curve.

a $\frac{dr}{d\theta} = 18\theta - 6\theta^2$

When $\frac{dr}{d\theta} = 0$:

$18\theta - 6\theta^2 = 0$

$6\theta(3 - \theta) = 0$

$\theta = 0$ or $\theta = 3$

The maximum and minimum values of r occur when $\frac{dr}{d\theta} = 0$.

$\frac{d^2r}{d\theta^2} = 18 - 12\theta$

Use the second derivative to decide which one gives the minimum and which the maximum value of r.

When $\theta = 0$:
$18 - 12(0) = 18\ (> 0)$ so there is a minimum at $(150, 0)$.

When $\theta = 0$, $r = 150$.

When $\theta = 3$:
$18 - 12(3) = -18\ (< 0)$ so there is a maximum at $(177, 3)$.

When $\theta = 3$, $r = 177$.

When $\theta = 2\pi, r = 9.21$.

Hence the minimum value of $r = 9.21$ occurs when $\theta = 2\pi$.

The maximum value of $r = 177$ occurs when $\theta = 3$.

The minimum value could occur at the end of the domain. You have already checked $\theta = 0$, so you just need to check $\theta = 2\pi$.

b

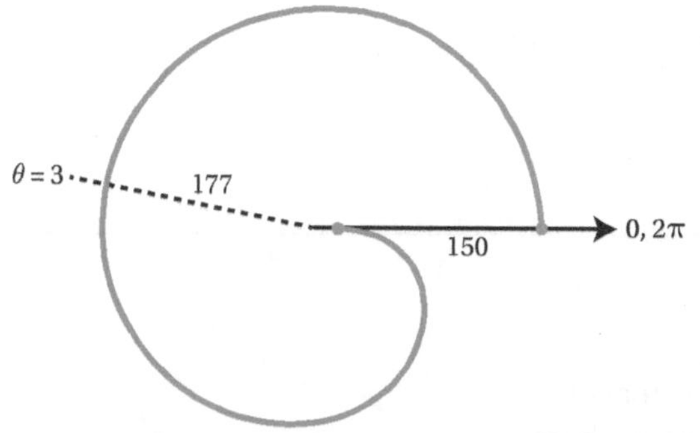

As θ increases from 0 to 2π, r increases from 150 to 177 (when $\theta = 3$, which is just below π) and then decreases to 9.21.

Polar equations often involve trigonometric functions. You may be able to use trigonometric graphs, rather than differentiation, to find the maximum and minimum values of r.

WORKED EXAMPLE 6.5

A curve has polar equation $r = 5 - 2\cos\theta$ for $0 \leqslant \theta \leqslant 2\pi$.

a Find the largest and smallest values of r.

b Hence sketch the curve.

a $-1 \leqslant \cos\theta \leqslant 1$

so $3 \leqslant 5 - 2\cos\theta \leqslant 7$

Start by considering the minimum and maximum values of $\cos\theta$: -1 when $\theta = \pi$, and 1 when $\theta = 0$ or 2π.

The largest value is $r = 7$ when $\theta = \pi$.

The smallest value is $r = 3$ when $\theta = 0, 2\pi$.

b

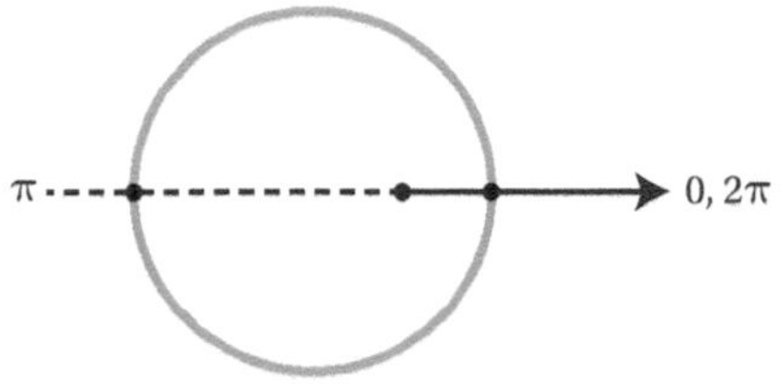

Tangents at the pole

The expression $r = \sin 2\theta$ changes from positive to negative, or vice versa, when $\theta = 0, \frac{\pi}{2}, \pi, \frac{3\pi}{2}, 2\pi$. Each of those θ values corresponds to a half-line, shown in red in the diagram. As the curve approaches each of the lines, r gets closer to zero (so points on the curve gets closer and closer to the pole). This means that each of the lines $\theta = 0, \frac{\pi}{2}, \pi, \frac{3\pi}{2}$ is a tangent to the curve at the pole.

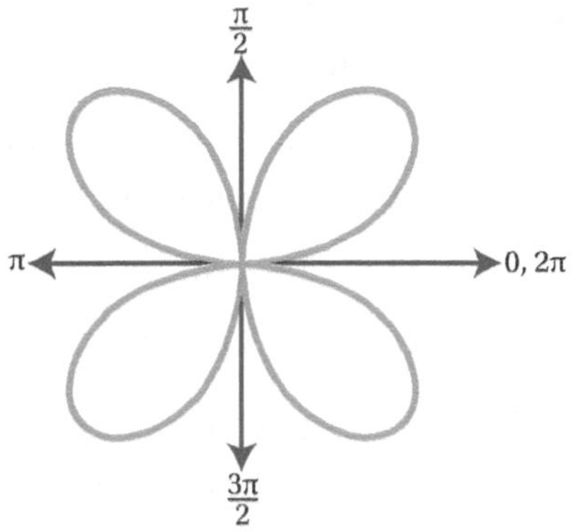

Key point 6.2

For a curve with polar equation $r = f(\theta)$, the line $\theta = \alpha$ is a **tangent at the pole** if $f(\alpha) = 0$ but $f(\alpha) > 0$ on one side of the line.

WORKED EXAMPLE 6.6

For the curve with polar equation $r = 2\cos 3\theta$, find the tangents at the pole and hence sketch the curve.

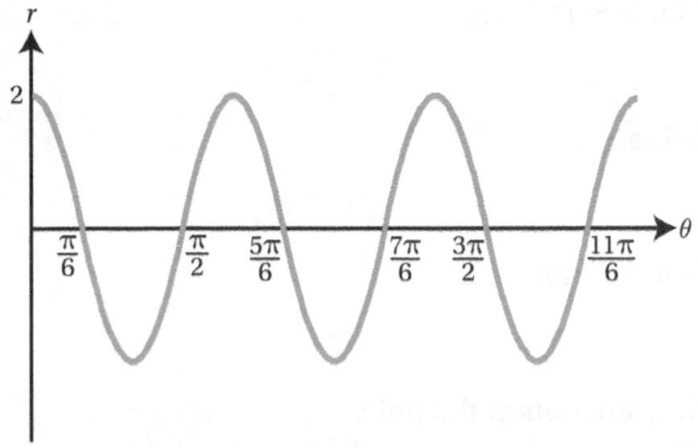

Sketch the graph of r against θ to see which values of θ produce negative values of r.

Continues on next page

The tangents at the pole are:

$\theta = \frac{\pi}{6}, \frac{\pi}{2}, \frac{5\pi}{6}, \frac{7\pi}{6}, \frac{3\pi}{2}$ and $\frac{11\pi}{6}$.

The value of $2\cos 3\theta$ passes through zero when $\theta = \frac{\pi}{6}, \frac{\pi}{2}, \frac{5\pi}{6}, \frac{7\pi}{6}, \frac{3\pi}{2}$ and $\frac{11\pi}{6}$.

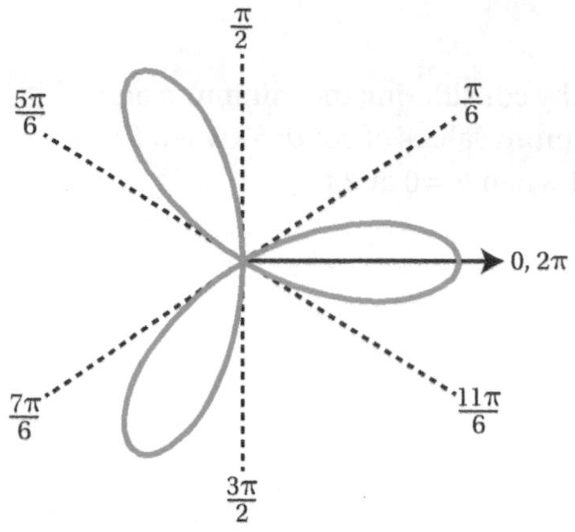

Between the tangents the value of $|r|$ increases from 0 to the maximum value of 2 and then decreases back to 0.

Notice that the curve is actually drawn twice as θ varies between 0 and 2π.

EXERCISE 6B

1 For each curve, find the minimum and maximum possible value of r, and the corresponding values of θ. Hence sketch the curve. (In all cases, $0 \leqslant \theta \leqslant 2\pi$.)

a **i** $r = 3 + 2\sin\theta$ **ii** $r = 5 + \cos\theta$

b **i** $r = 7 - 3\cos 2\theta$ **ii** $r = 5 - 2\sin 2\theta$

2 Find the equations of the tangents at the pole for each curve. Hence sketch the curve. (In all cases, $0 \leqslant \theta \leqslant 2\pi$.)

a **i** $r = 2\sin 3\theta$ **ii** $r = 3\cos 2\theta$

b **i** $r = 2 + 4\cos 3\theta$ **ii** $r = \sqrt{2} - 2\sin 4\theta$

3 Sketch the curves given in question 2 for $r \geqslant 0$.

4 A curve has equation $r = 3 + 2\sin\theta$, $0 \leqslant \theta \leqslant 2\pi$, in polar coordinates.

a Find the maximum and minimum values of r, and the corresponding values of θ.

b Sketch the curve.

5 Consider the curve with polar equation $r = 3\cos 4\theta$, $r \geqslant 0$ for $0 \leqslant \theta \leqslant 2\pi$.

a Find the equations of the tangents at the pole.

b State the set of values of θ for which the curve is not defined.

c Hence sketch the curve.

6 **a** Sketch the curve with polar equation $r = 3 - 2\cos^2\theta$, $0 \leqslant \theta \leqslant 2\pi$.

b State the largest and smallest values of r.

7 **a** Sketch the curve of $r = 1 + 2\sin 2\theta$, $0 \leqslant \theta \leqslant 2\pi$, labelling tangents at the pole.

b State the points where $|r|$ has its maximum value.

8 A curve has polar equation $r = \frac{3}{2 - \sin\theta}$, $0 \leqslant \theta \leqslant 2\pi$.

a Find the largest and smallest value of r and the values of θ at which they occur.

b Hence sketch the graph.

9 a Sketch the graph of $y = x\left(x - \frac{2\pi}{3}\right)\left(x - \frac{3\pi}{2}\right)$ for $0 \leqslant x \leqslant 2\pi$.

b Sketch the curve with polar equation $r = \theta\left(\theta - \frac{2\pi}{3}\right)\left(\theta - \frac{3\pi}{2}\right)$.

10 a Find the smallest and largest values of $y = 3x^2 - 6\pi x + 4\pi^2$ for $x \in [0, 2\pi]$.

b Sketch the curve with the polar equation $r = 3\theta^2 - 6\pi\theta + 4\pi^2$, for $0 \leqslant \theta \leqslant 2\pi$.

11 Sketch the curve with equation $r = -\theta(\theta - \pi)^2(\theta - 2\pi)$ for $0 \leqslant \theta \leqslant 2\pi$.

12 Sketch the curve with equation $r = \tan\theta$ for $0 \leqslant \theta \leqslant 2\pi$.

Section 3: Changing between polar and Cartesian coordinates

You can use trigonometry to find the Cartesian coordinates of a point with given polar coordinates.

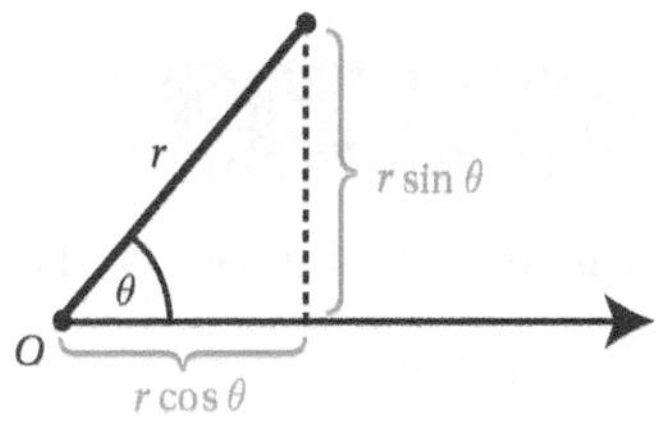

Key point 6.3

A point with polar coordinates (r, θ) has Cartesian coordinates $(r\cos\theta, r\sin\theta)$.

WORKED EXAMPLE 6.7

Points P and Q have polar coordinates $\left(4, \frac{\pi}{3}\right)$ and $\left(2, \frac{7\pi}{6}\right)$.

a Show points P and Q on the same diagram.
b Find the Cartesian coordinates of P and Q.

a

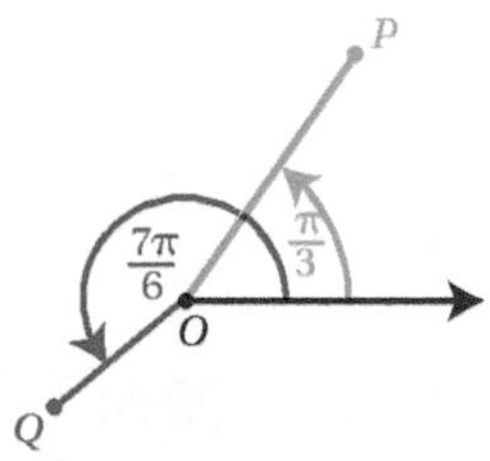

The first coordinate is the distance from the origin and the second coordinate is the angle.

$\frac{\pi}{3}$ is 60° and $\frac{7\pi}{6}$ is 210°.

b For P:

$x = 4\cos\frac{\pi}{3} = 2$

$y = 4\sin\frac{\pi}{3} = 2\sqrt{3}$

So the Cartesian coordinates of P are $(2, 2\sqrt{3})$.

Use $x = r\cos\theta$ and $y = r\sin\theta$.

Continues on next page

For Q:

$x = 2\cos\frac{7\pi}{6} = -\sqrt{3}$

$y = 2\sin\frac{7\pi}{6} = -1$

So the Cartesian coordinates of Q are $(-\sqrt{3}, -1)$.

To change from Cartesian to polar coordinates, consider the same diagram again.

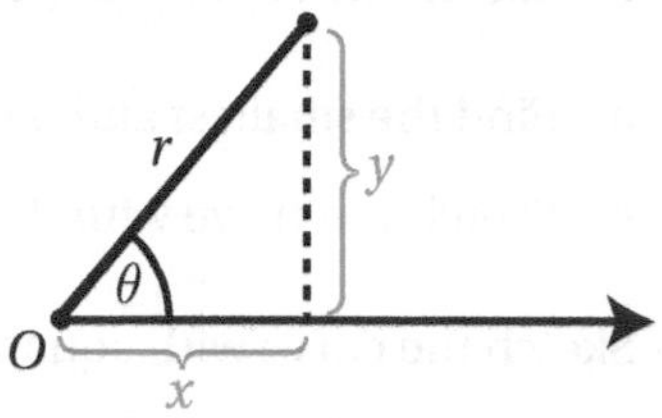

The value of r is the distance from the origin, so $r^2 = x^2 + y^2$. It is conventional to give r as a positive value. You need to be a little careful when finding the angle. Since $x = r\cos\theta$ and $y = r\sin\theta$, you can divide the two equations to get $\tan\theta = \frac{y}{x}$. However, there are two values $\theta \in [0, 2\pi)$ with the same value of $\tan\theta$; you need to consider the position of the point to decide which one is correct.

Key point 6.4

For a point with Cartesian coordinates (x, y) the polar coordinates satisfy:

- $r = \sqrt{x^2 + y^2}$
- $\tan\theta = \frac{y}{x}$

Rewind

This should remind you of the modulus and argument of a complex number – see Chapter 1, Section 3.

WORKED EXAMPLE 6.8

Find the polar coordinates of the points $A(-3, 5)$ and $B(4, -1)$.

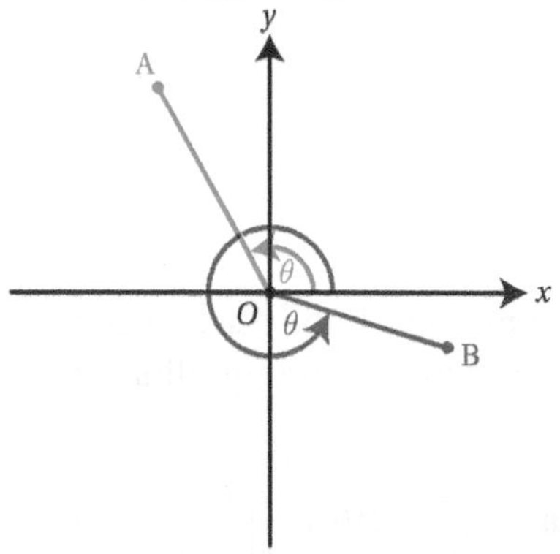

Start by plotting the points to see which angle to use.

For A:

$r = \sqrt{(-3)^2 + 5^2}$

$= \sqrt{34}$

Use $r = \sqrt{x^2 + y^2}$.

$\tan\theta = \frac{5}{-3}$

$\Rightarrow \theta = 2.11$ or $\theta = 5.25$

Find $\tan^{-1}\left(-\frac{5}{3}\right)$, then add π to get the two possible values between 0 and 2π.

Continues on next page

Hence the coordinates of A are $(\sqrt{34}, 2.11)$. — The angle for A is smaller than π.

For B:

$r = \sqrt{4^2 + (-1)^2}$

$= \sqrt{17}$

$\tan\theta = -\frac{1}{4}$

$\Rightarrow \theta = 2.90$ or $\theta = 6.04$

Hence the coordinates of B are $(\sqrt{17}, 6.04)$. — The angle for B is greater than π.

You can now convert equations of curves between polar and Cartesian forms.

WORKED EXAMPLE 6.9

Find the Cartesian equation of the curve with polar equation $r = 2\sin\theta$.

$r = \sqrt{x^2 + y^2}$ — Use $r^2 = x^2 + y^2$ and $y = r\sin\theta$.

$\sin\theta = \frac{y}{r}$

$\therefore \sin\theta = \frac{y}{\sqrt{x^2 + y^2}}$

Hence: — Substitute for $\sin\theta$ in the equation of the curve.

$\sqrt{x^2 + y^2} = 2\left(\frac{y}{\sqrt{x^2 + y^2}}\right)$

$x^2 + y^2 = 2y$ — Simplify if possible. In this case, multiply both sides by the 'square root' term.

WORKED EXAMPLE 6.10

Find the polar equation of the curve $(x^2 + y^2)^3 = 3xy$.

$(r^2)^3 = 3(r\cos\theta)(r\sin\theta)$ — Use $x = r\cos\theta$ and $y = r\sin\theta$.

You also know that $x^2 + y^2 = r^2$.

$r^6 = 3r^2 \sin\theta\cos\theta$

$r^4 = 3\sin\theta\cos\theta$

EXERCISE 6C

1 Each point is given in polar coordinates. Find the Cartesian coordinates.

a i $\left(5, \frac{\pi}{4}\right)$ ii $\left(3, \frac{\pi}{3}\right)$

b i $\left(\sqrt{2}, \frac{3\pi}{4}\right)$ ii $\left(\sqrt{3}, \frac{5\pi}{6}\right)$

c i $(6, 4.1)$ ii $(3, 5.7)$

2 Each point is given in Cartesian coordinates. Find the polar coordinates. Take $\theta \in [0, 2\pi)$.

a i $(5, 2)$ ii $(3, 4)$

b i $(0, 2)$ ii $(0, -3)$

c i $(-1, -5)$ ii $(4, -1)$

3 Find the polar equation of each curve.

a i $x^2 + y^2 = 3xy$ ii $(x^2 + y^2)^2 = 2x^2y$

b i $\frac{1}{x} + \frac{1}{y} = \frac{1}{5}$ ii $x^3 + y^3 = 3$

c i $y = 3x + 1$ ii $x^2 + y^2 = 6$

4 Find the Cartesian equation of each curve.

a i $r = 2\theta$ ii $r = 3\theta^2$

b i $r = 4\sin\theta$ ii $r = 2\cos\theta$

c i $r = 2\tan\theta$ ii $r^2 = \tan\theta$

5 A curve has polar equation $r = 3\cos\theta$.
Point Q, with polar coordinates $(2, \alpha)$ where $\pi < \alpha < 2\pi$, lies on the curve.

a Find the value of α.

b Find the Cartesian coordinates of Q.

c Find the Cartesian equation of the curve.

6 Find the polar equation of the circle $(x-1)^2 + (y-1)^2 = 2$, giving your answer in the form $r = \mathrm{f}(\theta)$.

7 Find the Cartesian equation of the curve with polar equation $r = 3\tan\theta$.

8 A curve has polar equation $r = \frac{3}{2 + \cos\theta}$.

Show that the Cartesian equation of the curve can be written as $3x^2 + 4y^2 + 6x = 9$.

9 Find the Cartesian equation of the curve with polar equation:

a $r = \cos^3\theta$ b $r = \frac{1}{\cos^3\theta}$.

Checklist of learning and understanding

- **Polar coordinates** (r, θ) describe the position of a point in terms of its distance from the **pole** and the angle measured anticlockwise from the **initial line**.
- The connection between polar and Cartesian coordinates is:
  - $x = r\cos\theta$, $y = r\sin\theta$
  - $r = \sqrt{x^2 + y^2}$, $\tan\theta = \dfrac{y}{x}$
- For a curve with equation given in polar coordinates:
  - when r has negative values, the curve will appear to be plotted 180° from the angle θ
  - there may be one or more tangents at the pole, given by the values of θ for which $r = 0$.

Mixed practice 6

1. Find the greatest distance from the pole of any point on the curve $r = 2(5 - 3\sin\theta)$, $0 \leqslant \theta \leqslant 2\pi$.

   Choose from these options.

   A 2 **B** 6 **C** 10 **D** 16

2. Find the polar equation of the curve $x^2 + y^2 = a(x - y)$.

3. Points A and B have polar coordinates $\left(7, \frac{\pi}{4}\right)$ and $\left(4, \frac{5\pi}{6}\right)$. Find:

   a the distance AB

   b the area of the triangle AOB.

4. Sketch the curve with polar equation $r = 1 - \cos 2\theta$, $0 \leqslant \theta \leqslant 2\pi$.

5. A curve has polar equation $r(1 - \sin\theta) = 4$. Find its Cartesian equation in the form $y = \text{f}(x)$.

   [©AQA 2007]

6. A curve has polar equation $r(4 - 3\cos\theta) = 4$. Find its Cartesian equation in the form $y^2 = \text{f}(x)$.

   [©AQA 2014]

7. Find a Cartesian equation for the curve $r = 3\cos^2\theta$, $0 \leqslant \theta \leqslant 2\pi$.

   Choose from these options.

   A $\sqrt{x^2 + y^2} = 3x^2$ **B** $\sqrt{x^2 + y^2} = 3(1 - y^2)$ **C** $(x^2 + y^2)^{\frac{3}{2}} = 3x^2$ **D** $(x^2 + y^2)^3 = 27x^2$

8. A curve has polar equation $r = 2 + 4\sin\theta$, $0 \leqslant \theta \leqslant 2\pi$

   a Find the equations of the tangents at the pole.

   b Sketch the curve.

   c Find the Cartesian equation of the curve.

9. **a** Sketch the curve with polar equation $r = 5 - 3\cos\theta$, $0 \leqslant \theta \leqslant 2\pi$.

   b Find the Cartesian coordinates of the point that is furthest away from the origin.

10. A curve has polar equation $r = 2 - 4\sin 2\theta$, $r \geqslant 0$ for $0 \leqslant \theta \leqslant 2\pi$.

    a Find the equations of the tangents at the pole.

    b State the polar coordinates of the points at greatest distance from the pole.

    c Hence sketch the graph.

11. A curve is defined by the polar equation $r = 3\theta$ for $0 \leqslant \theta \leqslant 2\pi$.

    a Sketch the curve.

    b Find the Cartesian coordinates of the point where the curve intersects the line $\theta = \frac{2\pi}{3}$.

12. Sketch the curve with polar equation $r = 5\cos 3\theta$, $0 \leqslant \theta \leqslant 2\pi$. Indicate the equations of the tangents at the pole, and give the polar coordinates of the point where the curve crosses the initial line.

13 **a** Show that $x^2 = 1 - 2y$ can be written in the form $x^2 + y^2 = (1 - y)^2$.

b A curve has Cartesian equation $x^2 = 1 - 2y$.

Find its polar equation in the form $r = f(\theta)$, given that $r > 0$.

[©AQA 2008]

14 The diagram shows the curve C with equation $r = 2\cos 2\theta$ and a circle of radius 1. The eight intersection points are connected to form two overlapping rectangles. Find the exact area of the shaded region.

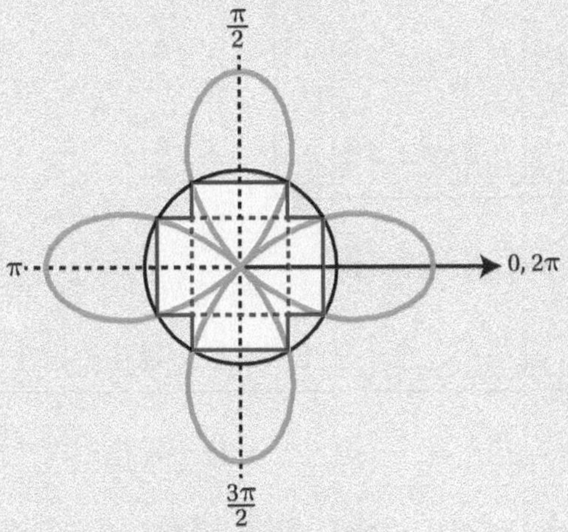

15 The diagram shows a sketch of a curve C, the pole O and the initial line.

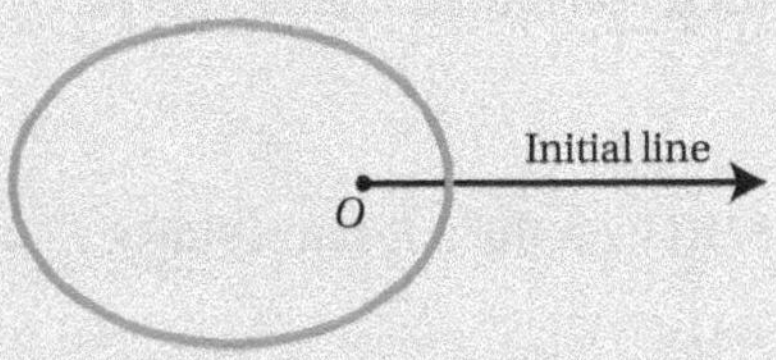

The curve C has polar equation $r = \frac{2}{3 + 2\cos\theta}$, $0 \leqslant \theta \leqslant 2\pi$.

a Verify that the point L with polar coordinates $(2, \pi)$ lies on C.

b The circle with polar equation $r = 1$ intersects C at the points M and N.

i Find the polar coordinates of M and N.

ii Find the area of triangle LMN.

c Find a Cartesian equation of C, giving your answer in the form $9y^2 = f(x)$.

[©AQA 2009]

FOCUS ON ... PROOF 1

Roots of real polynomials

In this section, you will prove that complex roots of real polynomials come in conjugate pairs. You will need to use the properties of complex conjugates.

Key point 1

- $(z \pm w)^* = z^* \pm w^*$
- $(zw)^* = z^* w^*$
- $\left(\frac{z}{w}\right)^* = \frac{z^*}{w^*}$
- $(z^n)^* = (z^*)^n$ for integer n

All these results are proved in a similar way; here is just one of them.

PROOF 6

Prove that $(zw)^* = z^* w^*$.

Let $z = x + \text{i}y$
$w = u + \text{i}v$

The best way to describe complex conjugates is to write them in terms of their real and imaginary parts.

Then: $z^* = x - \text{i}y$
$w^* = u - \text{i}v$

LHS $= (zw)^*$

Start with the left-hand side.

$= ((x + \text{i}y)(u + \text{i}v))^*$
$= ((xu - yv) + \text{i}(xv + yu))^*$
$= (xu - yv) - \text{i}(xv + yu)$

RHS $= z^* w^*$

If you get stuck, start on the right-hand side and try to meet in the middle.

$= (x - \text{i}y)(u - \text{i}v)$
$= xu - \text{i}xv - \text{i}yu - yv$
$= (xu - yv) - \text{i}(xv + yu)$

Therefore LHS = RHS

Every proof needs a conclusion.

You are now ready to prove the main result.

Key point 2

If $f(x)$ is a real polynomial and if $f(z) = 0$, then $f(z^*) = 0$.

Rewind

Remember that a **real polynomial** is a polynomial in which the coefficients are real numbers.

Key point 2 says that, if a number z is a root of $f(x)$, then so is its conjugate, z^*.

PROOF 7

Let $f(z) = a_n z^n + a_{n-1} z^{n-1} + \cdots + a_1 z + a_0$ — Write a general form of the polynomial. It is helpful if you label the coefficients with subscripts that correspond to the powers.

Then: $(a_k z^k)^* = (a_k)^* (z^k)^*$ — Take the complex conjugate of each term and use the results of Key point 1: $(zw)^* = z^* w^*$.

$= a_k (z^k)^*$ — Since a_k is real, $(a_k)^* = a_k$.

So: $f(z^*) = a_n (z^*)^n + \cdots + a_1 z^* + a_0$

$= (a_n z^n)^* + \cdots + (a_1 z)^* + (a_0)^*$ — Use the result from Proof 6. For the last term, remember that a_0 is real.

$= (a_n z^n + \cdots + a_1 z + a_0)^*$ — Use Key point 1 again: $z^* + w^* = (z + w)^*$.

$= (f(z))^*$

$= 0$, as required. — Remember that you started by assuming that $f(z) = 0$.

QUESTIONS

1. Prove the rest of the results in Key point 1.
2. Does the proof of Key point 2 work if z is a real number? What does the result say then?
3. Where in Proof 7 did you use the fact that $f(x)$ is a **real** polynomial? Does the result of Key point 2 still hold if some of the a_k are complex numbers?
4. You already know that you can use the roots of a polynomial to write it as a product of linear factors: if the roots of $f(x)$ are $z_1, z_2, \ldots, z_n$, then $f(x) = (x - z_1)(x - z_2)\ldots(x - z_n)$.

   However, some of the z_k may be complex. Key point 2 implies that, once you have found all the roots of a **real** polynomial, you can write it as a product of real linear and quadratic factors. Can you see why this is the case? You may want to start by looking at some examples.

Solving cubic equations

In this section, you will look at a famous historical example of problem solving. It shows you how mathematicians adapt their definitions and rules to enable them to solve a wider variety of problems.

The cubic formula

You know that the quadratic equation $ax^2 + bx + c = 0$ can be solved by applying the quadratic formula $x = \frac{-b \pm \sqrt{b^2 - 4ac}}{2a}$. Some quadratic equations have no real solutions, which is the case when the discriminant $b^2 - 4ac$ is negative.

A similar formula for the cubic equation was discovered by Cardano and Tartaglia in the 16th century. The formula is rather complicated, and here you will consider only the special case of cubic equations with no x^2 term.

The equation $x^3 + px + q = 0$ has solutions given by $x = \frac{-p}{3u} + u$, where $u = \sqrt[3]{-\frac{q}{2} \pm \sqrt{\frac{q^2}{4} + \frac{p^3}{27}}}$.

The quantity $\frac{q^2}{4} + \frac{p^3}{27}$ plays a role similar to the discriminant of a quadratic equation: it tells you how many solutions the equation has. Unfortunately, as you will see, there are examples where this discriminant is negative but the equation still has real solutions. This is where mathematicians needed to engage in some serious problem solving to make the formula work correctly.

Using the formula

Consider the equation $x^3 + 9x - 26 = 0$.

Here $p = 9$, $q = -26$, so $u = \sqrt[3]{13 \pm \sqrt{196}}$

$= -1 \text{ or } 3$

When $u = -1$, $x = \frac{-9}{-3} - 1$

$= 2$

When $u = 3$, $x = \frac{-9}{9} + 3$

$= 2$

So, in this case, both possible values of u give the same solution for x.

Plotting the graph confirms that $x = 2$ is the only real solution of the equation.

The negative discriminant

Now consider the equation $x^3 - 3x = 0$. Here $p = -3$, $q = 0$.

Then $u = \sqrt[3]{0 \pm \sqrt{0 + \frac{-27}{27}}}$

$= \sqrt[3]{\pm\sqrt{-1}}$

This appears to be a dead end – you are looking for real solutions, so you can't take the square root of −1. However, you can see (by factorising) that there should be three real solutions: 0, $\sqrt{3}$ and $-\sqrt{3}$. Is there any way you can make the formula work to find those three solutions?

Problem solving often requires perseverance and 'thinking outside the box'. So what happens if you start working with complex numbers and carry on with the calculation?

You now need $u = \sqrt[3]{\pm i}$. There are in fact three different complex numbers for which the cube is i:

$u_1 = -i$, $u_2 = \frac{\sqrt{3}}{2} + \frac{1}{2}i$ and $u_3 = -\frac{\sqrt{3}}{2} + \frac{1}{2}i$.

When $u = -i$:

$$\begin{aligned} x &= \frac{1}{-i} + (-i) \\ &= i - i \\ &= 0 \end{aligned}$$

When $u = \frac{\sqrt{3}}{2} + \frac{1}{2}i$:

$$\begin{aligned} x &= \frac{2}{\sqrt{3}+i} + \left(\frac{\sqrt{3}}{2} + \frac{1}{2}i\right) \\ &= \frac{2\left(\sqrt{3}-i\right)}{3+1} + \left(\frac{\sqrt{3}}{2} + \frac{1}{2}i\right). \\ &= \frac{\sqrt{3}}{2} - \frac{1}{2}i + \frac{\sqrt{3}}{2} + \frac{1}{2}i \\ &= \sqrt{3} \end{aligned}$$

When $u = \frac{\sqrt{3}}{2} + \frac{1}{2}i$, a similar calculation gives the third solution, $x = -\sqrt{3}$.

⏭ Fast forward

In Further Mathematics Student Book 2 you will learn how to find cube roots of a complex number. Here, you are just checking that the three roots given are correct.

⏭ Fast forward

In Further Mathematics Student Book 2 you will also use complex numbers to solve differential equations and prove trigonometric identities.

QUESTIONS

1. Apply the formula to solve the equation $x^3 + 6x - 20 = 0$.
2. Find the three possible values for $u = \sqrt[3]{-i}$. Confirm that these lead to the same three solutions of the equation $x^3 - 3x = 0$.
3. The fact that this formula uses complex numbers to find real solutions was one of the arguments that persuaded mathematicians to accept complex numbers. Did it change your opinion on whether complex numbers 'exist'?

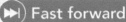

FOCUS ON ... MODELLING 1

Complex numbers and radios

You may have been told that complex numbers are used in electronics. The building blocks of electronic circuits are resistors, capacitors and inductors. Normally each of these needs to be analysed in a different way using quite tricky differential equations.

- For a resistor: $V = IR$, where V is the voltage, I is the current and R is the resistance. This is called Ohm's law.
- For a capacitor, $I = C\frac{dV}{dt}$ where C is the capacitance.
- For an inductor, $V = L\frac{dI}{dt}$ where L is the inductance.

However, if you use complex numbers and assume a sinusoidal input voltage with (angular) frequency ω, you can treat all three in the same way, as 'complex resistors', using the concept of impedance, Z. The impedance of a resistor is R, the impedance of a capacitor is $\frac{1}{i\omega C}$ and the impedance of an inductor is $i\omega L$. Then you can use a general version of Ohm's law, which says that $V = IZ$.

This diagram shows a model of a simple circuit which can act as a radio.

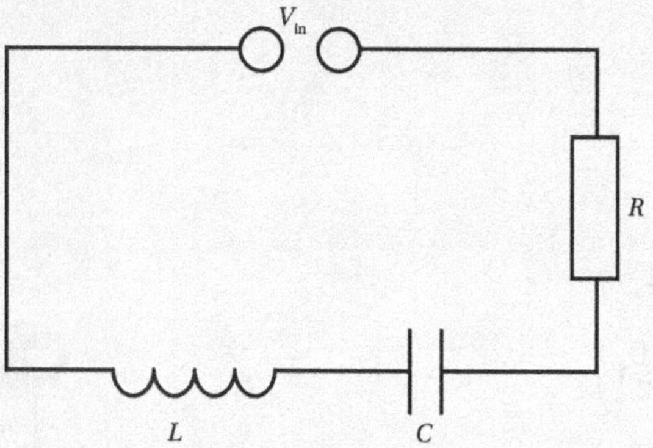

The total impedance for this circuit is:

$$\begin{aligned} Z &= R + i\omega L + \frac{1}{i\omega C} \\ &= R + i\omega\left(L - \frac{1}{\omega^2 C}\right) \end{aligned}$$

If the input voltage is V_{in}, then the resulting current is:

$$\begin{aligned} I = \frac{V_{in}}{Z} &= \frac{V_{in}}{R + i\omega\left(L - \frac{1}{\omega^2 C}\right)} \\ &= \frac{V_{in}}{R^2 + \omega^2\left(L - \frac{1}{\omega^2 C}\right)^2}\left(R - i\omega\left(L - \frac{1}{\omega^2 C}\right)\right) \end{aligned}$$

If you plot the modulus of the current against ω, your graph should look like this:

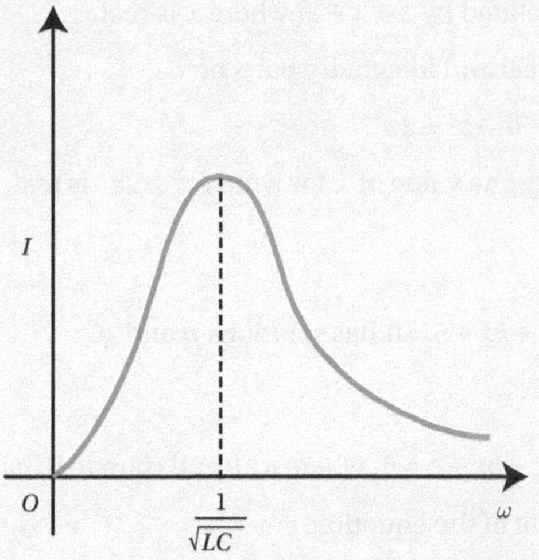

There is a maximum that occurs when the denominator of the expression above is minimised. This happens when $\omega = \frac{1}{\sqrt{LC}}$.

Now think about what you have found. There is a natural frequency that has a higher current than any other frequency. This phenomenon is called **resonance**. If the input voltage has this frequency, the circuit will produce a large current that can be used to power a speaker.

QUESTIONS

1. Radio 1 has a frequency of 98 MHz or 6.15×10^8 in the units required for ω. If an inductor has $L = 10^{-5}$ H (a fairly typical value), find the required capacitance for a circuit to 'tune in' to Radio 1.

2. **a** What is the current in the circuit when it is being driven at the resonant frequency?

   b What would happen if $R = 0$?

   c Why would $R = 0$ be a poor modelling assumption?

3. Use technology to sketch the current against ω. Explore the effect on the graph of changing C while keeping the same resonant frequency.

   Why might smaller values of C produce a better radio?

4. **a** The argument of the expression for the current gives the 'phase difference' between the input voltage and the current – a measure of the delay between the input and the circuit responding.

   Find an expression for this phase difference.

   b What is the value of the phase difference when the circuit is at resonance?

CROSS-TOPIC REVIEW EXERCISE 1

1 Solve the equation $3z^* - 2i = 2z + 5$.

2 The complex number z is defined by $z = x + 2i$ where x is real.

a Find, in terms of x, the real and imaginary parts of:

i z^2 ii $z^2 + 2z^*$.

b Show that there is exactly one value of x for which $z^2 + 2z^*$ is real.

[©AQA 2009]

3 The quadratic equation $2x^2 + kx + 5 = 0$ has solutions p and q.

Express $p^2 + q^2$ in terms of k.

4 The cubic equation $x^3 + ax^2 + bx + c = 0$, where a, b and c are real, has roots $(2 + 3i)$ and 4.

a Write down the other root of the equation.

b Find the values of a, b and c.

5 Use the substitution $x = u - 2$ to find the exact value of the real root of the equation $x^3 + 6x^2 + 12x - 2 = 0$.

6 The parabola P_1 with equation $y^2 = 4ax$ is transformed by a stretch with scale factor 3 parallel to the x-axis to form the curve P_2.

a Find the equation of P_2.

The parabola P_1 is now transformed by a stretch with scale factor $k > 0$ parallel to the y-axis to form the curve P_3.

b Given that P_2 and P_3 are the same curve, find the value of k.

7 A curve has equation $y = \dfrac{3x-1}{x+2}$.

a Write down the equations of the two asymptotes to the curve.

b Sketch the curve, indicating the coordinates of the points where the curve intersects the coordinate axes.

c Hence, or otherwise, solve the inequality $0 < \dfrac{3x-1}{x+2} < 3$.

[©AQA 2007]

8 Solve the equation $3\sinh x = 2\cosh x$, giving your answer in the form $x = \ln\sqrt{a}$.

9 Solve the equation $3\sinh^2 x + 2\sinh x - 8 = 0$, giving your answers in terms of natural logarithms.

10 The Cartesian equation of a circle is $(x+4)^2 + (y-7)^2 = 65$.

Using the origin O as the pole and the positive x-axis as the initial line, find the polar equation of this circle, giving your answer in the form $r = p\sin\theta + q\cos\theta$.

11 a Given that $z = x + iy$, where $x, y \in \mathbb{R}$, find $|z - i|$ in terms of x and y.

b Sketch on an Argand diagram the locus of points satisfying $|z - i| = |z + 1|$.

12 a By writing $z = x + \mathrm{i}y$, or otherwise, solve the equation $z^2 = \mathrm{i} - 1$.

b Solve the quadratic equation $w^2 + 2\mathrm{i}w = \mathrm{i}$.

13 A polynomial is defined by $\mathrm{f}(x) = x^3 + x + 10$.

a Find an integer root of the equation $\mathrm{f}(x) = 0$.

b Solve the equation $\mathrm{f}(x) = 0$.

The integer root of $\mathrm{f}(x) = 0$ is p and the complex roots are z_1 and z_2.

c On an Argand diagram, shade the locus of points z which satisfy $|p| \leqslant |z| < |z_1|$.

d Calculate $|\arg(z_1) - \arg(z_2)|$.

14 a Show that $(1+\mathrm{i})^3 = 2\mathrm{i} - 2$.

b The cubic equation $z^3 - (5+\mathrm{i})z^2 + (9+4\mathrm{i})z + k(1+\mathrm{i}) = 0$, where k is a real constant, has roots α, β and γ.

It is given that $\alpha = 1 + \mathrm{i}$.

i Find the value of k.

ii Show that $\beta + \gamma = 4$.

iii Find the values of β and γ.

[©AQA 2011]

15 The equation $2x^2 + 3x - 6 = 0$ has roots α and β.

a Write down the value of $\alpha + \beta$ and the value of $\alpha\beta$.

b Hence show that $\alpha^3 + \beta^3 = -\frac{135}{8}$.

c Find a quadratic equation, with integer coefficients, whose roots are $\alpha + \frac{\alpha}{\beta^2}$ and $\beta + \frac{\beta}{\alpha^2}$.

[©AQA 2013]

16 An ellipse E has equation

$$\frac{x^2}{3} + \frac{y^2}{4} = 1$$

a Sketch the ellipse E, showing the coordinates of the points of intersection of the ellipse with the coordinate axes.

b The ellipse E is stretched with scale factor 2 parallel to the y-axis.

Find and simplify the equation of the curve after the stretch.

c The original ellipse, E, is translated by the vector $\begin{bmatrix} a \\ b \end{bmatrix}$. The equation of the translated ellipse is $4x^2 + 3y^2 - 8x + 6y = 5$.

Find the values of a and b.

[© AQA 2009]

17 Show that $\sinh(\ln \sin x) = -\dfrac{\cos x}{2\tan x}$.

18 **a** Show that $9\sinh x - \cosh x = 4e^{x} - 5e^{-x}$.

b Given that $9\sinh x - \cosh x = 8$, find the exact value of $\tanh x$.

19 Let $z = \cos\dfrac{\pi}{3} + i\sin\dfrac{\pi}{3}$ and $w = \cos\dfrac{\pi}{4} + i\sin\dfrac{\pi}{4}$.

a Find zw in the form $r(\cos\theta + i\sin\theta)$.

b Find zw in the form $x + iy$, where x and y are expressed in terms of surds.

c Show zw on an Argand diagram and hence find the exact value of $\tan\dfrac{\pi}{12}$.

20 **a** Two points, A and B, on an Argand diagram are represented by the complex numbers $2 + 3i$ and $-4 - 5i$ respectively. Given that the points A and B are at the ends of a diameter of a circle C_1, express the equation of C_1 in the form $|z - z_0| = k$.

b A second circle, C_2, is represented on the Argand diagram by the equation $|z - 5 + 4i| = 4$. Sketch on one Argand diagram both C_1 and C_2.

c The points representing the complex numbers z_1 and z_2 lie on C_1 and C_2 respectively and are such that $|z_1 - z_2|$ has its maximum value. Find this maximum value, giving your answer in the form $a + b\sqrt{5}$.

[©AQA 2009]

21 The diagram shows a parabola P which has equation $y = \dfrac{1}{8}x^2$, and another parabola Q which is the image of P under a reflection in the line $y = x$.

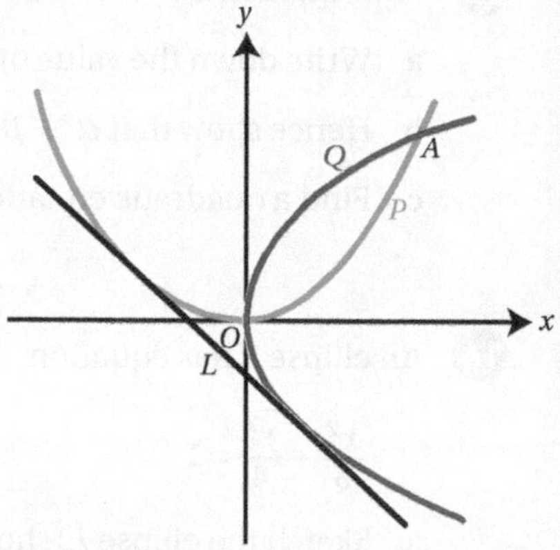

The parabolas P and Q intersect at the origin and again at the point A.

The line L is a tangent to both P and Q.

a **i** Find the coordinates of the point A.

ii Write down an equation for Q.

iii Give a reason why the gradient of L must be -1.

b **i** Given that the line $y = -x + c$ intersects the parabola P at two distinct points, show that $c > -2$.

ii Find the coordinates of the points at which the line L touches the parabolas P and Q.

(No credit will be given for solutions based on differentiation.)

[© AQA 2011]

22 A curve has equation $y = \frac{x^2 - 2x + 1}{x^2 - 2x - 3}$.

a Find the equations of the three asymptotes of the curve.

b **i** Show that if the line $y = k$ intersects the curve then $(k-1)x^2 - 2(k-1)x - (3k+1) = 0$.

ii Given that the equation $(k-1)x^2 - 2(k-1)x - (3k+1) = 0$ has real roots, show that $k^2 - k \geqslant 0$.

iii Hence show that the curve has only one stationary point and find its coordinates. (No credit will be given for solutions based on differentiation.)

c Sketch the curve and its asymptotes.

[©AQA 2013]

23 Find the set of values of k for which

$$2 \sinh x + 3 \cosh x = k$$

has at least one solution.

24 The diagram shows a sketch of a curve and a circle.

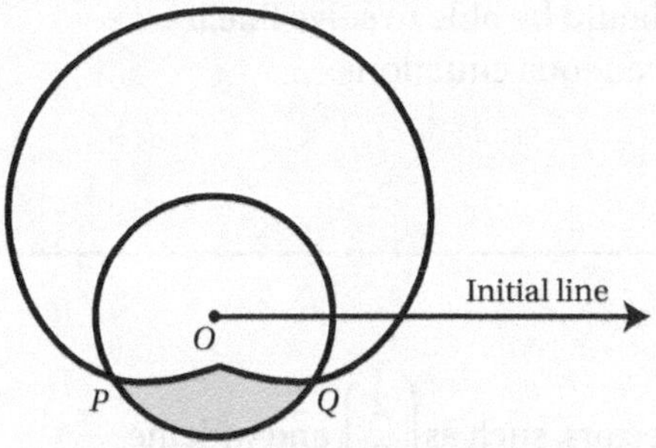

The polar equation of the curve is $r = 3 + 2\sin\theta,\ 0 \le \theta \le 2\pi$

The circle, whose polar equation is $r = 2$, intersects the curve at the points P and Q, as shown in the diagram.

a Find the polar coordinates of P and the polar coordinates of Q.

b A straight line, drawn from the point P through the pole O, intersects the curve again at the point A.

i Find the polar coordinates of A.

ii Find, in surd form, the length of AQ.

iii Hence, or otherwise, explain why the line AQ is a tangent to the circle $r = 2$.

[©AQA 2013]

7 Matrices

In this chapter you will learn how to:

- add, subtract and perform scalar multiplication with conformable matrices
- use and interpret zero and identity matrices
- calculate the determinant of a 2×2 matrix
- find and interpret the inverse of a 2×2 matrix
- use matrices to solve two simultaneous equations.

Before you start...

A Level Mathematics Student Book 1, Chapter 15	You should know how to add, subtract and perform scalar multiplication of vectors.	1 Calculate: $\begin{pmatrix} 2 \\ 3 \end{pmatrix} - 2\begin{pmatrix} 1 \\ -4 \end{pmatrix}$
GCSE	You should be able to solve linear simultaneous equations.	2 Solve these simultaneous equations. $\begin{cases} 2x - 3y = 5 \\ x + y = -1 \end{cases}$

What are matrices?

You are already familiar with column vectors, such as $\begin{pmatrix} 2 \\ 3 \end{pmatrix}$, and with the rules on how to add, subtract and multiply vectors. A matrix is an extension of the idea of a column vector into a rectangular array of numbers.

Matrices provide an efficient way of dealing with large amounts of information. They are used widely in many branches of advanced mathematics, from calculus to probability, and form the basis of many computer programs including, for example, those used in weather forecasting and quantum mechanics.

In this chapter you will look at the basic techniques for working with matrices, and apply some of these to solve two linear simultaneous equations.

Rewind

In A Level Mathematics Student Book 2, Chapter 16, vectors are extended into 3 dimensions.

Section 1: Addition, subtraction and scalar multiplication

A **matrix** (plural: **matrices**) is a rectangular array of **elements**, which may be numerical or algebraic. For example:

$$\mathbf{A} = \begin{pmatrix} 1 & -3 & x \\ 2.5 & a^2 - \pi & 4 \end{pmatrix}$$

is a matrix with two rows and three columns, for a total of six elements.

Key point 7.1

An $m \times n$ matrix is a rectangular array of elements with m rows and n columns. A matrix that has the same number of rows as columns is called a **square matrix**.

Matrices are generally designated by a bold or underlined upper-case letter (**A** in print and either $\underline{A}$ or A when handwritten).

There are two special matrices with which you will need to be familiar.

Key point 7.2

- A **zero matrix**, denoted by **Z**, has every element equal to zero.
- An **identity matrix**, denoted by **I**, is a square matrix with 1 on each element of the lead diagonal (upper left to lower right) and zeros everywhere else.

The 2×2 identity matrix is $\begin{pmatrix} 1 & 0 \\ 0 & 1 \end{pmatrix}$,

the 3×3 identity matrix is $\begin{pmatrix} 1 & 0 & 0 \\ 0 & 1 & 0 \\ 0 & 0 & 1 \end{pmatrix}$.

Fast forward

You will see the use of the identity matrix later in this chapter.

WORKED EXAMPLE 7.1

$\mathbf{A} = \begin{pmatrix} 1 & 3 & 2 \\ 6 & 3 & -4 \end{pmatrix}$, $\mathbf{B} = \begin{pmatrix} 2 \\ 4 \end{pmatrix}$ and $\mathbf{C} = \begin{pmatrix} -6 & 2 \end{pmatrix}$.

Write down the dimensions of the matrices **A**, **B** and **C**.

A is a 2×3 matrix.	There are 2 rows and 3 columns.
B is a 2×1 matrix.	There are 2 rows and 1 column.
C is a 1×2 matrix.	There is 1 row and 2 columns.

Notice that a column matrix such as **B** in Worked example 7.1 is a column vector.

In some circumstances you need to flip a matrix around its lead diagonal, exchanging rows for columns. This is called **transposing** a matrix.

Key point 7.3

The **transpose** of a matrix **A** is a new matrix, $\mathbf{A}^T$, in which the rows of $\mathbf{A}^T$ are the columns of **A**.

WORKED EXAMPLE 7.2

$\mathbf{A} = \begin{pmatrix} 1 & 3 & 2 \\ 6 & 3 & -4 \end{pmatrix}$, $\mathbf{B} = \begin{pmatrix} 2 & -8 \\ 4 & 3 \end{pmatrix}$ and $\mathbf{C} = (-6 \ \ 2)$.

Write down the transpose matrices $\mathbf{A}^{\mathrm{T}}$, $\mathbf{B}^{\mathrm{T}}$ and $\mathbf{C}^{\mathrm{T}}$.

$\mathbf{A}^{\mathrm{T}} = \begin{pmatrix} 1 & 6 \\ 3 & 3 \\ 2 & -4 \end{pmatrix}$

A is a 2×3 matrix so $\mathbf{A}^{\mathrm{T}}$ is a 3×2 matrix. The first row of **A** becomes the first column of $\mathbf{A}^{\mathrm{T}}$ and so on.

$\mathbf{A} = \begin{pmatrix} 1 & 3 & 2 \\ 6 & 3 & -4 \end{pmatrix}$

$\mathbf{B}^{\mathrm{T}} = \begin{pmatrix} 2 & 4 \\ -8 & 3 \end{pmatrix}$

B is a 2×2 matrix so $\mathbf{B}^{\mathrm{T}}$ is also a 2×2 matrix. The first row of **B** becomes the first column of $\mathbf{B}^{\mathrm{T}}$ and so on.

$\mathbf{B} = \begin{pmatrix} 2 & -8 \\ 4 & 3 \end{pmatrix}$

$\mathbf{C}^{\mathrm{T}} = \begin{pmatrix} -6 \\ 2 \end{pmatrix}$

C is a 1×2 row matrix so $\mathbf{C}^{\mathrm{T}}$ is a 2×1 column matrix.

Addition and subtraction

The rules for matrix addition and subtraction are equivalent to the rules for vector addition and subtraction, with which you are already familiar.

Key point 7.4

You can only add (or subtract) two matrices with the same dimensions.

Take each position in the matrix in turn and add (or subtract) the components for that position.

Matrices that have the appropriate dimensions for an operation are **conformable**. For addition and subtraction, two matrices are only conformable if they have identical dimensions.

WORKED EXAMPLE 7.3

The 2×2 matrix $\mathbf{A} = \begin{pmatrix} 1 & 3 \\ 2 & -5 \end{pmatrix}$.

For each part, determine: **i** $\mathbf{A}+\mathbf{B}$ and **ii** $\mathbf{A}-\mathbf{B}$ or explain why the sum or difference does not exist.

a $\mathbf{B}_1 = \begin{pmatrix} 3 & -4 \\ 1 & 2 \end{pmatrix}$ **b** $\mathbf{B}_2 = \begin{pmatrix} 4 & 5 & -3 \\ 2 & 22 & -1 \end{pmatrix}$ **c** $\mathbf{B}_3 = \begin{pmatrix} 2 & 1 \\ 2 & -3 \\ 2 & 3 \end{pmatrix}$ **d** $\mathbf{B}_4 = \begin{pmatrix} x & 2x \\ -2 & x^2+1 \end{pmatrix}$

Continues on next page

a i $\mathbf{A}+\mathbf{B}_1=\begin{pmatrix}1 & 3\\2 & -5\end{pmatrix}+\begin{pmatrix}3 & -4\\1 & 2\end{pmatrix}$

$=\begin{pmatrix}1+3 & 3-4\\2+1 & -5+2\end{pmatrix}$

$=\begin{pmatrix}4 & -1\\3 & -3\end{pmatrix}$

The dimensions match so you can add the matrices.

For each position in the matrix take the element in **A** and add the element in $\mathbf{B}_1$.

ii $\mathbf{A}-\mathbf{B}_1=\begin{pmatrix}1 & 3\\2 & -5\end{pmatrix}-\begin{pmatrix}3 & -4\\1 & 2\end{pmatrix}$

$=\begin{pmatrix}1-3 & 3-(-4)\\2-1 & -5-2\end{pmatrix}$

$=\begin{pmatrix}-2 & 7\\1 & -7\end{pmatrix}$

The dimensions match so you can subtract the matrices.

For each position in the matrix take the element in **A** and subtract the element in $\mathbf{B}_1$.

b Cannot add or subtract a 2×2 matrix and a 2×3 matrix.

The dimensions do not match so you cannot add or subtract.

c Cannot add or subtract a 2×2 matrix and a 3×2 matrix.

The dimensions do not match so you cannot add or subtract.

d i $\mathbf{A}+\mathbf{B}_4=\begin{pmatrix}1 & 3\\2 & -5\end{pmatrix}+\begin{pmatrix}x & 2x\\-2 & x^2+1\end{pmatrix}$

$=\begin{pmatrix}1+x & 3+2x\\2-2 & -5+x^2+1\end{pmatrix}$

$=\begin{pmatrix}1+x & 3+2x\\0 & x^2-4\end{pmatrix}$

The dimensions match so you can add the matrices.

For each position in the matrix take the element in **A** and add the element in $\mathbf{B}_4$.

ii $\mathbf{A}-\mathbf{B}_4=\begin{pmatrix}1 & 3\\2 & -5\end{pmatrix}-\begin{pmatrix}x & 2x\\-2 & x^2+1\end{pmatrix}$

$=\begin{pmatrix}1-x & 3-2x\\2-(-2) & -5-(x^2+1)\end{pmatrix}$

$=\begin{pmatrix}1-x & 3-2x\\4 & -6-x^2\end{pmatrix}$

The dimensions match so you can subtract the matrices.

For each position in the matrix take the element in **A** and subtract the element in $\mathbf{B}_4$.

Also, just as with vectors, you can multiply a matrix by a scalar value.

Tip

Remember that in the examination you may use a calculator to perform any operations on purely numerical matrices, without showing further working.

Key point 7.5

To multiply a matrix by a scalar, multiply each element by that scalar.

WORKED EXAMPLE 7.4

$\mathbf{A} = \begin{pmatrix} 1 & a \\ 2 & 2 \end{pmatrix}$, $\mathbf{B} = \begin{pmatrix} b & 3 \\ -4 & -5 \end{pmatrix}$ and $\mathbf{C} = \begin{pmatrix} 2 & 2 \\ 4 & 0 \end{pmatrix}$.

a Write down $\frac{1}{2}\mathbf{C}$.

b p and q are scalar constants; write down $p\mathbf{A}$ and $q\mathbf{B}$.

c Given $\frac{1}{2}\mathbf{C} = p\mathbf{A} + q\mathbf{B}$, find the values of a, b, p and q.

a $\frac{1}{2}\mathbf{C} = \frac{1}{2}\begin{pmatrix} 2 & 2 \\ 4 & 0 \end{pmatrix} = \begin{pmatrix} 1 & 1 \\ 2 & 0 \end{pmatrix}$

Multiply each element of the matrix by the scalar quantity.

b $p\mathbf{A} = \begin{pmatrix} p & ap \\ 2p & 2p \end{pmatrix}$, $q\mathbf{B} = \begin{pmatrix} bq & 3q \\ -4q & -5q \end{pmatrix}$

Multiply each element of the matrix by the scalar quantity.

c $p\mathbf{A} + q\mathbf{B} = \begin{pmatrix} p + bq & ap + 3q \\ 2p - 4q & 2p - 5q \end{pmatrix}$

For each position in the matrix take the element in $p\mathbf{A}$ and add the element in $q\mathbf{B}$.

$= \begin{pmatrix} 1 & 1 \\ 2 & 0 \end{pmatrix}$

$$\begin{cases} p + bq = 1 & (1) \\ ap + 3q = 1 & (2) \\ 2p - 4q = 2 & (3) \\ 2p - 5q = 0 & (4) \end{cases}$$

Comparing each element gives a set of simultaneous equations.

$(3) - (4): q = 2$

$(3): 2p - 8 = 2$

$p = 5$

(3) and (4) only involve unknowns p and q. Solve these, then substitute into (1) and (2) to solve for a and b.

Substituting $p = 5$, $q = 2$ into (1) and (2):

$$\begin{cases} 5 + 2b = 1 & (1) \\ 5a + 6 = 1 & (2) \end{cases}$$

$(1): b = -2$

$(2): a = -1$

$a = -1, b = -2, p = 5, q = 2$

EXERCISE 7A

1 **a** State the dimensions of each matrix.

i $\begin{pmatrix} 1 & 2 \\ 1 & 3 \end{pmatrix}$ **ii** $\begin{pmatrix} 1 & 5 \\ 2 & 3 \\ 1 & -3 \end{pmatrix}$ **iii** $\begin{pmatrix} 2 & 6 & 0 \\ 4 & 1 & 0 \end{pmatrix}$ **iv** $\begin{pmatrix} 1 & 2 & -4 \\ 8 & -3 & 3 \\ -1 & 7 & 22 \\ -5 & -2 & 0 \end{pmatrix}$

b For each matrix in **a**, write down its transpose matrix.

2 $\mathbf{A} = \begin{pmatrix} 1 & 2 \\ 1 & 3 \end{pmatrix}$, $\mathbf{B} = \begin{pmatrix} 4 & 2 \\ 6 & -8 \end{pmatrix}$, $\mathbf{C} = \begin{pmatrix} 3 & 3 \\ 0 & -2 \end{pmatrix}$, $\mathbf{D} = \begin{pmatrix} 2 & 1 \\ 3 & -4 \end{pmatrix}$,

$\mathbf{E} = \begin{pmatrix} 1 & 5 \\ 2 & 3 \\ 1 & -3 \end{pmatrix}$, $\mathbf{F} = \begin{pmatrix} -4 & 0 \\ 2 & 1 \\ 3 & 1 \end{pmatrix}$.

Calculate the resultant matrix, if there is a valid calculation.

a **i** $\mathbf{A}+\mathbf{B}$ **ii** $\mathbf{C}+\mathbf{D}$ **iii** $\mathbf{E}+\mathbf{F}$

b **i** $\mathbf{B}-\mathbf{C}$ **ii** $\mathbf{D}-\mathbf{E}$ **iii** $\mathbf{A}-\mathbf{C}$

c **i** $2\mathbf{A}$ **ii** $-3\mathbf{C}$ **iii** $4\mathbf{E}$

d **i** $2\mathbf{A}-3\mathbf{C}$ **ii** $\mathbf{B}-2\mathbf{D}$ **iii** $3\mathbf{E}-\mathbf{F}$

3 Add the two given matrices where possible.

a $\begin{pmatrix} 3 & -1 \\ 4 & 5 \end{pmatrix}$ and $\begin{pmatrix} -2 & 5 \\ 1 & 1 \end{pmatrix}$ **b** $\begin{pmatrix} -1 \\ 2 \end{pmatrix}$ and $\begin{pmatrix} 3 & 3 \\ 1 & 2 \end{pmatrix}$

c $\begin{pmatrix} -4 \\ 2 \end{pmatrix}$ and $(1 \ -3)$ **d** $\begin{pmatrix} 4 & -1 \\ 3 & 3 \end{pmatrix}$ and $\begin{pmatrix} 1 & 1 \\ -2 & 4 \end{pmatrix}$

e $\begin{pmatrix} 3 & -1 & 2 \\ 0 & 4 & 2 \end{pmatrix}$ and $\begin{pmatrix} 1 & 1 \\ -2 & 3 \end{pmatrix}$ **f** $\begin{pmatrix} 6 & 0 & 1 \\ 1 & 2 & 1 \\ 7 & 1 & 5 \end{pmatrix}$ and $\begin{pmatrix} 4 & -2 & 3 \\ -1 & 0 & 0 \\ 1 & 0 & 2 \end{pmatrix}$

4 Given that $\mathbf{A} = \begin{pmatrix} 1 & -1 \\ 3 & 2 \end{pmatrix}$, $\mathbf{B} = \begin{pmatrix} 1 & 0 \\ 2 & -2 \end{pmatrix}$ and $\mathbf{C} = \begin{pmatrix} 3 & -1 \\ 0 & 5 \end{pmatrix}$, calculate:

a $\mathbf{A}+2\mathbf{B}$ **b** $3\mathbf{A}-2\mathbf{I}$ **c** $3\mathbf{C}-\mathbf{B}+2\mathbf{I}$ **d** $\mathbf{A}-\mathbf{B}+2\mathbf{C}$

5 Find the values of x and y that satisfy each matrix equation, or state that there is no solution.

a $\begin{pmatrix} 1 & x \\ -1 & 5 \end{pmatrix} + \begin{pmatrix} 2 & 2 \\ y & 7 \end{pmatrix} = \begin{pmatrix} 3 & 5 \\ 7 & 12 \end{pmatrix}$ **b** $\begin{pmatrix} x & 2 \\ 3 & 5 \end{pmatrix} + \begin{pmatrix} 3 & y \\ 7 & 3 \end{pmatrix} = \begin{pmatrix} 5 & 7 \\ 10 & 8 \end{pmatrix}$

c $\begin{pmatrix} 2x & 3 \\ 5 & y \end{pmatrix} + \begin{pmatrix} -y & -5 \\ 3 & 3x \end{pmatrix} = \begin{pmatrix} 7 & -2 \\ 8 & 8 \end{pmatrix}$ **d** $\begin{pmatrix} 3 & 2x \\ -1 & y \end{pmatrix} + \begin{pmatrix} -1 & y \\ 5 & 2x \end{pmatrix} = \begin{pmatrix} 2 & 7 \\ 4 & 5 \end{pmatrix}$

6 Given that $\mathbf{A} = \begin{pmatrix} 4 & -1 \\ 3 & 3 \end{pmatrix}$ and $\mathbf{B} = \begin{pmatrix} 3 & -5 \\ 1 & 7 \end{pmatrix}$ find the matrix $\mathbf{X}$.

a $2\mathbf{A}+\mathbf{X}=\mathbf{B}$ **b** $3\mathbf{X}-4\mathbf{A}=\mathbf{B}$ **c** $\mathbf{X}-3\mathbf{I}=5\mathbf{A}$ **d** $2\mathbf{I}-3\mathbf{X}=\mathbf{B}$

7 $\mathbf{A}=\begin{pmatrix} a & 3 \\ -2 & 1 \end{pmatrix}$ and $\mathbf{B}=\begin{pmatrix} 2 & b \\ 1 & 3 \end{pmatrix}$.

Given that $\mathbf{A}+s\mathbf{B}=t\mathbf{I}$, find the values of a, b, s and t.

8 $\mathbf{A}=\begin{pmatrix} 2x^2 & x \\ -x^2 & -3 \end{pmatrix}$ and $\mathbf{B}=\begin{pmatrix} 6 & -1 \\ x & x+2 \end{pmatrix}$.

Given $\mathbf{A}+x\mathbf{B}=y\mathbf{I}$, find all possible values of y and x.

9 $\mathbf{A}=\begin{pmatrix} 1 & 2a \\ -a & 3 \end{pmatrix}$ and $\mathbf{B}=\begin{pmatrix} 4 & b+1 \\ 3b & 1 \end{pmatrix}$.

Given $c\mathbf{A}+d\mathbf{B}=\mathbf{I}$, find the values of a, b, c and d.

10 Explain why, for any two matrices **A** and **B** that have the same dimensions, $(\mathbf{A}+\mathbf{B})^{\mathrm{T}}=\mathbf{A}^{\mathrm{T}}+\mathbf{B}^{\mathrm{T}}$.

Section 2: Matrix multiplication

Matrices have a variety of uses and most involve multiplication.

Matrix $\mathbf{A}=\begin{pmatrix} 1 & 3 \\ 2 & 5 \end{pmatrix}$ and matrix $\mathbf{B}=\begin{pmatrix} -1 & 4 \\ 0 & -2 \end{pmatrix}$.

To find the product $\mathbf{C}=\mathbf{AB}$, let matrix $\mathbf{C}=\begin{pmatrix} c_{11} & c_{12} \\ c_{21} & c_{22} \end{pmatrix}$, so that:

$$\underset{\mathbf{A}}{\begin{pmatrix} 1 & 3 \\ 2 & 5 \end{pmatrix}}\underset{\mathbf{B}}{\begin{pmatrix} -1 & 4 \\ 0 & -2 \end{pmatrix}}=\underset{\mathbf{C}}{\begin{pmatrix} c_{11} & c_{12} \\ c_{21} & c_{22} \end{pmatrix}}$$

Multiply each pair of values together and calculate the total of these products to find c_{ij}.

So, to work out c_{11} you look at the **first row** of **A** and the **first column** of **B**:

$$\underset{\mathbf{A}}{\begin{pmatrix} 1 & 3 \\ 2 & 5 \end{pmatrix}}\quad\underset{\mathbf{B}}{\begin{pmatrix} -1 & 4 \\ 0 & -2 \end{pmatrix}}$$

$$c_{11}=1\times(-1)+3\times 0=-1$$

To work out c_{12} you look at the **first row** of **A** and the **second column** of **B**:

$$\underset{\mathbf{A}}{\begin{pmatrix} 1 & 3 \\ 2 & 5 \end{pmatrix}}\quad\underset{\mathbf{B}}{\begin{pmatrix} -1 & 4 \\ 0 & -2 \end{pmatrix}}$$

$$c_{12}=1\times 4+3\times(-2)=-2$$

Continuing in this fashion, you calculate each element of **C** to find $\mathbf{C}=\begin{pmatrix} -1 & -2 \\ -2 & -2 \end{pmatrix}$.

For multiplication, the left-hand matrix must have the same number of columns as the right-hand matrix has rows. If this is not the case, the matrices are not conformable for multiplication, and the product is not defined.

Key point 7.6

When an $m \times n$ matrix **A** and an $n \times p$ matrix **B** are multiplied together, the result is an $m \times p$ matrix $\mathbf{C} = \mathbf{AB}$.

To find the element in row i and column j of **C**:

1 take row i of **A** and column j of **B**
2 multiply each pair of corresponding values and add all the products together.

WORKED EXAMPLE 7.5

The 2×2 matrix $\mathbf{A} = \begin{pmatrix} 1 & 3 \\ 2 & -5 \end{pmatrix}$.

For each of the matrices **B**, determine **i AB** and **ii BA** or explain why the product does not exist.

a $\mathbf{B}_1 = \begin{pmatrix} 3 & -4 \\ 1 & 2 \end{pmatrix}$ **b** $\mathbf{B}_2 = \begin{pmatrix} 4 & 5 & -3 \\ 2 & 22 & -1 \end{pmatrix}$ **c** $\mathbf{B}_3 = \begin{pmatrix} 2 & 1 \\ 2 & -3 \\ 2 & 3 \end{pmatrix}$ **d** $\mathbf{B}_4 = \mathbf{A}$

a i $\mathbf{AB}_1 = \begin{pmatrix} 1 & 3 \\ 2 & -5 \end{pmatrix}\begin{pmatrix} 3 & -4 \\ 1 & 2 \end{pmatrix}$

$= \begin{pmatrix} 1 \times 3 + 3 \times 1 & 1 \times (-4) + 3 \times 2 \\ 2 \times 3 + (-5) \times 1 & 2 \times (-4) + (-5) \times 2 \end{pmatrix}$

$= \begin{pmatrix} 6 & 2 \\ 1 & -18 \end{pmatrix}$

There is the same number of columns in **A** ($2 \times \mathbf{2}$) as rows in **B** ($\mathbf{2} \times 2$) so the product exists.

For the **top left** element of the product, multiply the **top** row of **A** with the **left** column of **B**. Continue in the same way for each other element.

ii $\mathbf{B}_1\mathbf{A} = \begin{pmatrix} 3 & -4 \\ 1 & 2 \end{pmatrix}\begin{pmatrix} 1 & 3 \\ 2 & -5 \end{pmatrix}$

$= \begin{pmatrix} 3 \times 1 + (-4) \times 2 & 3 \times 3 + (-4) \times (-5) \\ 1 \times 1 + 2 \times 2 & 1 \times 3 + 2 \times (-5) \end{pmatrix}$

$= \begin{pmatrix} -5 & 29 \\ 5 & -7 \end{pmatrix}$

There is the same number of columns in **B** ($2 \times \mathbf{2}$) as rows in **A** ($\mathbf{2} \times 2$) so the product exists.

b i $\mathbf{AB}_2 = \begin{pmatrix} 1 & 3 \\ 2 & -5 \end{pmatrix}\begin{pmatrix} 4 & 5 & -3 \\ 2 & 22 & -1 \end{pmatrix}$

$= \begin{pmatrix} 1 \times 4 + 3 \times 2 & 1 \times 5 + 3 \times 22 & 1 \times (-3) + 3 \times (-1) \\ 2 \times 4 + (-5) \times 2 & 2 \times 5 + (-5) \times 22 & 2 \times (-3) + (-5) \times (-1) \end{pmatrix}$

$= \begin{pmatrix} 10 & 71 & -6 \\ -2 & -100 & -1 \end{pmatrix}$

There is the same number of columns in **A** ($2 \times \mathbf{2}$) as rows in **B** ($\mathbf{2} \times 3$) so the product exists.

Continues on next page

ii $\mathbf{B}_2\mathbf{A} = \begin{pmatrix} 4 & 5 & -3 \\ 2 & 22 & -1 \end{pmatrix}\begin{pmatrix} 1 & 3 \\ 2 & -5 \end{pmatrix}$

Not conformable: $\mathbf{B}_2$ has 3 columns but $\mathbf{A}$ has 2 rows.

The number of columns in $\mathbf{B}$ $(2\times\mathbf{3})$ is different from the number of rows in $\mathbf{A}$ $(\mathbf{2}\times 2)$ so the product does not exist.

c i $\mathbf{AB}_3 = \begin{pmatrix} 1 & 3 \\ 2 & -5 \end{pmatrix}\begin{pmatrix} 2 & 1 \\ 2 & -3 \\ 2 & 3 \end{pmatrix}$

Not conformable: $\mathbf{A}$ has 2 columns but $\mathbf{B}_3$ has 3 rows.

The number of columns in $\mathbf{A}$ $(2\times\mathbf{2})$ is different from the number of rows in rows in $\mathbf{B}$ $(\mathbf{3}\times 2)$ so the product does not exist.

ii $\mathbf{B}_3\mathbf{A} = \begin{pmatrix} 2 & 1 \\ 2 & -3 \\ 2 & 3 \end{pmatrix}\begin{pmatrix} 1 & 3 \\ 2 & -5 \end{pmatrix}$

$$= \begin{pmatrix} 2\times 1+1\times 2 & 2\times 3+1\times(-5) \\ 2\times 1+(-3)\times 2 & 2\times 3+(-3)\times(-5) \\ 2\times 1+3\times 2 & 2\times 3+3\times(-5) \end{pmatrix}$$

$$= \begin{pmatrix} 4 & 1 \\ -4 & 21 \\ 8 & -9 \end{pmatrix}$$

There is the same number of columns in $\mathbf{B}$ $(3\times\mathbf{2})$ as rows in $\mathbf{A}$ $(\mathbf{2}\times 2)$ so the product exists.

d $\mathbf{B}_4\mathbf{A} = \mathbf{AB}_4 = \mathbf{A}^2 = \begin{pmatrix} 1 & 3 \\ 2 & -5 \end{pmatrix}\begin{pmatrix} 1 & 3 \\ 2 & -5 \end{pmatrix}$

$$= \begin{pmatrix} 1\times 1+3\times 2 & 1\times 3+3\times(-5) \\ 2\times 1+(-5)\times 2 & 2\times 3+(-5)\times(-5) \end{pmatrix}$$

$$= \begin{pmatrix} 7 & -12 \\ -8 & 31 \end{pmatrix}$$

There is the same number of columns in $\mathbf{A}$ $(2\times\mathbf{2})$ as rows in $\mathbf{A}$ $(\mathbf{2}\times 2)$ so the product exists.

In part **d** of Worked example 7.5, you used the standard index notation $\mathbf{A}^2$ for a matrix $\mathbf{A}$ multiplied by itself. Similarly $\mathbf{A}^3 = \mathbf{AAA}$ and onwards, for higher integer powers.

Notice that, as shown in part **a** of Worked example 7.5, matrix multiplication is not generally **commutative**, which means that **AB** does not necessarily equal **BA**. However, some matrices will commute with each other and, as you will see, there are some matrices that commute with all other matrices of suitable dimension.

Focus on...

Focus on... Modelling 2 shows how you can use matrices to represent networks. The main result that is explored involves raising these matrices to powers.

Key point 7.7

Two matrices **A** and **B** commute (are commutative) if **AB** = **BA**.

Although matrix multiplication is not generally commutative, it is **associative**; that is, a multiple product **ABC** can be calculated as (**AB**)**C** or as **A**(**BC**), the result is still the same.

WORKED EXAMPLE 7.6

$\mathbf{A}=\begin{pmatrix}1 & 2\\ 4 & -2\end{pmatrix}$, $\mathbf{B}=\begin{pmatrix}3 & 1\\ -1 & -1\end{pmatrix}$, $\mathbf{C}=\begin{pmatrix}-2 & -1\\ 1 & 1\end{pmatrix}$

a Calculate **X** = **AB** and **Y** = **BC**.

b Calculate **XC** and **AY** and explain your findings.

a $\mathbf{X}=\mathbf{AB}$ — Standard matrix multiplication.

$=\begin{pmatrix}1 & 2\\ 4 & -2\end{pmatrix}\begin{pmatrix}3 & 1\\ -1 & -1\end{pmatrix}$

$=\begin{pmatrix}1 & -1\\ 14 & 6\end{pmatrix}$

$\mathbf{Y}=\mathbf{BC}$

$=\begin{pmatrix}3 & 1\\ -1 & -1\end{pmatrix}\begin{pmatrix}-2 & -1\\ 1 & 1\end{pmatrix}$

$=\begin{pmatrix}-5 & -2\\ 1 & 0\end{pmatrix}$

b $\mathbf{XC}=\begin{pmatrix}1 & -1\\ 14 & 6\end{pmatrix}\begin{pmatrix}-2 & -1\\ 1 & 1\end{pmatrix}$

$=\begin{pmatrix}-3 & -2\\ -22 & -8\end{pmatrix}$

$\mathbf{AY}=\begin{pmatrix}1 & 2\\ 4 & -2\end{pmatrix}\begin{pmatrix}-5 & -2\\ 1 & 0\end{pmatrix}$

$=\begin{pmatrix}-3 & -2\\ -22 & -8\end{pmatrix}$

$\mathbf{XC}=(\mathbf{AB})\mathbf{C}$
$\mathbf{AY}=\mathbf{A}(\mathbf{BC})$

The products give the same result because matrix multiplication is associative.

EXERCISE 7B

1 State whether the two matrices can be multiplied, and the dimensions of their product, when they are multiplied in the order given.

a $\begin{pmatrix} 1 & 1 & 2 \\ 0 & 3 & 5 \end{pmatrix}$ and $\begin{pmatrix} 3 & 1 \\ 3 & 1 \\ 3 & 2 \end{pmatrix}$

b $\begin{pmatrix} 3 & -1 \\ 1 & 0 \end{pmatrix}$ and $\begin{pmatrix} 1 & 1 \\ 2 & 1 \\ 2 & 2 \end{pmatrix}$

c $\begin{pmatrix} 1 & 0 & 0 \\ 0 & 1 & 0 \end{pmatrix}$ and $\begin{pmatrix} 0 & 1 \\ 1 & 0 \end{pmatrix}$

d $(3 \ \ -2)$ and $\begin{pmatrix} 0 & 0 \\ 0 & 0 \end{pmatrix}$

e $\begin{pmatrix} 1 & 0 \\ 0 & 1 \end{pmatrix}$ and $\begin{pmatrix} 3 & -1 & 1 \\ 2 & 1 & 3 \end{pmatrix}$

f $(1 \ \ -1 \ \ 2)$ and $\begin{pmatrix} 3 \\ 1 \\ 4 \end{pmatrix}$

g $\begin{pmatrix} 3 & 1 \\ 2 & 1 \end{pmatrix}$ and $\begin{pmatrix} 2 & -2 \\ 2 & 1 \end{pmatrix}$

h $\begin{pmatrix} 1 & a \\ 2 & 2 \end{pmatrix}$ and $\begin{pmatrix} -a & 2a \\ 1 & 2 \end{pmatrix}$

i $\begin{pmatrix} 2 & 1 & 1 \\ 3 & -1 & 2 \\ 1 & 0 & 4 \end{pmatrix}$ and $\begin{pmatrix} 3 & 0 & 1 \\ 1 & 2 & 2 \\ 0 & 0 & 5 \end{pmatrix}$

j $\begin{pmatrix} 1 & a & 0 \\ 2 & 2 & a \\ 1 & 0 & 5 \end{pmatrix}$ and $\begin{pmatrix} 2 & 1 & 1 \\ -1 & a & 5 \\ 2 & 0 & a \end{pmatrix}$

k $\begin{pmatrix} 2 \\ 1 \\ 3 \end{pmatrix}$ and $(-1 \ \ 1 \ \ 3)$

2 Multiply the matrices from Question **1** when possible. Use a calculator to check your answers.

3 Determine whether the matrices commute.

a $\begin{pmatrix} 2 & 0 \\ 0 & 3 \end{pmatrix}$ and $\begin{pmatrix} 4 & 1 \\ 2 & 5 \end{pmatrix}$

b $\begin{pmatrix} 1 & 5 \\ 0 & 1 \end{pmatrix}$ and $\begin{pmatrix} 1 & -3 \\ 0 & 1 \end{pmatrix}$

c $\begin{pmatrix} 3 & -1 \\ 2 & 3 \end{pmatrix}$ and $\begin{pmatrix} 2 & 0 \\ 0 & 2 \end{pmatrix}$

d $\begin{pmatrix} 2 & 3 \\ -3 & 2 \end{pmatrix}$ and $\begin{pmatrix} -1 & 5 \\ -5 & -1 \end{pmatrix}$

e $\begin{pmatrix} 1 & 2 \\ 2 & 3 \end{pmatrix}$ and $\begin{pmatrix} 1 & 3 \\ 0 & 2 \end{pmatrix}$

f $\begin{pmatrix} 2 & 0 \\ -3 & 3 \end{pmatrix}$ and $\begin{pmatrix} 1 & -5 \\ 0 & 1 \end{pmatrix}$

4 Find the values of the unknown scalars.

a $\begin{pmatrix} 2 & -3 \\ 1 & 0 \end{pmatrix}\begin{pmatrix} 1 & x \\ y & 2 \end{pmatrix} = \begin{pmatrix} -4 & 0 \\ 1 & 3 \end{pmatrix}$

b $\begin{pmatrix} a & 2 \\ -3 & b \end{pmatrix}\begin{pmatrix} 1 & 2 \\ -2 & 3 \end{pmatrix} = \begin{pmatrix} 0 & 14 \\ -5 & -3 \end{pmatrix}$

c $\begin{pmatrix} x & y \\ 2 & -1 \end{pmatrix}\begin{pmatrix} 3 & 1 \\ 1 & 2 \end{pmatrix} = \begin{pmatrix} 8 & 11 \\ 5 & 0 \end{pmatrix}$

d $\begin{pmatrix} 2 & 1 \\ -1 & 2 \end{pmatrix}\begin{pmatrix} 1 & p \\ -3 & q \end{pmatrix} = \begin{pmatrix} -1 & 0 \\ -7 & -5 \end{pmatrix}$

5 Given that $\mathbf{A} = \begin{pmatrix} 7 & -1 \\ 0 & 5 \end{pmatrix}$ and $\mathbf{B} = \begin{pmatrix} 3 & 3 \\ 1 & 6 \end{pmatrix}$ work out these matrices.

a $\mathbf{A} + 3\mathbf{B} - 2\mathbf{I}$ b $\mathbf{AB}$ c $\mathbf{A}^3$

6 Given that $\mathbf{A} = (2 \quad -3)$, $\mathbf{B} = \begin{pmatrix} 1 \\ 5 \end{pmatrix}$ and $\mathbf{C} = (-1 \quad 7)$, work out these matrices.

a $\mathbf{A} + \mathbf{C}$ b $\mathbf{AB}$ c $\mathbf{BC}$

7 Given that $\mathbf{M} = \begin{pmatrix} 2 & -1 & 5 \\ 0 & 1 & 4 \\ -4 & 0 & 2 \end{pmatrix}$ and $\mathbf{N} = \begin{pmatrix} 3 & 0 & -1 \\ 2 & 2 & 5 \\ -1 & 3 & 0 \end{pmatrix}$, complete these matrix calculations.

a $3\mathbf{M} - 2\mathbf{I}$ b $\mathbf{MN}$ c $\mathbf{N}^2$

8 Given that $\mathbf{P} = \begin{pmatrix} 2 & a \\ 3a & -1 \end{pmatrix}$ and $\mathbf{Q} = \begin{pmatrix} 2a & -2 \\ a & 3 \end{pmatrix}$, find, in terms of a:

a $\mathbf{P} - 3\mathbf{Q}$ b $\mathbf{PQ}$.

9 $\mathbf{A} = (2 \quad 1 \quad 3)$, $\mathbf{B} = \begin{pmatrix} p & 2 \\ 1 & 2p \\ -1 & 2 \end{pmatrix}$ and $\mathbf{C} = \begin{pmatrix} 3p \\ -2 \end{pmatrix}$

a Find, in terms of p:

i $\mathbf{AB}$ ii $\mathbf{BC}$.

b Explain why $\mathbf{B}^2$ does not exist.

10 Find the value of b such that the matrices $\begin{pmatrix} 3 & 1 \\ 0 & 3 \end{pmatrix}$ and $\begin{pmatrix} -5 & 2 \\ 0 & b \end{pmatrix}$ commute.

11 Prove that the matrices $\begin{pmatrix} a & b \\ -b & a \end{pmatrix}$ and $\begin{pmatrix} c & d \\ -d & c \end{pmatrix}$ commute for all values of a, b, c and d.

12 Given that the matrices $\begin{pmatrix} 2 & 3 \\ 3 & -2 \end{pmatrix}$ and $\begin{pmatrix} c & d \\ d & -c \end{pmatrix}$ commute, find an expression for d in terms of c.

13 $\mathbf{A} = \begin{pmatrix} a_{11} & a_{12} \\ a_{21} & a_{22} \end{pmatrix}$, $\mathbf{B} = \begin{pmatrix} b_{11} & b_{12} \\ b_{21} & b_{22} \end{pmatrix}$ and $\mathbf{C} = \begin{pmatrix} c_{11} & c_{12} \\ c_{21} & c_{22} \end{pmatrix}$

a Calculate $\mathbf{X} = \mathbf{AB}$ and $\mathbf{Y} = \mathbf{BC}$.

b Show that $\mathbf{XC} = \mathbf{AY}$.

14 Prove that for two 2×2 matrices $\mathbf{A}$ and $\mathbf{B}$, $(\mathbf{AB})^{\mathrm{T}} = \mathbf{B}^{\mathrm{T}}\mathbf{A}^{\mathrm{T}}$.

Explain whether this is the case for matrices of other dimensions for which the product exists.

Section 3: Determinants and inverses of 2×2 matrices

In this section you will answer the question: 'Given two matrices, **A** and **B**, find a matrix **X** such that **AX** = **B**.'

In the AS Level course you will focus on 2×2 matrices.

Fast forward

In Further Mathematics Student Book 2 you will extend this to 3 × 3 matrices.

The identity matrix

In arithmetic, an **identity** is a number that, for a given operation, produces no changes. So, for addition in the real numbers, zero is the identity because for any real value x:

$$x+0=0+x=x$$

For 2×2 matrix addition, the zero matrix has the property:

$$\mathbf{Z}+\mathbf{A}=\mathbf{A}+\mathbf{Z}=\mathbf{A}$$

The identity for multiplication of the real numbers is 1, because, for any real value x:

$$1\times x=x\times 1=x$$

In matrix multiplication, the identity matrix, which was defined in Section 1, has the property:

$$\mathbf{IA}=\mathbf{AI}=\mathbf{A}$$

You can see this in the case of 2×2 matrices by considering the general matrix $\mathbf{A}=\begin{pmatrix} a & b \\ c & d \end{pmatrix}$.

$$\begin{aligned}\mathbf{IA}&=\begin{pmatrix} 1 & 0 \\ 0 & 1 \end{pmatrix}\begin{pmatrix} a & b \\ c & d \end{pmatrix}\\ &=\begin{pmatrix} 1\times a+0\times c & 1\times b+0\times d \\ 0\times a+1\times c & 0\times b+1\times d \end{pmatrix}\\ &=\begin{pmatrix} a & b \\ c & d \end{pmatrix}\\ &=\mathbf{A}\end{aligned}$$

And similarly, you can show that $\mathbf{AI}=\mathbf{A}$.

This result can be generalised for multiplication of the $n\times n$ identity matrix with any $n\times n$ matrix.

Tip

In the definition of the identity, you need to require that both $\mathbf{IA}=\mathbf{A}$ and $\mathbf{BI}=\mathbf{B}$. This is because, generally, matrices don't commute. If $\mathbf{M}$ is a square $n\times n$ matrix, then $\mathbf{I}$ commutes with $\mathbf{M}$: $\mathbf{MI}=\mathbf{IM}$.

Key point 7.8

The $n\times n$ identity matrix $\mathbf{I}$ has the unique property that:

- for any $n\times m$ matrix $\mathbf{A}$, $\mathbf{IA}=\mathbf{A}$
- for any $m\times n$ matrix $\mathbf{B}$, $\mathbf{BI}=\mathbf{B}$.

Focus on...

See Focus on...Proof 2 to prove that the identity matrix is unique.

Determinants

When you compare real numbers, you can say that one is greater or less than another. If you want to compare vectors you can use their magnitudes. Although you cannot meaningfully write that $\begin{pmatrix} 3 \\ 1 \end{pmatrix}>\begin{pmatrix} 2 \\ 2 \end{pmatrix}$ or that $\begin{pmatrix} 3 \\ 1 \end{pmatrix}<\begin{pmatrix} 2 \\ 2 \end{pmatrix}$, you can compare their magnitudes, as given by the modulus of the vector.

$$\left|\begin{pmatrix} 3 \\ 1 \end{pmatrix}\right|=\sqrt{10}>\left|\begin{pmatrix} 2 \\ 2 \end{pmatrix}\right|=\sqrt{8}$$

There is a similar idea for square matrices, called the **determinant**, which, in many respects, can be considered the magnitude of the matrix. However, unlike a vector modulus, the determinant can take positive or negative values.

Key point 7.9

The determinant of a 2×2 matrix $\mathbf{A} = \begin{pmatrix} a & b \\ c & d \end{pmatrix}$ is written as det $\mathbf{A}$ or $|\mathbf{A}|$ or $\begin{vmatrix} a & b \\ c & d \end{vmatrix}$, and equals $ad - bc$.

Tip

In words, the determinant is the product of the **lead diagonal** elements less the product of the **reverse diagonal** elements.

WORKED EXAMPLE 7.7

Calculate the determinant of the matrix $\mathbf{A} = \begin{pmatrix} 1 & 5 \\ 2 & -2 \end{pmatrix}$.

$|\mathbf{A}| = \left|\begin{pmatrix} 1 & 5 \\ 2 & -2 \end{pmatrix}\right|$

$= (1 \times (-2)) - (5 \times 2)$

$= -2 - 10$

$= -12$

For a matrix $\mathbf{A} = \begin{pmatrix} a & b \\ c & d \end{pmatrix}$

$|\mathbf{A}| = ad - bc$

You can prove some useful properties for 2×2 matrix determinants algebraically.

Key point 7.10

For two 2×2 matrices $\mathbf{A}$ and $\mathbf{B}$ and a scalar k:

- det $\mathbf{AB}$ = det $\mathbf{BA}$ = det $\mathbf{A}$ × det $\mathbf{B}$
- $\det(k\mathbf{A}) = k^2 \det \mathbf{A}$

The proof of the first result is shown here. You can prove the second result in a similar way.

PROOF 8

Let $\mathbf{A} = \begin{pmatrix} a & b \\ c & d \end{pmatrix}$ and $\mathbf{B} = \begin{pmatrix} p & q \\ r & s \end{pmatrix}$.

You need to introduce some variables so you can do the calculation.

Then det $\mathbf{A} = ad - bc$

det $\mathbf{B} = ps - qr$

First find the individual determinants of **A** and **B**.

Continues on next page

$$\mathbf{AB} = \begin{pmatrix} ap+br & aq+bs \\ cp+dr & cq+ds \end{pmatrix}$$

$$\mathbf{BA} = \begin{pmatrix} ap+cq & bp+dq \\ ar+cs & br+ds \end{pmatrix}$$

Use the rule for multiplying matrices to find **AB** and **BA**. Remember that they are not the same!

$$\det \mathbf{AB} = (ap+br)(cq+ds) - (aq+bs)(cp+dr)$$
$$= acpq + adps + bcqr + bdrs - (acpq + adqr + bcps + bdrs)$$
$$= adps + bcqr - adqr - bcps$$

Expand and simplify the expression for det **AB**.

$$\det \mathbf{A} \times \det \mathbf{B} = (ad-bc)(ps-qr)$$
$$= adps - adqr + bcqr - bcps$$

It's not obvious how to simplify this, so look at $(\det \mathbf{A}) \times (\det \mathbf{B})$.

Hence: $\det \mathbf{AB} = \det \mathbf{A} \times \det \mathbf{B}$

The two expressions contain the same four terms.

$$\det \mathbf{BA} = (ap+cq)(br+ds) - (bp+dq)(ar+cs)$$
$$= (apbr + apds + cqbr + cqds) - (bpar + bpcs + dqar + dqcs)$$
$$= apds + cqbr - bpcs - bpcs$$

Since **BA** is not the same as **AB** you also need to expand det **BA**.

This is again the same as $(\det \mathbf{A}) \times (\det \mathbf{B})$.

So $\det \mathbf{AB} = \det \mathbf{BA} = \det \mathbf{A} \times \det \mathbf{B}$

Write a conclusion, summarising what you have proved.

EXERCISE 7C

1 Find the determinant of each matrix.

a **i** $\begin{pmatrix} 3 & -1 \\ 7 & 4 \end{pmatrix}$ **ii** $\begin{pmatrix} 1 & 1 \\ -3 & 2 \end{pmatrix}$ **b** **i** $\begin{pmatrix} 2 & -3 \\ 1 & 5 \end{pmatrix}$ **ii** $\begin{pmatrix} 3 & -2 \\ 1 & 3 \end{pmatrix}$

c **i** $\begin{pmatrix} 2a & a \\ 3 & -1 \end{pmatrix}$ **ii** $\begin{pmatrix} 1 & -a \\ 5 & 2a \end{pmatrix}$ **d** **i** $\begin{pmatrix} a & -3 \\ 2 & 5a \end{pmatrix}$ **ii** $\begin{pmatrix} 1 & 2a \\ -a & 3 \end{pmatrix}$

2 Find the determinants of the matrices $\mathbf{A} = \begin{pmatrix} 2 & -3 \\ 1 & 5 \end{pmatrix}$, $\mathbf{B} = \begin{pmatrix} 3 & -2 \\ 1 & 3 \end{pmatrix}$ and **AB**.

Confirm that det AB = det A det B.

3 The matrix $\mathbf{A} = \begin{pmatrix} 2a & 5 \\ a+2 & a \end{pmatrix}$ has determinant 2.

Find the possible values of a.

4 The matrix $\mathbf{M} = \begin{pmatrix} 6x & x+1 \\ x^2 & 2 \end{pmatrix}$ has determinant 0.

Find the possible values of x.

Inverse matrices

Remember that the aim, at the start of this section, was to find a matrix **X** such that **AX** = **B**. Effectively, you are trying to 'undo' matrix multiplication.

For real numbers, you can reverse multiplication by division or, equivalently, multiplication by the reciprocal: if $ax = b$ then $x = \frac{b}{a} = \frac{1}{a} \times b$.

The matrix equivalent of a reciprocal is called the **inverse matrix.**

Key point 7.11

The matrix $\mathbf{A}^{-1}$ is the inverse of a square matrix **A** and has the property that

$$\mathbf{AA}^{-1} = \mathbf{A}^{-1}\mathbf{A} = \mathbf{I}$$

You can use a calculator to find the inverse of a matrix with numerical entries, but you should also know how to use the formula.

Key point 7.12

Given a matrix $\mathbf{A} = \begin{pmatrix} a & b \\ c & d \end{pmatrix}$ with $ad - bc \neq 0$, the inverse is

$$\mathbf{A}^{-1} = \frac{1}{ad-bc}\begin{pmatrix} d & -b \\ -c & a \end{pmatrix}$$

Tip

You can remember how to find $\mathbf{A}^{-1}$ as:

- swap the elements on the leading diagonal.
- change the signs of the elements on the reverse diagonal.
- divide by the determinant.

You can prove that this formula gives the inverse by checking that $\mathbf{AA}^{-1} = \mathbf{A}^{-1}\mathbf{A} = \mathbf{I}$.

PROOF 9

$$\mathbf{AA}^{-1} = \begin{pmatrix} a & b \\ c & d \end{pmatrix} \times \frac{1}{ad-bc}\begin{pmatrix} d & -b \\ -c & a \end{pmatrix}$$

Check that $\mathbf{AA}^{-1} = \mathbf{I}$.

$$= \begin{pmatrix} a & b \\ c & d \end{pmatrix}\begin{pmatrix} \frac{d}{ad-bc} & \frac{-b}{ad-bc} \\ \frac{-c}{ad-bc} & \frac{a}{ad-bc} \end{pmatrix}$$

Firstly multiply the second matrix by the scalar (by multiplying each element by the scalar).

$$= \begin{pmatrix} \frac{ad-bc}{ad-bc} & \frac{-ab+ba}{ad-bc} \\ \frac{cd-dc}{ad-bc} & \frac{-bc+da}{ad-bc} \end{pmatrix}$$

$$= \begin{pmatrix} 1 & 0 \\ 0 & 1 \end{pmatrix} = \mathbf{I}$$

This is the required result: $\mathbf{AA}^{-1} = \mathbf{I}$.

as required.

A similar calculation shows that $\mathbf{A}^{-1}\mathbf{A} = \mathbf{I}$.

In the formula for the inverse of **A**, the expression at the bottom of the fraction is the determinant of **A**. This means that if det **A** = 0, then the matrix **A** has no inverse.

Key point 7.13

A matrix **A** with det **A** = 0 is called **singular** and has no inverse.

WORKED EXAMPLE 7.8

For each matrix, find the inverse matrix or establish that there is no inverse.

Where there is an inverse, verify that the product with the original matrix is **I**.

a $\mathbf{A}=\begin{pmatrix}4 & 1\\ 6 & 2\end{pmatrix}$ **b** $\mathbf{B}=\begin{pmatrix}2 & 4\\ -1 & 3\end{pmatrix}$ **c** $\mathbf{C}=\begin{pmatrix}1 & -2\\ -3 & 6\end{pmatrix}$ **d** $\mathbf{D}=\begin{pmatrix}0 & 1\\ 1 & 0\end{pmatrix}$

a $\det\mathbf{A} = 4\times2-1\times6=2$

Find det **A**; if it is non-zero, then $\mathbf{A}^{-1}$ exists.

$$\mathbf{A}^{-1}=\frac{1}{2}\begin{pmatrix}2 & -1\\ -6 & 4\end{pmatrix}$$

$$=\begin{pmatrix}1 & -0.5\\ -3 & 2\end{pmatrix}$$

Swap the elements on the lead diagonal.
Change the sign of the elements on the reverse diagonal.
Divide by the determinant of the original matrix.

$$\mathbf{AA}^{-1}=\begin{pmatrix}4 & 1\\ 6 & 2\end{pmatrix}\begin{pmatrix}1 & -0.5\\ -3 & 2\end{pmatrix}$$

Check the answer by multiplication to find $\mathbf{AA}^{-1}=\mathbf{I}$.

$$=\begin{pmatrix}4\times1+1\times(-3) & 4\times(-0.5)+1\times2\\ 6\times1+2\times(-3) & 6\times(-0.5)+2\times2\end{pmatrix}$$

$$=\begin{pmatrix}1 & 0\\ 0 & 1\end{pmatrix}$$

$$=\mathbf{I}$$

b $\det\mathbf{B}=2\times3-4\times(-1)=10$

Find det **B**; if it is non-zero, then $\mathbf{B}^{-1}$ exists.

$$\mathbf{B}^{-1}=\frac{1}{10}\begin{pmatrix}3 & -4\\ 1 & 2\end{pmatrix}$$

$$=\begin{pmatrix}0.3 & -0.4\\ 0.1 & 0.2\end{pmatrix}$$

Continues on next page

$$\mathbf{BB}^{-1} = \begin{pmatrix} 2 & 4 \\ -1 & 3 \end{pmatrix}\begin{pmatrix} 0.3 & -0.4 \\ 0.1 & 0.2 \end{pmatrix}$$

Check the answer by multiplication to find $\mathbf{BB}^{-1} = \mathbf{I}$.

$$= \begin{pmatrix} 2\times0.3+4\times0.1 & 2\times(-0.4)+4\times0.2 \\ (-1)\times0.3+3\times0.1 & (-1)\times(-0.4)+3\times0.2 \end{pmatrix}$$

$$= \begin{pmatrix} 1 & 0 \\ 0 & 1 \end{pmatrix}$$

$$= \mathbf{I}$$

c $\det \mathbf{C} = 1\times6-2\times(-3) = 0$

Find $\det \mathbf{C}$; if it is zero, then $\mathbf{C}^{-1}$ does not exist.

C is singular and has no inverse.

d $\det \mathbf{D} = 0\times0-1\times1 = -1$

Find $\det \mathbf{D}$; if it is non-zero, then $\mathbf{D}^{-1}$ exists.

$$\mathbf{D}^{-1} = -\begin{pmatrix} 0 & -1 \\ -1 & 0 \end{pmatrix} = \begin{pmatrix} 0 & 1 \\ 1 & 0 \end{pmatrix}$$

$$\mathbf{DD}^{-1} = \begin{pmatrix} 0 & 1 \\ 1 & 0 \end{pmatrix}\begin{pmatrix} 0 & 1 \\ 1 & 0 \end{pmatrix}$$

Check the answer by multiplication to find $\mathbf{DD}^{-1} = \mathbf{I}$.

$$= \begin{pmatrix} 0\times0+1\times1 & 0\times1+1\times0 \\ 1\times0+0\times1 & 1\times1+0\times0 \end{pmatrix}$$

$$= \begin{pmatrix} 1 & 0 \\ 0 & 1 \end{pmatrix}$$

$$= \mathbf{I}$$

Using the inverse

You can use the inverse matrix to 'undo' matrix multiplication. Because matrix multiplication is generally not commutative, you need to be careful to multiply the matrices in the correct order.

If **A** is a non-singular matrix and if $\mathbf{AX} = \mathbf{B}$, then you can multiply both sides by $\mathbf{A}^{-1}$ on the left.

$$\mathbf{A}^{-1}(\mathbf{AX}) = \mathbf{A}^{-1}\mathbf{B}$$
$$\Rightarrow (\mathbf{A}^{-1}\mathbf{A})\mathbf{X} = \mathbf{A}^{-1}\mathbf{B}$$
$$\Rightarrow \mathbf{IX} = \mathbf{A}^{-1}\mathbf{B}$$
$$\Rightarrow \mathbf{X} = \mathbf{A}^{-1}\mathbf{B}$$

Tip

If **A** is singular (i.e. if $\det \mathbf{A} = 0$), then there is no solution as the inverse matrix, $\mathbf{A}^{-1}$, doesn't exist.

However, if you want to solve the equation $\mathbf{XA} = \mathbf{B}$, you need to multiply by $\mathbf{A}^{-1}$ on the right, so that **A** and $\mathbf{A}^{-1}$ are next to each other.

$$\mathbf{XAA}^{-1} = \mathbf{BA}^{-1}$$
$$\Rightarrow \mathbf{X} = \mathbf{BA}^{-1}$$

WORKED EXAMPLE 7.9

$\mathbf{A} = \begin{pmatrix} 4 & 1 \\ 7 & 2 \end{pmatrix}$, $\mathbf{B} = \begin{pmatrix} 3 & -4 \\ 0 & 8 \end{pmatrix}$.

a Find **X** such that **AX** = **B**. **b** Find **Y** such that **YA** = **B**.

$\det \mathbf{A} = 4 \times 2 - 1 \times 7 = 1$

$\mathbf{A}^{-1} = \begin{pmatrix} 2 & -1 \\ -7 & 4 \end{pmatrix}$

You need to multiply through by $\mathbf{A}^{-1}$ in order to find **X** and **Y**.

Calculate det **A** and then use it to find $\mathbf{A}^{-1}$.

a $\mathbf{AX} = \mathbf{B}$

$\mathbf{A}^{-1}\mathbf{AX} = \mathbf{A}^{-1}\mathbf{B}$

You need to cancel **A** from the left-hand side of the product, so you must multiply by $\mathbf{A}^{-1}$ on the left on both sides of the equation.

Then use $\mathbf{A}^{-1}\mathbf{A} = \mathbf{I}$ and $\mathbf{IX} = \mathbf{X}$.

$\mathbf{X} = \mathbf{A}^{-1}\mathbf{B} = \begin{pmatrix} 2 & -1 \\ -7 & 4 \end{pmatrix}\begin{pmatrix} 3 & -4 \\ 0 & 8 \end{pmatrix}$

$= \begin{pmatrix} 6 & -16 \\ -21 & 60 \end{pmatrix}$

b $\mathbf{YA} = \mathbf{B}$

$\mathbf{YAA}^{-1} = \mathbf{BA}^{-1}$

You need to cancel **A** from the right-hand side of the product, so you must multiply by $\mathbf{A}^{-1}$ on the right on both sides of the equation.

Then use $\mathbf{AA}^{-1} = \mathbf{I}$ and $\mathbf{YI} = \mathbf{Y}$.

$\mathbf{Y} = \mathbf{BA}^{-1} = \begin{pmatrix} 3 & -4 \\ 0 & 8 \end{pmatrix}\begin{pmatrix} 2 & -1 \\ -7 & 4 \end{pmatrix}$

$= \begin{pmatrix} 34 & -19 \\ -56 & 32 \end{pmatrix}$

You will often need to use the inverse of a product of two matrices. With real numbers, the reciprocal of a product is the product of the reciprocals: $\frac{1}{ab} = \frac{1}{a} \times \frac{1}{b}$. Because matrix multiplication is not commutative, you need to be a little more careful.

WORKED EXAMPLE 7.10

For non-singular square matrices **A** and **B**:

a show that $(\mathbf{AB})^{-1}$ exists **b** write down $(\mathbf{AB})^{-1}$ in terms of $\mathbf{A}^{-1}$ and $\mathbf{B}^{-1}$.

a If **A** and **B** are non-singular, then det **A** and det **B** are both non-zero.

Since $\det(\mathbf{AB}) = \det \mathbf{A} \times \det \mathbf{B}$, it follows that $\det(\mathbf{AB})$ is also non-zero.

$\therefore$ **AB** is non-singular and has an inverse.

Continues on next page

b By definition of an inverse:

$(\mathbf{AB})(\mathbf{AB})^{-1} = \mathbf{I}$

$\mathbf{AB}(\mathbf{AB})^{-1} = \mathbf{I}$

$\mathbf{A}^{-1}\mathbf{AB}(\mathbf{AB})^{-1} = \mathbf{A}^{-1}\mathbf{I}$

$\mathbf{B}(\mathbf{AB})^{-1} = \mathbf{A}^{-1}$

$\mathbf{B}^{-1}\mathbf{B}(\mathbf{AB})^{-1} = \mathbf{B}^{-1}\mathbf{A}^{-1}$

$(\mathbf{AB})^{-1} = \mathbf{B}^{-1}\mathbf{A}^{-1}$

Multiply by inverses on the left of each side to find $(\mathbf{AB})^{-1}$.

Multiply by $\mathbf{A}^{-1}$ on the left.

$\mathbf{A}^{-1}\mathbf{A} = \mathbf{I}$

Now multiply by $\mathbf{B}^{-1}$ on the left.

$\mathbf{B}^{-1}\mathbf{B} = \mathbf{I}$

Key point 7.14

The inverse of a product consists of the inverses of the individual matrices, but listed in the opposite order:

$$(\mathbf{AB})^{-1} = \mathbf{B}^{-1}\mathbf{A}^{-1}$$
$$(\mathbf{ABC})^{-1} = \mathbf{C}^{-1}\mathbf{B}^{-1}\mathbf{A}^{-1}$$

EXERCISE 7D

1 Find the inverses of the matrices from Exercise 7C, Question 1. State any values of a for which the inverse does not exist.

2 Find all possible values of x for which these matrices are singular.

a i $\begin{pmatrix} x & -3 \\ 1 & 2 \end{pmatrix}$ ii $\begin{pmatrix} x & 1 \\ 1 & 5 \end{pmatrix}$ b i $\begin{pmatrix} 2x & -x \\ 5 & 2 \end{pmatrix}$ ii $\begin{pmatrix} 3 & -1 \\ x & 3x \end{pmatrix}$

c i $\begin{pmatrix} 4x & 1 \\ 2 & 2x \end{pmatrix}$ ii $\begin{pmatrix} 1 & x \\ 2x & 8 \end{pmatrix}$ d i $\begin{pmatrix} 2x & -x \\ 4 & x \end{pmatrix}$ ii $\begin{pmatrix} 1 & x \\ 5x & x \end{pmatrix}$

3 Given that $\mathbf{A} = \begin{pmatrix} 1 & 1 \\ 3 & -1 \end{pmatrix}$ and $\mathbf{B} = \begin{pmatrix} -1 & 1 \\ 2 & 3 \end{pmatrix}$, find the matrix $\mathbf{X}$ such that:

a $\mathbf{AX} = \mathbf{B}$ b $\mathbf{XA} = \mathbf{B}$ c $\mathbf{BX} = \begin{pmatrix} 3 & 1 \\ 1 & 5 \end{pmatrix}$

d $\mathbf{XB} = \begin{pmatrix} 1 & -1 \\ 3 & 2 \end{pmatrix}$ e $\mathbf{AX} = \begin{pmatrix} 1 \\ 4 \end{pmatrix}$ f $\mathbf{XB} = (-3 \quad 1)$

4 If $\mathbf{A} = \begin{pmatrix} 3 & -1 \\ 2 & 1 \end{pmatrix}$ find the matrix $\mathbf{B}$ such that $\mathbf{BA} = \begin{pmatrix} -1 & 1 \\ 3 & 5 \end{pmatrix}$.

5 Given that $\mathbf{A} = \begin{pmatrix} -3 & 1 \\ 1 & 8 \end{pmatrix}$ and $\mathbf{B} = \begin{pmatrix} 2 & 1 \\ 1 & 5 \end{pmatrix}$:

a show that $\mathbf{A}$ is non-singular b find $\mathbf{A}^{-1}\mathbf{B}$.

6 If $\mathbf{A} = \begin{pmatrix} 1 & 0 \\ 2 & 1 \end{pmatrix}$, find the matrix **X** such that $\mathbf{AX} = \begin{pmatrix} 3 \\ -2 \end{pmatrix}$.

7 Find the values of k for which the matrix $\begin{pmatrix} 3k & 1 \\ 9 & k \end{pmatrix}$ is singular.

8 Let $\mathbf{A} = \begin{pmatrix} 5 & 3c \\ -c & 1 \end{pmatrix}$.

a Show that **A** is non-singular for all values of c.

b Find $\mathbf{A}^{-1}$ in terms of c.

9 Given that $\mathbf{M} = \begin{pmatrix} 1 & -3 \\ 1 & 5 \end{pmatrix}$, find the matrix **X** such that $\mathbf{M}^{-1}\mathbf{XM} = \begin{pmatrix} 4 & 0 \\ 0 & -2 \end{pmatrix}$.

10 **A** and **B** are non-singular matrices with inverses $\mathbf{A}^{-1}$ and $\mathbf{B}^{-1}$.

a Show that $\frac{1}{3}\mathbf{A}^{-1}$ is the inverse of $3\mathbf{A}$.

b Simplify $(\mathbf{A}^{-1}\mathbf{B})^{-1}(3\mathbf{A})^{-1}$.

11 Given that **A** and **B** are non-singular matrices, simplify $(\mathbf{A}^{-1}\mathbf{B}^{-1}\mathbf{A})^{-1}$.

12 Given that **P** and **Q** are non-singular matrices, simplify:

a $\mathbf{P}(\mathbf{QP})^{-1}$

b $(2\mathbf{PQ})(3\mathbf{PQ})^{-1}$.

13 Prove that, if the non-singular matrices **A** and **B** commute, then so do $\mathbf{A}^{-1}$ and $\mathbf{B}^{-1}$.

14 For non-singular square matrix **A**, find in terms of det **A**:

a $\det(\mathbf{A}^{-1})$

b $\det(\mathbf{A}^{T})$ where **A** is a 2×2 matrix

c $\det(\mathbf{A}^{n})$ where n is an integer.

Section 4: Linear simultaneous equations

You already know how to solve simultaneous equations by elimination and substitution. In this section, you will learn how to represent two simultaneous equations as a matrix equation, which you can then solve using the methods from Section 3. In Further Mathematics Student Book 2 you will see this technique extended to systems of three equations, where it can prove more useful than solving by elimination or substitution.

Fast forward

In Further Mathematics Student Book 2 you will see the same idea applied to three simultaneous equations with a 3 × 3 matrix.

When you multiply a column vector by a square matrix you get another column vector. For example:

$$\begin{pmatrix} 3 & 2 \\ 2 & -7 \end{pmatrix}\begin{pmatrix} x \\ y \end{pmatrix} = \begin{pmatrix} 3x+2y \\ 2x-7y \end{pmatrix}$$

The two components of this new vector look like expressions that appear in simultaneous equations. So, for example, the system of equations:

$$\begin{cases} 3x+2y=5 \\ 2x-7y=20 \end{cases}$$

can be written as $\begin{pmatrix} 3 & 2 \\ 2 & -7 \end{pmatrix}\begin{pmatrix} x \\ y \end{pmatrix} = \begin{pmatrix} 5 \\ 20 \end{pmatrix}$

This is now an equation of the form $\mathbf{AX} = \mathbf{B}$, where $\mathbf{X} = \begin{pmatrix} x \\ y \end{pmatrix}$ so you can solve it by finding $\mathbf{A}^{-1}$.

WORKED EXAMPLE 7.11

Use a matrix method to these simultaneous equations.

$$\begin{cases} 3x+2y=5 \\ 2x-7y=20 \end{cases}$$

$$\begin{pmatrix} 3 & 2 \\ 2 & -7 \end{pmatrix}\begin{pmatrix} x \\ y \end{pmatrix}=\begin{pmatrix} 5 \\ 20 \end{pmatrix}$$

Rewrite the problem as a matrix multiplication of the form $\mathbf{A}\begin{pmatrix} x \\ y \end{pmatrix}=\mathbf{B}$

$$\begin{pmatrix} 3 & 2 \\ 2 & -7 \end{pmatrix}^{-1}=\frac{1}{-25}\begin{pmatrix} -7 & -2 \\ -2 & 3 \end{pmatrix}$$

Find the inverse of the square matrix.

$$\begin{pmatrix} x \\ y \end{pmatrix}=\begin{pmatrix} 3 & 2 \\ 2 & -7 \end{pmatrix}^{-1}\begin{pmatrix} 5 \\ 20 \end{pmatrix}$$

$\begin{pmatrix} x \\ y \end{pmatrix}=\mathbf{A}^{-1}\mathbf{B}$

$$=\frac{1}{-25}\begin{pmatrix} -7 & -2 \\ -2 & 3 \end{pmatrix}\begin{pmatrix} 5 \\ 20 \end{pmatrix}$$

$$=-\frac{1}{25}\begin{pmatrix} -75 \\ 50 \end{pmatrix}$$

$$=\begin{pmatrix} 3 \\ -2 \end{pmatrix}$$

$x=3, y=-2$

Sometimes simultaneous equations don't have a unique solution. For example, the system $\begin{cases} x+y=1 \\ x+y=2 \end{cases}$ doesn't have any solutions, while the system $\begin{cases} x+y=1 \\ 2x+2y=2 \end{cases}$ has infinitely many.

The system will have no unique solution whenever the determinant is zero (as it isn't possible to find the inverse matrix).

Tip

Remember that you are allowed to perform all numerical matrix multiplication on your calculator.

Key point 7.15

Two simultaneous equations can be written as a matrix problem:

$$\mathbf{A}\begin{pmatrix} x \\ y \end{pmatrix}=\mathbf{B}$$

where **A** contains the coefficients of the system.
If det **A** = 0, there is no unique solution to the system

Fast forward

You will see in Further Mathematics Student Book 2 that it is particularly useful with systems of three equations to use the determinant to establish whether or not the system has a unique solution.

WORKED EXAMPLE 7.12

Find the values of a for which the simultaneous equations:

$$\begin{cases} ax - 7y = 4 \\ 6x + (1-5a)y = 5 \end{cases}$$

do not have a unique solution.

$$\begin{pmatrix} a & -7 \\ 6 & 1-5a \end{pmatrix}\begin{pmatrix} x \\ y \end{pmatrix} = \begin{pmatrix} 4 \\ 5 \end{pmatrix}$$

Write the simultaneous equations as a matrix equation.

$$\det\begin{pmatrix} a & -7 \\ 6 & 1-5a \end{pmatrix} = 0$$

Since there isn't a unique solution, the determinant is zero.

$$a(1-5a)-(-7)6=0$$
$$a-5a^2+42=0$$
$$5a^2-a-42=0$$
$$(5a+14)(a-3)=0$$
$$a=-\frac{14}{5} \text{ or } 3$$

Find an expression for the determinant and solve the equation for a.

EXERCISE 7E

1. By rewriting each set of simultaneous equations as a matrix problem, solve for x and y.

   a **i** $\begin{cases} x+2y=3 \\ 4x-2y=-8 \end{cases}$ **ii** $\begin{cases} 7x-3y=13 \\ 2x+4y=11 \end{cases}$

   b **i** $\begin{cases} 4y-x=2 \\ 3x-2y=-1 \end{cases}$ **ii** $\begin{cases} 4x+9y=12 \\ -8x+15y=31 \end{cases}$

   c **i** $\begin{cases} x-7y=17 \\ 2x+3y=-17 \end{cases}$ **ii** $\begin{cases} 3x-4y+2=0 \\ x-2y-8=0 \end{cases}$

2. Find the values of k for which these simultaneous equations do not have a unique solution.

   $$\begin{cases} 8x+(k+1)y=3 \\ 4kx+6y=1 \end{cases}$$

3. **a** Find the value of k for which these simultaneous equations do not have a unique solution.

   $$\begin{cases} kx+2y=1 \\ (k-2)x+3y=-1 \end{cases}$$

   b Assuming that the equations have a unique solution, find the solution in terms of k.

4 This system of equations has a unique solution.

$$\begin{cases} ax + 3y = 2 \\ 5x + (a-2)y = 1 \end{cases}$$

a Find the set of possible values of a.

b Find the solution in terms of a.

Checklist of learning and understanding

- An $m \times n$ matrix **A** is a rectangular array of numerical or algebraic elements with m rows and n columns.
- Addition (and subtraction) of conformable matrices **A** and **B** involves adding (or subtracting) elements of matching position.
- Scalar multiplication of a matrix is equivalent to multiplying each element by the scalar.
- When an $m \times n$ matrix **A** and an $n \times p$ matrix **B** are multiplied together, the result is an $m \times p$ matrix $\mathbf{C} = \mathbf{AB}$.
- To find the element in row i and column j of **C**:
  - take row i of **A** and column j of **B**
  - multiply each pair of corresponding values and add all the products together.
- In general, matrices are not commutative, i.e. $\mathbf{AB} \neq \mathbf{BA}$.
- The zero matrix **Z** has all elements equal to 0.
- The identity matrix **I** is a square $(n \times n)$ matrix with elements equal to 1 on the lead diagonal and 0 elsewhere.
  - For suitably sized **I**, $\mathbf{AI} = \mathbf{A}$, $\mathbf{IA} = \mathbf{A}$ for any matrix **A**.
- The inverse matrix of satisfies $\mathbf{AA}^{-1} = \mathbf{A}^{-1}\mathbf{A} = \mathbf{I}$.
  - A matrix is singular if its determinant is zero. Otherwise it is non-singular.
  - Only non-singular matrices have inverse matrices.
  - For two non-singular matrices **A** and **B**, $(\mathbf{AB})^{-1} = \mathbf{B}^{-1}\mathbf{A}^{-1}$.
  - Inverse matrices can be used to solve equations.
    - If $\mathbf{AX} = \mathbf{B}$ then $\mathbf{X} = \mathbf{A}^{-1}\mathbf{B}$.
    - If $\mathbf{XA} = \mathbf{B}$ then $\mathbf{X} = \mathbf{BA}^{-1}$.
- The determinant of a 2×2 matrix $\mathbf{A} = \begin{pmatrix} a & b \\ c & d \end{pmatrix}$ is $|\mathbf{A}| = ad - bc$
- The inverse of a non-singular 2×2 matrix $\mathbf{A} = \begin{pmatrix} a & b \\ c & d \end{pmatrix}$ is $\mathbf{A}^{-1} = \frac{1}{|\mathbf{A}|}\begin{pmatrix} d & -b \\ -c & a \end{pmatrix}$
- Two simultaneous equations can be written as a matrix problem: $\mathbf{A}\begin{pmatrix} x \\ y \end{pmatrix} = \mathbf{B}$, where **A** contains the coefficients of the system.
  - If $\det \mathbf{A} = 0$, there is no unique solution to the system.

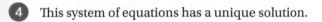

Mixed practice 7

1. The matrix **A** is given by $\mathbf{A} = \begin{pmatrix} 3 & -1 \\ 5 & -2 \end{pmatrix}$.

Find $\mathbf{A}^{-1}$. Choose from these options.

A $\begin{pmatrix} -2 & 1 \\ -5 & 3 \end{pmatrix}$ B $\begin{pmatrix} 2 & -1 \\ 5 & -3 \end{pmatrix}$ C $\begin{pmatrix} -3 & 5 \\ -1 & 2 \end{pmatrix}$ D $\begin{pmatrix} 3 & -5 \\ 1 & -2 \end{pmatrix}$

2. $\mathbf{A} = \begin{pmatrix} 2 & -3 \\ 0 & 4 \end{pmatrix}$ and $\mathbf{B} = \begin{pmatrix} 2k & -k \\ k & 1 \end{pmatrix}$

Find, in terms of k, the matrix **AB**. Choose from these options.

A $\begin{pmatrix} 4k & 3k \\ 0 & 4 \end{pmatrix}$ B $\begin{pmatrix} 2+2k & -3-k \\ k & 5 \end{pmatrix}$

C $\begin{pmatrix} k & -2k-3 \\ 4k & 4 \end{pmatrix}$ D $\begin{pmatrix} 4k & -2k \\ -2k & 3k+4 \end{pmatrix}$

3. Matrices **A**, **B** and **C** are given by:

$\mathbf{A} = \begin{pmatrix} a & -1 \\ 2 & 1 \end{pmatrix}$, $\mathbf{B} = \begin{pmatrix} 3 & b \\ 0 & -4 \end{pmatrix}$ and $\mathbf{C} = \begin{pmatrix} 1 & 4 \\ 2 & c \end{pmatrix}$.

If $\mathbf{AB} + k\mathbf{C} = \mathbf{I}$, find a, b, c and k.

4. Matrices **A** and **B** are given by $\mathbf{A} = \begin{pmatrix} 2k & 0 \\ k-1 & -1 \end{pmatrix}$ and $\mathbf{B} = \begin{pmatrix} 2 & 0 \\ 3 & -1 \end{pmatrix}$ where k is a constant.

a Given that **A** and **B** are commutative, find k.

b Show, by choosing matrices **C** and **D**, that matrix multiplication is not always commutative.

5. $\mathbf{A} = \begin{pmatrix} 2 & 3 \\ -2 & -5 \end{pmatrix}$ and $\mathbf{C} = \begin{pmatrix} 1 & 2 \\ 0 & 3 \end{pmatrix}$. Given that $\mathbf{AB} = \mathbf{C}$, find **B**.

6. Matrix **A** is given by $\mathbf{A} = \begin{pmatrix} a & -2 \\ 2 & 4 \end{pmatrix}$.

a Given that **A** is singular, calculate a.

b Given that **A** is non-singular, express $\mathbf{A}^{-1}$ in terms of a.

7. The matrices **A** and **B** are given by

$\mathbf{A} = \begin{pmatrix} 0 & 2 \\ 2 & 0 \end{pmatrix}$, $\mathbf{B} = \begin{pmatrix} 2 & 0 \\ 0 & -2 \end{pmatrix}$

a Calculate the matrix **AB**.

b Show that $\mathbf{A}^2$ is of the form $k\mathbf{I}$, where k is an integer and **I** is the 2×2 identity matrix.

c Show that $(\mathbf{AB})^2 \neq \mathbf{A}^2\mathbf{B}^2$.

[©AQA 2008]

8 $\mathbf{M}=\begin{pmatrix}3 & 5\\ 7 & k\end{pmatrix}$

a **i** Find the value of k for which **M** is singular.

ii Given that **M** is non-singular, find $\mathbf{M}^{-1}$, the inverse of **M**, in terms of k.

b When $k=12$, use $\mathbf{M}^{-1}$ to solve the simultaneous equations $\begin{cases}3x+5y=9\\ 7x+12y=11\end{cases}$

9 **a** Given that $\mathbf{ABC}=\mathbf{I}$, prove that $\mathbf{B}^{-1}=\mathbf{CA}$.

$\mathbf{A}=\begin{pmatrix}4 & 5\\ -2 & -3\end{pmatrix}$ and $\mathbf{C}=\begin{pmatrix}0 & -1\\ -3 & 4\end{pmatrix}$

b Find **B**.

10 It is given that $\mathbf{A}=\begin{pmatrix}1 & 4\\ 3 & 1\end{pmatrix}$ and that **I** is the 2×2 identity matrix.

a Show that $(\mathbf{A}-\mathbf{I})^2=k\mathbf{I}$ for some integer k.

b Given further that $\mathbf{B}=\begin{pmatrix}1 & 3\\ p & 1\end{pmatrix}$, find the integer p such that $(\mathbf{A}-\mathbf{B})^2=(\mathbf{A}-\mathbf{I})^2$

[©AQA 2010]

11 Let $\mathbf{X}=\begin{pmatrix}3 & x\\ -1 & 7\end{pmatrix}$

a Determine $\mathbf{XX}^{\mathrm{T}}$.

b Show that $\det(\mathbf{XX}^{\mathrm{T}}-\mathbf{X}^{\mathrm{T}}\mathbf{X})\leqslant 0$ for all real values of x.

c Find the value of x for which the matrix $(\mathbf{XX}^{\mathrm{T}}-\mathbf{X}^{\mathrm{T}}\mathbf{X})$ is singular.

[©AQA 2012]

12 **a** Show that for two 2×2 matrices **A** and **B**, where neither **A** nor **B** is the zero matrix Z, if $\mathbf{AB}=\mathbf{Z}$ then both **A** and **B** must be singular.

b Find two matrices **C** and **D**, where no element of either matrix is zero, for which $\mathbf{CD}=\mathbf{Z}$.

13 Matrix $\mathbf{A}=\begin{pmatrix}1 & 2\\ x & y\end{pmatrix}$. Given that **A** is non-singular and $\mathbf{A}+\mathbf{A}^{-1}=\mathbf{I}$, find x and y.

14 **A** and **B** are two 2×2 matrices.

a Show that if $\det(\mathbf{A}+\mathbf{B})=0=\det(\mathbf{A}-\mathbf{B})$, then $\det\mathbf{A}+\det\mathbf{B}=0$.

b Give an example of two matrices **A** and **B** for which $\det\mathbf{A}+\det\mathbf{B}=0$ but $\det(\mathbf{A}+\mathbf{B})\neq0$.

8 Matrix transformations

In this chapter you will learn how to:

- interpret matrices as linear transformations in two and three dimensions
- find a matrix representing a combined transformation
- find invariant points and invariant lines of a linear transformation.

Before you start…

GCSE	You should understand simple transformations: rotation, reflection, enlargement.	1 Triangle PQR is given by $P(1, 2)$, $Q(1, 4)$, $R(2, 2)$. Find the image of PQR under each transformation. a Rotation 90° anticlockwise about the origin. b Reflection through the line $y = -x$. c Enlargement centred at the origin, scale factor 2.
A Level Mathematics Student Book 1, Chapter 15	You should be able to use position vectors to represent points in 2 and 3 dimensions.	2 Write down the position vectors of points $A(5, 3, -2)$ and $B(0, -1, 1)$.
A Level Mathematics Student Book 1, Chapter 10	You should be able to solve trigonometric equations.	3 Find the value of θ in $[0°, 360°]$ for which $\sin\theta = \frac{1}{2}$ and $\cos\theta = -\frac{\sqrt{3}}{2}$.
Chapter 7	You should be able to multiply conformable matrices.	4 Calculate: $\begin{pmatrix} 2 & -1 \\ 0 & 3 \end{pmatrix}\begin{pmatrix} 2 & 5 \\ -4 & 1 \end{pmatrix}$
Chapter 7	You should know how to find determinants and inverses of 2×2 matrices.	5 For the matrix $\mathbf{A} = \begin{pmatrix} 2 & 5 \\ -4 & -3 \end{pmatrix}$ find: a det $\mathbf{A}$ b $\mathbf{A}^{-1}$
Chapter 7	You should be able to identify a singular matrix.	6 Find the value of k so that the matrix $\begin{pmatrix} 1 & 2 \\ 5 & k \end{pmatrix}$ is singular.

Why are matrices useful?

In Chapter 7 you learned about the structure and arithmetic of matrices, how to calculate inverses and determinants and how to solve equations involving matrices. In this chapter, you'll use these techniques to describe geometric transformations with matrices. This allows visual information to be recorded mathematically, which is an essential technique in applications such as computer game design.

Section 1: Matrices as linear transformations

You are familiar with transformations in two dimensions, such as reflections, rotations or enlargements, where the points (making up the **object**) are moved under a consistent rule to new (**image**) points. If you represent the points by their position vectors, then this rule can be written as a matrix equation.

For example, consider the rotation through 90° anticlockwise about the origin, applied to the triangle with vertices $P(1, 3)$, $Q(2, 3)$ and $R(1, 1)$, as shown in the diagram.

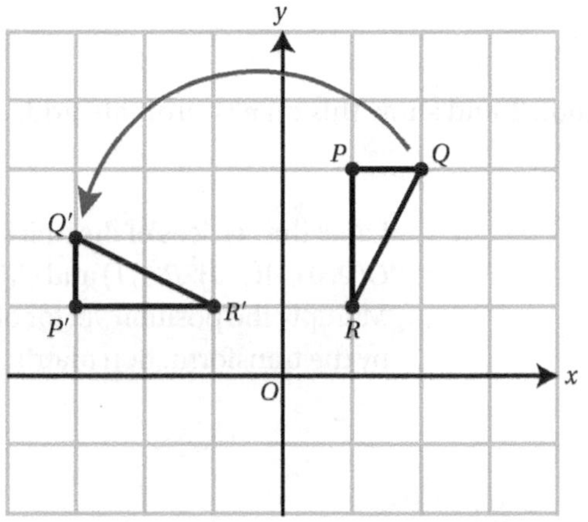

The image points have coordinates $P'(-3, 1)$, $Q'(-3, 2)$ and $R'(-1, 1)$. Look at how the transformation affects the position vectors:

$$\begin{pmatrix} 1 \\ 3 \end{pmatrix} \mapsto \begin{pmatrix} -3 \\ 1 \end{pmatrix}$$

$$\begin{pmatrix} 2 \\ 3 \end{pmatrix} \mapsto \begin{pmatrix} -3 \\ 2 \end{pmatrix}$$

$$\begin{pmatrix} 1 \\ 1 \end{pmatrix} \mapsto \begin{pmatrix} -1 \\ 1 \end{pmatrix}$$

It looks as if the rule can be written as:

$$\begin{pmatrix} x \\ y \end{pmatrix} \mapsto \begin{pmatrix} -y \\ x \end{pmatrix}$$

This can be expressed as a matrix equation:

$$\begin{pmatrix} x' \\ y' \end{pmatrix} = \begin{pmatrix} 0 & -1 \\ 1 & 0 \end{pmatrix}\begin{pmatrix} x \\ y \end{pmatrix}$$

The matrix $\mathbf{A} = \begin{pmatrix} 0 & -1 \\ 1 & 0 \end{pmatrix}$ represents a 90° anticlockwise rotation about the origin. You can use this matrix to find the image of any other point after the rotation. For example, the point $\begin{pmatrix} 5 \\ -8 \end{pmatrix}$ would be rotated to the point $\begin{pmatrix} 0 & -1 \\ 1 & 0 \end{pmatrix}\begin{pmatrix} 5 \\ -8 \end{pmatrix} = \begin{pmatrix} 8 \\ 5 \end{pmatrix}$.

You can identify a transformation represented by a matrix by looking at the image of the **unit square**, which is the square with coordinates $(0, 0)$, $(1, 0)$, $(1, 1)$ and $(0, 1)$.

WORKED EXAMPLE 8.1

A transformation $\mathbf{T}$ is given by $\mathbf{T} = \begin{pmatrix} 0 & -1 \\ -1 & 0 \end{pmatrix}$.

Find the image of the unit square under transformation $\mathbf{T}$ and show this on a coordinate grid. Hence describe the transformation.

$O: \begin{pmatrix} 0 & -1 \\ -1 & 0 \end{pmatrix}\begin{pmatrix} 0 \\ 0 \end{pmatrix} = \begin{pmatrix} 0 \\ 0 \end{pmatrix}$

$A: \begin{pmatrix} 0 & -1 \\ -1 & 0 \end{pmatrix}\begin{pmatrix} 1 \\ 0 \end{pmatrix} = \begin{pmatrix} 0 \\ -1 \end{pmatrix}$

$B: \begin{pmatrix} 0 & -1 \\ -1 & 0 \end{pmatrix}\begin{pmatrix} 1 \\ 1 \end{pmatrix} = \begin{pmatrix} -1 \\ -1 \end{pmatrix}$

$C: \begin{pmatrix} 0 & -1 \\ -1 & 0 \end{pmatrix}\begin{pmatrix} 0 \\ 1 \end{pmatrix} = \begin{pmatrix} -1 \\ 0 \end{pmatrix}$

Label the vertices of the unit square $O(0, 0)$, $A(1, 0)$, $B(1, 1)$ and $C(0, 1)$. Multiply the position vector of each vertex by the transformation matrix.

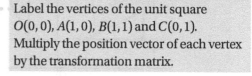

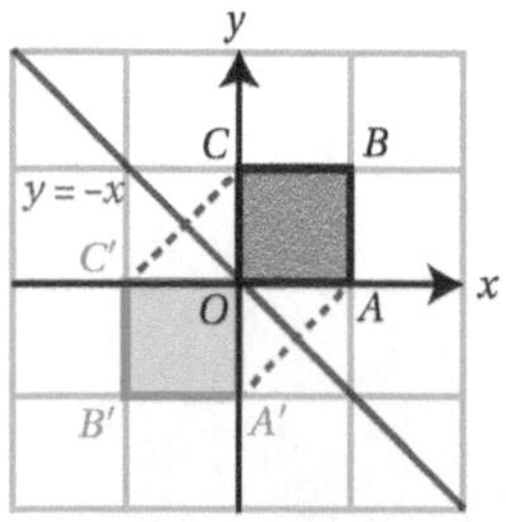

The transformation represents a reflection in the line $y = -x$.

Looking at the images of A and C helps identify the line of reflection.

Rather than considering all four points on the unit square, you can find the matrix to represent a given transformation by just looking at (1, 0) and (0, 1).

For a transformation $\mathbf{A} = \begin{pmatrix} a & b \\ c & d \end{pmatrix}$:

$$\mathbf{A}\begin{pmatrix} 1 \\ 0 \end{pmatrix} = \begin{pmatrix} a \\ c \end{pmatrix} \text{ and } \mathbf{A}\begin{pmatrix} 0 \\ 1 \end{pmatrix} = \begin{pmatrix} b \\ d \end{pmatrix}.$$

Notice that a matrix transformation never changes the position of the origin, as $\mathbf{A}\begin{pmatrix} 0 \\ 0 \end{pmatrix} = \begin{pmatrix} 0 \\ 0 \end{pmatrix}$.

Key point 8.1

The first column of a transformation matrix is the image of $\begin{pmatrix} 1 \\ 0 \end{pmatrix}$ and the second column is the image of $\begin{pmatrix} 0 \\ 1 \end{pmatrix}$.

WORKED EXAMPLE 8.2

Find the matrix representing the enlargement with scale factor 3 and centre at the origin.

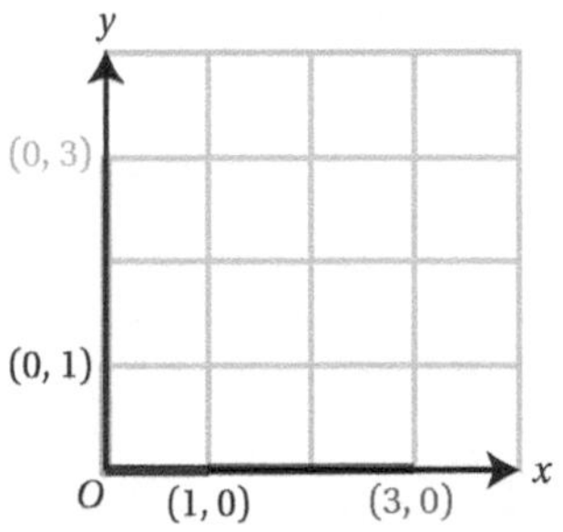

$\begin{pmatrix} 1 \\ 0 \end{pmatrix} \rightarrow \begin{pmatrix} 3 \\ 0 \end{pmatrix}$ and $\begin{pmatrix} 0 \\ 1 \end{pmatrix} \rightarrow \begin{pmatrix} 0 \\ 3 \end{pmatrix}$ — The images of the unit vectors give the columns of the transformation matrix.

So the matrix is $\begin{pmatrix} 3 & 0 \\ 0 & 3 \end{pmatrix}$

Using this method you can find a matrix for any transformation in two dimensions.

The most common transformations are summarised in Key point 8.2.

Key point 8.2

Transformation matrices

Rotation about the origin through the given angle	**90°**	**180°**	**270°**	
	$\begin{pmatrix} 0 & -1 \\ 1 & 0 \end{pmatrix}$	$\begin{pmatrix} -1 & 0 \\ 0 & -1 \end{pmatrix}$	$\begin{pmatrix} 0 & 1 \\ -1 & 0 \end{pmatrix}$	
Reflection in the given mirror line	x-axis	y-axis	$y = x$	$y = -x$
	$\begin{pmatrix} 1 & 0 \\ 0 & -1 \end{pmatrix}$	$\begin{pmatrix} -1 & 0 \\ 0 & 1 \end{pmatrix}$	$\begin{pmatrix} 0 & 1 \\ 1 & 0 \end{pmatrix}$	$\begin{pmatrix} 0 & -1 \\ -1 & 0 \end{pmatrix}$
Enlargement with centre at the origin	scale factor k			
	$\begin{pmatrix} k & 0 \\ 0 & k \end{pmatrix}$			

Tip

Note that by convention a positive angle means the rotation is anticlockwise. If the rotation is clockwise this will either be stated or a negative angle will be given.

Fast forward

In Section 2 rotations are generalised to any angle about the origin, and reflections to any mirror line through the origin.

The determinant

In some cases it can be difficult to distinguish between a rotation and a reflection. In Worked example 8.1 it could have been difficult to tell whether it was a reflection in the line $y = -x$ or a rotation through 180°.

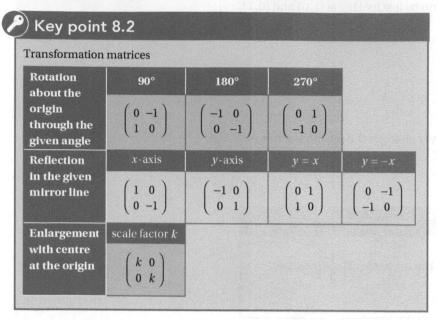

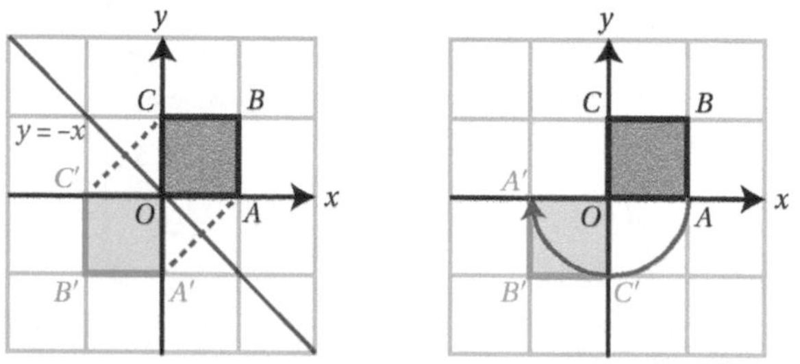

The image of the unit square is the same in both cases, but the points are arranged in a different order. Reading anticlockwise around the shape, the square $OABC$ is mapped to $OC'B'A'$ by the reflection, and to $OA'B'C'$ by the rotation.

A usfeul way of distinguishing between the two is to use the determinant of the transformation matrix **A**.

Table 8.1: Summary of the determinant value for transformation matrix **A**

Rotation	Reflection	Enlargement
det $\mathbf{A} = 1$	det $\mathbf{A} = -1$	det $\mathbf{A} = k^2$

The transformations that reverse the orientation have a negative determinant. Rotations and reflections do not change the size of the shape, while enlargements do.

Fast forward

In Further Mathematics Student Book 2, you'll see:

- rotation preserving the orientation
- reflection reversing the orientation of the shape.

Successive transformations

When two transformations are applied in succession, the overall result is another transformation. You can find its matrix as shown here.

Suppose the original point (the object) has the position vector $\mathbf{X} = \begin{pmatrix} x \\ y \end{pmatrix}$.

If the matrix representing the first transformation is **A**, then the first image is **AX**. After applying the second transformation with matrix **B**, the final image is **B**(**AX**).

Since matrix multiplication is associative, this is the same as (**BA**)**X**. Hence the matrix representing the combined transformation is **BA**.

Key point 8.3

For two transformations **A** and **B**, if **C** represents **A** followed by **B**, then transformation **C** is given by **C** = **BA**.

Fast forward

In Further Mathematics Student Book 2 you'll see how the determinant of a matrix representing an enlargement relates to the area of the image of a shape.

Common error

Notice that the matrix for the transformation that is applied first is written on the right.

WORKED EXAMPLE 8.3

$\mathbf{R} = \begin{pmatrix} 0 & 1 \\ 1 & 0 \end{pmatrix}$ represents a reflection in the line $y = x$ and $\mathbf{S} = \begin{pmatrix} 0 & 1 \\ -1 & 0 \end{pmatrix}$ represents a rotation through 270° about the origin.

Transformation $\mathbf{T}_1$ is defined as **R** followed by **S**, and $\mathbf{T}_2$ is defined by **S** followed by **R**.

B is the image of the point $A(2, 3)$ under $\mathbf{T}_1$ and C is the image of A under $\mathbf{T}_2$.

a Find the coordinates of B and C.

b Describe the transformations $\mathbf{T}_1$ and $\mathbf{T}_2$.

a $\mathbf{T}_1 = \mathbf{SR}$ — **R** followed by **S** is represented by **SR**.

$$= \begin{pmatrix} 0 & 1 \\ -1 & 0 \end{pmatrix}\begin{pmatrix} 0 & 1 \\ 1 & 0 \end{pmatrix}$$

$$= \begin{pmatrix} 1 & 0 \\ 0 & -1 \end{pmatrix}$$

$$\mathbf{T}_1\begin{pmatrix} 2 \\ 3 \end{pmatrix} = \begin{pmatrix} 1 & 0 \\ 0 & -1 \end{pmatrix}\begin{pmatrix} 2 \\ 3 \end{pmatrix}$$

$$= \begin{pmatrix} 2 \\ -3 \end{pmatrix}$$

B has coordinates $(2, -3)$.

$\mathbf{T}_2 = \mathbf{RS}$ — **S** followed by **R** is represented by **RS**.

$$= \begin{pmatrix} 0 & 1 \\ 1 & 0 \end{pmatrix}\begin{pmatrix} 0 & 1 \\ -1 & 0 \end{pmatrix}$$

$$= \begin{pmatrix} -1 & 0 \\ 0 & 1 \end{pmatrix}$$

Continues on next page

$$\mathbf{T}_2\begin{pmatrix}2\\3\end{pmatrix} = \begin{pmatrix}-1 & 0\\0 & 1\end{pmatrix}\begin{pmatrix}2\\3\end{pmatrix}$$

$$= \begin{pmatrix}-2\\3\end{pmatrix}$$

C has coordinates $(-2, 3)$.

b

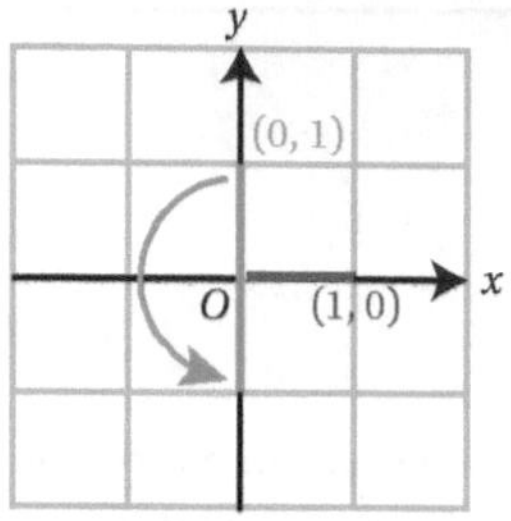

Looking at the columns of $\mathbf{T}_1$, the unit vector **i** is unchanged and the vector **j** is reversed.

$\mathbf{T}_1$ is a reflection in the x-axis.

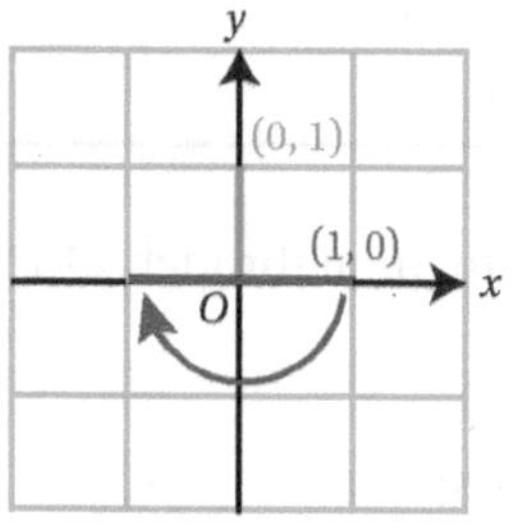

$\mathbf{T}_2$ leaves the **j** vector unchanged and reverses the direction of **i**.

$\mathbf{T}_2$ is a reflection in the y-axis.

Inverse transformations

You can also use a matrix to reverse the effect of a transformation. If a transformation has matrix **A**, and if the image of the object **X** under this transformation is $\mathbf{X}' = \mathbf{AX}$, then you can recover the original object by using the inverse matrix: $\mathbf{A}^{-1}\mathbf{X}' = \mathbf{X}$.

Key point 8.4

If a linear transformation is represented by a matrix **A**, then its inverse transformation is represented by the matrix $\mathbf{A}^{-1}$.

Notice that all the transformations that you have met so far have had non-zero determinants, so their inverse matrices exist.

When you have two successive transformations, the inverse transformations are applied in reverse order. This corresponds to the inverse of a product of two matrices: $(\mathbf{AB})^{-1} = \mathbf{B}^{-1}\mathbf{A}^{-1}$.

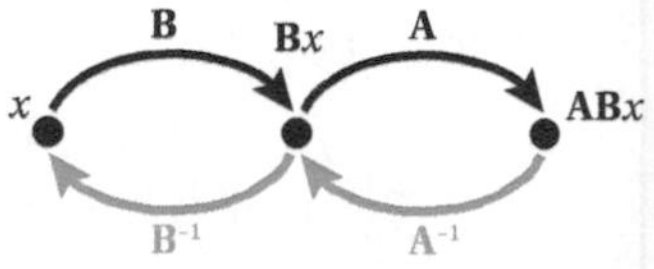

WORKED EXAMPLE 8.4

Point P is rotated 180° about the origin, and then reflected in the y-axis. The final image has coordinates $(-3, 7)$.

Find the coordinates of P.

The matrix for the 180° rotation is:

$$\mathbf{A} = \begin{pmatrix} -1 & 0 \\ 0 & -1 \end{pmatrix}$$

and for the reflection:

$$\mathbf{B} = \begin{pmatrix} -1 & 0 \\ 0 & 1 \end{pmatrix}$$

You are looking for the inverse of the combined transformation. Start by writing down the individual transformation matrices.

The combined transformation is:

$$\mathbf{C} = \mathbf{BA}$$

$$= \begin{pmatrix} -1 & 0 \\ 0 & 1 \end{pmatrix}\begin{pmatrix} -1 & 0 \\ 0 & -1 \end{pmatrix}$$

$$= \begin{pmatrix} 1 & 0 \\ 0 & -1 \end{pmatrix}$$

Find the combined transformation. Remember that the second transformation is written on the left.

You know that $\mathbf{C}\begin{pmatrix} x \\ y \end{pmatrix} = \begin{pmatrix} -3 \\ 7 \end{pmatrix}$, where (x, y) are the coordinates of P.

Hence:

$$\begin{pmatrix} x \\ y \end{pmatrix} = \mathbf{C}^{-1}\begin{pmatrix} -3 \\ 7 \end{pmatrix}$$

You can multiply on the left by the inverse matrix.

$$\mathbf{C}^{-1} = \frac{1}{-1}\begin{pmatrix} -1 & 0 \\ 0 & 1 \end{pmatrix}$$

$$= \begin{pmatrix} 1 & 0 \\ 0 & -1 \end{pmatrix}$$

The inverse of $\begin{pmatrix} a & b \\ c & d \end{pmatrix}$ is $\frac{1}{ad-bc}\begin{pmatrix} d & -b \\ -c & a \end{pmatrix}$.

So:

$$\begin{pmatrix} x \\ y \end{pmatrix} = \begin{pmatrix} 1 & 0 \\ 0 & -1 \end{pmatrix}\begin{pmatrix} -3 \\ 7 \end{pmatrix}$$

$$= \begin{pmatrix} -3 \\ -7 \end{pmatrix}$$

The coordinates of P are $(-3, -7)$.

EXERCISE 8A

1 Draw the image of the unit square under the transformation represented by each matrix.

a $\begin{pmatrix} 1 & 2 \\ 1 & 0 \end{pmatrix}$ b $\begin{pmatrix} -1 & 1 \\ 0 & 1 \end{pmatrix}$ c $\begin{pmatrix} 3 & 0 \\ 0 & 3 \end{pmatrix}$ d $\begin{pmatrix} 0 & 2 \\ -3 & 0 \end{pmatrix}$

2 Each diagram shows the image of the unit square under a linear transformation. Write down the matrix representing the transformation.

a

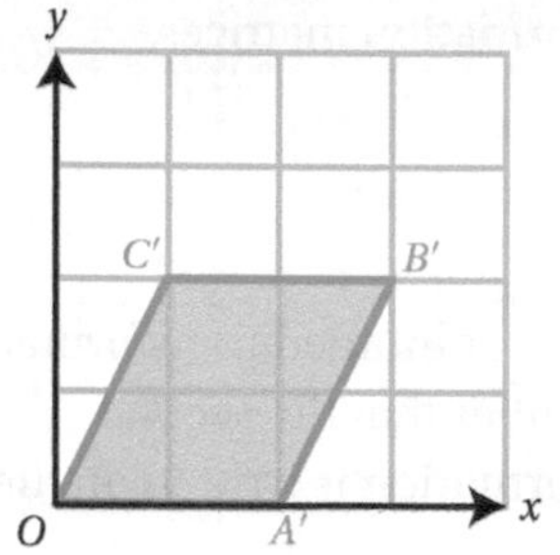

b

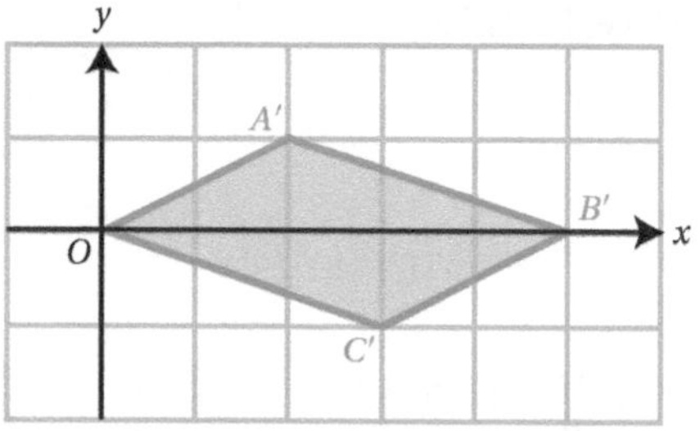

c

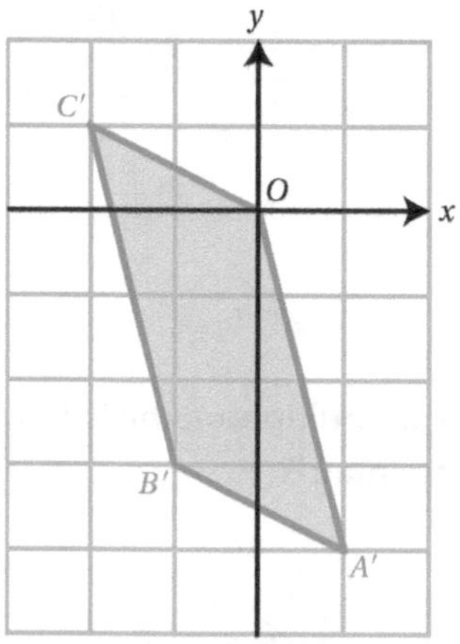

d

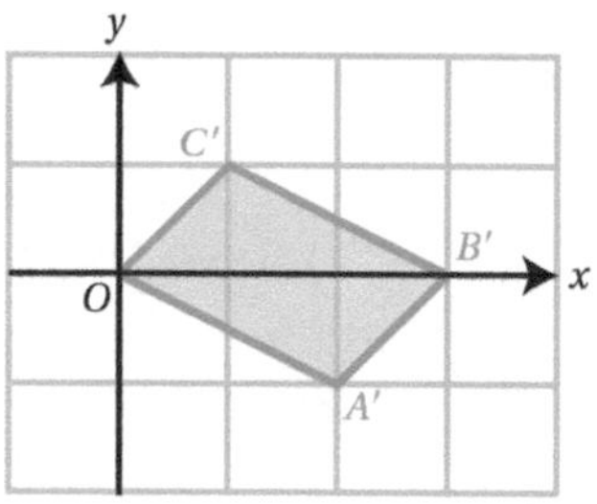

3 By considering the determinant, classify each transformation as a rotation, reflection or neither. Where appropriate, state the angle of rotation or the equation of the line of symmetry.

a $\begin{pmatrix} 0 & -1 \\ -1 & 0 \end{pmatrix}$ b $\begin{pmatrix} 0 & 1 \\ 1 & 0 \end{pmatrix}$ c $\begin{pmatrix} 1 & -1 \\ -1 & 2 \end{pmatrix}$

d $\begin{pmatrix} -1 & 0 \\ 0 & 1 \end{pmatrix}$ e $\begin{pmatrix} -1 & 0 \\ 0 & -1 \end{pmatrix}$ f $\begin{pmatrix} 1 & 1 \\ -1 & 1 \end{pmatrix}$

4 Find the image of the given point after the given transformation (or sequence of transformations).

a i Point $(3, 5)$; reflection in the line $y = x$

ii Point $(3, 5)$; reflection in the y-axis

b i Point $(-1, 3)$; rotation $270°$ about the origin

ii Point $(-1, 3)$; rotation $180°$ about the origin

c i Point $(4, -7)$; an enlargement with scale factor 3 followed by a rotation 90° about the origin

ii Point $(4, -7)$; reflection in the line $y = x$ followed by an enlargement with scale factor 4

d i Point $(4, 1)$; rotation 180° about the origin followed by an enlargement with scale factor $\frac{1}{2}$

ii Point $(4, 1)$; enlargement with scale factor $\frac{1}{3}$ followed by a reflection in the x-axis

5 You are given a transformation (or a sequence of transformations) and the coordinates of the image. Find the coordinates of the original object.

a i Rotation 90° about the origin; image $(3, 6)$

ii Rotation 270° about the origin; image $(3, 6)$

b i Reflection in the x-axis; image $(7, 2)$

ii Reflection in the line $y = -x$; image $(7, 2)$

c i Enlargement with scale factor 3 followed by a reflection in the line $y = x$; image $(-5, 2)$

ii Reflection in the y-axis followed by an enlargement with scale factor 5; image $(-5, 2)$

d i Rotation through 90° about the origin followed by a reflection in the x-axis; image $(6, 2)$

ii Reflection in the line $y = x$ followed by a rotation through 180° about the origin; image $(6, 2)$

6 a Find the matrix for the resulting transformation when a rotation 90° about the origin is followed by a rotation 180° about the origin. Describe the transformation.

b Repeat part **a** for some other pairs of rotations, including combining a rotation with itself. What do you notice?

c Find the matrix for the resulting transformation when a reflection in the line $y = x$ is followed by a reflection in the y-axis. Describe the transformation.

d Repeat part **c** for other pairs of reflections. What do you notice?

7 What is the inverse transformation of:

a reflection in the line $y = x$

b rotation through 90° about the origin?

8 The triangle with vertices $P(1, 1)$, $Q(6, 1)$ and $R(4, -2)$ is rotated through 90° anticlockwise about the origin. Find the coordinates of the vertices of the image.

9 The triangle with vertices $A(-1, 3)$, $B(2, 4)$ and $C(2, 3)$ is reflected in the line $y = -x$.

Find the coordinates of the vertices of the image. Show the original and the image triangle on a grid.

10 Transformation **T** is defined as a reflection in the line $y = x$ followed by an enlargement with scale factor 3.

a Find the matrix representing **T**.

b The image of the point P under the transformation **T** has coordinates $(3, -9)$. Find the coordinates of P.

c The rectangle with vertices $(0, 0)$, $(2, 0)$, $(2, 5)$ and $(0, 5)$ is transformed using the transformation **T**. Find the area of the image.

Section 2: Further transformations in 2D

There are five types of basic transformation that can be produced by matrices. In Section 1, you worked with enlargements and special cases of rotations and reflections. In this section, you will extend your understanding to general rotations and reflections and also meet stretches.

Rotations

The origin can never move under a matrix transformation, so it has to be the centre of rotation.

A rotation through angle θ (anticlockwise) about the origin moves the point $(1, 0)$ to the point $(\cos\theta, \sin\theta)$ and $(0, 1)$ to $(-\sin\theta, \cos\theta)$, so the matrix for this transformation is $\mathbf{A} = \begin{pmatrix} \cos\theta & -\sin\theta \\ \sin\theta & \cos\theta \end{pmatrix}$

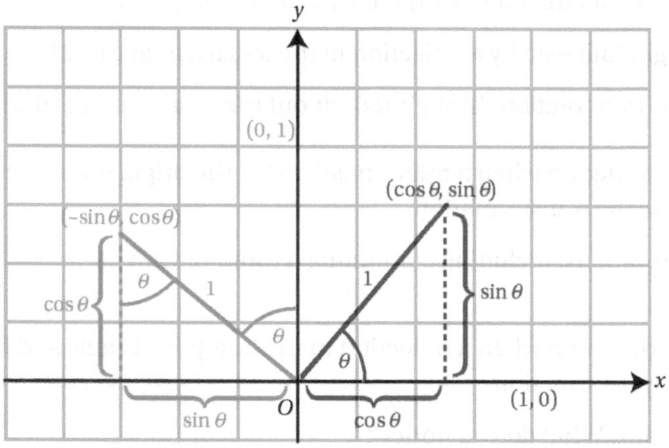

Tip

Remember that the columns of the transformation matrices correspond to the images of $\begin{pmatrix} 1 \\ 0 \end{pmatrix}$ and $\begin{pmatrix} 0 \\ 1 \end{pmatrix}$; this allows you to write down the matrix for each of these transformations.

Rewind

Check that using $\theta = 90°$, $180°$ and $270°$ gives the three rotation matrices from Section 1.

WORKED EXAMPLE 8.5

Find the image of the point $(-4, 2)$ after a rotation through $75°$ about the origin.

The matrix is:

$$\mathbf{A} = \begin{pmatrix} \cos 75° & -\sin 75° \\ \sin 75° & \cos 75° \end{pmatrix}$$

Use the general expression to find the rotation matrix.

$$= \begin{pmatrix} 0.259 & -0.966 \\ 0.966 & 0.259 \end{pmatrix}$$

The image of $(-4, 2)$ is:

$$\begin{pmatrix} 0.259 & -0.966 \\ 0.966 & 0.259 \end{pmatrix}\begin{pmatrix} -4 \\ 2 \end{pmatrix} = \begin{pmatrix} -2.97 \\ -3.35 \end{pmatrix}$$

Reflections

Since the origin cannot be affected by a transformation matrix, all lines of reflection must pass through the origin. Such lines have

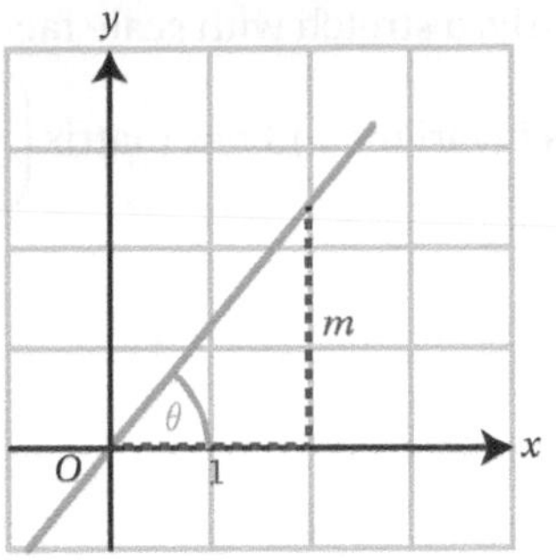

equations of the form $y = mx$. The gradient can also be defined by the angle the line makes with the x-axis: if this angle is θ then the gradient of the line is $m = \tan\theta$.

Considering the images of the two unit vectors, it can be shown that the matrix for the reflection is $\mathbf{A} = \begin{pmatrix} \cos 2\theta & \sin 2\theta \\ \sin 2\theta & -\cos 2\theta \end{pmatrix}$.

WORKED EXAMPLE 8.6

Line l has equation $y = \sqrt{3}x$.

a Find the angle that the line makes with the x-axis.
b Write down the matrix that describes a reflection in the line $y = \sqrt{3}x$.

a

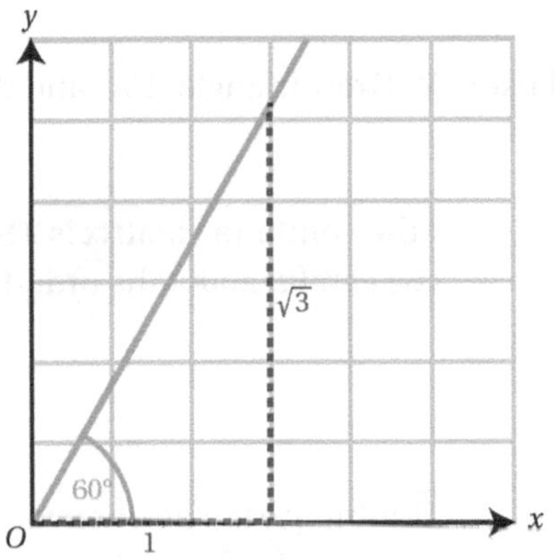

The line is at an angle $\arctan\sqrt{3} = 60°$ above the x-axis.

b Use $\theta = 60°$ (so $2\theta = 120°$):

$$\mathbf{A} = \begin{pmatrix} \cos 120° & \sin 120° \\ \sin 120° & -\cos 120° \end{pmatrix}$$

$$= \begin{pmatrix} -\frac{1}{2} & \frac{\sqrt{3}}{2} \\ \frac{\sqrt{3}}{2} & \frac{1}{2} \end{pmatrix}$$

The special lines from Section 1 are also covered by this general formula. For example, the matrix $\begin{pmatrix} 0 & -1 \\ -1 & 0 \end{pmatrix}$ from Worked example 8.1 corresponds to $\theta = 135°$ (since $\cos 270° = 0$ and $\sin 270° = -1$), which is the line $y = -x$.

Stretches

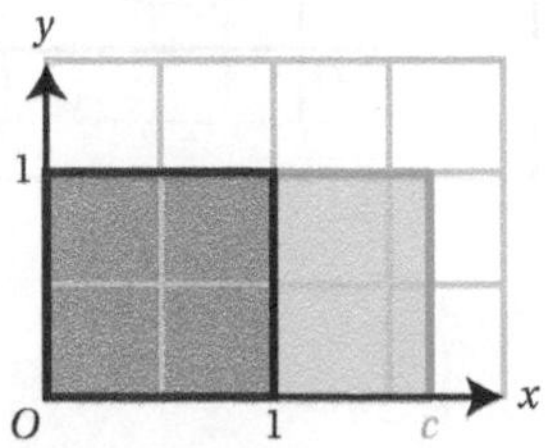

A stretch with scale factor c parallel to the x-axis leaves the y-axis unchanged (this is called the invariant axis). Hence its matrix is $\begin{pmatrix} c & 0 \\ 0 & 1 \end{pmatrix}$.

Similarly, a stretch with scale factor d parallel to the y-axis leaves the x-axis invariant and has matrix $\begin{pmatrix} 1 & 0 \\ 0 & d \end{pmatrix}$.

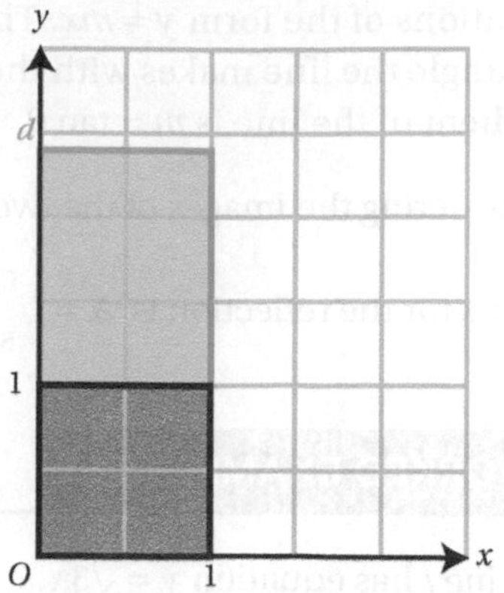

WORKED EXAMPLE 8.7

Transformation **S** is a stretch with scale factor 3 with the x-axis invariant. Transformation **T** is a stretch with scale factor 5 with the y-axis invariant. Transformation **V** is the result of **S** followed by **T**.

a Find the matrix representing **V**.

b The triangle with vertices $A(1,1)$, $B(2,1)$ and $C(1,3)$ is transformed using **V**. Draw triangle ABC and its image on a grid.

a $\mathbf{TS} = \begin{pmatrix} 5 & 0 \\ 0 & 1 \end{pmatrix}\begin{pmatrix} 1 & 0 \\ 0 & 3 \end{pmatrix}$

$= \begin{pmatrix} 5 & 0 \\ 0 & 3 \end{pmatrix}$

The combined matrix is **TS** – be careful about the order!

b A': $\begin{pmatrix} 5 & 0 \\ 0 & 3 \end{pmatrix}\begin{pmatrix} 1 \\ 1 \end{pmatrix} = \begin{pmatrix} 5 \\ 3 \end{pmatrix}$

B': $\begin{pmatrix} 5 & 0 \\ 0 & 3 \end{pmatrix}\begin{pmatrix} 2 \\ 1 \end{pmatrix} = \begin{pmatrix} 10 \\ 3 \end{pmatrix}$

C': $\begin{pmatrix} 5 & 0 \\ 0 & 3 \end{pmatrix}\begin{pmatrix} 1 \\ 3 \end{pmatrix} = \begin{pmatrix} 5 \\ 9 \end{pmatrix}$

Multiply each position vector by the transformation matrix.

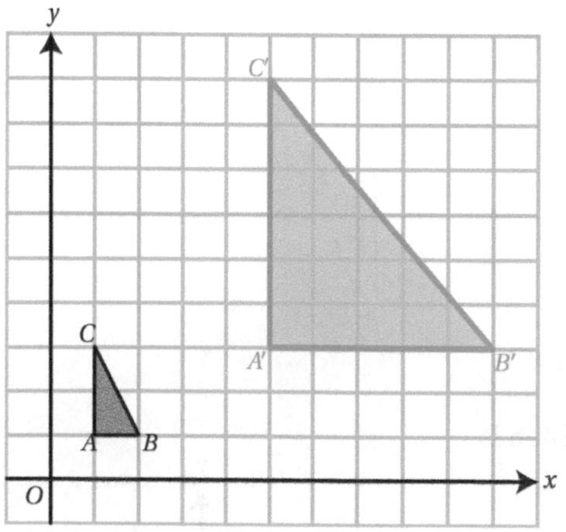

Enlargements

An enlargement (centred at the origin) is a combination of a vertical and horizontal stretch with the same scale factors. Its matrix is $\begin{pmatrix} k & 0 \\ 0 & k \end{pmatrix}$, where k is the scale factor.

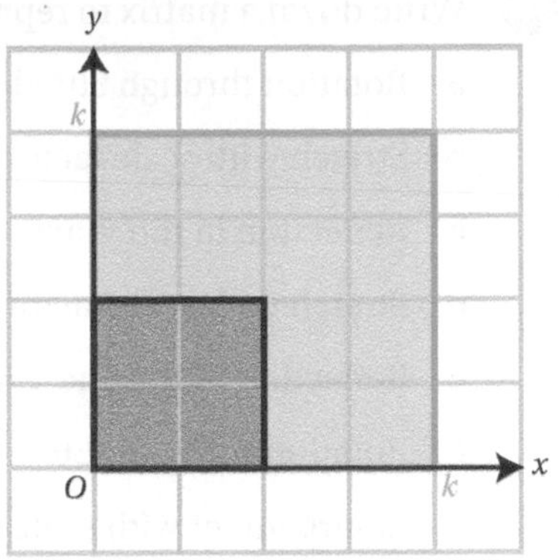

The table in Key point 8.5 summarises all the 2D linear transformations.

Key point 8.5

2D linear transformations

Transformation	Matrix	Determinant
Rotation about O, angle θ (anticlockwise)	$\begin{pmatrix} \cos\theta & -\sin\theta \\ \sin\theta & \cos\theta \end{pmatrix}$	1
Reflection in the line $y = (\tan\theta)x$	$\begin{pmatrix} \cos 2\theta & \sin 2\theta \\ \sin 2\theta & -\cos 2\theta \end{pmatrix}$	−1
Stretch, scale factor c, parallel to the x-axis	$\begin{pmatrix} c & 0 \\ 0 & 1 \end{pmatrix}$	c
Stretch, scale factor d, parallel to the y-axis	$\begin{pmatrix} 1 & 0 \\ 0 & d \end{pmatrix}$	d
Enlargement, centre O, scale factor k	$\begin{pmatrix} k & 0 \\ 0 & k \end{pmatrix}$	k^2

The matrices for rotation and reflection will be given in your formula book.

Tip

Note that only the matrices for rotation and reflection are given in the formula book.

EXERCISE 8B

1 Find the angle of (anticlockwise) rotation represented by each matrix.

a $\begin{pmatrix} \frac{1}{2} & -\frac{\sqrt{3}}{2} \\ \frac{\sqrt{3}}{2} & \frac{1}{2} \end{pmatrix}$ **b** $\begin{pmatrix} 0.6 & -0.8 \\ 0.8 & 0.6 \end{pmatrix}$ **c** $\begin{pmatrix} -\frac{\sqrt{2}}{2} & -\frac{\sqrt{2}}{2} \\ \frac{\sqrt{2}}{2} & -\frac{\sqrt{2}}{2} \end{pmatrix}$ **d** $\begin{pmatrix} -\frac{3}{5} & \frac{4}{5} \\ -\frac{4}{5} & -\frac{3}{5} \end{pmatrix}$

2 Find the equation of the line of symmetry for the reflection represented by each matrix.

a $\begin{pmatrix} \frac{1}{2} & \frac{\sqrt{3}}{2} \\ \frac{\sqrt{3}}{2} & -\frac{1}{2} \end{pmatrix}$ **b** $\begin{pmatrix} 0 & 1 \\ 1 & 0 \end{pmatrix}$ **c** $\begin{pmatrix} \frac{3}{4} & \frac{\sqrt{7}}{4} \\ \frac{\sqrt{7}}{4} & -\frac{3}{4} \end{pmatrix}$ **d** $\begin{pmatrix} \frac{\sqrt{3}}{2} & -\frac{1}{2} \\ -\frac{1}{2} & -\frac{\sqrt{3}}{2} \end{pmatrix}$

3 Describe each stretch.

a $\begin{pmatrix} 6 & 0 \\ 0 & 1 \end{pmatrix}$ **b** $\begin{pmatrix} 1 & 0 \\ 0 & 2 \end{pmatrix}$ **c** $\begin{pmatrix} -3 & 0 \\ 0 & 1 \end{pmatrix}$ **d** $\begin{pmatrix} -\frac{3}{2} & 0 \\ 0 & 1 \end{pmatrix}$

4 Write down a matrix to represent each transformation in two dimensions.

a Rotation through 60° about the origin

b Stretch with scale factor 3 parallel to the x-axis

c Reflection in the y-axis

d Stretch with scale factor $\frac{1}{2}$ parallel to the y-axis

e Reflection in the line $y = -2x$

f Rotation through 150° clockwise about the origin

g Enlargement with scale factor 5 centred at the origin

5 Find the image of the point $(3, -1)$ under each transformation.

a Reflection in the line $y = -x$

b Rotation through 45° clockwise about the origin

c Reflection in the line $y = \frac{1}{2}x$

d Stretch with scale factor 4 parallel to the y-axis

e Rotation through 120° about the origin

f Enlargement with scale factor $\frac{1}{3}$ centred at the origin

g Stretch with scale factor $\frac{1}{3}$ parallel to the x-axis

6 For each transformation, find the point that has image $(-1, 2)$.

a Reflection in the line $y = x$

b Rotation through 90° about the origin

c Reflection in the line $y = -\sqrt{3}x$

d Enlargement with scale factor 2 with the centre at the origin

e Stretch with scale factor 2 parallel to the y-axis

f Rotation through 60° about the origin

g Stretch with scale factor $\frac{1}{2}$ parallel to the x-axis

7 Find the image of the point $(2, -3)$ under a rotation through 135° about the origin.

8 The vertices of a rectangle are $A(-2, 1)$, $B(3, 1)$, $C(3, -2)$ and $D(-2, -2)$. $PQRS$ is the image of $ABCD$ under the stretch with scale factor 3 parallel to the x-axis.

a Find the coordinates of P, Q, R and S.

b Find the ratio (area of $PQRS$) : (area of $ABCD$).

9 a Find a 2×2 matrix representing a reflection in the line $y = 3x$.

b Find the coordinates of the point that has image $(2, 2)$ after reflection.

10 a Write down the 2×2 matrix representing a stretch with scale factor 1.5 parallel to the y-axis.

b A stretch with scale factor 1.5 parallel to the y-axis is applied to a triangle. The image has vertices with coordinates $(0, 0)$, $(1, 3)$ and $(-3, 1)$. Find the coordinates of the vertices of the original triangle.

Section 3: Invariant points and invariant lines

You already know that every linear transformation maps the origin to itself; you say that O is an **invariant point**. Some transformations, such as rotations, have no other invariant points: all other points are moved by the transformation. In this section, you will learn how to determine whether a transformation has any invariant points.

Consider a reflection in the line $l_1 : y = 2x$; it is represented by matrix $\mathbf{R} = \begin{pmatrix} -0.6 & 0.8 \\ 0.8 & 0.6 \end{pmatrix}$.

If a point with position vector $\begin{pmatrix} u \\ v \end{pmatrix}$ is invariant, then its image will have the same position vector:

$$\mathbf{R}\begin{pmatrix} u \\ v \end{pmatrix} = \begin{pmatrix} 0.8v - 0.6u \\ 0.6v + 0.8u \end{pmatrix} = \begin{pmatrix} u \\ v \end{pmatrix}$$

This is a system of linear equations:

$$\begin{cases} 0.8v - 0.6u = u \\ 0.6v + 0.8u = v \end{cases} \Leftrightarrow \begin{cases} 0.8v - 1.6u = 0 \\ -0.4v + 0.8u = 0 \end{cases}$$

You can see that the equations are the same, and equivalent to $v = 2u$. This means that every invariant point must lie on the line $y = 2x$. This makes sense: the points on the reflection line don't move and all the other points do. You say that $y = 2x$ is a **line of invariant points**.

Key point 8.6

An invariant point is any point that is unaffected by the transformation, which means that the image of the point is the point itself: $\mathbf{A}\begin{pmatrix} u \\ v \end{pmatrix} = \begin{pmatrix} u \\ v \end{pmatrix}$.

There may be infinitely many invariant points, forming a line of invariant points. In that case, this equation will have infinitely many solutions of the form $v = ku$.

WORKED EXAMPLE 8.8

Determine whether $\mathbf{B} = \begin{pmatrix} 3 & -2 \\ 2 & -1 \end{pmatrix}$ has any lines of invariant points.

You are looking for points (u, v) such that:

$$\begin{pmatrix} 3 & -2 \\ 2 & -1 \end{pmatrix}\begin{pmatrix} u \\ v \end{pmatrix} = \begin{pmatrix} u \\ v \end{pmatrix}$$

Invariant points satisfy $\mathbf{A}\begin{pmatrix} u \\ v \end{pmatrix} = \begin{pmatrix} u \\ v \end{pmatrix}$.

$$\Rightarrow \begin{cases} 3u - 2v = u \\ 2u - v = v \end{cases}$$

Rewrite this as a system of equations.

$$\Rightarrow \begin{cases} 2u - 2v = 0 \\ 2u - 2v = 0 \end{cases}$$

Continues on next page

Both equations are equivalent to $u - v = 0$ or $v = u$.

Check whether both equations are, in fact, the same.

Hence $y = x$ is the line of invariant points.

If so, they give the line of invariant points.

WORKED EXAMPLE 8.9

Show that the transformation represented by the matrix $\mathbf{M} = \begin{pmatrix} -4 & 1 \\ 3 & -2 \end{pmatrix}$ has no invariant points other than the origin.

Invariant points satisfy:

$$\begin{pmatrix} -4 & 1 \\ 3 & -2 \end{pmatrix}\begin{pmatrix} u \\ v \end{pmatrix} = \begin{pmatrix} u \\ v \end{pmatrix}$$

$$\Rightarrow \begin{cases} -4u + v = u \\ 3u - 2v = v \end{cases}$$

Write the condition for invariant points and turn it into simultaneous equations.

The only solution is $u = 0, v = 0$.

Solve the equations.

Hence the only invariant point is $(0, 0)$.

Consider again the reflection in the line $y = 2x$. All the points on the reflection line are invariant. Any other point moves under the reflection. However, it stays on the same line perpendicular to the reflection line. This means that this perpendicular line, considered as a whole, is invariant.

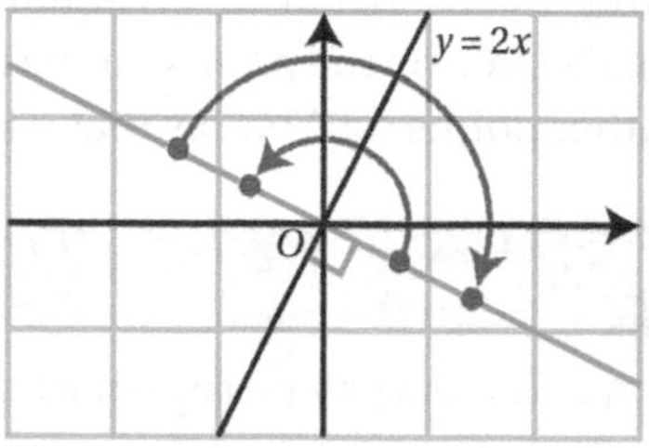

Key point 8.7

An invariant line l is a line for which the image of any point on l is also on l (though is not necessarily the same point).

Any line of invariant points is an invariant line. However, the opposite is not true; there can be an invariant line that contains non-invariant points (for example, any line perpendicular to the line of reflection).

WORKED EXAMPLE 8.10

Show that $y = -\frac{1}{2}x + 3$ is an invariant line for the reflection in $y = 2x$.

Any point on the line $y = -\frac{1}{2}x + 3$ has coordinates of the form $\left(u, -\frac{1}{2}u + 3\right)$.

You are going to find the image of a general point on the line $y = -\frac{1}{2}x + 3$.

Continues on next page

Its image is:

You already know the matrix representing the reflection in $y = 2x$.

$$\begin{pmatrix} u' \\ v' \end{pmatrix} = \begin{pmatrix} -0.6 & 0.8 \\ 0.8 & 0.6 \end{pmatrix}\begin{pmatrix} u \\ -\frac{1}{2}u + 3 \end{pmatrix}$$

$$= \begin{pmatrix} -0.6u - 0.4u + 2.4 \\ 0.8u - 0.3u + 1.8 \end{pmatrix}$$

$$= \begin{pmatrix} -u + 2.4 \\ 0.5u + 1.8 \end{pmatrix}$$

Then:

To check whether this image lies on the same line, you need to check that $v' = -\frac{1}{2}u' + 3$.

$$-\frac{1}{2}u' + 3 = -\frac{1}{2}(-u + 2.4) + 3$$
$$= 0.5u + 1.8$$
$$= v'$$

Hence (u', v') also lies on the line $y = -\frac{1}{2}x + 3$, and so this line is invariant.

WORKED EXAMPLE 8.11

Show that an enlargement has an infinite number of invariant lines through the origin.

The matrix for an enlargement is $\begin{pmatrix} k & 0 \\ 0 & k \end{pmatrix}$.

Invariant lines: $v = mu$

Lines through the origin have the form $y = mx$.

$$\begin{pmatrix} u' \\ v' \end{pmatrix} = \begin{pmatrix} k & 0 \\ 0 & k \end{pmatrix}\begin{pmatrix} u \\ mu \end{pmatrix}$$

Write an equation for the image of any point on the line.

$$\Rightarrow \begin{cases} u' = ku \\ v' = kmu \end{cases}$$

Hence $v' = mu'$ for all m, so every line through the origin is invariant.

If line is invariant, then the coordinates of the image also satisfy $y = mx$.

WORKED EXAMPLE 8.12

A transformation is represented by the matrix $\mathbf{A} = \begin{pmatrix} -4 & 1 \\ 3 & -2 \end{pmatrix}$.

a Find the equations of invariant lines through the origin.

b Show that there are no lines of invariant points.

Continues on next page

a The image of (u, mu):

Any line through the origin has the form $y = mx$.

$$\begin{pmatrix} u' \\ v' \end{pmatrix} = \begin{pmatrix} -4 & 1 \\ 3 & -2 \end{pmatrix}\begin{pmatrix} u \\ mu \end{pmatrix} = \begin{pmatrix} -4u + mu \\ 3u - 2mu \end{pmatrix}$$

For an invariant line:

If the line is invariant, the image also lies on the line.

$$3u - 2mu = m(-4u + mu)$$

If the whole line is invariant, then this is true for all u.

$$u(m^2 - 2m - 3) = 0$$

$$m^2 - 2m - 3 = 0$$

$$m = -1 \text{ or } m = 3$$

The invariant lines are $y = -x$ and $y = 3x$.

b For an invariant point:

If a point is invariant, then $(u', v') = (u, v)$.

$$\begin{pmatrix} u' \\ v' \end{pmatrix} = \begin{pmatrix} -4 & 1 \\ 3 & -2 \end{pmatrix}\begin{pmatrix} u \\ v \end{pmatrix} = \begin{pmatrix} u \\ v \end{pmatrix}$$

Solve to find u, v.

$$\Rightarrow \begin{cases} -4u + v = u \\ 3u - 2v = v \end{cases}$$

$$\Rightarrow \begin{cases} -5u + v = 0 \\ 3u - 3v = 0 \end{cases}$$

$$\Rightarrow u = 0, v = 0$$

Hence the origin is the only invariant point.

EXERCISE 8C

1 For each matrix, find any lines of invariant points and any other invariant lines through the origin.

a $\begin{pmatrix} 1 & 1 \\ 0 & 4 \end{pmatrix}$ **b** $\begin{pmatrix} 2 & 5 \\ 0 & 4 \end{pmatrix}$ **c** $\begin{pmatrix} 1 & 4 \\ 2 & -1 \end{pmatrix}$ **d** $\begin{pmatrix} 3 & -2 \\ 4 & -3 \end{pmatrix}$

2 **R** is a reflection through the line $y = 2x$.

S is a 90° rotation anticlockwise about the origin.

T is a stretch, scale factor 3, parallel to the y-axis.

In each case, find any lines of invariant points and any other invariant lines through the origin.

a **R** followed by **S** **b** $\mathbf{S}^{-1}\mathbf{RS}$ **c** **TR**

d **RT** **e** **TS** **f** $\mathbf{S}^{-1}\mathbf{TS}$

3 Transformation **A** is given by $\mathbf{A} = \begin{pmatrix} 3 & -1 \\ 6 & a \end{pmatrix}$.

a Given that $y = 2x$ is an invariant line under transformation **A**, find the value of a.

b For this value a, determine whether there is a second invariant line.

4 Transformations **A** and **B** are given by $\mathbf{A} = \begin{pmatrix} 3 & 1 \\ 2 & 4 \end{pmatrix}$, $\mathbf{B} = \begin{pmatrix} 5 & -3 \\ -1 & 7 \end{pmatrix}$.

a Show that they have a common invariant line.

b Find any invariant lines of the product **AB**.

5 Transformation **A** is given by $\mathbf{A} = \begin{pmatrix} -3 & -2 \\ 9 & 8 \end{pmatrix}$.

a Find any invariant lines of **A** and determine whether there is a line of invariant points.

Transformation $\mathbf{B} = \mathbf{A}^2$.

b Write down the equations of any invariant lines of **B**.

6 **R** is a reflection in the line $y = 2x$.

S is a clockwise rotation about the origin by angle θ.

Find a condition on θ for which $y = x$ is an invariant line of $\mathbf{S}^{-1}\mathbf{RS}$.

7 Transformation **A** is given by $\mathbf{A} = \begin{pmatrix} 3 & 1 \\ 0 & 1 \end{pmatrix}$.

a Find any invariant lines of **A** and determine whether any are lines of invariant points.

Transformation **B** is given by $\mathbf{B} = \begin{pmatrix} -1 & b \\ 7 & -4 \end{pmatrix}$.

b Given that $y = 7x$ is an invariant line of the product **AB**, find the value of b.

8 Matrix $\mathbf{T} = \begin{pmatrix} 3 & p & -1 \\ 1 & 2 & q \\ 0 & -1 & -1 \end{pmatrix}$ represents a three-dimensional linear transformation, for which the line $\frac{x}{2} = \frac{y}{3} = -z$ is invariant.

a By considering the product $\mathbf{T}\begin{pmatrix} 2 \\ 3 \\ -1 \end{pmatrix}$, find the values of p and q.

b For these values of p and q:

i find $\mathbf{T}^{-1}$ or show that **T** is singular

ii describe the image of the line $x = 0$, $y = -z$.

9 For this question, two distinct lines are said to be **twinned** under a transformation if each line is the image of the other.

a Under transformation $\mathbf{T} = \begin{pmatrix} 4 & 1 \\ a & b \end{pmatrix}$, line $y = -2x$ is twinned with $y = 3x$. Calculate the values a and b.

b For transformation $\mathbf{S} = \begin{pmatrix} 1 & 2 \\ -4 & -1 \end{pmatrix}$, find the line that is twinned with $y = 3x$.

Section 4: Transformations in 3D

When you move from the plane to three-dimensional space, some of the basic transformations need to be described in a different way. For example, while in two dimensions a rotation is about a point, in three dimensions it is about an **axis of rotation**. Similarly, you can't reflect a three-dimensional object in a line, but you can use a **plane of reflection**.

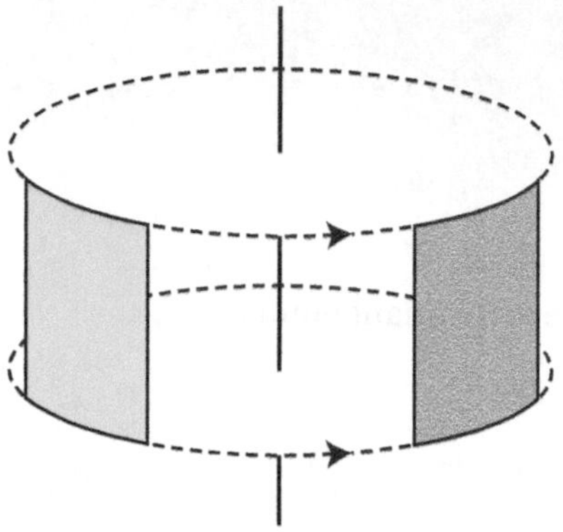

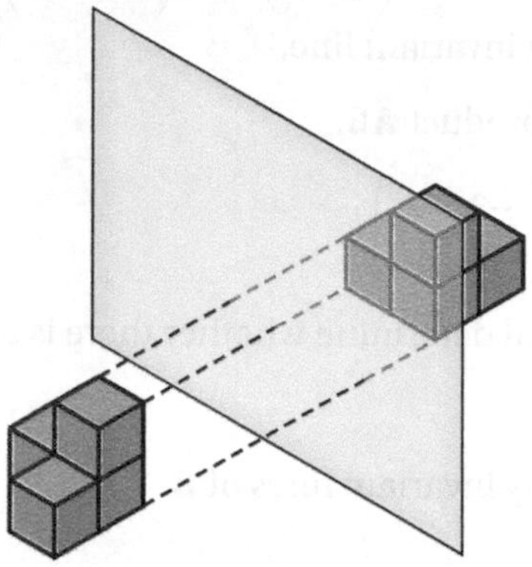

In two dimensions, it was useful to consider the image of the unit square. In three dimensions, you can look at the **unit cube**. This is the cube with edge length one unit, with one vertex at the origin and edges parallel to the coordinate axes. Three of the edges are given by the unit vectors,

$$\mathbf{i}=\begin{pmatrix}1\\0\\0\end{pmatrix}, \mathbf{j}=\begin{pmatrix}0\\1\\0\end{pmatrix} \text{ and } \mathbf{k}=\begin{pmatrix}0\\0\\1\end{pmatrix}.$$

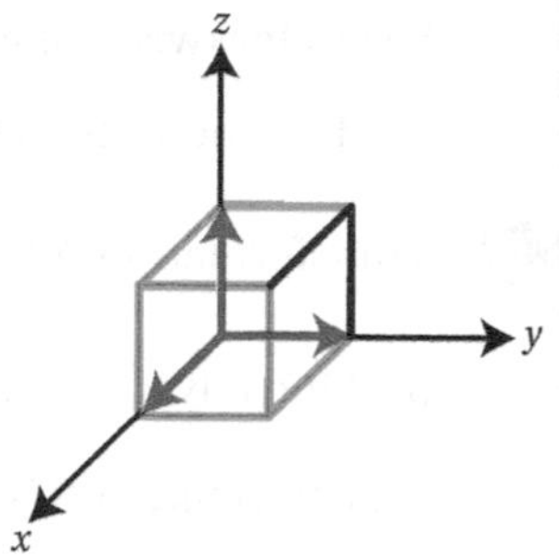

The three columns of a transformation matrix give the image positions for these three unit vectors, and so outline the three edges of the image of the unit cube.

In each of the two types of transformation you will consider, at least one of the unit direction vectors will be unaffected. By fixing the row and column of that direction vector in the matrix, you can use your understanding of 2D transformations to define the 3D matrix.

Reflections

In this course, you only need to consider three possible planes of reflection: the x–y plane, the y–z plane and the x–z plane. The x–y plane contains the x and y axes. Every point in this plane has z-coordinate equal to zero, so this plane can also be described as the $z = 0$ plane. The other two planes are the $x = 0$ and the $y = 0$ planes.

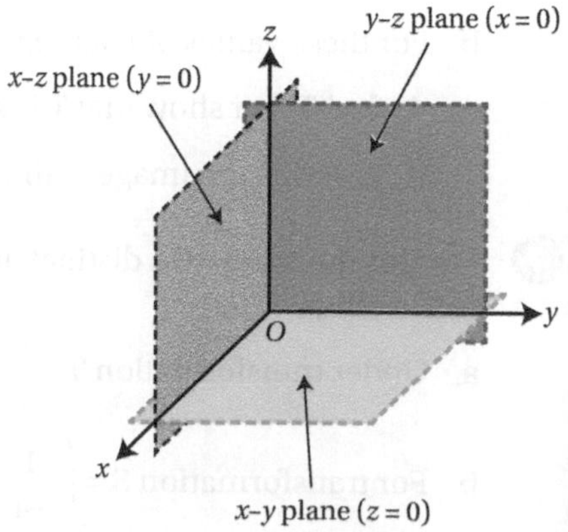

WORKED EXAMPLE 8.13

Transformation **R** represents a reflection in the plane $x = 0$.

Write down the 3×3 matrix for **R**.

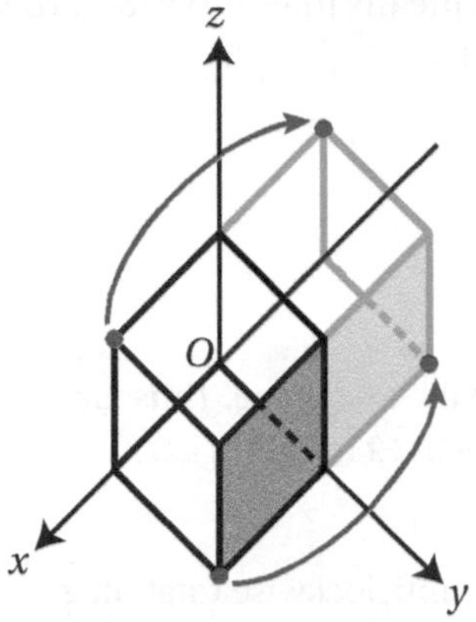

Sketch the unit cube being reflected in the plane ($x = 0$ is the y–z plane).

Images of unit vectors:

$$\begin{pmatrix} 1 \\ 0 \\ 0 \end{pmatrix} \mapsto \begin{pmatrix} -1 \\ 0 \\ 0 \end{pmatrix}$$

$$\begin{pmatrix} 0 \\ 1 \\ 0 \end{pmatrix} \mapsto \begin{pmatrix} 0 \\ 1 \\ 0 \end{pmatrix}$$

$$\begin{pmatrix} 0 \\ 0 \\ 1 \end{pmatrix} \mapsto \begin{pmatrix} 0 \\ 0 \\ 1 \end{pmatrix}$$

The vector i has its direction reversed.

The unit vectors **j** and **k** lie in the plane of reflection, so are unaffected by the reflection.

Hence:

$$\mathbf{R} = \begin{pmatrix} -1 & 0 & 0 \\ 0 & 1 & 0 \\ 0 & 0 & 1 \end{pmatrix}$$

The images of the unit vectors give the columns of the transformation matrix.

Rotations

You only need to work with rotations about one of the coordinate axes. The axis of rotation is unaffected by the transformation, so one of the columns in the transformation matrix will be just the unit vector. To find the rest of the matrix, you need to think about the other two unit vectors being rotated in two dimensions.

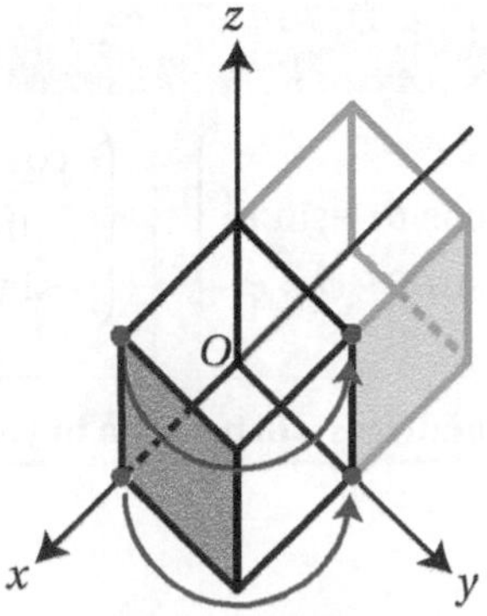

WORKED EXAMPLE 8.14

Transformation S represents the rotation 60° about the y-axis. Find the 3 × 3 matrix for S.

The axis of rotation is along direction **j**, so

$$\begin{pmatrix} 0 \\ 1 \\ 0 \end{pmatrix} \mapsto \begin{pmatrix} 0 \\ 1 \\ 0 \end{pmatrix}$$

$$\mathbf{S} = \begin{pmatrix} ? & 0 & ? \\ 0 & 1 & 0 \\ ? & 0 & ? \end{pmatrix}$$

The unit vector **j** lies along the axis of rotation so is unaffected by the rotation. This means that the second row and column are as shown in red.

A 60° rotation about the origin in the $x-z$ plane has 2D matrix:

$$\begin{pmatrix} \frac{1}{2} & \frac{\sqrt{3}}{2} \\ -\frac{\sqrt{3}}{2} & \frac{1}{2} \end{pmatrix}$$

Filling in the gaps with this rotation matrix gives:

$$\mathbf{S} = \begin{pmatrix} \frac{1}{2} & 0 & \frac{\sqrt{3}}{2} \\ 0 & 1 & 0 \\ -\frac{\sqrt{3}}{2} & 0 & \frac{1}{2} \end{pmatrix}$$

The **i** and **k** vectors are rotated in the x–z plane. Thus the remaining four elements form the 2 × 2 matrix of a 60° rotation.

You need to be a little careful: the anticlockwise rotation goes from z towards x, so the image of the unit vector **k** has both components positive, while the image of **i** has one negative component. This means that the negative entry (from − sin 60°) is in the first column rather than the third.

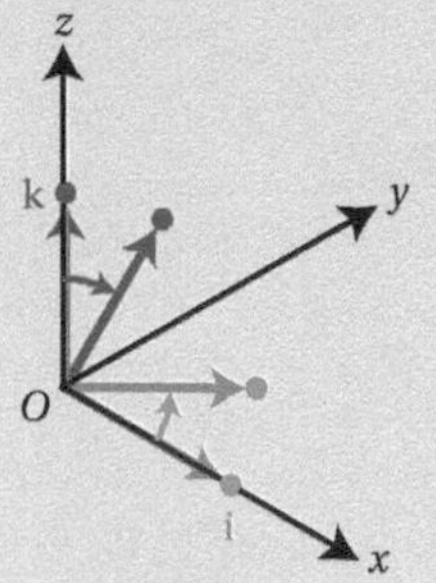

The three 3D rotation matrices you need to know are summarised in Key point 8.8.

Key point 8.8

Rotation about the given axis

x-axis	y-axis	z-axis
$\begin{pmatrix} 1 & 0 & 0 \\ 0 & \cos\theta & -\sin\theta \\ 0 & \sin\theta & \cos\theta \end{pmatrix}$	$\begin{pmatrix} \cos\theta & 0 & \sin\theta \\ 0 & 1 & 0 \\ -\sin\theta & 0 & \cos\theta \end{pmatrix}$	$\begin{pmatrix} \cos\theta & -\sin\theta & 0 \\ \sin\theta & \cos\theta & 0 \\ 0 & 0 & 1 \end{pmatrix}$

These matrices will be given in your formula book.

Tip

Remember that a positive angle means the rotation is anticlockwise.

The three 3D reflection matrices you need to know are summarised in Key point 8.9.

Key point 8.9

Reflection in a given plane.

$x = 0$	$y = 0$	$z = 0$
$\begin{pmatrix} -1 & 0 & 0 \\ 0 & 1 & 0 \\ 0 & 0 & 1 \end{pmatrix}$	$\begin{pmatrix} 1 & 0 & 0 \\ 0 & -1 & 0 \\ 0 & 0 & 1 \end{pmatrix}$	$\begin{pmatrix} 1 & 0 & 0 \\ 0 & 1 & 0 \\ 0 & 0 & -1 \end{pmatrix}$

Other transformations

As in two dimensions, combining transformations corresponds to multiplying their matrices.

WORKED EXAMPLE 8.15

Find the matrix of the resulting transformation when a rotation through 90° about the x-axis is followed by a rotation through 90° about the y-axis.

Let $\mathbf{A}$ represent the rotation about the x-axis and $\mathbf{B}$ the rotation about the y-axis. Then:

Start by writing down the two standard rotation matrices.

$$\mathbf{A} = \begin{pmatrix} 1 & 0 & 0 \\ 0 & 0 & -1 \\ 0 & 1 & 0 \end{pmatrix} \text{ and } \mathbf{B} = \begin{pmatrix} 0 & 0 & 1 \\ 0 & 1 & 0 \\ -1 & 0 & 0 \end{pmatrix}$$

The combined transformation is:

The combined transformation corresponds to the matrix product.

$$\mathbf{C} = \mathbf{BA}$$

$$= \begin{pmatrix} 0 & 0 & 1 \\ 0 & 1 & 0 \\ -1 & 0 & 0 \end{pmatrix}\begin{pmatrix} 1 & 0 & 0 \\ 0 & 0 & -1 \\ 0 & 1 & 0 \end{pmatrix}$$

$$= \begin{pmatrix} 0 & 1 & 0 \\ 0 & 0 & -1 \\ -1 & 0 & 0 \end{pmatrix}$$

This doesn't seem to be any of the three standard rotations, and you don't need to be able to interpret such a compound transformation. However, in some simple cases, you can recognise the transformation by considering the images of the three unit vectors.

WORKED EXAMPLE 8.16

Describe transformation **T** given by $\begin{pmatrix} 1 & 0 & 0 \\ 0 & 1 & 0 \\ 0 & 0 & 4 \end{pmatrix}$.

Vector **i** is unaffected by **T**:

$$\mathbf{T} = \begin{pmatrix} 1 & 0 & 0 \\ 0 & 1 & 0 \\ 0 & 0 & 4 \end{pmatrix}$$

Find which of the first, second or third row and column match the identity matrix; reduce the matrix to a 2×2 and identify the transformation in the plane.

The transformation in the y–z plane shows that **T** is a stretch parallel to the z-axis, scale factor 4.

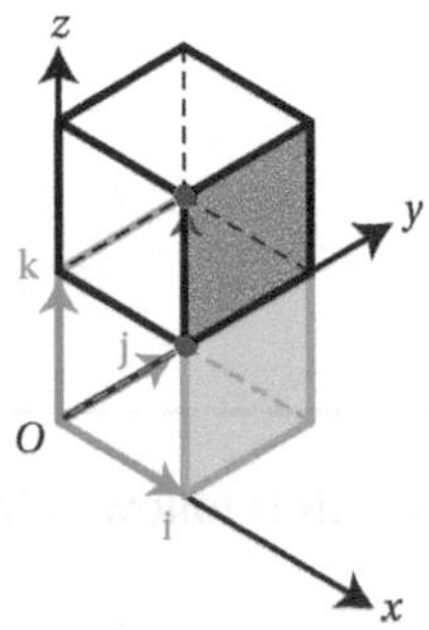

Looking at the images of the unit vectors confirms that this is a stretch parallel to the z-axis.

EXERCISE 8D

1 Write down a matrix to represent each transformation in three dimensions.

- **a** Rotation through 90° about the x-axis
- **b** Rotation through 30° about the z-axis
- **c** Reflection in the plane $x = 0$
- **d** Reflection in the plane $y = 0$

2 Find the image of the point $(2, -1, 1)$ under each transformation in three dimensions.

- **a** Rotation through 60° about the y-axis
- **b** Rotation through 180° about the z-axis
- **c** Reflection in the plane $z = 0$
- **d** Reflection in the plane $y = 0$

3 Point A is reflected in the plane $x = 0$. The coordinates of the image are $(2, -5, 1)$. Find the coordinates of A.

4 Transformation **T** is a reflection in the plane $y = 0$. Transformation **S** is a rotation through 180° about the x-axis. Transformation **M** is **T** followed by **S**.

- **a** Write down the matrices representing **T** and **S**.
- **b** Find the matrix representing **M**. Hence describe the transformation **M**.

5 **R** represents a rotation through 90° clockwise about the x-axis. Triangle ABC is given by $A(1,1,1)$, $B(1,3,1)$, $C(1,1,4)$.

Find the image of triangle ABC under transformation **R**.

6 $(4,2,-1)$ are the coordinates of the image of point A after a rotation through 135° about the z-axis. Find the coordinates of the object point A.

7 Find the coordinates of the image when the point $(-3,1,1)$ is rotated 30° clockwise about the y-axis.

8 A rotation through 90° about the z-axis is followed by the reflection in the plane $x = 0$. Find the matrix representing the resulting transformation.

9 A rotation through 90° about the x-axis is followed by a rotation through 90° clockwise about the y-axis. Find the matrix representing the resulting transformation.

10 Point P is rotated through 90° about the z-axis and then reflected in the plane $y = 0$. The coordinates of the image are $(3,-1,4)$. Find the coordinates of P.

11 Transformation **T** represents a 90° rotation about the x-axis followed by a 90° rotation about the y-axis.

a Find matrix **T**.

b Show that $\mathbf{T}^3 = \mathbf{I}$.

Checklist of learning and understanding

- A matrix can represent a linear transformation: the image of a point with position vector $\begin{pmatrix} u \\ v \end{pmatrix}$ is $\mathbf{A}\begin{pmatrix} u \\ v \end{pmatrix}$.
- The first column of a transformation matrix is the image of $\begin{pmatrix} 1 \\ 0 \end{pmatrix}$ and the second column is the image of $\begin{pmatrix} 0 \\ 1 \end{pmatrix}$.
- For two transformations **A** and **B**, if **C** represents **A** followed by **B**, then transformation **C** is given by $\mathbf{C} = \mathbf{BA}$.
- If a linear transformation is represented by a matrix **A**, then its inverse transformation is represented by the matrix $\mathbf{A}^{-1}$.
- Two-dimensional linear transformations include:
  - rotation about the origin
  - reflection through line $y = mx$
  - stretch parallel to the x-axis or y-axis
  - enlargement centred at the origin.
- Three-dimensional linear transformations include:
  - rotation about the x-axis, y-axis or z-axis
  - reflection through the plane $x = 0$, $y = 0$ or $z = 0$.
- An invariant line is one whose image coincides with the original line.
- An invariant point is one whose image is the same point.
  - The origin is always an invariant point.
  - Invariant points may form a line of invariant points.

Mixed practice 8

1 A reflection is represented by the matrix $\begin{pmatrix} -1 & 0 \\ 0 & 1 \end{pmatrix}$.

State the equation of the line of invariant points. Choose from these options.

A $x = 0$ **B** $y = 0$ **C** $y = x$ **D** $y = -x$

2 **R** is a rotation clockwise through 120° about the z-axis.

S is the matrix $\frac{1}{2}\begin{pmatrix} -1 & -\sqrt{3} & 0 \\ \sqrt{3} & -1 & 0 \\ 0 & 0 & -2 \end{pmatrix}$.

a $\mathbf{T} = \mathbf{RS}$. Calculate matrix **T**.

b Describe the transformation represented by **T**.

3 For matrices **A** and **B**, find any invariant lines of the form $y = mx$ and determine whether any are lines of invariant points.

a $\mathbf{A} = \begin{pmatrix} 4 & 1 \\ -27 & -8 \end{pmatrix}$ **b** $\mathbf{B} = \begin{pmatrix} 4 & 1 \\ -1 & 2 \end{pmatrix}$

4 **a** Describe the single transformation given by each matrix.

i $\mathbf{A} = \begin{pmatrix} -0.5 & 0 \\ 0 & -0.5 \end{pmatrix}$ **ii** $\mathbf{B} = \begin{pmatrix} 1 & 0 \\ 0 & 4 \end{pmatrix}$

b Write down the 3×3 matrix describing a rotation through 60° about the x-axis.

5 The matrix $\begin{pmatrix} 1 & 0 & 0 \\ 0 & -0.6 & -0.8 \\ 0 & 0.8 & -0.6 \end{pmatrix}$ represents a rotation.

a State the axis of rotation.

b Find the angle of rotation, giving your answer to the nearest degree.

[©AQA 2014]

6 Find the transformation represented by the matrix $\frac{1}{5}\begin{pmatrix} 3 & 4 \\ 4 & -3 \end{pmatrix}$.
Choose from these options.

A Rotation anticlockwise about O

B Rotation anticlockwise about O

C Reflection in the line $y = 2x$

D Reflection in the line $y = \frac{1}{2}x$

7 Transformation **A** is a 90° rotation clockwise about the origin and transformation **B** is a stretch, scale factor 3, parallel to the y-axis.

a Write down the 2×2 matrices for **A** and **B**.

b Matrix **C** is such that $\mathbf{BC} = \mathbf{AC}$. Find det **C**.

8 The matrices **A** and **B** are defined by:

$$\mathbf{A} = \begin{pmatrix} \frac{1}{\sqrt{2}} & -\frac{1}{\sqrt{2}} \\ \frac{1}{\sqrt{2}} & \frac{1}{\sqrt{2}} \end{pmatrix}, \mathbf{B} = \begin{pmatrix} \frac{1}{\sqrt{2}} & \frac{1}{\sqrt{2}} \\ \frac{1}{\sqrt{2}} & -\frac{1}{\sqrt{2}} \end{pmatrix}$$

Describe fully the geometrical transformation represented by each of the following matrices.

a **A** **b** **B** **c** $\mathbf{A}^2$ **d** $\mathbf{B}^2$ **e** **AB**

[©AQA 2010]

9 The transformation represented by matrix $\mathbf{A} = \begin{pmatrix} -a & a^2 \\ -2 & 2 \end{pmatrix}$ has line of invariant points $y = bx$.

Find a and b.

10 Under transformation **T**, the images of points $P(1, 3)$ and $Q(-2, 5)$ are $P'(3, 6)$ and $Q'(-6, 10)$.

a By considering $\mathbf{T}\begin{pmatrix} 1 & -2 \\ 3 & 5 \end{pmatrix}$ or otherwise, calculate the matrix representing **T**.

b **T** is the result of an enlargement, scale factor $k > 0$, and a transformation with matrix **S** where $\det \mathbf{S} = 1.5$.

i Find k.

ii Describe the transformation **S**.

11 **a** Write down the 2×2 matrix corresponding to each transformation:

i a reflection in the line $y = -x$

ii a stretch parallel to the y-axis of scale factor 7.

b Hence find the matrix corresponding to the combined transformation of a reflection in the line $y = -x$ followed by a stretch parallel to the y-axis of scale factor 7.

c The matrix **A** is defined by $\begin{pmatrix} -3 & -\sqrt{3} \\ -\sqrt{3} & 3 \end{pmatrix}$.

i Show that $\mathbf{A}^2 = k\mathbf{I}$, where k is a constant and **I** is the 2×2 identity matrix.

ii Show that the matrix **A** corresponds to a combination of an enlargement and a reflection. State the scale factor of the enlargement and state the equation of the line of reflection in the form $y = (\tan \theta)x$.

[©AQA 2014]

12 Transformation **A** is given by matrix $\mathbf{A} = \begin{pmatrix} -2 & 6 \\ 3 & -5 \end{pmatrix}$.

a A rectangle $OPQR$ with $P = (2, 1)$ and $R(-2, 4)$ is transformed by **A**.

i Write down the coordinates of Q.

ii Find the image $OP'Q'R'$ of $OPQR$ under transformation **A**.

b **i** Write down the equation of the line of invariant points for **A**.

iii Determine whether there is a second invariant line for transformation **A** and, if so, find its equation.

13 Matrices **A** and **B** are given by:

$$\mathbf{A} = \begin{pmatrix} 8-3k & -1 \\ 12-9k & 1 \end{pmatrix}, \mathbf{B} = \begin{pmatrix} 9-2k & -2 \\ 0 & 3-2k \end{pmatrix}$$

a Find, in terms of k, the invariant lines of **A**.

b Given that $\det \mathbf{A} = \det \mathbf{B} + 1$:

i find the possible values of k

ii show that **A** and **B** have a common invariant line

iii for each value of k, find the equations of any other invariant lines of **A**.

14 Matrices **A** and **B** are given by $\mathbf{A} = \begin{pmatrix} -1 & 2 \\ 0 & a \end{pmatrix}$ and $\mathbf{B} = \begin{pmatrix} -2 & 2 \\ b & -2 \end{pmatrix}$.

a Given **A** and **B** commute under multiplication, find a and b.

b For a transformation given by matrix **T**, where $\det \mathbf{T} = 1$, $\mathbf{AB} = k\mathbf{T}$:

i find the matrix **T**, and state the equation of the invariant line

ii describe the second linear transformation that comprises **AB**.

15 a The matrix **A** is defined by $\mathbf{A} = \begin{pmatrix} -1 & s \\ t & 2 \end{pmatrix}$.

Given that the image of the point $(2, 1)$ under the transformation represented by **A** is $(1, 10)$, find the value of s and the value of t.

b The matrix **B** is defined by $\mathbf{B} = \begin{pmatrix} \sqrt{2} & -\sqrt{2} \\ \sqrt{2} & \sqrt{2} \end{pmatrix}$.

i Show that $\mathbf{B}^4 = k\mathbf{I}$, where k is an integer and **I** is the 2×2 identity matrix.

ii Describe the transformation represented by the matrix **B** as a combination of two geometrical transformations.

iii Find the matrix $\mathbf{B}^{15}$.

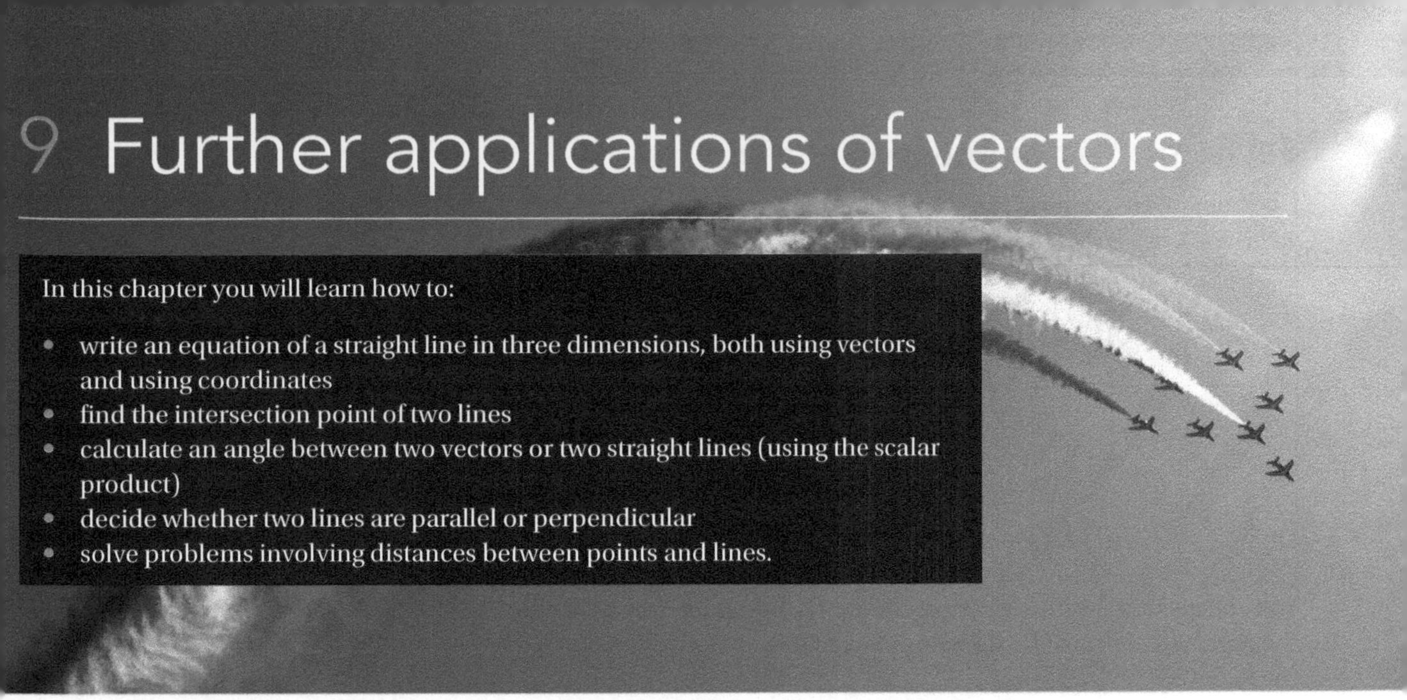

9 Further applications of vectors

In this chapter you will learn how to:

- write an equation of a straight line in three dimensions, both using vectors and using coordinates
- find the intersection point of two lines
- calculate an angle between two vectors or two straight lines (using the scalar product)
- decide whether two lines are parallel or perpendicular
- solve problems involving distances between points and lines.

Before you start...

A Level Mathematics Student Book 1, Chapter 15	You should know how to describe vectors using components in two and three dimensions.	1 a Write $\begin{pmatrix} 3 \\ -5 \end{pmatrix}$ using **i** and **j** base vectors. b Write $3\mathbf{i} - 2\mathbf{k}$ as a three-dimensional column vector.
	You should recognise when two vectors are equal.	2 Find the values of a and b such that $\begin{pmatrix} 3a-2 \\ b+1 \end{pmatrix} = \begin{pmatrix} b-3 \\ 2a+1 \end{pmatrix}$.
	You should recognise when two vectors are parallel.	3 Find the values of p and q such that the vectors $2\mathbf{i} + p\mathbf{j} - \mathbf{k}$ and $q\mathbf{i} + 3\mathbf{j} + 5\mathbf{k}$ are parallel.
A Level Mathematics Student Book 1, Chapter 10	You should be able to solve trigonometric equations to find an unknown angle, working in both degrees and radians.	4 Find two possible values of $\theta \in [0°, 360°]$ such that $12 - 5\cos\theta = 14$.

Describing lines in three dimensions

Many situations involve lines in three-dimensional space: for example, representing flight paths of aeroplanes or describing the motion of a character in a computer game. In order to solve problems involving lines in space, you need a way of deciding whether a point lies on a given straight line.

Suppose you are given two points, A and B. These two points determine a unique straight line (a 'straight line' means a line extending indefinitely in both directions). How can you check whether a third point lies on the same line, or describe all points on this line? You can use vectors to answer this question.

Section 1: Vector equation of a line

Consider, as an example, the straight line passing through points $A(-1, 1, 4)$, $B(1, 4, 2)$ and another point $C(5, 10, -2)$. Then:

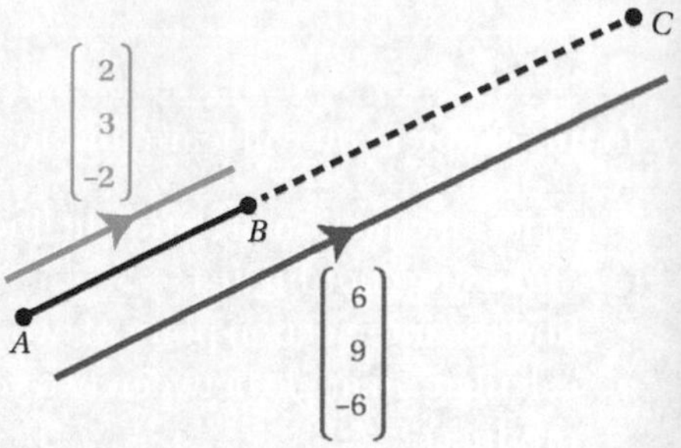

$$\overrightarrow{AC} = \begin{pmatrix} 5 \\ 10 \\ -2 \end{pmatrix} - \begin{pmatrix} -1 \\ 1 \\ 4 \end{pmatrix} = \begin{pmatrix} 6 \\ 9 \\ -6 \end{pmatrix}$$

$$\overrightarrow{AB} = \begin{pmatrix} 1 \\ 4 \\ 2 \end{pmatrix} - \begin{pmatrix} -1 \\ 1 \\ 4 \end{pmatrix} = \begin{pmatrix} 2 \\ 3 \\ -2 \end{pmatrix}$$

$$\therefore \overrightarrow{AC} = 3\overrightarrow{AB}$$

This means that the line AC is parallel to AB. But, since they both contain the point A, they must be the same straight line: in other words, C lies on the line AB.

Rewind

Remember that a scalar is just a number.

You can now characterise all the points on the line AB. Using the above idea, a point R lies on AB if AR and AB are parallel. But you know that this can be expressed using vectors by saying that $\overrightarrow{AR} = \lambda\overrightarrow{AB}$ for some value of the scalar λ.

Rewind

See A Level Mathematics Student Book 1, Chapter 15, for a reminder of vector algebra.

So $\overrightarrow{AR} = \begin{pmatrix} 2\lambda \\ 3\lambda \\ -2\lambda \end{pmatrix}$

But you also know that $\overrightarrow{AR} = \mathbf{r} - \mathbf{a}$, where $\mathbf{r}$ and $\mathbf{a}$ are the position vectors of R and A, respectively.

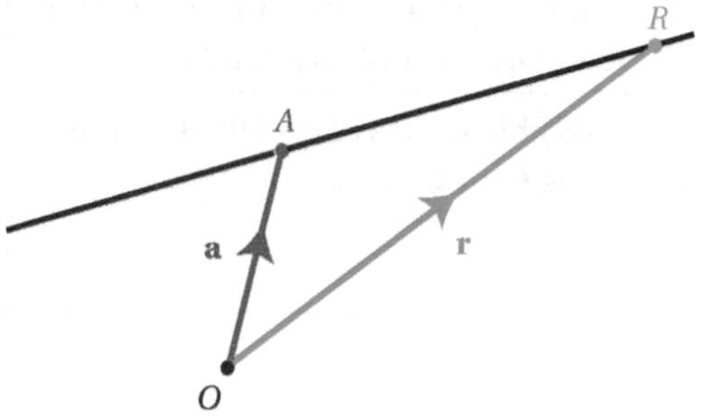

This means that $\mathbf{r} = \begin{pmatrix} -1 \\ 1 \\ 4 \end{pmatrix} + \begin{pmatrix} 2\lambda \\ 3\lambda \\ -2\lambda \end{pmatrix}$ is the position vector of a general point on the line AB. In other words, R has coordinates $(-1+2\lambda, 1+3\lambda, 4-2\lambda)$ for some value of λ. Different values of λ correspond to different points on the line; for example, $\lambda = 0$ corresponds to point A, $\lambda = 1$ to point B and $\lambda = 3$ to point C.

The line is parallel to the vector $\begin{pmatrix} 2 \\ 3 \\ -2 \end{pmatrix}$, so this vector determines the direction of the line. The expression for the position vector of $\mathbf{r}$ is usually written as $\mathbf{r} = \begin{pmatrix} -1 \\ 1 \\ 4 \end{pmatrix} + \lambda \begin{pmatrix} 2 \\ 3 \\ -2 \end{pmatrix}$, so it is easy to identify the direction vector.

Key point 9.1

The vector equation of a line is:

$$\mathbf{r} = \mathbf{a} + \lambda \mathbf{d}$$

where:

- $\mathbf{r}$ is the position vector of a general point on the line
- $\mathbf{d}$ is the **direction vector** of the line
- $\mathbf{a}$ is the position vector of any known point on the line.

Fast forward

In Further Mathematics Student Book 2 you will see that there is more than one possible form for the vector equation of a line.

WORKED EXAMPLE 9.1

Write down a vector equation of the line passing through the point (−1, 1, 2) in the direction of the vector $\begin{pmatrix} 2 \\ 2 \\ 1 \end{pmatrix}$.

$$\underline{r} = \begin{pmatrix} -1 \\ 1 \\ 2 \end{pmatrix} + \lambda \begin{pmatrix} 2 \\ 2 \\ 1 \end{pmatrix}$$

The equation of the line is $\mathbf{r} = \mathbf{a} + \lambda \mathbf{d}$, where $\mathbf{a}$ is the position vector of a point on the line and $\mathbf{d}$ is the direction vector.

You can also use the vector equation for lines in two dimensions.

For example, a line with direction vector $\begin{pmatrix} 1 \\ 3 \end{pmatrix}$ has gradient 3, because an increase of 1 unit in x produces an increase of 3 units in y.

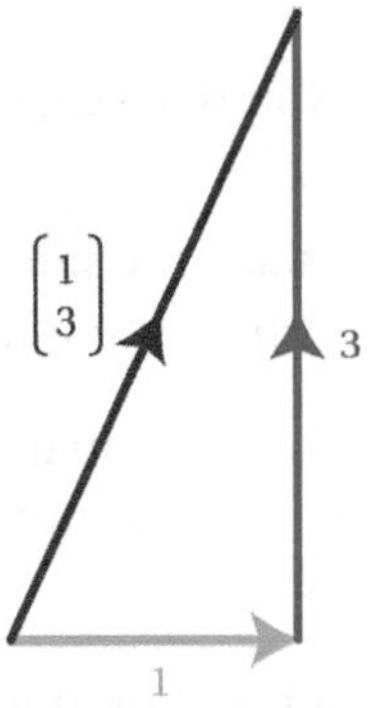

In three dimensions, you cannot use a single number to replace the direction vector, as you can to describe the gradient in two dimensions.

For example, for a line with direction vector $\begin{pmatrix} 2 \\ 2 \\ 1 \end{pmatrix}$, an increase of 2 units in x produces an increase of 2 units in y and an increase of 1 unit in z.

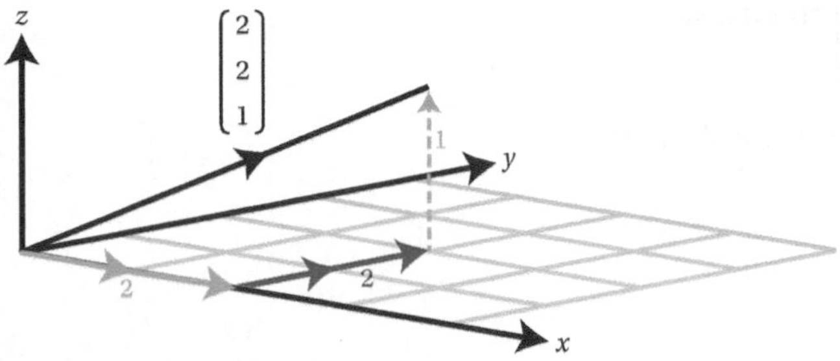

Worked example 9.2 shows how to find a vector equation of the line when two points on the line are known.

WORKED EXAMPLE 9.2

Find a vector equation of the line through points $A(-1, 1, 2)$ and $B(3, 5, 4)$.

$\underline{a} = \begin{pmatrix} -1 \\ 1 \\ 2 \end{pmatrix}$ The line passes through $A(-1, 1, 2)$.

$\underline{d} = \overrightarrow{AB} = \underline{b} - \underline{a} = \begin{pmatrix} 4 \\ 4 \\ 2 \end{pmatrix}$ A direction vector for the line is given by the vector $\overrightarrow{AB}$ (or $\overrightarrow{BA}$).

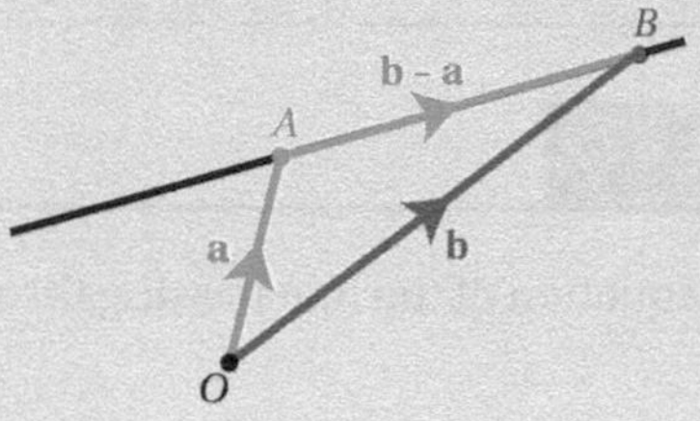

$\underline{r} = \begin{pmatrix} -1 \\ 1 \\ 2 \end{pmatrix} + \lambda \begin{pmatrix} 4 \\ 4 \\ 2 \end{pmatrix}$ The vector equation is $\mathbf{r} = \mathbf{a} + \lambda\mathbf{d}$.

If instead of **a** you had used the position vector of point B in the formula, then you would have got the equation $\mathbf{r} = \begin{pmatrix} 3 \\ 5 \\ 4 \end{pmatrix} + \lambda \begin{pmatrix} 4 \\ 4 \\ 2 \end{pmatrix}$.

This equation represents the same line, but the values of the parameter λ corresponding to particular points will be different.

For example, with the first equation point A has $\lambda = 0$ and point B has $\lambda = 1$, while with the second equation point A has $\lambda = -1$ and point B has $\lambda = 0$.

The direction vector is not unique either, since you are only interested in its direction and not its magnitude. Hence $\begin{pmatrix} 2 \\ 2 \\ 1 \end{pmatrix}$ or $\begin{pmatrix} -6 \\ -6 \\ -3 \end{pmatrix}$ could also be used as direction vectors for the line in Worked example 9.2, as they are all in the same direction. So yet another form of the equation of the same line would be $\mathbf{r}=\begin{pmatrix}-1\\1\\2\end{pmatrix}+\lambda\begin{pmatrix}-6\\-6\\-3\end{pmatrix}$. With this equation, point A has $\lambda=0$ and point B has $\lambda=-\frac{2}{3}$.

When there is more than one line in a question, you should use different letters for the parameters.

Tip

The most commonly used letters for the parameter in the equation of a line are λ (lambda) and μ (mu) or s and t.

WORKED EXAMPLE 9.3

a Show that the equations $\mathbf{r}=\begin{pmatrix}-1\\1\\2\end{pmatrix}+\lambda\begin{pmatrix}2\\2\\1\end{pmatrix}$ and $\mathbf{r}=\begin{pmatrix}5\\7\\5\end{pmatrix}+\mu\begin{pmatrix}6\\6\\3\end{pmatrix}$ represent the same straight line.

b Show that the equation $\mathbf{r}=\begin{pmatrix}-5\\-3\\1\end{pmatrix}+t\begin{pmatrix}-4\\-4\\-2\end{pmatrix}$ represents a different straight line.

You need to show that the two lines have parallel direction vectors (the lines will be parallel) and one common point (then they will be the same line).

a $\begin{pmatrix}6\\6\\3\end{pmatrix}=3\begin{pmatrix}2\\2\\1\end{pmatrix}$ — Two vectors are parallel if one is a scalar multiple of the other.

The direction vectors are parallel.

Show that (5, 7, 5) lies on the first line. — You know that the point (5, 7, 5) lies on the second line; now check it lies on the first line.

$-1+2\lambda=5$

$\lambda=3$ — Find the value of λ which gives the first coordinate.

Checking the other coordinates: — This value of λ must give the other two coordinates.

$\begin{cases}1+3\times2=7\\2+3\times1=5\end{cases}$

$\therefore$ (5, 7, 5) lies on the line.

Hence the two lines are the same.

b $\begin{pmatrix}-4\\-4\\-2\end{pmatrix}=-2\begin{pmatrix}2\\2\\1\end{pmatrix}$ — Check whether the direction vectors are parallel.

So the line is parallel to the other two.

$-1+2\lambda=-5$

$\lambda=-2$ — Check whether (−5, − 3, 1) lies on the first line. Find the value of λ which gives the first coordinate.

Continues on next page

Checking the other two coordinates:

$$\begin{cases} 1+(-2)\times 2=-3 \\ 2+(-2)\times 1=0\neq 1 \end{cases}$$

This value of λ must give the other two coordinates.

$\therefore (-5, -3, 1)$ does not lie on the line.

Hence the second line is not the same as the first line.

WORK IT OUT 9.1

Find a vector equation of the line that passes through the points $A(-3, 4, 2)$ and $B(5, 1, 1)$.

Which is the correct solution? Identify the errors made in the incorrect solutions.

Solution 1	Solution 2	Solution 3
$\mathbf{b}-\mathbf{a}=8\mathbf{i}-3\mathbf{j}-\mathbf{k}$ $\mathbf{r}=8\mathbf{i}-3\mathbf{j}-\mathbf{k}+t(-3\mathbf{i}+4\mathbf{j}+2\mathbf{k})$	$\mathbf{r}=\begin{pmatrix}-3\\4\\2\end{pmatrix}+\lambda\begin{pmatrix}5\\1\\1\end{pmatrix}$	$\mathbf{a}-\mathbf{b}=-8\mathbf{i}+3\mathbf{j}+\mathbf{k}$ $\mathbf{r}=5\mathbf{i}+\mathbf{j}+\mathbf{k}+t(-8\mathbf{i}+3\mathbf{j}+\mathbf{k})$

Sometimes, you know that a point lies on a given line, but you do not know its precise coordinates. Worked example 9.4 shows you how to work with a general point on the line (with an unknown value of λ).

WORKED EXAMPLE 9.4

Point $B(3, 5, 4)$ lies on the line with equation $\mathbf{r}=\begin{pmatrix}-1\\1\\2\end{pmatrix}+\lambda\begin{pmatrix}2\\2\\1\end{pmatrix}$.

Find the possible positions of a point Q on the line such that $BQ=15$.

$$\underline{q}=\begin{pmatrix}-1\\1\\2\end{pmatrix}+\lambda\begin{pmatrix}2\\2\\1\end{pmatrix}=\begin{pmatrix}-1+2\lambda\\1+2\lambda\\2+\lambda\end{pmatrix}$$

You know that Q lies on the line, so it has the position vector $\begin{pmatrix}-1\\1\\2\end{pmatrix}+\lambda\begin{pmatrix}2\\2\\1\end{pmatrix}$ for some value of λ.

Continues on next page

$\overrightarrow{BQ} = \underline{q} - \underline{b}$

$= \begin{pmatrix} -1+2\lambda \\ 1+2\lambda \\ 2+\lambda \end{pmatrix} - \begin{pmatrix} 3 \\ 5 \\ 4 \end{pmatrix}$

$= \begin{pmatrix} 2\lambda-4 \\ 2\lambda-4 \\ \lambda-2 \end{pmatrix}$

You can express vector $\overrightarrow{BQ}$ in terms of λ and then set its magnitude equal to 15.

$|\overrightarrow{BQ}| = 15$

$(2\lambda-4)^2 + (2\lambda-4)^2 + (\lambda-2)^2 = 15^2$

$9\lambda^2 - 36\lambda - 189 = 0$

$\lambda = -3$ or 7

It is easier to work without the square root, so square the magnitude equation.

$\underline{q} = \begin{pmatrix} -7 \\ -5 \\ -1 \end{pmatrix}$ or $\begin{pmatrix} 13 \\ 15 \\ 9 \end{pmatrix}$

Now find the position vector of Q, using $\mathbf{q} = \begin{pmatrix} -1+2\lambda \\ 1+2\lambda \\ 2+\lambda \end{pmatrix}$.

EXERCISE 9A

1 Find the vector equation of each line in the given direction through each given point.

a **i** Direction $\begin{pmatrix} 1 \\ 4 \end{pmatrix}$; point $(4, -1)$ **ii** Direction $\begin{pmatrix} 2 \\ -3 \end{pmatrix}$; point $(4, 1)$

b **i** Point $(1, 0, 5)$; direction $\begin{pmatrix} 1 \\ 3 \\ -3 \end{pmatrix}$ **ii** Point $(-1, 1, 5)$; direction $\begin{pmatrix} 3 \\ -2 \\ 2 \end{pmatrix}$

c **i** Point $(4, 0)$; direction $2\mathbf{i} + 3\mathbf{j}$ **ii** Point $(0, 2)$; direction $\mathbf{i} - 3\mathbf{j}$

d **i** Direction $\mathbf{i} - 3\mathbf{k}$; point $(0, 2, 3)$ **ii** Direction $2\mathbf{i} + 3\mathbf{j} - \mathbf{k}$; point $(4, -3, 0)$

2 Find the vector equation of each line through the two given points.

a **i** $(4, 1)$ and $(1, 2)$ **ii** $(2, 7)$ and $(4, -2)$

b **i** $(-5, -2, 3)$ and $(4, -2, 3)$ **ii** $(1, 1, 3)$ and $(10, -5, 0)$

3 Decide whether or not each given point lies on the given line.

a **i** Line $\mathbf{r} = \begin{pmatrix} 2 \\ 1 \\ 5 \end{pmatrix} + t\begin{pmatrix} -1 \\ 2 \\ 2 \end{pmatrix}$; point (0, 5, 9) **ii** Line $\mathbf{r} = \begin{pmatrix} -1 \\ 0 \\ 3 \end{pmatrix} + t\begin{pmatrix} 4 \\ 1 \\ 5 \end{pmatrix}$; point (−1, 0, 3)

b **i** Line $\mathbf{r} = \begin{pmatrix} 4 \\ 0 \\ 3 \end{pmatrix} + t\begin{pmatrix} 4 \\ 0 \\ 3 \end{pmatrix}$; point (0, 0, 0) **ii** Line $\mathbf{r} = \begin{pmatrix} -1 \\ 5 \\ 1 \end{pmatrix} + t\begin{pmatrix} 0 \\ 0 \\ 7 \end{pmatrix}$; point (−1, 3, 8)

4 Determine whether the two equations describe the same straight line.

a $\mathbf{r} = \begin{pmatrix} 3 \\ -1 \\ 2 \end{pmatrix} + \lambda\begin{pmatrix} 1 \\ 1 \\ 3 \end{pmatrix}$ and $\mathbf{r} = \begin{pmatrix} 4 \\ 0 \\ 5 \end{pmatrix} + \mu\begin{pmatrix} 2 \\ 1 \\ 1 \end{pmatrix}$

b $\mathbf{r} = \begin{pmatrix} 4 \\ 1 \\ 2 \end{pmatrix} + \lambda\begin{pmatrix} 3 \\ 4 \\ 1 \end{pmatrix}$ and $\mathbf{r} = \begin{pmatrix} 7 \\ 5 \\ 1 \end{pmatrix} + \mu\begin{pmatrix} 3 \\ 4 \\ 1 \end{pmatrix}$

c $\mathbf{r} = 3\mathbf{i} + 2\mathbf{j} - 2\mathbf{k} + t(2\mathbf{i} - 5\mathbf{j} + 3\mathbf{k})$ and $\mathbf{r} = 3\mathbf{i} + 2\mathbf{j} - 2\mathbf{k} + s(-4\mathbf{i} + 10\mathbf{j} - 6\mathbf{k})$

d $\mathbf{r} = 3\mathbf{i} + 2\mathbf{j} - 2\mathbf{k} + t(2\mathbf{i} - 5\mathbf{j} + 3\mathbf{k})$ and $\mathbf{r} = -\mathbf{i} + 12\mathbf{j} - 6\mathbf{k} + s(-4\mathbf{i} + 10\mathbf{j} - 6\mathbf{k})$

5 A line passes through the point $A(3, -1, 4)$ and has direction vector $5\mathbf{i} - \mathbf{j} + 2\mathbf{k}$.

a Write down the vector equation of the line.

b Point B has coordinates (−7, 1, 0). Show that B lies on the line.

c Find the exact distance AB.

6 **a** Find the vector equation of the line through the points with coordinates (4, −1, 5) and (7, 7, 2).

b Determine whether the point with coordinates (10, 15, 1) lies on this line.

7 Line l_1 has vector equation $\mathbf{r} = \begin{pmatrix} 3 \\ -1 \\ 1 \end{pmatrix} + \lambda\begin{pmatrix} 1 \\ -6 \\ 2 \end{pmatrix}$. Find the vector equation of the line parallel to l_1 which passes through the point $P(4, 1, 7)$.

8 Find the vector equation of the line that passes through the point (4, −1, 2) which is parallel to the line with equation $\mathbf{r} = (3 + 2\lambda)\mathbf{i} + (5 - \lambda)\mathbf{j} + (3\lambda)\mathbf{k}$.

9 **a** Show that the points $A(4, -1, -8)$ and $B(2, 1, -4)$ lie on the line l with equation $\mathbf{r} = \begin{pmatrix} 2 \\ 1 \\ -4 \end{pmatrix} + t\begin{pmatrix} -1 \\ 1 \\ 2 \end{pmatrix}$.

b Find the coordinates of the point C on the line l such that $AB = BC$.

10 **a** Find the vector equation of line l through points $P(7, 1, 2)$ and $Q(3, -1, 5)$.

b Point R lies on l and $PR = 3PQ$. Find the possible coordinates of R.

11 **a** Write down the vector equation of the line l through the point $A(2, 1, 4)$ parallel to the vector $2\mathbf{i} - 3\mathbf{j} + 6\mathbf{k}$.

b Calculate the magnitude of the vector $2\mathbf{i} - 3\mathbf{j} + 6\mathbf{k}$.

c Find the possible coordinates of point P on l such that $AP = 35$.

Section 2: Cartesian equation of a line

In Section 1, you worked with vector equations of lines in both two and three dimensions. You already know that, in two dimensions, you can also write the equation of a straight line in the form $y = mx + c$. In this section, you will look at how the two forms of the equation are related.

WORKED EXAMPLE 9.5

A straight line passes through the point $(3, -5)$ and has gradient $-\frac{2}{7}$.

Find the vector equation of the line.

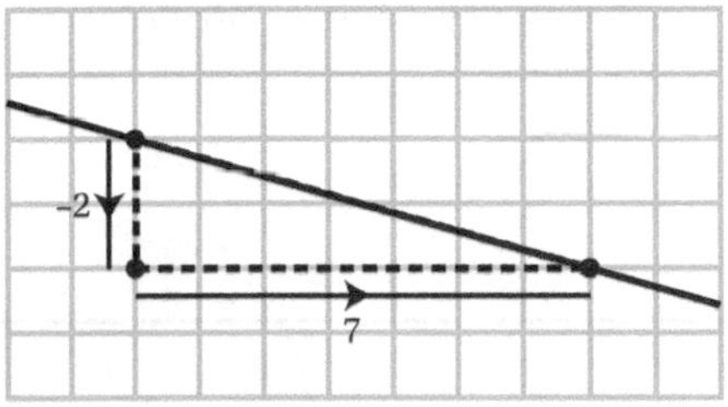

Draw a gradient triangle to identify the direction vector: 7 units to the right and 2 units down.

The direction vector is $\underline{d} = \begin{pmatrix} 7 \\ -2 \end{pmatrix}$

The line goes through point $\underline{a} = \begin{pmatrix} 3 \\ -5 \end{pmatrix}$ so the equation is:

$\underline{r} = \begin{pmatrix} 3 \\ -5 \end{pmatrix} + \lambda \begin{pmatrix} 7 \\ -2 \end{pmatrix}$

The vector equation of the line is $\mathbf{r} = \mathbf{a} + \lambda\mathbf{d}$.

Key point 9.2

In two dimensions, a line with gradient $\frac{p}{q}$ has direction vector $\begin{pmatrix} q \\ p \end{pmatrix}$.

WORKED EXAMPLE 9.6

A straight line has equation $\mathbf{r} = \begin{pmatrix} 1 \\ 2 \end{pmatrix} + \lambda \begin{pmatrix} 5 \\ 3 \end{pmatrix}$.

Write the equation of the line in the form $ax + by + c = 0$.

The line has gradient $\frac{3}{5}$ and passes through the point (1, 2).

You can find the gradient from the direction vector. The vector equation also shows the coordinates of one point on the line.

$$y - 2 = \frac{3}{5}(x - 1)$$

Use $y - y_1 = m(x - x_1)$.

$$5y - 10 = 3x - 3$$

$$3x - 5y + 7 = 0$$

The equation of the line in the form $ax + by + c = 0$ is called a **Cartesian equation.** This means that the equation is in terms of x- and y-coordinates, rather than in terms of position vectors.

There is another way to change from a vector to a Cartesian equation. The key is to realise that the position vector **r** gives the coordinates of a point on the line: $\mathbf{r} = \begin{pmatrix} x \\ y \end{pmatrix}$.

Tip

The Cartesian equation of a line in two dimensions can be written as $ax + by + c = 0$, $ax + by = c$ or $y = mx + c$.

WORKED EXAMPLE 9.7

A line has vector equation $\mathbf{r} = \begin{pmatrix} 3 \\ -4 \end{pmatrix} + \lambda \begin{pmatrix} -2 \\ 5 \end{pmatrix}$.

By expressing x and y in terms of λ, write the equation of the line in the form $ax + by = c$.

$$\begin{pmatrix} x \\ y \end{pmatrix} = \begin{pmatrix} 3 \\ -4 \end{pmatrix} + \lambda \begin{pmatrix} -2 \\ 5 \end{pmatrix}$$

The position vector is related to coordinates: $\mathbf{r} = \begin{pmatrix} x \\ y \end{pmatrix}$.

$$\Rightarrow \begin{cases} x = 3 - 2\lambda & (1) \\ y = -4 + 5\lambda & (2) \end{cases}$$

From (1):

$$\lambda = \frac{3 - x}{2}$$

Express λ from the first equation and substitute into the second.

Substituting into (2):

$$y = -4 + 5\left(\frac{3 - x}{2}\right)$$

$$2y = -8 + 15 - 5x$$

$$5x + 2y = 7$$

In three dimensions, there is no equivalent of the gradient – you cannot use a single number to replace the direction vector. But you can still use the method from Worked example 9.7 to find a Cartesian equation of the line.

WORKED EXAMPLE 9.8

A line has vector equation $\mathbf{r} = \begin{pmatrix} 1 \\ 4 \\ -1 \end{pmatrix} + \lambda \begin{pmatrix} 3 \\ 2 \\ 5 \end{pmatrix}$. Express y and z in terms of x.

$$\begin{pmatrix} x \\ y \\ z \end{pmatrix} = \begin{pmatrix} 1 \\ 4 \\ -1 \end{pmatrix} + \lambda \begin{pmatrix} 3 \\ 2 \\ 5 \end{pmatrix}$$

Write $\mathbf{r} = \begin{pmatrix} x \\ y \\ z \end{pmatrix}$.

$$\Rightarrow \begin{cases} x = 1 + 3\lambda & (1) \\ y = 4 + 2\lambda & (2) \\ z = -1 + 5\lambda & (3) \end{cases}$$

From (1):

$$\lambda = \frac{x-1}{3}$$

Express λ from the first equation and substitute into the second and third.

Substituting into (2) and (3):

$$y = \frac{2}{3}x + \frac{10}{3}, \ z = \frac{5}{3}x - \frac{8}{3}$$

It is possible to combine these two equations. Make λ the subject of all three equations and equate them to each other.

$$\begin{cases} \lambda = \dfrac{x-1}{3} \\ \lambda = \dfrac{y-4}{2} \\ \lambda = \dfrac{z+1}{5} \end{cases}$$

$$\Rightarrow \frac{x-1}{3} = \frac{y-4}{2} = \frac{z+1}{5}$$

Rewind

Writing x, y and z in terms of λ gives the parametric equation of the line. These are covered in A Level Mathematics Student Book 2, Chapter 12.

Key point 9.3

To find the Cartesian equation of a line given its vector equation:

- write $\begin{pmatrix} x \\ y \\ z \end{pmatrix}$ in terms of λ, giving three equations
- make λ the subject of each equation
- equate the three expressions for λ to get an equation of the form $\dfrac{x-a}{k} = \dfrac{y-b}{m} = \dfrac{z-c}{n}$.

Sometimes a Cartesian equation cannot be written in this form, and you need to make a slight adjustment to the procedure.

WORKED EXAMPLE 9.9

Find the Cartesian equation of the line with vector equation $\mathbf{r} = \begin{pmatrix} 1 \\ \frac{1}{2} \\ -3 \end{pmatrix} + \lambda \begin{pmatrix} \frac{1}{3} \\ 5 \\ 0 \end{pmatrix}$.

$$\begin{cases} x = 1 + \frac{1}{3}\lambda \quad \Rightarrow \lambda = \frac{x-1}{\frac{1}{3}} \\ y = \frac{1}{2} + 5\lambda \quad \Rightarrow \lambda = \frac{y-\frac{1}{2}}{5} \\ z = -3 \end{cases}$$

You need an equation involving x, y and z. Remembering that $\mathbf{r} = \begin{pmatrix} x \\ y \\ z \end{pmatrix}$, you can express λ in terms of x, y and z.

$$\frac{x-1}{\frac{1}{3}} = \frac{y-\frac{1}{2}}{5},\ z = -3$$

Equate the expressions for λ from the first two equations. However, the third equation does not contain λ, so you have to leave it as a separate equation.

$$\frac{3x-3}{1} = \frac{2y-1}{10},\ z = -3$$

It will look neater if you rewrite the equation without 'fractions within fractions'.

You can reverse this procedure to go from a Cartesian to a vector equation. A vector equation is convenient if you need to identify the direction vector of the line, or to solve problems involving intersections of lines.

Key point 9.4

To find a vector equation of a line, from a Cartesian equation in the form $\frac{x-a}{k} = \frac{y-b}{m} = \frac{z-c}{n}$:

- set each of the three expressions equal to λ
- express x, y and z in terms of λ
- write $\mathbf{r} = \begin{pmatrix} x \\ y \\ z \end{pmatrix}$ to obtain $\mathbf{r}$ in terms of λ.

Tip

The Cartesian equation can sometimes be read from the vector equation: if the vector equation is $\mathbf{r} = \begin{pmatrix} a \\ b \\ c \end{pmatrix} + \lambda \begin{pmatrix} k \\ m \\ n \end{pmatrix}$ then the Cartesian equation is $\frac{x-a}{k} = \frac{y-b}{m} = \frac{z-c}{n}$. However, if any of the components of the direction vector is 0, you should work through the procedure described in Key point 9.3.

You can adapt this procedure even when the Cartesian equation is in a different form, as in Worked example 9.10.

WORKED EXAMPLE 9.10

Find a vector equation of the line with Cartesian equation $x = -2$, $\frac{3y+1}{4} = \frac{2-z}{5}$, and hence write down the direction vector of the line.

$$\begin{cases} \frac{3y+1}{4} = \lambda \\ \frac{2-z}{5} = \lambda \end{cases}$$

You need to introduce a parameter λ. As the two expressions involving y and z are equal, you can set them both equal to λ.

$$\begin{cases} x = -2 \\ y = \frac{4\lambda - 1}{3} \\ z = 2 - 5\lambda \end{cases}$$

Now express x, y and z in terms of λ.

$$\mathbf{r} = \begin{pmatrix} -2 \\ -\frac{1}{3} \\ 2 \end{pmatrix} + \lambda \begin{pmatrix} 0 \\ \frac{4}{3} \\ -5 \end{pmatrix}$$

The vector equation is an equation for $\mathbf{r} = \begin{pmatrix} x \\ y \\ z \end{pmatrix}$ in terms of λ. You usually separate the expression into a part without λ and a part involving λ.

The direction vector is $\begin{pmatrix} 0 \\ \frac{4}{3} \\ -5 \end{pmatrix}$,

Now identify the direction vector.

or $\begin{pmatrix} 0 \\ 4 \\ -15 \end{pmatrix}$.

You can change the magnitude of the direction vector so that it does not contain fractions, in this case by multiplying by a scalar.

If you want to check whether a given point lies on a given line, you can do this by substituting the numbers for the x, y and z coordinates.

WORKED EXAMPLE 9.11

Determine whether the point $A(3, -2, 2)$ lies on the line with equation $\frac{x+1}{2} = \frac{4-y}{3} = \frac{2z}{3}$.

Substituting $x = 3$, $y = -2$, $z = 2$:

$$\begin{cases} \frac{x+1}{2} = \frac{3+1}{2} = 2 \\ \frac{4-y}{3} = \frac{4+2}{3} = 2 \\ \frac{2z}{3} = \frac{2\times 2}{3} = \frac{4}{3} \end{cases}$$

If the point lies on the line, the coordinates should satisfy the Cartesian equation. This means that all three expressions should be equal.

The point does not lie on the line.

The third equality is not satisfied.

EXERCISE 9B

1 Write down the Cartesian equation of each line.

a **i** $\mathbf{r}=\begin{pmatrix}3\\-1\end{pmatrix}+\lambda\begin{pmatrix}-7\\4\end{pmatrix}$ **ii** $\mathbf{r}=\begin{pmatrix}-1\\5\end{pmatrix}+\lambda\begin{pmatrix}2\\3\end{pmatrix}$

b **i** $\mathbf{r}=\begin{pmatrix}4\\-1\\5\end{pmatrix}+\lambda\begin{pmatrix}2\\-1\\7\end{pmatrix}$ **ii** $\mathbf{r}=\begin{pmatrix}1\\7\\2\end{pmatrix}+\lambda\begin{pmatrix}-1\\1\\2\end{pmatrix}$

c **i** $\mathbf{r}=\begin{pmatrix}-1\\5\\0\end{pmatrix}+\lambda\begin{pmatrix}0\\-2\\2\end{pmatrix}$ **ii** $\mathbf{r}=\begin{pmatrix}3\\0\\6\end{pmatrix}+\lambda\begin{pmatrix}7\\1\\0\end{pmatrix}$

2 Find a vector equation of each line.

a **i** $y=\frac{3}{5}x+2$ **ii** $y=-\frac{4}{3}x-1$

b **i** $3x-5y=17$ **ii** $2x+3y+4=0$

3 Write down a vector equation of each line.

a **i** $\frac{x-2}{5}=\frac{y-2}{3}=\frac{z+1}{7}$ **ii** $\frac{x+1}{4}=\frac{y-6}{-1}=\frac{z-5}{3}$

b **i** $\frac{x+1}{3}=\frac{y}{-7}=\frac{z-1}{-5}$ **ii** $\frac{x-3}{2}=\frac{y+1}{-4}=\frac{z}{5}$

c **i** $\frac{x-11}{3}=\frac{y+1}{6},\ z=-2$ **ii** $\frac{x+1}{5}=\frac{3-z}{2},\ y=1$

4 Determine whether each pair of lines is parallel, the same line, or neither.

a $\mathbf{r}=\begin{pmatrix}1\\1\\2\end{pmatrix}+\lambda\begin{pmatrix}-1\\1\\3\end{pmatrix}$ and $4\mathbf{i}+\mathbf{j}-2\mathbf{k}+t(5\mathbf{i}+2\mathbf{j}+\mathbf{k})$

b $\mathbf{r}=\begin{pmatrix}\frac{13}{2}\\-7\\1\end{pmatrix}+t\begin{pmatrix}2\\-3\\-\frac{2}{3}\end{pmatrix}$ and $\frac{2x-1}{4}=\frac{y-2}{-3}=\frac{6-3z}{2}$

c $\frac{x-5}{7}=\frac{y-2}{-1}=4-z$ and $\mathbf{r}=\begin{pmatrix}2\lambda+1\\4\\5-\lambda\end{pmatrix}$

d $x=2t+1,\ y=1-4t,\ z=3$ and $\mathbf{r}=\begin{pmatrix}8\\-13\\3\end{pmatrix}+s\begin{pmatrix}-1\\2\\0\end{pmatrix}$

5 Find a vector equation of the line $5x + 3y = 30$.

6 Determine whether the point with coordinates (2, 4, 5) lies on the line with equation $\frac{2x-1}{3} = \frac{y+1}{5} = \frac{3-z}{2}$.

7 **a** Find the Cartesian equation of the line with vector equation $\mathbf{r} = (3\lambda + 1)\mathbf{i} + (4 - 2\lambda)\mathbf{j} + (3\lambda - 1)\mathbf{k}$.

b Find the unit vector in the direction of the line.

8 Line l has Cartesian equation $\frac{x-4}{3} = \frac{2y+1}{4}$, $z = -2$. Point $M(-2, p, q)$ lies on the line.

a Find the values of p and q.

b Point N also lies on the line and the distance $MN = \sqrt{52}$. Find the possible coordinates of N.

Section 3: Intersections of lines

Suppose two lines have vector equations $\mathbf{r}_1 = \mathbf{a} + \lambda\mathbf{d}_1$ and $\mathbf{r}_2 = \mathbf{b} + \mu\mathbf{d}_2$.
If they intersect, then there is a point that lies on both lines.
Remembering that the position vector of a point on the line is given by the vector $\mathbf{r}$, this means that you need to find the values of λ and μ which make $\mathbf{r}_1 = \mathbf{r}_2$.

WORKED EXAMPLE 9.12

Find the coordinates of the point of intersection of the lines $\mathbf{r} = \begin{pmatrix} 0 \\ -4 \\ 1 \end{pmatrix} + \lambda\begin{pmatrix} 1 \\ 2 \\ 1 \end{pmatrix}$ and $\mathbf{r} = \begin{pmatrix} 1 \\ 3 \\ 5 \end{pmatrix} + \mu\begin{pmatrix} 4 \\ -2 \\ -2 \end{pmatrix}$.

$$\begin{pmatrix} 0 \\ -4 \\ 1 \end{pmatrix} + \lambda\begin{pmatrix} 1 \\ 2 \\ 1 \end{pmatrix} = \begin{pmatrix} 1 \\ 3 \\ 5 \end{pmatrix} + \mu\begin{pmatrix} 4 \\ -2 \\ -2 \end{pmatrix}$$

You need to make $\mathbf{r}_1 = \mathbf{r}_2$.

$$\Rightarrow \begin{pmatrix} 0+\lambda \\ -4+2\lambda \\ 1+\lambda \end{pmatrix} = \begin{pmatrix} 1+4\mu \\ 3-2\mu \\ 5-2\mu \end{pmatrix}$$

$$\Rightarrow \begin{cases} 0+\lambda = 1+4\mu \\ -4+2\lambda = 3-2\mu \\ 1+\lambda = 5-2\mu \end{cases}$$

If two vectors are equal, then all their components are equal.

$$\Rightarrow \begin{cases} \lambda - 4\mu = 1 & (1) \\ 2\lambda + 2\mu = 7 & (2) \\ \lambda + 2\mu = 4 & (3) \end{cases}$$

Continues on next page

(3) − (1):

$6\mu = 3$

$\mu = \frac{1}{2}$

$\therefore \lambda = 3$

You know how to solve two simultaneous equations in two variables. Pick any two of the three equations. Subtract the first from the third to eliminate λ.

Check in (2):

$2 \times 3 + 2 \times \frac{1}{2} = 7$

So the lines intersect.

The values of λ and μ you have found must also satisfy the second equation.

$$\underline{r_1} = \begin{pmatrix} 0 \\ -4 \\ 1 \end{pmatrix} + 3\begin{pmatrix} 1 \\ 2 \\ 1 \end{pmatrix} = \begin{pmatrix} 3 \\ 2 \\ 4 \end{pmatrix}$$

The position of the intersection point is given by the vector $\mathbf{r}_1$ (or $\mathbf{r}_2$ as they should be the same).

The lines intersect at the point (3, 2, 4).

In a plane, two different straight lines either intersect or are parallel. However, in three dimensions it is possible to have lines that are not parallel but do not intersect, like the red and the blue lines shown in the diagram.

Tip

You might be able to use your calculator to solve simultaneous equations.

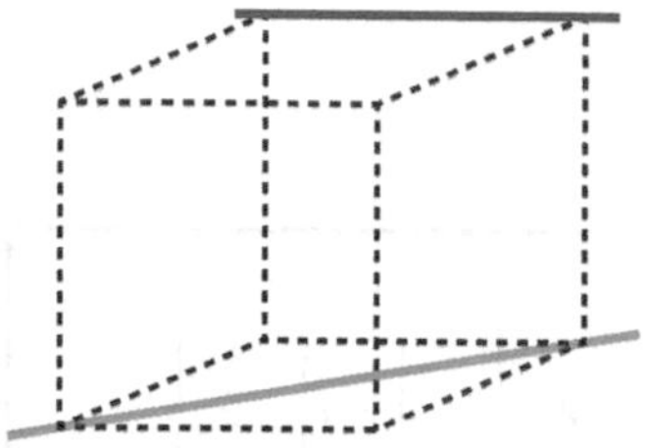

Such lines are called **skew lines**.

When two lines are skew, it is not possible to find values of λ and μ such that $\mathbf{r}_1 = \mathbf{r}_2$.

WORKED EXAMPLE 9.13

Show that the lines $\mathbf{r} = \begin{pmatrix} -4 \\ 1 \\ 3 \end{pmatrix} + t\begin{pmatrix} 1 \\ 1 \\ 4 \end{pmatrix}$ and $\mathbf{r} = \begin{pmatrix} 2 \\ 1 \\ 1 \end{pmatrix} + \lambda\begin{pmatrix} 2 \\ -3 \\ 2 \end{pmatrix}$ do not intersect.

$$\begin{pmatrix} -4 \\ 3 \\ 3 \end{pmatrix} + t\begin{pmatrix} 1 \\ 1 \\ 4 \end{pmatrix} = \begin{pmatrix} 2 \\ 1 \\ 1 \end{pmatrix} + \lambda\begin{pmatrix} 2 \\ -3 \\ 2 \end{pmatrix}$$

Make the two position vectors equal and try to solve the three equations to find λ and t.

$$\Rightarrow \begin{cases} -4 + t = 2 + 2\lambda \\ 3 + t = 1 - 3\lambda \\ 3 + 4t = 1 + 2\lambda \end{cases}$$

$$\Rightarrow \begin{cases} t - 2\lambda = 6 & (1) \\ t + 3\lambda = -2 & (2) \\ 4t - 2\lambda = -2 & (3) \end{cases}$$

Continues on next page

$(1)-(2) \Rightarrow \lambda = -\frac{8}{5}, t = \frac{14}{5}$ — You can find t and λ from the first two equations.

Check in (3): — The values found must also satisfy the third equation.

$4\times\frac{14}{5} - 2\times\left(-\frac{8}{5}\right) = \frac{72}{5} \neq -2$

The two lines do not intersect. — This tells us that it is impossible to find t and λ to make $\mathbf{r}_1 = \mathbf{r}_2$.

It is also possible to check whether a given line crosses one of the coordinate axes. In Worked example 9.14, you use the Cartesian equation of a line.

WORKED EXAMPLE 9.14

a Find the coordinates of the point where the line with equation $\frac{x-6}{2} = \frac{y+1}{7} = \frac{z+9}{-3}$ intersects the y-axis.

b Show that the line does not intersect the z-axis.

a Substitute $x = 0, y = k, z = 0$: — A point on the y-axis has coordinates $(0, k, 0)$. Substitute these into the Cartesian equation.

$$\frac{0-6}{2} = \frac{k+1}{7} = \frac{0+9}{-3}$$

$$-3 = \frac{k+1}{7} = -3$$

$k+1 = -21$ — Solve for k.

$k = -22$

The point of intersection is $(0, -22, 0)$.

b Substitute $x = 0, y = 0$: — A point on the z-axis has coordinates $(0, 0, m)$.

$$\frac{0-6}{2} \neq \frac{0+1}{7}$$

The line does not intersect the z-axis. — The first equality is not satisfied.

EXERCISE 9C

1 Determine whether each pair of lines intersect and, if they do, find the coordinates of the point of intersection.

a i $\mathbf{r} = \begin{pmatrix} 6 \\ 1 \\ 2 \end{pmatrix} + \lambda \begin{pmatrix} -1 \\ 2 \\ 1 \end{pmatrix}$ and $\mathbf{r} = \begin{pmatrix} 2 \\ 1 \\ -14 \end{pmatrix} + \mu \begin{pmatrix} 2 \\ -2 \\ 3 \end{pmatrix}$

ii $\mathbf{r} = \begin{pmatrix} 4 \\ -1 \\ 2 \end{pmatrix} + \lambda \begin{pmatrix} 1 \\ 2 \\ -4 \end{pmatrix}$ and $\mathbf{r} = \begin{pmatrix} 6 \\ -2 \\ 0 \end{pmatrix} + \mu \begin{pmatrix} 3 \\ -4 \\ 0 \end{pmatrix}$

b **i** $\mathbf{r}=\begin{pmatrix}1\\2\\3\end{pmatrix}+t\begin{pmatrix}-1\\1\\2\end{pmatrix}$ and $\mathbf{r}=\begin{pmatrix}-4\\-4\\-11\end{pmatrix}+s\begin{pmatrix}5\\1\\2\end{pmatrix}$

ii $\mathbf{r}=\begin{pmatrix}4\\0\\2\end{pmatrix}+t\begin{pmatrix}2\\0\\1\end{pmatrix}$ and $\mathbf{r}=\begin{pmatrix}-1\\2\\3\end{pmatrix}+s\begin{pmatrix}1\\-2\\-2\end{pmatrix}$

c **i** $\frac{x+1}{2}=\frac{y-6}{-2}=\frac{z+7}{1}$ and $\frac{x-2}{-1}=\frac{y-5}{3}=\frac{z}{5}$

ii $\frac{x-2}{-3}=\frac{y+1}{2}=\frac{z-1}{2}$ and $\frac{x+1}{1}=\frac{y}{6}=\frac{z-1}{3}$

2 Show that the lines with equations $\mathbf{r}=(3+2t)\mathbf{i}+(3-t)\mathbf{j}+(6+3t)\mathbf{k}$ and $\mathbf{r}=(2-3t)\mathbf{i}+(2t)\mathbf{j}+6\mathbf{k}$ are skew.

3 Show that the lines $\mathbf{r}=\begin{pmatrix}7\\1\\5\end{pmatrix}+t\begin{pmatrix}2\\2\\1\end{pmatrix}$ and $\mathbf{r}=\begin{pmatrix}4\\-6\\-3\end{pmatrix}+s\begin{pmatrix}-3\\1\\5\end{pmatrix}$ intersect, and find the coordinates of the intersection point.

4 Determine whether the line with equation $\mathbf{r}=\begin{pmatrix}3\\-7\\2\end{pmatrix}+\lambda\begin{pmatrix}-1\\1\\5\end{pmatrix}$ crosses the x-axis.

5 The line with equation $\frac{x-p}{3}=\frac{y+2}{4}=\frac{z-12}{8}$ crosses the y-axis. Find the value of p.

6 Show that the lines with equations $\mathbf{r}=\begin{pmatrix}1\\1\\2\end{pmatrix}+\lambda\begin{pmatrix}1\\0\\3\end{pmatrix}$, $\mathbf{r}=\begin{pmatrix}-2\\3\\-1\end{pmatrix}+\mu\begin{pmatrix}1\\-1\\0\end{pmatrix}$ and $\mathbf{r}=\begin{pmatrix}2\\-1\\-1\end{pmatrix}+t\begin{pmatrix}-1\\2\\3\end{pmatrix}$ form a triangle, and find its area.

Section 4: The scalar product

Angles between lines

The diagram shows two vectors, **a** and **b**, directed away from the intersection point. The angle between these vectors, θ, can be calculated using the scalar product (also called the dot product).

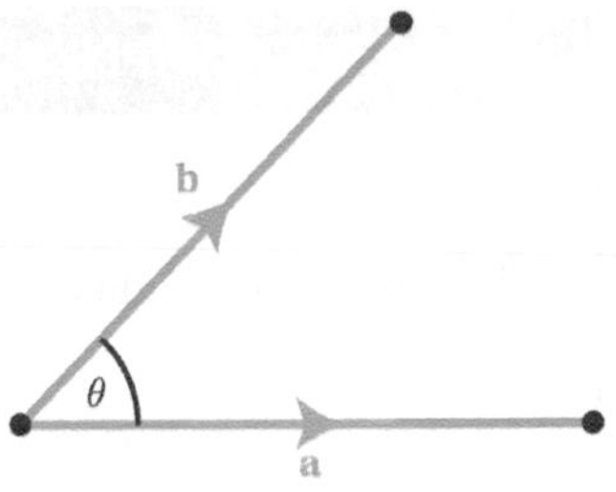

Key point 9.5

The scalar product of two vectors, $\mathbf{a} \cdot \mathbf{b}$, is defined as

$$\mathbf{a} \cdot \mathbf{b} = |\mathbf{a}||\mathbf{b}|\cos\theta$$

where θ is the angle between the vectors **a** and **b**.

This definition leads to a formula for finding the scalar product from the components of the two vectors.

Key point 9.6

If $\mathbf{a} = \begin{pmatrix} a_1 \\ a_2 \\ a_3 \end{pmatrix}$ and $\mathbf{b} = \begin{pmatrix} b_1 \\ b_2 \\ b_3 \end{pmatrix}$, then:

$$\mathbf{a} \cdot \mathbf{b} = a_1b_1 + a_2b_2 + a_3b_3$$

You can now find the angle between two vectors. One of the most common uses is to find the angle between two lines, which you can see, from the diagram, is the same as the angle between their direction vectors.

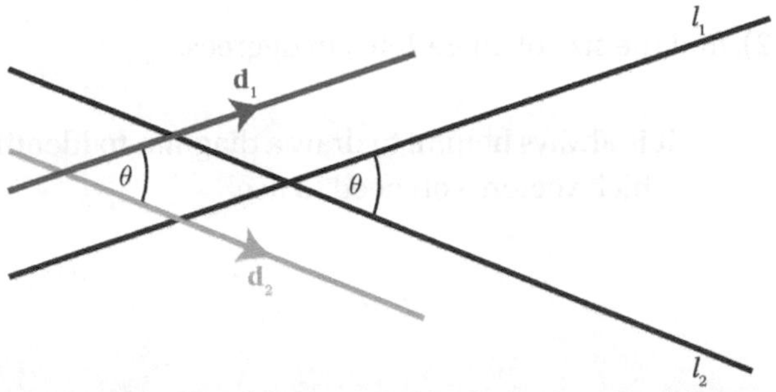

Key point 9.7

The angle between two lines is equal to the angle between their direction vectors.

Common error

When taking the scalar product of two vectors to find the angle θ between them, you need both vectors to be directed away from the angle; otherwise the angle you find will be $180° - \theta$.

WORKED EXAMPLE 9.15

Find the acute angle between lines with equations $\mathbf{r} = \begin{pmatrix} 4 \\ 1 \\ -2 \end{pmatrix} + t\begin{pmatrix} 1 \\ -1 \\ 3 \end{pmatrix}$ and $\mathbf{r} = \begin{pmatrix} 4 \\ 1 \\ -2 \end{pmatrix} + \lambda\begin{pmatrix} -1 \\ 4 \\ 1 \end{pmatrix}$.

$$\cos\theta = \frac{\mathbf{a} \bullet \mathbf{b}}{|\mathbf{a}||\mathbf{b}|}$$

Rearrange the formula from Key point 9.5.

$$\underline{a} = \begin{pmatrix} 1 \\ -1 \\ 3 \end{pmatrix}, \underline{b} = \begin{pmatrix} -1 \\ 4 \\ 1 \end{pmatrix}$$

Take **a** and **b** to be the direction vectors of the two lines.

$$\underline{a} \bullet \underline{b} = \begin{pmatrix} 1 \\ -1 \\ 3 \end{pmatrix} \bullet \begin{pmatrix} -1 \\ 4 \\ 1 \end{pmatrix}$$
$$= 1\times(-1) + (-1)\times 4 + 3\times 1$$
$$= -2$$

Find the dot product.

$$\cos\theta = \frac{-2}{\sqrt{(1+1+9)}\sqrt{1+16+1}}$$
$$= -\frac{2}{\sqrt{11}\sqrt{18}}$$

Now use the formula to calculate the angle.

The acute angle $= 180° - 98.2°$
$= 81.8°$
$\therefore \theta = 81.8°$

The angle found is obtuse but the question asked for the acute angle.

WORKED EXAMPLE 9.16

Given points $A(3, -5, 2)$, $B(4, 1, 1)$ and $C(-1, 1, 2)$, find the size of angle BAC, in degrees.

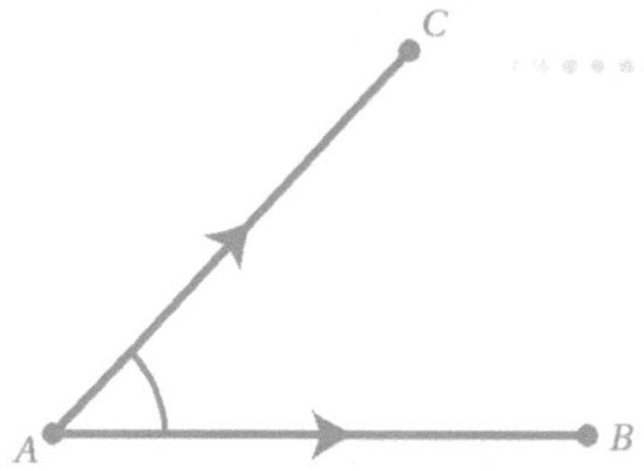

It is always helpful to draw a diagram to identify which vectors you need to use.

Let $\theta = \angle BAC$.

$$\cos\theta = \frac{\overrightarrow{AB} \bullet \overrightarrow{AC}}{|\overrightarrow{AB}||\overrightarrow{AC}|}$$

You can see that the required angle is between vectors $\overrightarrow{AB}$ and $\overrightarrow{AC}$.

Note that you need both vectors to be directed away from the angle you want to find.

$$\overrightarrow{AB} = \begin{pmatrix} 4 \\ 1 \\ 1 \end{pmatrix} - \begin{pmatrix} 3 \\ -5 \\ 2 \end{pmatrix} = \begin{pmatrix} 1 \\ 6 \\ -1 \end{pmatrix}$$

$\overrightarrow{AB} = \mathbf{b} - \mathbf{a}$ and $\overrightarrow{AC} = \mathbf{c} - \mathbf{a}$.

Continues on next page

$$\overrightarrow{AC} = \begin{pmatrix} -1 \\ 1 \\ 2 \end{pmatrix} - \begin{pmatrix} 3 \\ -5 \\ 2 \end{pmatrix} = \begin{pmatrix} -4 \\ 6 \\ 0 \end{pmatrix}$$

$$\overrightarrow{AB} \cdot \overrightarrow{AC} = \begin{pmatrix} 1 \\ 6 \\ -1 \end{pmatrix} \cdot \begin{pmatrix} -4 \\ 6 \\ 0 \end{pmatrix}$$

Find the dot product.

$$= 1 \times (-4) + 6 \times 6 + (-1) \times 0$$
$$= 32$$

$$\cos\theta = \frac{32}{\sqrt{1^2+6^2+1^2}\sqrt{4^2+6^2+0^2}}$$

Use $\cos\theta = \frac{\overrightarrow{AB} \cdot \overrightarrow{AC}}{|\overrightarrow{AB}||\overrightarrow{AC}|}$

$$= \frac{32}{\sqrt{38}\sqrt{52}}$$
$$= 0.7199$$

$$\theta = \cos^{-1}(0.7199) = 44.0°$$

WORK IT OUT 9.2

Find the angle between the lines $\mathbf{r} = \begin{pmatrix} 1 \\ 3 \\ 3 \end{pmatrix} + \lambda \begin{pmatrix} -1 \\ 1 \\ 2 \end{pmatrix}$ and $\mathbf{r} = \begin{pmatrix} 1 \\ -1 \\ 0 \end{pmatrix} + \mu \begin{pmatrix} 1 \\ 0 \\ 3 \end{pmatrix}$.

Which is the correct solution? Identify the errors made in the incorrect solutions.

Solution 1	Solution 2	Solution 3
$\mathbf{a} = \begin{pmatrix} 1 \\ 1 \\ 3 \end{pmatrix}, \mathbf{b} = \begin{pmatrix} 1 \\ -1 \\ 0 \end{pmatrix}$	$\mathbf{a} = \begin{pmatrix} -1 \\ 1 \\ 2 \end{pmatrix}, \mathbf{b} = \begin{pmatrix} 1 \\ 0 \\ 3 \end{pmatrix}$	$\mathbf{a} = \begin{pmatrix} 1 \\ 3 \\ 3 \end{pmatrix} + \begin{pmatrix} -1 \\ 1 \\ 2 \end{pmatrix} = \begin{pmatrix} 0 \\ 4 \\ 5 \end{pmatrix}$
$\cos\theta = \frac{1-1+0}{\sqrt{11}\sqrt{2}}$	$\cos\theta = \frac{-1+0+6}{\sqrt{6}\sqrt{10}}$	$\mathbf{b} = \begin{pmatrix} 1 \\ -1 \\ 0 \end{pmatrix} + \begin{pmatrix} 1 \\ 0 \\ 3 \end{pmatrix} = \begin{pmatrix} 2 \\ -1 \\ 3 \end{pmatrix}$
$= 0$	$= 0.645$	$\cos\theta = \frac{(0-4+15)}{\sqrt{41}\sqrt{14}}$
$\Rightarrow \theta = 90°$	$\Rightarrow \theta = 49.8°$	$= 0.459$
		$\Rightarrow \theta = 62.7°$

Perpendicular lines

The formula in Key point 9.5 is very convenient for checking whether two vectors are perpendicular. If $\theta = 90°$, then $\cos\theta = 0$, so the numerator of the fraction in the formula must be zero. This means that you don't even need to calculate the magnitudes of the two vectors.

Key point 9.8

Two vectors **a** and **b** are perpendicular if $\mathbf{a} \cdot \mathbf{b} = 0$.

Two lines are perpendicular if their direction vectors satisfy $\mathbf{d}_1 \cdot \mathbf{d}_2 = 0$.

WORKED EXAMPLE 9.17

Given that $\mathbf{p} = \begin{pmatrix} 4 \\ -1 \\ 2 \end{pmatrix}$ and $\mathbf{q} = \begin{pmatrix} 2 \\ 1 \\ 1 \end{pmatrix}$ find the value of the scalar t such that $\mathbf{p} + t\mathbf{q}$ is perpendicular to $\begin{pmatrix} 3 \\ 5 \\ 1 \end{pmatrix}$.

$$(\underline{p} + t\underline{q}) \cdot \begin{pmatrix} 3 \\ 5 \\ 1 \end{pmatrix} = 0$$

Two vectors are perpendicular if their scalar product is equal to 0.

$$\underline{p} + t\underline{q} = \begin{pmatrix} 4+2t \\ -1+t \\ 2+t \end{pmatrix}$$

You need to find the components of $\mathbf{p} + t\mathbf{q}$ in terms of t and then form an equation.

So:

$$\begin{pmatrix} 4+2t \\ -1+t \\ 2+t \end{pmatrix} \cdot \begin{pmatrix} 3 \\ 5 \\ 1 \end{pmatrix} = 0$$

$$3(4+2t) + 5(-1+t) + 1(2+t) = 0$$

$$9 + 12t = 0$$

$$t = -\frac{3}{4}$$

WORKED EXAMPLE 9.18

Prove that the lines $2x + 1 = \frac{y+1}{2} = \frac{5-z}{4}$ and $\frac{x-2}{4} = \frac{y}{6} = \frac{2z-6}{7}$ are perpendicular.

$$2x + 1 = \frac{y+1}{2} = \frac{5-z}{4}$$

$$\Leftrightarrow \frac{x + \frac{1}{2}}{\frac{1}{2}} = \frac{y+1}{2} = \frac{z-5}{-4}$$

In order to identify the two direction vectors, you need to rearrange the equations into the form $\frac{x-a}{k} = \frac{y-b}{m} = \frac{z-c}{n}$.

$$\therefore \underline{d}_1 = \begin{pmatrix} \frac{1}{2} \\ 2 \\ -4 \end{pmatrix}$$

$$\frac{x-2}{4} = \frac{y}{6} = \frac{2z-6}{7}$$

Continues on next page

$\Leftrightarrow \frac{x-2}{4} = \frac{y}{6} = \frac{z-3}{\frac{7}{2}}$

$\therefore \underline{d}_2 = \begin{pmatrix} 4 \\ 6 \\ \frac{7}{2} \end{pmatrix}$

$\underline{d}_1 \cdot \underline{d}_2 = \left(\frac{1}{2}\right)(4) + (2)(6) + (-4)\left(\frac{7}{2}\right)$

$= 2 + 12 - 14$

$= 0$

Hence the two lines are perpendicular.

Use the scalar product to show that $\mathbf{d}_1$ and $\mathbf{d}_2$ are perpendicular.

Distance from a point to a line

You can use perpendicular lines to find the shortest distance from a point to a line.

Focus on...

See Focus on ... Problem solving 2 for alternatives to the method in Worked example 9.19.

WORKED EXAMPLE 9.19

Line l has equation $\mathbf{r} = \begin{pmatrix} 3 \\ -1 \\ 0 \end{pmatrix} + \lambda \begin{pmatrix} 1 \\ -1 \\ 1 \end{pmatrix}$ and point A has coordinates (3, 9, −2).

a Find the coordinates of point B on l so that AB is perpendicular to l.
b Hence find the shortest distance from A to l.
c Find the coordinates of the reflection of the point A in l.

a

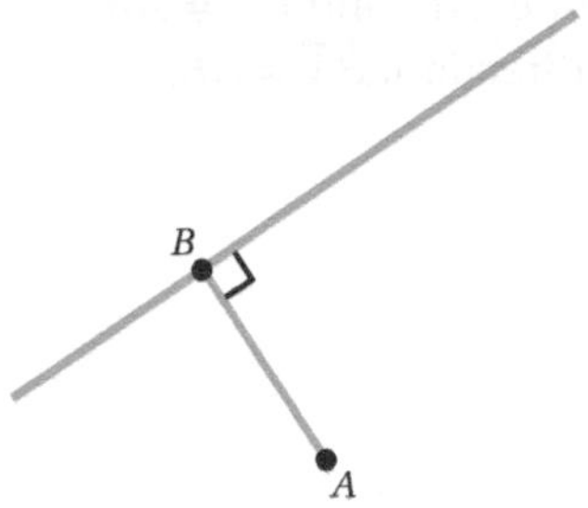

Draw a diagram. The line AB needs to be perpendicular to the direction vector of l.

$\overrightarrow{AB} \cdot \begin{pmatrix} 1 \\ -1 \\ 1 \end{pmatrix} = 0$

$\overrightarrow{OB} = \underline{r} = \begin{pmatrix} 3+\lambda \\ -1-\lambda \\ \lambda \end{pmatrix}$

You know that B lies on l, so its position vector is given by the equation for $\mathbf{r}$.

Continues on next page

$$\overrightarrow{AB} = \begin{pmatrix} 3+\lambda \\ -1-\lambda \\ \lambda \end{pmatrix} - \begin{pmatrix} 3 \\ 9 \\ -2 \end{pmatrix}$$

$$= \begin{pmatrix} \lambda \\ -10-\lambda \\ \lambda+2 \end{pmatrix}$$

$$\begin{pmatrix} \lambda \\ -10-\lambda \\ \lambda+2 \end{pmatrix} \bullet \begin{pmatrix} 1 \\ -1 \\ 1 \end{pmatrix} = 0$$

You can now find the value of λ for which the two lines are perpendicular.

$$(\lambda) + (10+\lambda) + (\lambda+2) = 0$$

$$\lambda = -4$$

$$\underline{r} = \begin{pmatrix} 3 \\ -1 \\ 0 \end{pmatrix} - 4\begin{pmatrix} 1 \\ -1 \\ 1 \end{pmatrix} = \begin{pmatrix} -1 \\ 3 \\ -4 \end{pmatrix}$$

Using this value of λ in the equation of the line gives the position vector of B.

B has coordinates $(-1, 3, -4)$.

b $\overrightarrow{AB} = \begin{pmatrix} -1 \\ 3 \\ -4 \end{pmatrix} - \begin{pmatrix} 3 \\ 9 \\ -2 \end{pmatrix} = \begin{pmatrix} -4 \\ -6 \\ -2 \end{pmatrix}$

The shortest distance from a point to a line is the perpendicular distance, in other words, the distance AB.

$$|\overrightarrow{AB}| = \sqrt{16+36+4}$$

$$= 2\sqrt{14}$$

c

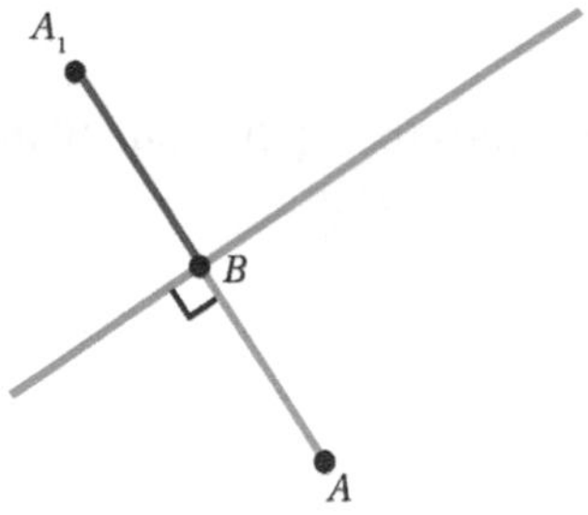

The reflection A_1 lies on the line (AB) and $BA_1 = AB$. As they are also in the same direction, $\overrightarrow{BA_1} = \overrightarrow{AB}$.

$$\overrightarrow{BA_1} = \overrightarrow{AB}$$

$$\underline{a}_1 - \underline{b} = \overrightarrow{AB}$$

$$\underline{a}_1 = \begin{pmatrix} -4 \\ -6 \\ -2 \end{pmatrix} + \begin{pmatrix} -1 \\ 3 \\ -4 \end{pmatrix}$$

So A_1 has coordinates $(-5, -3, -6)$.

The final part of Worked example 9.19 illustrates the power of vectors: as they contain both distance and direction information, just one equation ($\overrightarrow{BA_1} = \overrightarrow{AB}$) was able to express both the fact that A_1 is on the line AB and that $BA_1 = AB$.

Distance between lines

The approach used in Worked example 9.19 to find the shortest distance from a point to a line can also be used to find the shortest distance between two parallel lines. You can just pick any known point on the second line to play the role of the point A.

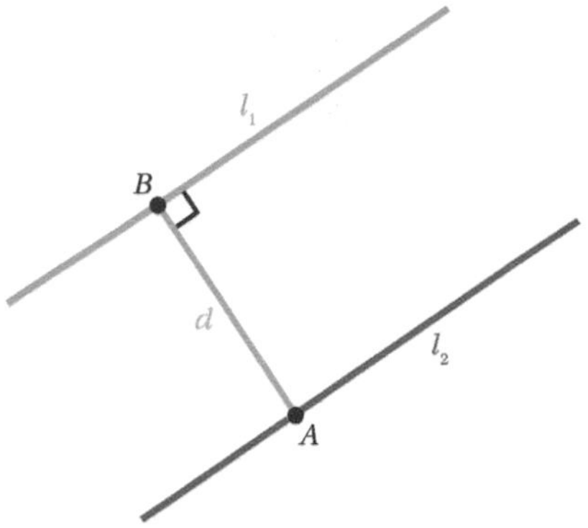

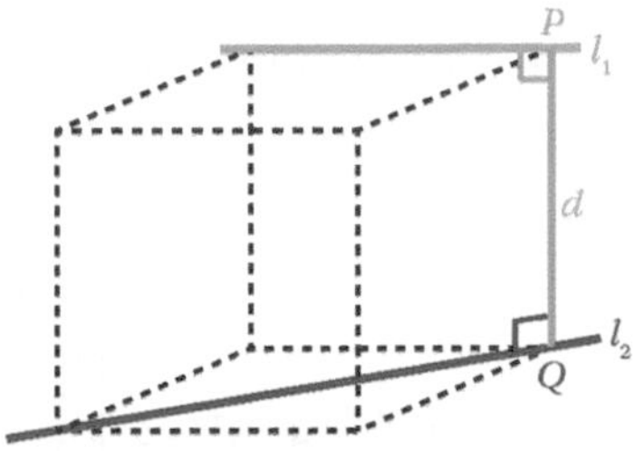

However, a different approach is needed to find the shortest distance between two skew lines. You can't now pick a fixed point on the second line; instead you need a general point on the first line, P, and a general point on the second line, Q. The distance between P and Q will be shortest when $\overrightarrow{PQ}$ is perpendicular to both lines.

Focus on...

See Focus on ... Problem solving 2 question 2 for alternatives to the method in Worked example 9.20 for finding the shortest distance between skew lines.

WORKED EXAMPLE 9.20

Find the shortest distance between the skew lines l_1: $\mathbf{r} = \begin{pmatrix} 1 \\ 0 \\ 2 \end{pmatrix} + \lambda \begin{pmatrix} 1 \\ -1 \\ 1 \end{pmatrix}$ and l_2: $\mathbf{r} = \begin{pmatrix} 0 \\ 11 \\ 0 \end{pmatrix} + \mu \begin{pmatrix} 2 \\ 0 \\ -1 \end{pmatrix}$.

Let P be a general point on l_1 and Q a general point on l_2.

Then

$$\underline{p} = \begin{pmatrix} 1+\lambda \\ -\lambda \\ 2+\lambda \end{pmatrix}$$

and

$$\underline{q} = \begin{pmatrix} 2\mu \\ 11 \\ -\mu \end{pmatrix}$$

The position vector of a general point on a line is given by the equation of that line.

$$\overrightarrow{PQ} = \underline{q} - \underline{p}$$

$$= \begin{pmatrix} 2\mu \\ 11 \\ -\mu \end{pmatrix} - \begin{pmatrix} 1+\lambda \\ -\lambda \\ 2+\lambda \end{pmatrix}$$

$$= \begin{pmatrix} 2\mu-1-\lambda \\ 11+\lambda \\ -\mu-2-\lambda \end{pmatrix}$$

You can then find a general vector, $\overrightarrow{PQ}$, connecting l_1 and l_2.

Continues on next page

If $\overrightarrow{PQ}$ is perpendicular to both lines then

$$\begin{pmatrix} 2\mu-1-\lambda \\ 11+\lambda \\ -\mu-2-\lambda \end{pmatrix} \cdot \begin{pmatrix} 1 \\ -1 \\ 1 \end{pmatrix} = 0$$

This connecting vector will be shortest when it is perpendicular to both lines.

The dot product of $\overrightarrow{PQ}$ with the direction vector of each line must be zero.

$$(2\mu-1-\lambda)-(11+\lambda)+(-\mu-2-\lambda)=0$$

$$\mu-3\lambda=14 \qquad (1)$$

and

$$\begin{pmatrix} 2\mu-1-\lambda \\ 11+\lambda \\ -\mu-2-\lambda \end{pmatrix} \cdot \begin{pmatrix} 2 \\ 0 \\ 1 \end{pmatrix} = 0$$

$$2(2\mu-1-\lambda)+0-(-\mu-2-\lambda)=0$$

$$5\mu-\lambda=0 \qquad (2)$$

Solving (1) and (2) simultaneously:

$\lambda=-5,\ \mu=-1$

$$\therefore \overrightarrow{PQ} = \begin{pmatrix} -2-1+5 \\ 11-5 \\ 1-2+5 \end{pmatrix} = \begin{pmatrix} 2 \\ 6 \\ 4 \end{pmatrix}$$

Substitute these values of λ and μ back into the expression for $\overrightarrow{PQ}$.

$$d = \left|\overrightarrow{PQ}\right| = \sqrt{2^2+6^2+4^2} = 2\sqrt{14}$$

The distance is the length of $\overrightarrow{PQ}$.

EXERCISE 9D

1 Calculate the scalar product for each pair of vectors.

a **i** $\begin{pmatrix} 5 \\ 1 \\ 2 \end{pmatrix}$ and $\begin{pmatrix} 1 \\ -2 \\ 3 \end{pmatrix}$ **ii** $\begin{pmatrix} 3 \\ 0 \\ 2 \end{pmatrix}$ and $\begin{pmatrix} 0 \\ -1 \\ 1 \end{pmatrix}$

b **i** $2\mathbf{i}+2\mathbf{j}-\mathbf{k}$ and $\mathbf{i}-\mathbf{j}+3\mathbf{k}$ **ii** $3\mathbf{i}+\mathbf{j}$ and $\mathbf{i}-2\mathbf{k}$

c **i** $\begin{pmatrix} 3 \\ 2 \end{pmatrix}$ and $\begin{pmatrix} -1 \\ 4 \end{pmatrix}$ **ii** $\mathbf{i}-\mathbf{j}$ and $2\mathbf{i}+3\mathbf{j}$

2 Calculate the angle between the pairs of vectors from Question 1.

Give your answers in radians.

3 The angle between vectors **a** and **b** is θ.

Find the exact value of $\cos\theta$ in each case.

a **i** $\mathbf{a} = 2\mathbf{i} + 3\mathbf{j} - \mathbf{k}$ and $\mathbf{b} = \mathbf{i} - 2\mathbf{j} + \mathbf{k}$ **ii** $\mathbf{a} = \mathbf{i} - 3\mathbf{j} + 3\mathbf{k}$ and $\mathbf{b} = \mathbf{i} + 5\mathbf{j} - 2\mathbf{k}$

b **i** $\mathbf{a} = \begin{pmatrix} 2 \\ 2 \\ 3 \end{pmatrix}$ and $\mathbf{b} = \begin{pmatrix} 1 \\ 1 \\ -2 \end{pmatrix}$ **ii** $\mathbf{a} = \begin{pmatrix} 5 \\ 1 \\ -3 \end{pmatrix}$ and $\mathbf{b} = \begin{pmatrix} 2 \\ -1 \\ 2 \end{pmatrix}$

c **i** $\mathbf{a} = -2\mathbf{k}$ and $\mathbf{b} = 4\mathbf{i}$ **ii** $\mathbf{a} = 5\mathbf{i}$ and $\mathbf{b} = 3\mathbf{j}$

4 Which pairs of vectors are perpendicular?

a **i** $\begin{pmatrix} 2 \\ 1 \\ -3 \end{pmatrix}$ and $\begin{pmatrix} 1 \\ -2 \\ 2 \end{pmatrix}$ **ii** $\begin{pmatrix} 3 \\ -1 \\ 2 \end{pmatrix}$ and $\begin{pmatrix} 2 \\ 6 \\ 0 \end{pmatrix}$

b **i** $5\mathbf{i} - 2\mathbf{j} + \mathbf{k}$ and $3\mathbf{i} + 4\mathbf{j} - 7\mathbf{k}$ **ii** $\mathbf{i} - 3\mathbf{k}$ and $2\mathbf{i} + \mathbf{j} + \mathbf{k}$

5 Find the acute angle between each pair of lines, giving your answers in degrees.

a **i** $\mathbf{r} = \begin{pmatrix} 5 \\ -1 \\ 2 \end{pmatrix} + \lambda \begin{pmatrix} 2 \\ 2 \\ 3 \end{pmatrix}$ and $\mathbf{r} = \begin{pmatrix} 1 \\ 1 \\ 0 \end{pmatrix} + \mu \begin{pmatrix} 4 \\ -1 \\ 3 \end{pmatrix}$ **ii** $\mathbf{r} = \begin{pmatrix} 4 \\ 0 \\ 2 \end{pmatrix} + \lambda \begin{pmatrix} 2 \\ -1 \\ 1 \end{pmatrix}$ and $\mathbf{r} = \begin{pmatrix} 1 \\ 0 \\ 2 \end{pmatrix} + \mu \begin{pmatrix} -5 \\ 1 \\ 3 \end{pmatrix}$

b **i** $\mathbf{r} = \begin{pmatrix} 2 \\ 0 \\ 1 \end{pmatrix} + t \begin{pmatrix} -1 \\ 0 \\ 0 \end{pmatrix}$ and $\mathbf{r} = \begin{pmatrix} 1 \\ 3 \\ 3 \end{pmatrix} + s \begin{pmatrix} 4 \\ 0 \\ 2 \end{pmatrix}$ **ii** $\mathbf{r} = \begin{pmatrix} 6 \\ 6 \\ 2 \end{pmatrix} + t \begin{pmatrix} -1 \\ 0 \\ 3 \end{pmatrix}$ and $\mathbf{r} = \begin{pmatrix} 1 \\ 0 \\ 0 \end{pmatrix} + s \begin{pmatrix} 4 \\ -1 \\ 2 \end{pmatrix}$

6 **a** The vertices of a triangle have position vectors $\mathbf{a} = \begin{pmatrix} 1 \\ 1 \\ 3 \end{pmatrix}$, $\mathbf{b} = \begin{pmatrix} 2 \\ -1 \\ 1 \end{pmatrix}$ and $\mathbf{c} = \begin{pmatrix} 5 \\ 1 \\ 2 \end{pmatrix}$.

Find, in degrees, the angles of the triangle.

b Find, in degrees, the angles of the triangle with vertices (2, 1, 2), (4, −1, 5) and (7, 1, −2).

7 Points A and B have position vectors $\overrightarrow{OA} = \begin{pmatrix} 2 \\ 2 \\ 3 \end{pmatrix}$ and $\overrightarrow{OB} = \begin{pmatrix} -1 \\ 7 \\ 2 \end{pmatrix}$. Find the angle between $\overrightarrow{AB}$ and $\overrightarrow{OA}$.

8 Given four points with coordinates $A(2, -1, 3)$, $B(1, 1, 2)$, $C(6, -1, 2)$ and $D(7, -3, 3)$, find the acute angle between $\overrightarrow{AC}$ and $\overrightarrow{BD}$.

9 The line l_1 has equation $\mathbf{r} = \begin{pmatrix} 3 \\ -1 \\ -5 \end{pmatrix} + s \begin{pmatrix} 6 \\ 4 \\ -3 \end{pmatrix}$ and the line l_2 has equation $\mathbf{r} = \begin{pmatrix} 0 \\ 2 \\ 1 \end{pmatrix} + t \begin{pmatrix} -4 \\ 3 \\ -1 \end{pmatrix}$.

a Find, to 1 decimal place, the acute angle between l_1 and l_2.

The line l_3 has equation $\mathbf{r} = \begin{pmatrix} 1 \\ 0 \\ -1 \end{pmatrix} + t \begin{pmatrix} -3 \\ 6 \\ 2 \end{pmatrix}$.

b Show that l_1 and l_3 are perpendicular.

10 Four points have coordinates $A(2, 4, 1)$, $B(k, 4, 2k)$, $C(k+4, 2k+4, 2k+2)$ and $D(6, 2k+4, 3)$.

a Show that $ABCD$ is a parallelogram for all values of k.

b When $k=1$ find the angles of the parallelogram.

c Find the value of k for which $ABCD$ is a rectangle.

11 The vertices of a triangle have position vectors $\mathbf{a} = \mathbf{i} - 2\mathbf{j} + 2\mathbf{k}$, $\mathbf{b} = 3\mathbf{i} - \mathbf{j} + 7\mathbf{k}$ and $\mathbf{c} = 5\mathbf{i}$.

a Show that the points are the vertices of a right-angled triangle.

b Calculate the other two angles of the triangle.

c Find the area of the triangle.

12 Line l has equation $\mathbf{r} = \begin{pmatrix} 4 \\ 2 \\ -1 \end{pmatrix} + \lambda \begin{pmatrix} 2 \\ -1 \\ 2 \end{pmatrix}$ and point P has coordinates $(7, 2, 3)$. Point C lies on l and PC is perpendicular to l. Find the coordinates of C.

13 The line l_1 passes through the point A with coordinates $(2, -4, 3)$ and the point B with coordinates $(4, -5, 7)$.

a Find the vector equation of l_1.

The line l_2 passes through the point C with coordinates $(5, 0, -2)$ and is parallel to l_1.

b Find the distance AC.

c Find the exact value of $\cos\theta$, where θ is the angle between AC and l_2.

d Hence find, to 3 significant figures, the distance between l_1 and l_2.

14 Two lines with equations $l_1 : \mathbf{r} = \begin{pmatrix} 0 \\ -1 \\ 2 \end{pmatrix} + \lambda \begin{pmatrix} 1 \\ 5 \\ 3 \end{pmatrix}$ and $l_2 : \mathbf{r} = \begin{pmatrix} 2 \\ 2 \\ 1 \end{pmatrix} + t \begin{pmatrix} -1 \\ 1 \\ 3 \end{pmatrix}$ intersect at point P.

a Find the coordinates of P.

b Find, in degrees, the acute angle between the two lines.

Point Q has coordinates $(-1, 5, 10)$.

c Show that Q lies on l_2.

d Find the distance PQ.

e Hence find the shortest distance from Q to the line l_1.

15 Find the shortest distance of the line with equation $\mathbf{r} = \begin{pmatrix} 1 \\ -2 \\ 2 \end{pmatrix} + \lambda \begin{pmatrix} 2 \\ 2 \\ 1 \end{pmatrix}$ from the origin.

16 Find the exact distance between the lines $\mathbf{r} = \begin{pmatrix} 1 \\ 1 \\ 2 \end{pmatrix} + s \begin{pmatrix} 1 \\ 0 \\ 2 \end{pmatrix}$ and $\mathbf{r} = \begin{pmatrix} 1 \\ -2 \\ 3 \end{pmatrix} + t \begin{pmatrix} 1 \\ 0 \\ 2 \end{pmatrix}$.

17 The line l has equation $\mathbf{r} = \begin{pmatrix} 5 \\ 1 \\ 2 \end{pmatrix} + \lambda \begin{pmatrix} 2 \\ -3 \\ 3 \end{pmatrix}$ and point P has coordinates (21, 5, 10).

a Find the coordinates of point M on l such that PM is perpendicular to l.

b Show that the point $Q(15, -14, 17)$ lies on l.

c Find the coordinates of point R on l such that $|PR| = |PQ|$.

18 Two lines have vector equations l_1: $\mathbf{r} = \begin{pmatrix} 1 \\ 3 \\ 1 \end{pmatrix} + \lambda \begin{pmatrix} 1 \\ -1 \\ 2 \end{pmatrix}$ and l_2: $\mathbf{r} = \begin{pmatrix} 5 \\ -1 \\ -6 \end{pmatrix} + \mu \begin{pmatrix} 1 \\ 1 \\ 3 \end{pmatrix}$.

The point A on l_1 and the point B on l_2 are such that $\overrightarrow{AB}$ is perpendicular to both lines.

a Show that $\mu - \lambda = 1$.

b Find a second equation linking λ and μ.

c Hence find the shortest distance between l_1 and l_2, giving your answer as an exact value.

Checklist of learning and understanding

- A **vector equation** of a straight line has the form $\mathbf{r} = \mathbf{a} + \lambda\mathbf{d}$, where **a** is one point on the line and **b** is the **direction vector.**
  - To find the equation of the line through two points with position vectors **a** and **b**, use the direction vector $\mathbf{d} = \mathbf{b} - \mathbf{a}$.
- To find the Cartesian equation of a line given its vector equation:
  - write $\begin{pmatrix} x \\ y \\ z \end{pmatrix}$ in terms of λ, giving three equations
  - make λ the subject of each equation
  - equate the three expressions for λ to get an equation of the form $\frac{x-a}{k} = \frac{y-b}{m} = \frac{z-c}{n}$.
- To find the intersection of two lines, solve simultaneous equations. If there is no solution, the lines are **skew**.
- The **scalar product** (or dot product) of two vectors can be calculated in two ways:
  - $\mathbf{a} \cdot \mathbf{b} = a_1b_1 + a_2b_2 + a_3b_3$
  - $\mathbf{a} \cdot \mathbf{b} = |\mathbf{a}||\mathbf{b}|\cos\theta$
  - Two vectors are perpendicular if $\mathbf{a} \cdot \mathbf{b} = 0$.
- The angle between two vectors can be found from $\cos\theta = \frac{\mathbf{a} \cdot \mathbf{b}}{|\mathbf{a}||\mathbf{b}|}$.

Mixed practice 9

1 The points A, B and C have position vectors $\mathbf{a} = \begin{pmatrix} 1 \\ 2 \\ 0 \end{pmatrix}$, $\mathbf{b} = \begin{pmatrix} 3 \\ -1 \\ -2 \end{pmatrix}$ and $\mathbf{c} = \begin{pmatrix} 4 \\ 0 \\ -5 \end{pmatrix}$ respectively.

Find the size of the angle ABC.

Choose from these options.

A 150° **B** 111° **C** 78.6° **D** 30.1°

2 Find the vector equation of the line passing through points (−1, 2, 5) and (7, 0, 3).

3 Two lines have equations $\mathbf{r} = (3\mathbf{i} - \mathbf{j} + 3\mathbf{k}) + t(2\mathbf{i} + \mathbf{j} + \mathbf{k})$ and $\mathbf{r} = (-2\mathbf{i} - 2\mathbf{j} - 2\mathbf{k}) + s(\mathbf{i} - \mathbf{j} + 3\mathbf{k})$. Given that the lines intersect:

a find the coordinates of the point of intersection

b find the acute angle between the lines.

4 Show that the lines with equations $\mathbf{r} = (5\mathbf{i} - 2\mathbf{j} + \mathbf{k}) + t(2\mathbf{i} - \mathbf{j} - \mathbf{k})$ and $\mathbf{r} = (\mathbf{i} + \mathbf{j} - \mathbf{k}) + s(-2 + \mathbf{i} + \mathbf{j} + 5\mathbf{k})$ are skew.

5 The vector $\mathbf{p} = x\mathbf{i} - 3\mathbf{j} + x\mathbf{k}$ and the vector $\mathbf{q} = 2x\mathbf{i} - 2\mathbf{j} - 7\mathbf{k}$. Given that $\mathbf{p}$ and $\mathbf{q}$ are perpendicular, find the possible values of x.

6 Find the acute angle between the skew lines $\frac{x+3}{1} = \frac{y-1}{2} = \frac{z-4}{-1}$ and $\frac{x-5}{2} = \frac{y-2}{-3} = \frac{z+3}{2}$.

7 **a** Find a vector equation of the line with Cartesian equation $\frac{2x-1}{4} = \frac{y+2}{3} = \frac{4-3z}{6}$.

b Determine whether the line intersects the x-axis.

c Find the angle the line makes with the x-axis.

8 **a** Find the coordinates of the point of intersection of the lines with Cartesian equations $\frac{x-2}{3} = \frac{y+1}{4} = \frac{z+1}{1}$ and $5 - x = \frac{y+2}{-3} = \frac{z-7}{2}$.

b Show that the line with equation $\mathbf{r} = \begin{pmatrix} 7 \\ 8 \\ -1 \end{pmatrix} + \lambda \begin{pmatrix} 1 \\ -1 \\ 2 \end{pmatrix}$ passes through the intersection point from part **a**.

9 Four points have coordinates $A(-1, 5, 4)$, $B(0, 1, 7)$, $C(-1, 1, 2)$ and $D(0, 0, 5)$.

Determine whether the lines AB and CD intersect.

10 Three points have coordinates $A(4, 1, 2)$, $B(1, 5, 1)$ and $C(p, p, 3)$.

a Find the value of p for which the triangle ABC has a right angle at B.

b For the value of p in part **a**, find the coordinates of point D on the side AC such that $AD = 2DC$.

11 Two lines are given by Cartesian equations:

$l_1: \frac{x-2}{3} = \frac{y+1}{-1} = \frac{z-2}{1}$

$l_2: \frac{x-5}{3} = 1 - y = z + 4$

a Show that l_1 and l_2 are parallel.

b Show that the point $A(14, -5, 6)$ lies on l_1.

c Find the coordinates of the point B on l_2 such that AB is perpendicular to the two lines.

d Hence find the distance between l_1 and l_2, giving your answer to 3 significant figures.

12 The points A, B and C have coordinates $(3, -2, 4)$, $(1, -5, 6)$ and $(-4, 5, -1)$ respectively.

The line l passes through A and has equation $\mathbf{r} = \begin{pmatrix} 3 \\ -2 \\ 4 \end{pmatrix} + \lambda \begin{pmatrix} 7 \\ -7 \\ 5 \end{pmatrix}$.

a Show that the point C lies on the line l.

b Find a vector equation of the line that passes through points A and B.

c The point D lies on the line through A and B such that the angle CDA is a right angle. Find the coordinates of D.

d The point E lies on the line through A and B such that the area of triangle ACE is three times the area of triangle ACD. Find the coordinates of the two possible positions of E.

[©AQA 2013]

13 The points A and B have coordinates $(2, 1, -1)$ and $(3, 1, -2)$ respectively. The angle OBA is θ, where O is the origin.

a **i** Find the vector $\overrightarrow{AB}$. **ii** Show that $\cos\theta = \frac{5}{2\sqrt{7}}$.

b The point C is such that $\overrightarrow{OC} = 2\overrightarrow{OB}$. The line l is parallel to $\overrightarrow{AB}$ and passes through the point C. Find a vector equation of l.

c The point D lies on l such that angle $ODC = 90°$. Find the coordinates of D.

[©AQA 2009]

14 Find the shortest distance from the point $(-1, 1, 2)$ to the line with equation $\mathbf{r} = \begin{pmatrix} 1 \\ 0 \\ 2 \end{pmatrix} + t \begin{pmatrix} -3 \\ 1 \\ 1 \end{pmatrix}$.

15 A line l_1 passes through the points $(-2, 3, 4)$ and $(1, 5, -3)$.

a Find a vector equation for l_1.

A line l_2 has equation $\mathbf{r} = \begin{pmatrix} 0 \\ 1 \\ -2 \end{pmatrix} + \lambda \begin{pmatrix} 3 \\ 2 \\ -7 \end{pmatrix}$.

b Find the exact distance between l_1 and l_2.

16 Two lines have equations l_1: $\mathbf{r} = \begin{pmatrix} 2 \\ -1 \\ 0 \end{pmatrix} + \lambda \begin{pmatrix} 1 \\ -2 \\ 2 \end{pmatrix}$ and l_2: $\mathbf{r} = \begin{pmatrix} 2 \\ -1 \\ 0 \end{pmatrix} + \mu \begin{pmatrix} 1 \\ 1 \\ 2 \end{pmatrix}$ and intersect at point P.

a Show that $Q(5, 2, 6)$ lies on l_2.

b R is a point on l_1 such that $|PR| = |PQ|$. Find the possible coordinates of R.

17 Two lines are given by l_1: $\mathbf{r} = \begin{pmatrix} -5 \\ 1 \\ 10 \end{pmatrix} + \lambda \begin{pmatrix} -3 \\ 0 \\ 4 \end{pmatrix}$ and l_2: $\mathbf{r} = \begin{pmatrix} 3 \\ 0 \\ -9 \end{pmatrix} + \mu \begin{pmatrix} 1 \\ 1 \\ 7 \end{pmatrix}$.

a l_1 and l_2 intersect at P. Find the coordinates of P.

b Show that the point $Q(5, 2, 5)$ lies on l_2.

c Find the coordinates of point M on l_1 such that QM is perpendicular to l_1.

d Find the area of the triangle PQM.

18 a Show that the lines l_1: $\mathbf{r} = -3\mathbf{i} + 3\mathbf{j} + 18\mathbf{k} + s(2\mathbf{i} - \mathbf{j} - 8\mathbf{k})$ and l_2: $\mathbf{r} = 5\mathbf{i} + 2\mathbf{k} + t(\mathbf{i} + \mathbf{j} - \mathbf{k})$ do not intersect.

b Points P and Q lie on l_1 and l_2 respectively, such that $\overrightarrow{PQ}$ is perpendicular to both lines.

i Show that $3t - 23s + 49 = 0$.

ii Find a second equation for s and t.

iii Hence find the shortest distance between l_1 and l_2, giving your answer as an exact value.

19 The coordinates of the points A and B are $(3, -2, 4)$ and $(6, 0, 3)$ respectively.

The line l_1 has equation $\mathbf{r} = \begin{pmatrix} 3 \\ -2 \\ 4 \end{pmatrix} + \lambda \begin{pmatrix} 2 \\ -1 \\ 3 \end{pmatrix}$.

a i Find the vector $\overrightarrow{AB}$.

ii Calculate the acute angle between $\overrightarrow{AB}$ and the line l_1, giving your answer to the nearest 0.1°.

b The point D lies on l_1 where $\lambda = 2$. The line l_2 passes through D and is parallel to AB.

i Find a vector equation of line l_2 with parameter μ.

ii The diagram shows a symmetrical trapezium $ABCD$, with angle DAB equal to angle ABC.

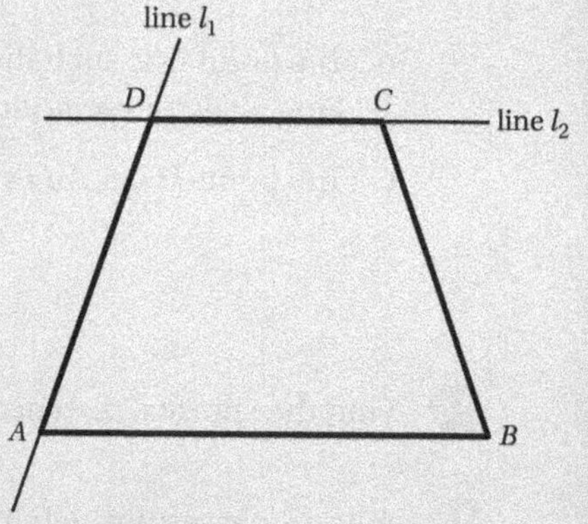

The point C lies on line l_2. The length of AD is equal to the length of BC.

Find the coordinates of C.

[© AQA 2011]

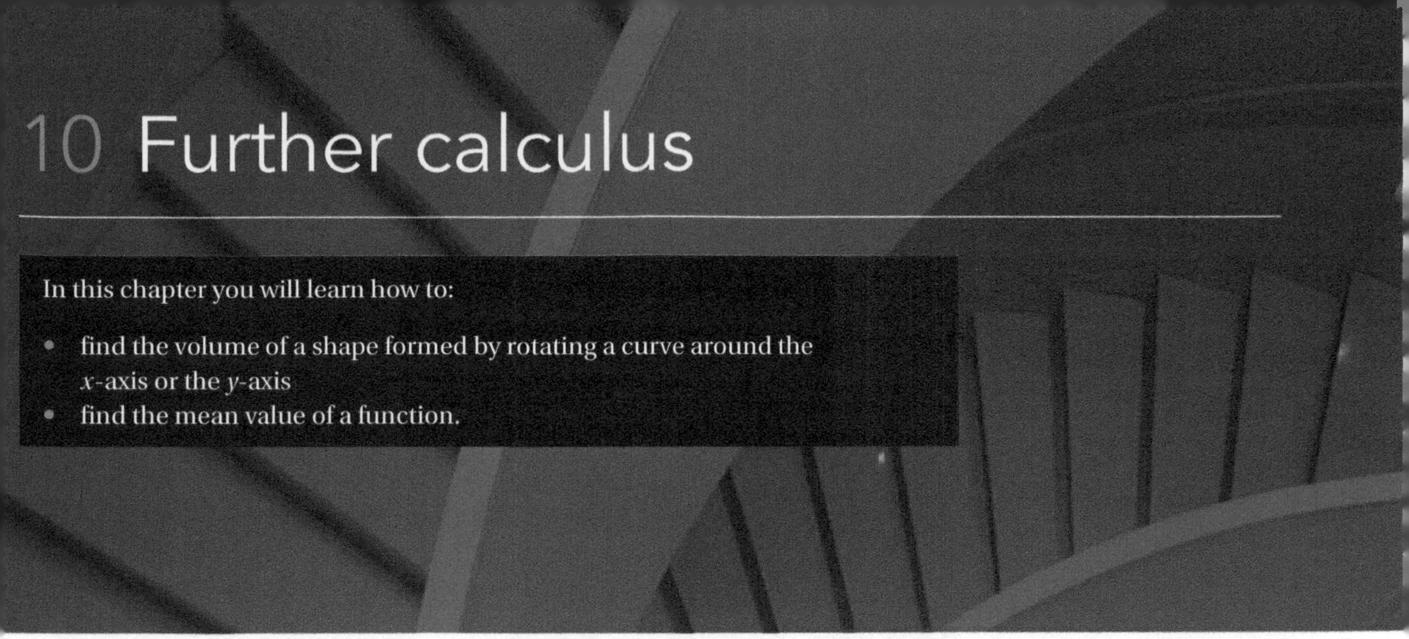

10 Further calculus

In this chapter you will learn how to:

- find the volume of a shape formed by rotating a curve around the x-axis or the y-axis
- find the mean value of a function.

Before you start...

GCSE	You should know the formula for the volume of a cylinder.	1 Find the exact volume of a cylinder with base radius 4 cm and height 10 cm.
A Level Mathematics Student Book 1, Chapter 14	You should know how to find the definite integral of a polynomial.	2 Evaluate $\int_1^3 (x^4 + 2)\,dx$.
A Level Mathematics Student Book 1, Chapter 16	You should know that displacement is found by $\int v\,dt$.	3 Find the displacement in the first 10 seconds of a particle with velocity $3x^3$ m s^{-1}.

What else can you do with calculus?

You have already seen several applications of calculus, such as finding tangents and normals to curves, optimisation, finding areas and converting between displacement, velocity and acceleration in kinematics. In this chapter, you will see two further applications - finding volumes and finding the mean value of a function.

Section 1: Volumes of revolution

In A Level Mathematics Student Book 1, Chapter 14, you saw that the area between a curve and the x-axis from $x = a$ to $x = b$ is given by $\int_a^b y\,dx$, as long as $y > 0$. In this section, you will use a similar formula to find the volume of a shape formed by rotating the curve about either the x-axis or the y-axis.

If a curve is rotated about the x-axis or the y-axis, the resulting shape is called a **solid of revolution** and the volume of that shape is referred to as the **volume of revolution**.

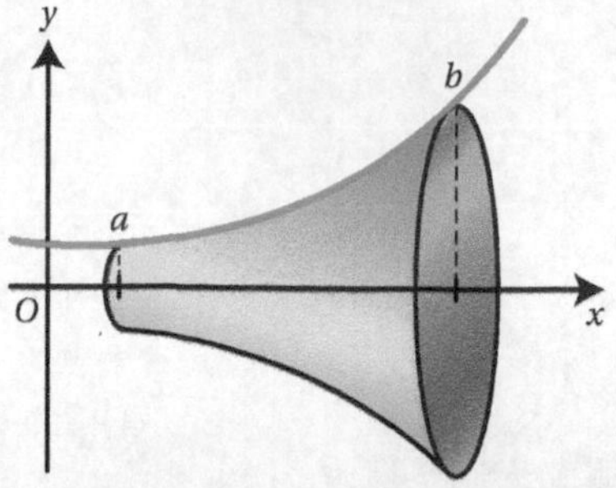

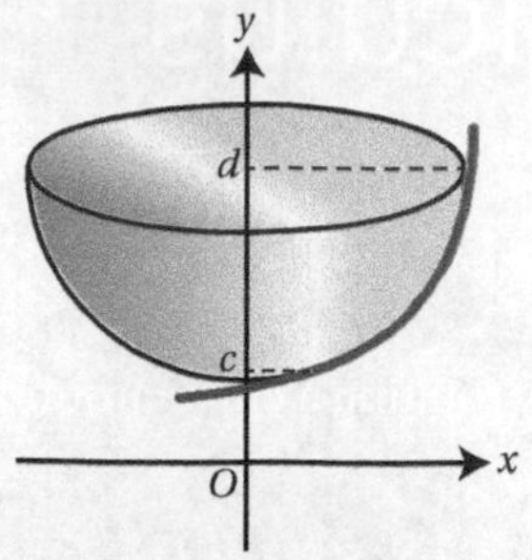

Key point 10.1

- When the curve $y = f(x)$ between $x = a$ and $x = b$ is rotated 360° about the x-axis, the volume of revolution is given by $V = \pi\int_a^b y^2 \,dx$.
- When the curve $y = f(x)$ between $y = c$ and $y = d$ is rotated 360° about the y-axis, the volume of revolution is given by $V = \pi\int_c^d x^2 \,dy$.

The proof of these results is very similar. The proof for rotation about the x-axis is given here.

PROOF 10

The solid can be split into small cylinders.

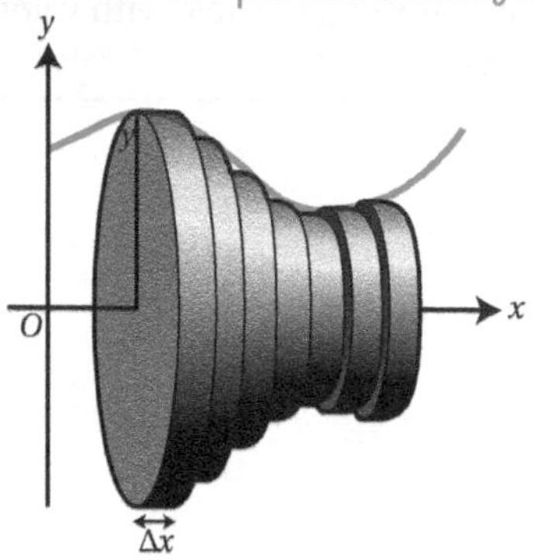

Draw an outline of a representative function to illustrate the argument.

The volume of each cylinder is $\pi y^2 \Delta x$.

The radius of each cylinder is the y-coordinate and the height is Δx.

The total volume is approximately:

$$V \approx \sum_a^b \pi y^2 \Delta x$$

You are starting at $x = a$ and stopping at $x = b$.
It is only approximate because the volume of revolution is not exactly the same as the total volume of the cylinders.

$$V = \lim_{\Delta x \to 0} \sum_a^b \pi y^2 \Delta x$$

$$= \int_a^b \pi y^2 \,dx$$

$$= \pi\int_a^b y^2 \,dx$$

However, as you make the cylinders smaller the volume gets more and more accurate. The sum then becomes an integral. You can leave π out of the integration and multiply by it at the end.

WORKED EXAMPLE 10.1

The graph of $y=\sqrt{x^2+1}$, $0\leqslant x\leqslant 3$, is rotated 360° about the x-axis.

Find, in terms of π, the volume of the solid generated.

$V=\pi\int_0^3 (x^2+1)\,dx$ — Use the formula: $V=\pi\int_a^b y^2\,dx$.

$=\pi\left[\frac{x^3}{3}+x\right]_0^3$ — Evaluate the definite integral.

$=\pi\left[\left(\frac{3^3}{3}+3\right)-0\right]$

$=12\pi$

To find the volume of revolution about the y-axis you will often have to rearrange the equation of the curve to find x in terms of y.

Common error

Remember that the limits of the integration need to be in terms of y and not x.

WORKED EXAMPLE 10.2

The part of the curve $y=\frac{1}{x}$ between $x=1$ and $x=4$ is rotated 360° about the y-axis.
Find the exact value of the volume of the solid generated.

When $x=1$, $y=\frac{1}{1}=1$ — Find the limits in terms of y.

When $x=4$, $y=\frac{1}{4}$

$y=\frac{1}{x}\Rightarrow x=\frac{1}{y}$ — Express x in terms of y.

$V=\pi\int_a^b x^2\,dy$ — Use the formula $V=\pi\int_a^b x^2\,dy$, substituting in $x=\frac{1}{y}$.

$=\pi\int_{\frac{1}{4}}^{1}\left(\frac{1}{y}\right)^2 dy$

$=\pi\int_{\frac{1}{4}}^{1} y^{-2}\,dy$

$=\pi\left[-y^{-1}\right]_{\frac{1}{4}}^{1}$

$=\pi[(-1)-(-4)]$

$=3\pi$

You might also be asked to find a volume of revolution of an area between two curves.

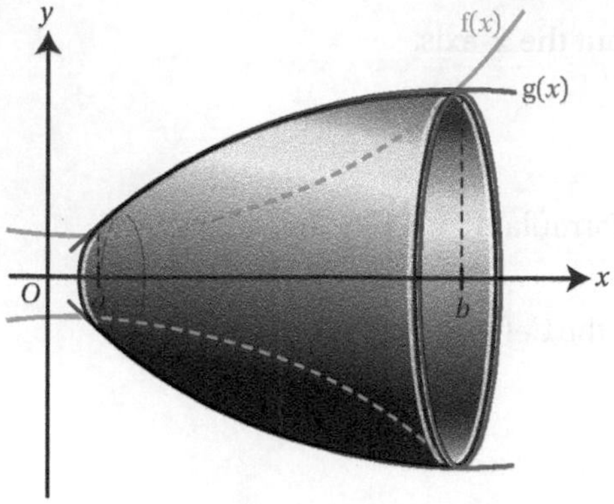

From the diagram you can see that the volume formed when the region R is rotated around the x-axis is given by the volume of revolution of $g(x)$ minus the volume of revolution of $f(x)$.

Tip

Remember that many calculators can find definite integrals.

Key point 10.2

The volume of revolution of the region between curves $g(x)$ and $f(x)$ is:

$$V = \pi\int_a^b (g(x)^2 - f(x)^2)\,dx$$

where $g(x)$ is above $f(x)$ and the curves intersect at $x = a$ and $x = b$.

Common error

Make sure that you square **each term** within the brackets and do not make the mistake of squaring the whole expression inside the brackets: the formula **is not** $\pi\int_a^b (g(x) - f(x))^2\,dx$.

WORKED EXAMPLE 10.3

Find the volume formed when the region enclosed by $y = x^2 + 6$ and $y = 8x - x^2$ is rotated through 360° about the x-axis.

For points of intersection:

$$x^2 + 6 = 8x - x^2$$
$$2x^2 - 8x + 6 = 0$$
$$x^2 - 4x + 3 = 0$$
$$(x-1)(x-3) = 0$$
$$x = 1 \text{ or } x = 3$$

First find the x-coordinates of the points where the curves meet, by equating the RHS of both equations and solving. This will give you the limits of integration.

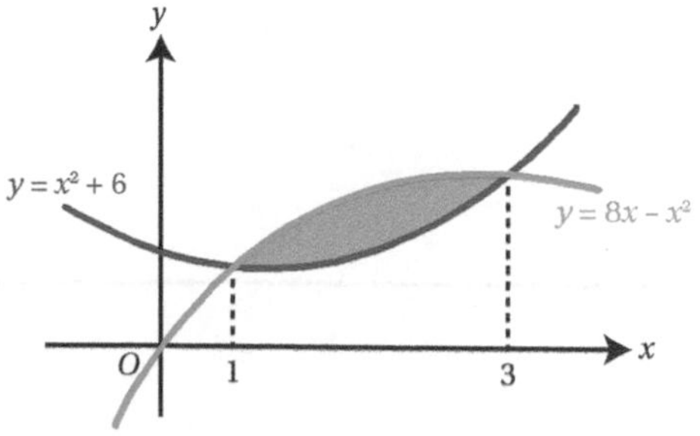

Sketch the graphs in the region concerned.

$y = 8x - x^2$ is above $y = x^2 + 6$.

Continues on next page

$$V = \pi\int_1^3 ((8x - x^2)^2 - (x^2+6)^2)\,dx$$

Apply the formula $V = \pi\int_a^b (g(x)^2 - f(x)^2)\,dx$.

$$= \pi\int_1^3 ((64x^2 - 16x^3 + x^4) - (x^4 + 12x^2 + 36))\,dx$$

$$= \pi\int_1^3 (52x^2 - 16x^3 - 36)\,dx$$

Expand and simplify.

$$= \pi\left[\frac{52}{3}x^3 - 4x^4 - 36x\right]_1^3$$

Then evaluate the definite integral.

$$= \pi\left[\left(\frac{52}{3}\times 3^3 - 4\times 3^4 - 36\times 3\right) - \left(\frac{52}{3}\times 1^3 - 4\times 1^4 - 36\times 1\right)\right]$$

$$= \frac{176}{3}\pi$$

Did you know?

There are also formulae to find the surface area of a solid formed by rotating a region around an axis. Some particularly interesting examples arise if you allow one end of the region to tend to infinity; for example, rotating the region formed by the lines $y = \frac{1}{x}$, $x = 1$ and the x-axis results in a solid called Gabriel's horn or Torricelli's trumpet.

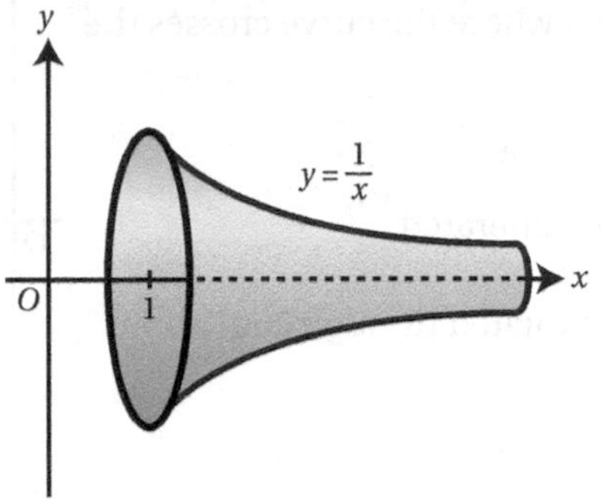

Areas and volumes can also be calculated using what are called improper integrals, and it ensues that it is possible to have a solid of finite volume but infinite surface area!

EXERCISE 10A

1 The part of the curve $y = f(x)$ for $a \leqslant y \leqslant b$ is rotated 360° about the x-axis. Find the exact volume of revolution formed in each case.

a **i** $f(x) = x^2$; $a = -1$, $b = 1$ **ii** $f(x) = x^3$; $a = 0$, $b = 2$

b **i** $f(x) = x^2 + 6$; $a = -1$, $b = 3$ **ii** $f(x) = 2x^3 + 1$; $a = 0$, $b = 1$

c **i** $f(x) = \frac{1}{x}$; $a = 1$, $b = 2$ **ii** $f(x) = \frac{1}{x^2}$; $a = 1$, $b = 4$

2 Find the exact volume of revolution formed when each curve, for $a \leqslant x \leqslant b$, is rotated through 2π radians about the x-axis.

a **i** $y = e^x$; $a = 0$, $b = 1$ **ii** $y = e^{-x}$; $a = 0$, $b = 3$

b **i** $y = e^{2x} + 1$; $a = 0$, $b = 1$ **ii** $y = e^{-x} + 2$; $a = 0$, $b = 2$

c **i** $y = \sqrt{\sin x}$; $a = 0$, $b = \pi$ **ii** $y = \sqrt{\cos x}$; $a = 0$, $b = \frac{\pi}{2}$

3 The part of the curve for $a \leqslant y \leqslant b$ is rotated 360° about the y-axis.

Find the exact volume of revolution formed in each case.

a **i** $y = 4x^2 + 1$; $a = 1, b = 17$ **ii** $y = \frac{x^2 - 1}{3}$; $a = 0, b = 5$

b **i** $y = x^3$; $a = 0, b = 8$ **ii** $y = x^4$; $a = 2, b = 8$

c **i** $y = \frac{1}{x^3}$; $a = 8, b = 27$ **ii** $y = \frac{1}{x^5}$; $a = 1, b = 32$

A **4** The part of the curve $y = \mathrm{f}(x)$ for $a \leqslant y \leqslant b$ is rotated 360° about the y-axis.

Find the exact volume of revolution formed in each case.

a **i** $\mathrm{f}(x) = \ln x + 1$; $a = 1, b = 3$ **ii** $\mathrm{f}(x) = \ln(2x - 1)$; $a = 0, b = 4$

b **i** $\mathrm{f}(x) = \frac{1}{x^2}$; $a = 1, b = 2$ **ii** $\mathrm{f}(x) = \frac{1}{x^2} + 2$; $a = 3, b = 5$

c **i** $\mathrm{f}(x) = \arcsin x$; $a = -\frac{\pi}{2}, b = \frac{\pi}{2}$ **ii** $\mathrm{f}(x) = \arcsin x$; $a = -\frac{\pi}{4}, b = \frac{\pi}{4}$

5 The diagram shows the region, R, bounded by the curve $y = \sqrt{x} - 2$, the x-axis and the line $x = 9$.

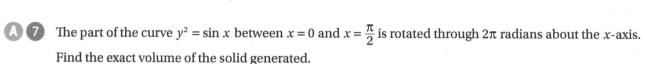

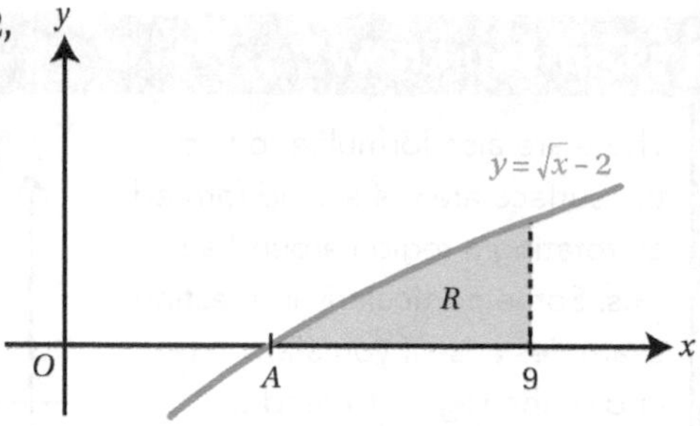

a Find the coordinates of the point A where the curve crosses the x-axis.

This region is rotated 360° about the x-axis.

b Find the exact volume of the solid generated.

6 The curve $y = 3x^2 + 1$, for $0 \leqslant x \leqslant 2$, is rotated through 360° about the y-axis.

Find the volume of revolution generated, correct to 3 s.f.

A **7** The part of the curve $y^2 = \sin x$ between $x = 0$ and $x = \frac{\pi}{2}$ is rotated through 2π radians about the x-axis.

Find the exact volume of the solid generated.

8 The curve $y = x^2$, for $0 < x < a$, is rotated through 180° about the x-axis.

The resulting volume is $\frac{16\pi}{5}$.

Find the value of a.

9 The region enclosed by the curve $y = x^2 - a^2$ and the x-axis is rotated 90° about the x-axis.

Find an expression for the volume of revolution formed.

A **10** The part of the curve $y = \sqrt{\frac{3}{x}}$ between $x = 1$ and $x = a$ is rotated through 2π radians about the x-axis. The volume of the resulting solid is $\pi \ln \frac{64}{27}$.

Find the exact value of a.

11 **a** Find the coordinates of the points of intersection of curves $y = x^2 + 3$ and $y = 4x + 3$.

b Find the volume of revolution generated when the region between the curves $y = x^2 + 3$ and $y = 4x + 3$ is rotated through 360° about the x-axis.

12 The region bounded by the curves $y = x^2 + 6$ and $y = 8x - x^2$ is rotated through 360° about the x-axis. Find the volume of the resulting shape.

13 **a** Find the coordinates of the points of intersection of the curves $y = 4\sqrt{x}$ and $y = x + 3$.

b The region between the curves $y = 4\sqrt{x}$ and $y = x + 3$ is rotated through 360° about the y-axis. Find the volume of the solid generated.

14 By rotating the circle $x^2 + y^2 = r^2$ around the x-axis, prove that the volume of a sphere of radius r is given by $\frac{4}{3}\pi r^3$.

15 By choosing a suitable function to rotate around the x-axis, prove that the volume of a circular cone with base radius r and height h is $\frac{\pi r^2 h}{3}$.

16 Find the volume of revolution when the region enclosed by the graphs of $y = e^x$, $y = 1$ and $x = 1$ is rotated through 360° about the line $y = 1$.

Section 2: Mean value of a function

Suppose an object travels between $t = 0$ s and $t = 3$ s with a velocity given by $v = t$. Its velocity–time graph looks like this.

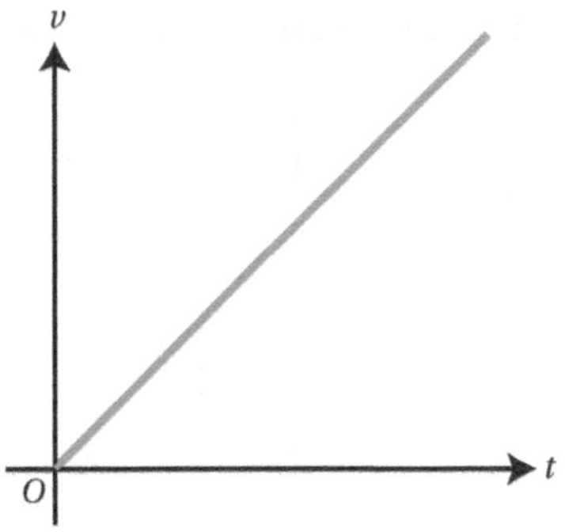

Its average velocity can be found from:

$$\frac{\text{initial velocity} + \text{final velocity}}{2} = \frac{0+3}{2}$$
$$= 1.5$$

Suppose, instead, the object has velocity given by $v = \frac{t^2}{3}$. Then you can compare the two velocity–time graphs.

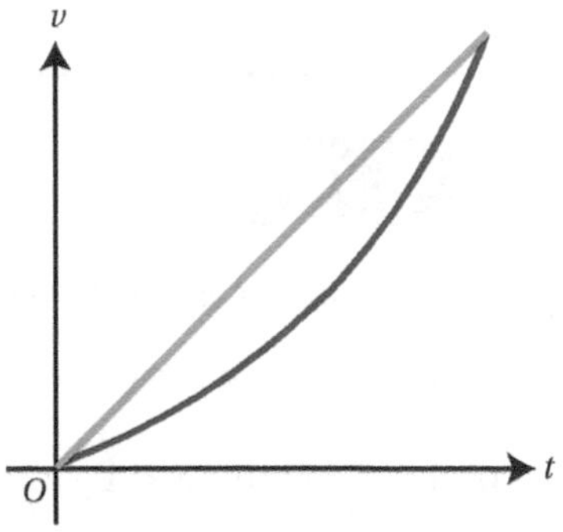

The formula $\frac{\text{initial velocity} + \text{final velocity}}{2}$ would give the same average velocity for the two graphs, which can't be correct because the red curve is underneath the blue line everywhere other than at the end points.

You need a measure of average that takes into account the value of the function everywhere.

One possibility is to use $\frac{\text{total distance}}{\text{time taken}}$.

You can then use the fact that total distance is the integral of velocity with respect to time.

For the blue line this gives:

$$\text{average velocity} = \frac{\int_0^3 t \, dt}{3}$$
$$= \frac{1}{3}\left[\frac{t^2}{2}\right]_0^3$$
$$= 1.5$$

For the red curve this gives:

$$\text{average velocity} = \frac{\int_0^3 \frac{t^2}{3}\,dt}{3}$$

$$= \frac{1}{3}\left[\frac{t^3}{9}\right]_0^3$$

$$= 1$$

This process can be generalised for any function.

Key point 10.4

The mean value of a function $f(x)$ between a and b is:

$$\frac{\int_a^b f(x)\,dx}{b-a}$$

WORKED EXAMPLE 10.4

Find the mean value of $x^2 - x$ between 3 and 4.

$$\text{Mean value} = \frac{\int_3^4 (x^2 - x)\,dx}{4-3}$$

Use the formula for the mean value of a function: $\frac{\int_a^b f(x)\,dx}{b-a}$

$$= \frac{1}{4-3}\left[\frac{x^3}{3} - \frac{x^2}{2}\right]_3^4$$

Notice that $x^2 - x$ varies between 6 and 12, so a mean of around 9 seems reasonable.

$$= \frac{40}{3} - \frac{9}{2}$$

$$= \frac{53}{6}$$

EXERCISE 10B

1 Find the mean value of each function between the given values of x.

a **i** x^2 for $0 < x < 1$ **ii** x^2 for $1 < x < 3$

b **i** $\sqrt{x}$ for $0 < x < 4$ **ii** $\frac{1}{x^2}$ for $1 < x < 5$

c **i** $x^3 + 1$ for $0 < x < 4$ **ii** $x^4 - x$ for $0 < x < 10$

A **2** Find the mean value of each function over the domain given.

a **i** $\sin x$ for $0 < x < \pi$ **ii** $\cos x$ for $0 < x < \pi$

b **i** e^x for $0 < x < 1$ **ii** $\frac{1}{x}$ for $1 < x < e$

c **i** $\sqrt{x+1}$ for $3 < x < 8$ **ii** $x \sin(x^2)$ for $0 < x < \sqrt{\pi}$

3 The velocity of a rocket is given by $v = 30\sqrt{t}$ where t is time, in seconds, and v is velocity, in metres per second.

Find the mean velocity in the first T seconds.

4 The mean value of the function $x^2 - x$ for $0 < x < a$ is zero.

Find the value of a.

5 $f(x) = x^2$ for $x \geqslant 0$.

a f_{mean} is the mean value of $f(x)$ between 0 and a. Find an expression for f_{mean} in terms of a.

b Given that $f(c) = f_{mean}$ find an expression for c in terms of a.

6 Show that the mean value of $\frac{1}{x^2}$ between 1 and a is inversely proportional to a.

7 An alternating current has time period 2. The power dissipated by the current through a resistor is given by $P = P_0 \sin^2(\pi t)$.

Find the ratio of the mean power of one complete period to the maximum power.

8 The mean value of $f(x)$ between a and b is F.

Prove that the mean value of $f(x)+1$ between a and b is $F+1$.

9 **a** Sketch the graph of $\frac{1}{2\sqrt{x}}$.

b Use the graph to explain why the mean value of the function between a and b is less than the mean of $f(a)$ and $f(b)$.

c Hence prove that, if $0 < a < b$, $\sqrt{b} - \sqrt{a} < \frac{1}{3}\left(\frac{b}{\sqrt{a}} - \frac{a}{\sqrt{b}}\right)$.

10 If f_{mean} is the mean value of $f(x)$ for $a < x < b$ and $f(a) < f(b)$, then $f(a) < f_{mean} < f(b)$.

Either prove this statement or disprove it using a counterexample.

Checklist of learning and understanding

- The volume of a shape formed by rotating a curve about the x-axis or the y-axis is known as the volume of revolution.
  - When the curve $y = f(x)$ between $x = a$ and $x = b$ is rotated 360° about the x-axis, the volume of revolution is given by

    $$V = \pi \int_a^b y^2 \, dx$$

  - When the curve $y = f(x)$ between $y = c$ and $y = d$ is rotated 360° about the y-axis, the volume of revolution is given by

    $$V = \pi \int_c^d x^2 \, dy$$

  - The volume of revolution of the region between curves $g(x)$ and $f(x)$ is:

    $$V = \pi \int_a^b \left(g(x)^2 - f(x)^2 \right) dx$$

    where $g(x)$ is above $f(x)$ and the curves intersect at $x = a$ and $x = b$.
- The mean value of a function $f(x)$ between a and b is:

  $$\frac{\int_a^b f(x) \, dx}{b - a}$$

Mixed practice 10

1 Find the volume of revolution when the curve $y = x^2$ for $1 < x < 2$ is rotated through 360° around the x-axis.

Choose from these options.

A $\frac{27\pi}{5}$ B $\frac{31\pi}{5}$ C $\frac{32\pi}{5}$ D 15π

2 Find the mean value of x^3 between 1 and 4.

Choose from these options.

A $\frac{85}{4}$ B $\frac{65}{3}$ C $\frac{65}{2}$ D $\frac{255}{4}$

3 The curve $y = \sqrt{x}$ between 0 and a is rotated through 360° about the x-axis. The resulting shape has a volume of 18π.

Find the value of a.

4 For $0 < x < a$, the mean value of x is equal to the mean value of x^2.

Find the value of a.

5 The mean value of $\frac{1}{\sqrt{x}}$ from 0 to b is 1.

Find the value of b.

6 The curve $x = \frac{y^2 - 1}{3}$, with $1 \leqslant y \leqslant 4$, is rotated through 360° about the y-axis.

Find the volume of revolution generated, correct to 3 s.f.

7 The diagram shows the curve with equation $y = \sqrt{100 - 4x^2}$, where $x \geqslant 0$.

Calculate the volume of the solid generated when the region bounded by the curve shown and the coordinate axes is rotated through 360° about the y-axis, giving your answer in terms of π.

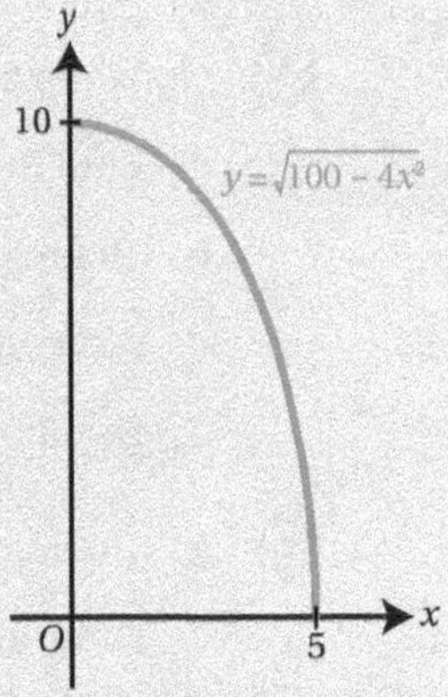

[©AQA 2009]

A 8 The diagram shows the curve with equation $y = \sqrt{(x-2)^5}$ for $x \geqslant 2$.

The shaded region R is bounded by the curve $y = \sqrt{(x-2)^5}$, the x-axis and the lines $x = 3$ and $x = 4$.

Find the exact value of the volume of the solid formed when the region R is rotated through 360° about the x-axis.

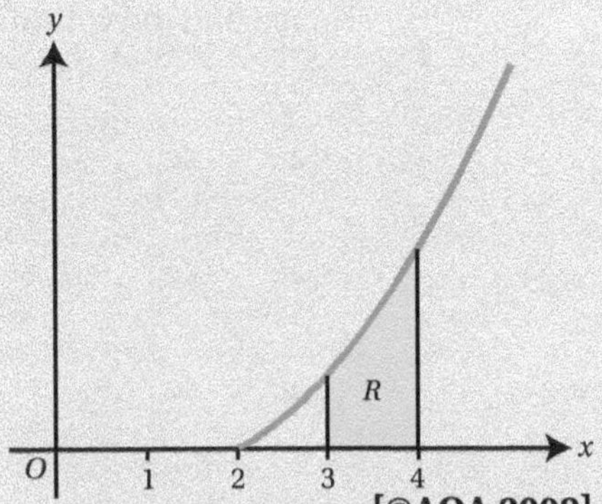

[©AQA 2009]

9 $f(x) = \frac{1}{x^2}$ for $x > 0$

a f_{mean} is the mean value of $f(x)$ between 1 and a. Find an expression for f_{mean} in terms of a.

b Given that $f(c) = f_{mean}$ find an expression for c in terms of a.

10 The region bounded by the curve $y = ax - x^2$ and the x-axis is rotated one full turn about the x-axis. Find, in terms of a, the resulting volume of revolution.

11 Prove that the mean value of x between a and b is the arithmetic mean of a and b.

A 12 The diagram shows the curve $y = \ln x$ and the line $y = -\frac{1}{e}x + 2$.

a Show that the two graphs intersect at (e, 1).

The shaded region is rotated through 360° about the y-axis.

b Find the exact value of the volume of revolution.

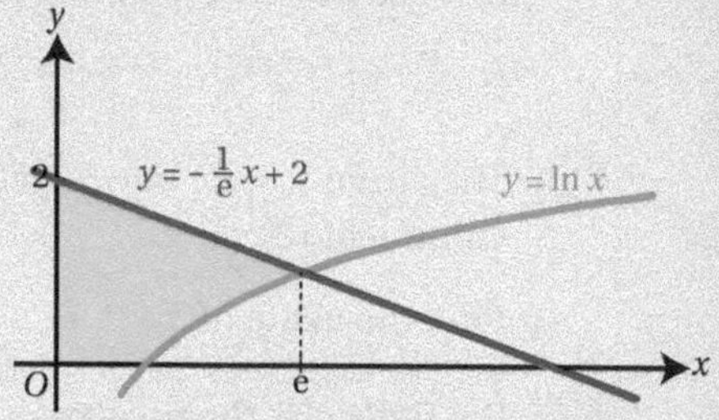

13 The region enclosed by $y = (x-1)(x-2)+1$ and the line $y = 1$ is rotated through 180° about the line $y = 1$.

Find the exact value of the resulting volume.

14 The part of the curve $y = x^2 + 3$ between $y = 3$ and $y = k$ $(k > 0)$ is rotated 360° about the y-axis. The volume of revolution formed is 25π.

Find the value of k.

15 Consider two curves with equations $y = x^2 - 8x + 12$ and $y = 12 + x - x^2$.

a Find the coordinates of the points of intersection of the two curves.

b The region enclosed by the curves is rotated through 360° about the x-axis.

Write down an integral expression for the volume of the solid generated.

c Evaluate the volume, giving your answer to the nearest integer.

16 **a** The region enclosed by $y = x^2$ and $y = \sqrt{x}$ is labelled R.

Draw a sketch showing R.

b Find the volume when R is rotated through 360° about the x-axis.

c Hence find the volume when R is rotated through 360° about the y-axis.

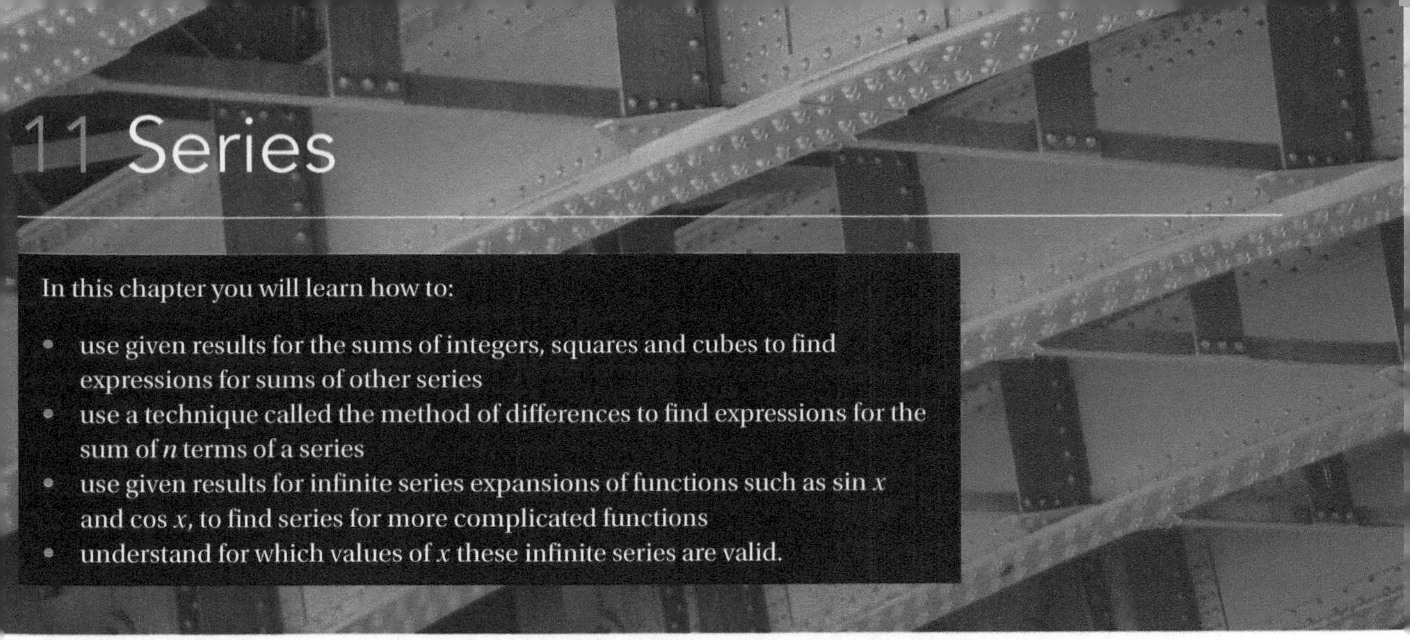

11 Series

In this chapter you will learn how to:

- use given results for the sums of integers, squares and cubes to find expressions for sums of other series
- use a technique called the method of differences to find expressions for the sum of n terms of a series
- use given results for infinite series expansions of functions such as $\sin x$ and $\cos x$, to find series for more complicated functions
- understand for which values of x these infinite series are valid.

Before you start...

GCSE	You should be able to use the nth term formula to generate terms of a sequence.	1 A sequence is defined by $u_n = n^2 + 3n - 1$. Find the first three terms.
GCSE	You should be able to simplify expressions by factorising.	2 Simplify $n(n+1)(2n+3)+n(n+1)(n-3)$.
A Level Mathematics Student Book 1, Chapter 9	You should be able to use the binomial expansion on $(x+y)^n$ for positive integer n.	3 Expand $(2+3x)^4$.

Summing sequences

If you sum the terms of a sequence, you get a **series**. The formula for the nth term of a sequence is useful because it allows you to find any term you wish without having to find all the terms that have come before. In the same way, it is useful to have a formula for the sum of the first n terms of a sequence.

In this chapter, you will look at methods for finding formulae for these sums and at forming infinite series for some common functions such as sine, cosine and the exponential function.

Section 1: Sigma notation

The sum of a sequence up to a certain point is called a **series**. You often use the symbol S_n to denote the sum of the first n terms of a sequence.

$$S_n = u_1 + u_2 + u_3 + \cdots + u_n$$

> **Tip**
>
> The expression u_n is the standard notation for the nth term of a sequence. So in the sequence 2, 5, 8, 11, 14, ..., $u_1 = 2$, $u_2 = 5$, $u_3 = 8$, and so on.

Instead of $S_n = u_1 + u_2 + u_3 + \cdots + u_n$, you will often see exactly the same idea expressed in a shorter form, using sigma notation:

Key point 11.1

This is the last value taken by r, where counting ends.

The value of r changes with each new term.

Greek capital sigma means 'add up' → $\sum_{r=1}^{r=n} f(r) = f(1) + f(2) + ... + f(n)$

This is the first value taken by r, where counting starts.

Tip

Don't be put off by this notation. If you are in any doubt, try writing out the first few terms in full.

There is nothing special about the letter r here; you could use any letter, but r and k are the most usual. Note that sometimes sigma is written without the r and k above and below it – you may just see the first and last values.

WORKED EXAMPLE 11.1

$$T_n = \sum_{2}^{n} r^2$$

Find the value of T_4.

$T_4 = 2^2 + 3^2 + 4^2$ — Substitute the starting value, $r = 2$, into the expression to be summed, r^2. You've not reached the end value, so put in $r = 3$. You've still not reached the end value, so put in $r = 4$.

$= 4 + 9 + 16$

$= 29$ — You've reached the end value, so stop and evaluate.

WORKED EXAMPLE 11.2

Write the series $\frac{1}{2} + \frac{1}{3} + \frac{1}{4} + \frac{1}{5} + \frac{1}{6}$ in sigma notation.

The general term is $\frac{1}{r}$. — Describe each term of the series using a general term in the variable r.

The series starts at $r = 2$. — Note the first value of r.

The series ends at $r = 6$. — Note the final value of r.

$\frac{1}{2} + \frac{1}{3} + \frac{1}{4} + \frac{1}{5} + \frac{1}{6} = \sum_{2}^{6} \frac{1}{r}$ — Summarise in sigma notation.

EXERCISE 11A

1 Evaluate each expression.

a **i** $\sum_{2}^{4} 3r$ **ii** $\sum_{5}^{7} (2r+1)$

b **i** $\sum_{3}^{6} 2^r - 1$ **ii** $\sum_{1}^{4} 1.5^r$

c **i** $\sum_{a=0}^{a=5} b(a+1)$ **ii** $\sum_{q=2}^{q=5} pq^2$

2 Write each expression in sigma notation.
Note that there is more than one correct answer.

a **i** $2+3+4+\cdots+43$ **ii** $6+8+10+\cdots+60$

b **i** $\frac{1}{4}+\frac{1}{8}+\frac{1}{16}+\cdots+\frac{1}{128}$ **ii** $2+\frac{2}{3}+\frac{2}{9}+\cdots+\frac{2}{243}$

c **i** $14a+21a+28a+\cdots+70a$ **ii** $0+1+2^b+3^b+\cdots+19^b$

Section 2: Using standard formulae

In general, it is difficult to find a formula for the first n terms of a series, but if you know the formula for a few 'standard' series, you can use them to establish formulae for many other series.

You can use the three standard formulae in Key point 11.2 without proof – unless you're asked to prove them with the method used in Section 3!

Key point 11.2

Formulae for the sums of integers, squares and cubes:

- $\sum_{r=1}^{n} r = \frac{1}{2}n(n+1)$
- $\sum_{r=1}^{n} r^2 = \frac{1}{6}n(n+1)(2n+1)$
- $\sum_{r=1}^{n} r^3 = \frac{1}{4}n^2(n+1)^2$

The second and third formulae will be given in your formula book.

Fast forward

The formulae in Key point 11.2 will be proved in Section 3 of this chapter.

You will also see in Chapter 12, Section 2, how these results can be proved more directly by means of a technique called **induction**.

Did you know?

The formula for the sum of the first n integers, $\frac{1}{2}n(n+1)$, is the nth term formula for the sequence of triangular numbers: 1, 3, 6, 10, 15, 21 ...

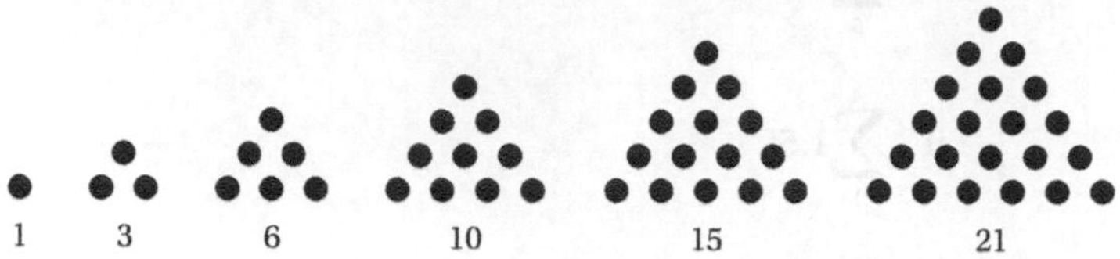

Before you use these results, notice how you can split up sums and take out constants. For example:

$$\sum_{r=1}^{n}(3r+2) = (3\times1+2)+(3\times2+2)+(3\times3+2)+(3\times4+2)+\cdots+(3n+2)$$

$$= 3(1+2+3+4+\cdots+n)+\underbrace{2+2+2+2+\cdots+2}_{n\text{ times}}$$

$$= 3\sum_{r=1}^{n} r+\sum_{r=1}^{n} 2$$

where $\sum_{r=1}^{n} 2 = 2n$.

Key point 11.3

You can manipulate series in several ways.

- $\sum(u_r+v_r)=\sum u_r+\sum v_r$
- $\sum cu_r = c\sum u_r$
- $\sum_{r=1}^{n} c = nc$

where c is a constant.

Common error

Remember that a constant, c, summed n times is nc and not just c. For example, $\sum_{r=1}^{n} 2 = 2n$ and not 2.

WORKED EXAMPLE 11.3

a Use the formula for $\sum_{r=1}^{n} r$ to show that $\sum_{r=1}^{n}(4r+3) = n(2n+5)$.

b Hence find $\sum_{r=8}^{20}(4r+3)$.

a $\sum_{r=1}^{n}(4r+3) = \sum_{r=1}^{n} 4r + \sum_{r=1}^{n} 3$

You need to rearrange the expression into a form to which you can apply the standard formulae. Start by splitting up the sum.

Continues on next page

$$= 4\sum_{r=1}^{n} r + \sum_{r=1}^{n} 3$$

Then take 4 out of the first sum as a factor.

$$= 4 \times \frac{1}{2}n(n+1) + 3n$$

$\sum_{r=1}^{n} r = \frac{1}{2}n(n+1)$ and $\sum_{r=1}^{n} 3 = 3n$

$$= n[2(n+1)+3]$$
$$= n(2n+5)$$

Notice that it's always a good idea to factorise first. In this case only n factorises, but in more complicated examples this will avoid having to expand and then factorise a higher order polynomial later.

b $\sum_{r=8}^{20}(4r+3) = \sum_{r=1}^{20}(4r+3) - \sum_{r=1}^{7}(4r+3)$

You can only use the formula in part **a** if the sum starts from $r = 1$. Therefore, work out the sum of the first 20 terms and subtract the sum of the first 7 terms.

$$= 20(2\times 20+5) - 7(2\times 7+5)$$
$$= 900 - 133$$
$$= 767$$

Now use the formula $n(2n+5)$ with $n = 20$ and $n = 7$.

WORKED EXAMPLE 11.4

a Use the formulae for $\sum_{r=1}^{n} r$, $\sum_{r=1}^{n} r^2$ and $\sum_{r=1}^{n} r^3$ to show that $\sum_{r=1}^{n} r(2r-5)(r+1) = \frac{1}{2}n(n+1)(n+2)(n-3)$.

b Hence find an expression for $\sum_{r=1}^{2n} r(2r-5)(r+1)$, simplifying your answer fully.

a $\sum_{r=1}^{n} r(2r-5)(r+1) = \sum_{r=1}^{n} r(2r^2-3r-5)$

Expand the brackets.

$$= \sum_{r=1}^{n}(2r^3 - 3r^2 - 5r)$$

$$= \sum_{r=1}^{n} 2r^3 - \sum_{r=1}^{n} 3r^2 - \sum_{r=1}^{n} 5r$$

Split up the series into separate sums.

$$= 2\sum_{r=1}^{n} r^3 - 3\sum_{r=1}^{n} r^2 - 5\sum_{r=1}^{n} r$$

Take out constants.

Continues on next page

$$= 2\left[\frac{1}{4}n^2(n+1)^2\right] - 3\left[\frac{1}{6}n(n+1)(2n+1)\right] - 5\left[\frac{1}{2}n(n+1)\right]$$

Substitute in the standard formulae.

$$= \frac{1}{2}n^2(n+1)^2 - \frac{1}{2}n(n+1)(2n+1) - 5\left[\frac{1}{2}n(n+1)\right]$$

Simplify the first two terms.

$$= \frac{1}{2}n(n+1)[n(n+1) - (2n+1) - 5]$$

Now factorise as many terms as possible. Note that this is much easier than expanding everything first.

$$= \frac{1}{2}n(n+1)[n^2 + n - 2n - 1 - 5]$$

$$= \frac{1}{2}n(n+1)[n^2 - n - 6]$$

$$= \frac{1}{2}n(n+1)(n+2)(n-3)$$

b $\sum_{r=1}^{2n} r(2r-5)(r+1) = \frac{1}{2}2n(2n+1)(2n+2)(2n-3)$

Substitute $2n$ for n in the formula found in part **a**.

$$= n(2n+1)(2n+2)(2n-3)$$

$$= 2n(2n+1)(n+1)(2n-3)$$

Simplify and factorise a 2 from the second bracket.

WORK IT OUT 11.1

Given that $\sum_{r=1}^{n}(r^2 - 2r) = \frac{n}{6}(n+1)(2n-5)$, find an expression for $\sum_{r=n+1}^{2n}(r^2 - 2r)$.

Which is the correct solution? Identify the errors made in the incorrect solutions.

Solution 1	$\sum_{r=n+1}^{2n}(r^2-2r) = \frac{2n}{6}(2n+1)(2(2n)-5)$ $= \frac{n}{3}(2n+1)(4n-5)$
Solution 2	$\sum_{r=n+1}^{2n}(r^2-2r) = \sum_{r=1}^{2n}(r^2-2r) - \sum_{r=1}^{n}(r^2-2r)$ $= \frac{2n}{6}(2n+1)(2(2n)-5) - \frac{n}{6}(n+1)(2n-5)$ $= \frac{n}{6}[2(2n+1)(2(2n)-5) - (n+1)(2n-5)]$ $= \frac{n}{6}[2(8n^2-6n-5) - (2n^2-3n-5)]$ $= \frac{n}{6}(14n^2-9n-5)$ $= \frac{n}{6}(14n+5)(n-1)$

Continues on next page

Solution 3

$$\sum_{r=n+1}^{2n}(r^2-2r)=\sum_{r=1}^{2n}(r^2-2r)-\sum_{r=1}^{n+1}(r^2-2r)$$

$$=\frac{2n}{6}(2n+1)(2(2n)-5)-\frac{n}{6}((n+1)+1)(2(n+1)-5)$$

$$=\frac{n}{6}[2(2n+1)(2(2n)-5)-(n+2)(2n-3)]$$

$$=\frac{n}{6}[2(8n^2-6n-5)-(2n^2-n-6)]$$

$$=\frac{n}{6}(14n^2-11n-4)$$

EXERCISE 11B

In this exercise, you can assume the formulae for $\sum_{r=1}^{n} r$, $\sum_{r=1}^{n} r^2$ and $\sum_{r=1}^{n} r^3$.

1 Evaluate each expression.

a **i** $\sum_{r=1}^{30} r^2$ **ii** $\sum_{r=1}^{20} r^3$ **b** **i** $\sum_{r=32}^{50} r^3$ **ii** $\sum_{r=25}^{100} r$

2 Find a formula for each series, giving your answer in its simplest form.

a **i** $\sum_{r=1}^{4n} r$ **ii** $\sum_{r=1}^{3n} r^2$ **b** **i** $\sum_{r=1}^{n-1} r^2$ **ii** $\sum_{r=1}^{n+1} r^3$

3 Show that $\sum_{r=1}^{n} r(3r-5)=n(n+1)(n-2)$.

4 Show that $\sum_{r=1}^{n} 3r(r-1)=n(n^2-1)$.

5 **a** Find an expression for $\sum_{r=1}^{n}(6r+7)$.

b Hence find the least value of n such that $\sum_{r=1}^{n}(6r+7)>2400$.

6 **a** Show that $\sum_{r=1}^{n}(r+1)(r+5)=\frac{n}{6}(n+7)(2n+7)$.

b Hence evaluate $\sum_{r=16}^{40}(r+1)(r+5)$.

7 Show that $\sum_{r=1}^{n} r^2(r-1)=\frac{n}{12}(n^2-1)(kn+2)$, where k is an integer to be found.

8 a Show that $\sum_{r=1}^{n} r(r^2-3) = \frac{n}{4}(n+1)(n-2)(n+3)$.

b Hence find a formula for $\sum_{r=1}^{2n} r(r^2-3)$, fully simplifying your answer.

9 a Show that $\sum_{r=1}^{n} r(r+1) = \frac{n}{3}(n+1)(n+2)$.

b Hence find, in the form $\ln 3^k$, the exact value of $2\ln 3 + 3\ln 3^2 + 4\ln 3^3 + \cdots + 20\ln 3^{19}$.

10 Show that the sum of the squares of the first n odd numbers is given by $S = \frac{n}{3}(an^2-1)$, where a is an integer to be found.

Section 3: Method of differences

If you are investigating a series, start by writing out a few terms to see if any patterns develop.

For example, for the series $\sum_{r=1}^{n}[r(r+1)-r(r-1)]$:

$$
\begin{aligned}
u_1 &= 1(2)-1(0)\\
u_2 &= 2(3)-2(1)\\
u_3 &= 3(4)-3(2)\\
u_4 &= 4(5)-4(3)\\
&\vdots
\end{aligned}
$$

You can see that each term shares a common element with the next; in the first term, this element is positive and in the next it is negative. Therefore, when you complete the sum, these common elements will cancel out.

This cancellation continues right through to the nth term.

$$
\begin{aligned}
&1(2)-1(0)\\
&+2(3)-2(1)\\
&+3(4)-3(2)\\
&+4(5)-4(3)\\
&\quad\vdots\\
&+(n-1)n-(n-1)(n-2)\\
&+n(n+1)-n(n-1)
\end{aligned}
$$

$$\therefore \sum_{r=1}^{n} r(r+1)-r(r-1) = n(n+1)-1(0) = n(n+1)$$

In fact, because:

$$r(r+1)-r(r-1) = r^2+r-r^2+r = 2r$$

you have just shown that:

$$\sum_{r=1}^{n} 2r = n(n+1) \Rightarrow \sum_{r=1}^{n} r = \frac{n}{2}(n+1)$$

which is the result for the sum of the first n integers that you met in Section 2.

This process for finding a formula for the sum of the first n terms of a sequence is called the **method of differences**.

Key point 11.4

Method of differences

If the general term of a series, u_r, can be written in the form $u_r = f(r+1) - f(r)$, then:

$$\sum_{r=1}^{n} u_r = f(n+1) - f(1)$$

Tip

The series won't always take exactly this form, so always write out several terms to see how the cancellation occurs.

WORKED EXAMPLE 11.5

a Show that $(2r+1)^3 - (2r-1)^3 \equiv 24r^2 + 2$.

b Hence show that $\sum_{r=1}^{n} r^2 = \frac{1}{6}n(n+1)(2n+1)$.

a $(2r+1)^3 - (2r-1)^3$

$= (2r)^3 + 3(2r)^2 1 + 3(2r)1^2 + 1^3$

$\quad - [(2r)^3 + 3(2r)^2(-1) + 3(2r)(-1)^2 + (-1)^3]$

Use the binomial expansion to expand the cubed brackets.

$= 8r^3 + 12r^2 + 6r + 1 - [8r^3 - 12r^2 + 6r - 1]$

$= 24r^2 + 2$

Simplify to give the result required.

b $\sum_{r=1}^{n} (24r^2 + 2) = \sum_{r=1}^{n} [(2r+1)^3 - (2r-1)^2]$

Sum both sides of the result from **a**.

RHS $= (3)^3 - (1)^3$

$\quad + (5)^3 - (3)^3$

$\quad + (7)^3 - (5)^3$

$\quad \vdots$

$\quad + (2n-1)^3 - (2n-3)^2$

$\quad + (2n+1)^3 - (2n-1)^2$

The RHS is a difference, so you expect cancellation.

Write out the first few terms ($r = 1, 2, 3, \ldots$) and the last couple of terms ($r = n-1, n$).

$= (2n+1)^3 - 1^3$

Everything cancels except the terms shown.

Continues on next page

$$\text{LHS} = 24\sum_{r=1}^{n} r^2 + 2n$$

For the LHS, remember that $\sum_{r=1}^{n} 2 = 2n$.

$$\therefore 24\sum_{r=1}^{n} r^2 + 2n = (2n+1)^3 - 1^3$$

Make the expressions for the LHS and RHS equal.

$$24\sum_{r=1}^{n} r^2 + 2n = (2n)^3 + 3(2n)^2 1 + 3(2n)1^2 + 1^3 - 1^3$$

You now need to make $\sum_{r=1}^{n} r^2$ the subject.

Start by expanding the RHS and then simplify.

$$24\sum_{r=1}^{n} r^2 + 2n = 8n^3 + 12n^2 + 6n$$

$$24\sum_{r=1}^{n} r^2 = 8n^3 + 12n^2 + 4n$$

$$6\sum_{r=1}^{n} r^2 = 2n^3 + 3n^2 + n$$

$$6\sum_{r=1}^{n} r^2 = n(2n^2 + 3n + 1)$$

Factorise the RHS.

$$6\sum_{r=1}^{n} r^2 = n(n+1)(2n+1)$$

$$\sum_{r=1}^{n} r^2 = \frac{1}{6}n(n+1)(2n+1)$$

Finally, divide by 6.

Sometimes the cancellation occurs two terms apart.

WORKED EXAMPLE 11.6

a Show that $\frac{1}{k+1} - \frac{1}{k+3} \equiv \frac{2}{(k+1)(k+3)}$.

b Find an expression for $\sum_{k=1}^{n} \frac{2}{(k+1)(k+3)}$.

c Hence find $\sum_{k=1}^{\infty} \frac{2}{(k+1)(k+3)}$.

Tip

In an A Level question you could be expected to split $\frac{2}{(k+1)(k+3)}$ into partial fractions rather than being given the identity to prove.

a
$$\frac{1}{k+1} - \frac{1}{k+3} \equiv \frac{k+3}{(k+1)(k+3)} - \frac{k+1}{(k+1)(k+3)}$$
$$\equiv \frac{k+3-k-1}{(k+1)(k+3)}$$
$$\equiv \frac{2}{(k+1)(k+3)}$$

Continues on next page

b $\displaystyle\sum_{k=1}^{n} \frac{2}{(k+1)(k+3)} = \sum_{k=1}^{n} \frac{1}{k+1} - \frac{1}{k+3}$ .... Sum both sides of the result in **a**.

$$= \frac{1}{2} - \cancel{\frac{1}{4}}$$

$$+ \frac{1}{3} - \cancel{\frac{1}{5}}$$ .... Writing out several terms $(k = 1, 2, 3, \ldots)$ shows that the cancellations in the series occur two terms apart.

$$+ \cancel{\frac{1}{4}} - \cancel{\frac{1}{6}}$$

$$+ \cancel{\frac{1}{5}} - \cancel{\frac{1}{7}}$$

$$\vdots$$

$$+ \cancel{\frac{1}{n-1}} - \cancel{\frac{1}{n+1}}$$

$$+ \cancel{\frac{1}{n}} - \frac{1}{n+2}$$ .... Continue the pattern of cancellation for the last few terms $(k = n-2, n-1, n)$.

$$+ \cancel{\frac{1}{n+1}} - \frac{1}{n+3}$$

$$= \frac{1}{2} + \frac{1}{3} - \frac{1}{n+2} - \frac{1}{n+3}$$ .... This leaves part of the first two terms and part of the last two. You could put this all over a common denominator and combine into one fraction but, as the question doesn't specifically require this, there is no need to do anything else.

c Let $n \to \infty$ in the result in part **b**. .... As the denominator tends to ∞, these fractions tend to zero.

$$\frac{1}{n+2} \to 0$$

$$\frac{1}{n+3} \to 0$$

$$\therefore \sum_{k=1}^{\infty} \frac{2}{(k+1)(k+3)} = \frac{1}{2} + \frac{1}{3} = \frac{5}{6}$$

EXERCISE 11C

1   **a**   Show that $(r+1)^2 - r^2 \equiv 2r + 1$.

    b   Hence show that $\displaystyle\sum_{r=1}^{n} 2r + 1 = n(n+2)$.

2   **a**   Show that $r^2(r+1)^2 - (r-1)^2 r^2 \equiv 4r^3$.

    b   Hence show that $\displaystyle\sum_{r=1}^{n} r^3 = \frac{1}{4} n^2 (n+1)^2$.

3 a Show that $\frac{1}{k+1} - \frac{1}{k+2} \equiv \frac{1}{(k+1)(k+2)}$.

b i Hence show that $\sum_{k=1}^{n} \frac{2}{(k+1)(k+2)} = \frac{n}{n+2}$.

ii Find $\sum_{k=11}^{24} \frac{2}{(k+1)(k+2)}$.

A **4** a Express $\frac{2}{(2r-1)(2r+1)}$ in partial fractions.

b Use the method of differences to show that $\sum_{r=1}^{n} \frac{2}{(2r-1)(2r+1)} = \frac{2n}{2n+1}$.

c Find $\sum_{r=1}^{\infty} \frac{2}{(2r-1)(2r+1)}$.

5 a Show that $(r+1)! - (r-1)! \equiv (r^2+r-1)(r-1)!$

b Use the method of differences to show that $\sum_{r=1}^{n} (r^2+r-1)(r-1)! = (n+2)n! - 2$.

6 a Show that $\frac{1}{k} - \frac{1}{(k+2)} \equiv \frac{2}{k(k+2)}$.

b Hence, find $\sum_{k=1}^{\infty} \frac{2}{k(k+2)}$.

7 a Show that $\frac{1}{2k+1} - \frac{1}{2k+3} \equiv \frac{2}{(2k+1)(2k+3)}$.

b Find an expression for $\sum_{k=1}^{n} \frac{2}{(2k+1)(2k+3)}$.

c Hence show that $\frac{1}{3\times5} + \frac{1}{5\times7} + \frac{1}{7\times9} + \ldots < \frac{1}{6}$.

A **8** Use the method of differences to find $\sum_{r=1}^{n} \frac{1}{(r+1)(r+2)(r+3)}$.

9 a Show that $\frac{1}{2k} - \frac{1}{k+1} + \frac{1}{2(k+2)} \equiv \frac{1}{k(k+1)(k+2)}$

b Use the method of differences to show that

$$\sum_{k=1}^{2n} \frac{1}{k(k+1)(k+2)} = \frac{n(an+b)}{c(n+1)(2n+1)}$$

where a, b and c are constants to be found.

c Find $\frac{1}{11\times12\times13} + \frac{1}{12\times13\times14} + \ldots + \frac{1}{20\times21\times22}$

10 a Use the method of differences to find $\sum_{k=1}^{n} \ln\left(1+\frac{1}{k}\right)$.

b Hence, prove that the series $\sum_{k=1}^{\infty} \ln\left(1+\frac{1}{k}\right)$ diverges.

Section 4: Maclaurin series

You can write many common functions as infinite series, called **Maclaurin series**. You need to be aware of the results in this section, which are given in the formula book.

Key point 11.5

Maclaurin series for some common functions and the values of x for which they are valid:

- $e^x = 1 + x + \frac{x^2}{2!} + \frac{x^3}{3!} + \cdots$ all $x \in \mathbb{R}$
- $\ln(1+x) = x - \frac{x^2}{2} + \frac{x^3}{3} - \frac{x^4}{4} + \cdots$ $-1 < x \leqslant 1$
- $\sin x = x - \frac{x^3}{3!} + \frac{x^5}{5!} - \frac{x^7}{7!} + \cdots$ all $x \in \mathbb{R}$
- $\cos x = 1 - \frac{x^2}{2!} + \frac{x^4}{4!} - \frac{x^6}{6!} + \cdots$ all $x \in \mathbb{R}$
- $(1+x)^n = 1 + nx + \frac{n(n-1)x^2}{2!} + \frac{n(n-1)(n-2)x^3}{3!} + \cdots$ $|x| < 1$

These will be given in your formula book.

You can use these standard results to find Maclaurin series of more complicated functions.

Did you know?

Maclaurin series are named after the 18th-century mathematician Colin Maclaurin, who also developed some of Newton's work on calculus, algebra and gravitation theory.

Tip

Don't overlook the information on the values of x for which these series are valid; this is a very important part of each result.

Fast forward

You will see where these results come from in Further Mathematics Student Book 2.

Rewind

Note that the last result in Key point 11.5 is the binomial expansion, which is covered in A Level Mathematics Student Book 2, Chapter 6.

WORKED EXAMPLE 11.7

a Use the Maclaurin series for $\cos x$ to find the first four terms in the series for $\cos(2x^3)$.

b State the values of x for which the series is valid.

a $\cos(2x^3) = 1 - \frac{(2x^3)^2}{2!} + \frac{(2x^3)^4}{4!} - \frac{(2x^3)^6}{6!} + \ldots$

Substitute $2x^3$ into the series for $\cos x$: $\cos x = 1 - \frac{x^2}{2!} + \frac{x^4}{4!} - \frac{x^6}{6!} + \cdots$

$= 1 - \frac{4x^6}{2!} + \frac{16x^{12}}{4!} - \frac{64x^{18}}{6!} + \ldots$

$= 1 - 2x^6 + \frac{2}{3}x^{12} - \frac{4}{45}x^{18} + \ldots$

Expand and simplify.

b Valid for all $x \in \mathbb{R}$.

Both $2x^3$ and $\cos x$ are valid for all $x \in \mathbb{R}$.

This process can be more complicated if it involves finding two separate Maclaurin series and then combining them.

WORKED EXAMPLE 11.8

a Use the Maclaurin series for $\sin x$ and e^x to find the series for $e^{\sin x}$ as far as the term in x^4.

b State the values of x for which your series is valid.

a $e^{\sin x} = e^{x - \frac{x^3}{3!} + \dots}$

Start by replacing $\sin x$ with its series.

$\approx e^x \times e^{-\frac{x^3}{3!}}$

Now split this into a product of two terms...

$= \left(1 + x + \frac{x^2}{2!} + \frac{x^3}{3!} + \frac{x^4}{4!} + \dots\right)\left(1 + \left(-\frac{x^3}{3!}\right) + \dots\right)$

...and form the series for each of them.

For $e^{-\frac{x^3}{3!}}$ substitute $-\frac{x^3}{3!}$ into the series for e^x.

$= 1 + x + \frac{x^2}{2} + \frac{x^3}{6} + \frac{x^4}{24} - \frac{x^3}{6} - \frac{x^4}{6} + \dots$

$= 1 + x + \frac{x^2}{2} - \frac{x^4}{8} + \dots$

Expand term by term and simplify.

b Valid for all $x \in \mathbb{R}$

The series for both e^x and $\sin x$ are valid for all $x \in \mathbb{R}$.

WORKED EXAMPLE 11.9

a Find the first three terms in the Maclaurin series for $\ln(2 - 3x)$.

b Hence find the Maclaurin series up to the term in x^3 for $\ln\left(\frac{\sqrt{1+2x}}{2-3x}\right)$.

c State the interval in which the expansion is valid.

a $\ln(2-3x) = \ln\left[2\left(1 - \frac{3x}{2}\right)\right]$

$= \ln\left[2\left(1 + \frac{-3x}{2}\right)\right]$

You know the series expansion for $\ln(1+x)$ so you need to write $\ln(2-3x)$ in this form.

Start by factorising 2.

$= \ln 2 + \ln\left(1 + \frac{-3x}{2}\right)$

Separate the 2, using $\ln(ab) = \ln a + \ln b$

$= \ln 2 + \left(\frac{-3x}{2}\right) - \frac{\left(\frac{-3x}{2}\right)^2}{2} + \frac{\left(\frac{-3x}{2}\right)^3}{3} + \dots$

Then substitute $\frac{-3x}{2}$ into the series of $\ln(1+x)$:

$\ln(1+x) = x - \frac{x^2}{2} + \frac{x^3}{3} - \frac{x^4}{4} + \dots$

$= \ln 2 + \frac{-3x}{2} - \frac{\frac{9x^2}{4}}{2} + \frac{\frac{-27x^3}{8}}{3} + \dots$

$= \ln 2 - \frac{3x}{2} - \frac{9x^2}{8} - \frac{9x^3}{8} + \dots$

Expand and simplify.

Continues on next page

b $\ln\left(\frac{\sqrt{1+2x}}{2-3x}\right) = \ln(\sqrt{1+2x}) - \ln(2-3x)$

$= \frac{1}{2}\ln(1+2x) - \ln(2-3x)$

Again, you need everything in the form of $\ln(1+x)$.

First, use the laws of logs.

$$\frac{1}{2}\ln(1+2x) = \frac{1}{2}\left((2x) - \frac{(2x)^2}{2} + \frac{(2x)^3}{3} + \cdots\right)$$

$$= \frac{1}{2}\left(2x - 2x^2 + \frac{4x^3}{3} + \cdots\right)$$

$$= x - x^2 + \frac{2x^3}{3} + \cdots$$

You know the series expansion for the second term from part **a**.

For the first term, substitute $2x$ into the series for $\ln(1+x)$.

So:

Now put both series together.

$$\ln\left(\frac{\sqrt{1+2x}}{2-3x}\right) = \left(x - x^2 + \frac{2x^3}{3} + \cdots\right)$$

$$-\left(\ln 2 - \frac{3x}{2} - \frac{9x^2}{8} - \frac{9x^3}{8} + \cdots\right)$$

$$= x - x^2 + \frac{2x^3}{3} + \cdots - \ln 2 + \frac{3x}{2} + \frac{9x^2}{8} + \frac{9x^3}{8} + \cdots$$

$$= -\ln 2 + \frac{5x}{2} + \frac{x^2}{8} + \frac{43x^3}{24} + \cdots$$

c Since $\ln(1+x)$ is valid when $-1 < x \leqslant 1$:

Find the interval of validity separately for each function.

- $\ln\left(1+\frac{-3x}{2}\right)$ is valid when $-1 < \frac{-3x}{2} \leqslant 1$

  This is when $-\frac{2}{3} \leqslant x < \frac{2}{3}$

- $\ln(1+2x)$ is valid when $-1 < 2x \leqslant 1$

  This is when $-\frac{1}{2} < x \leqslant \frac{1}{2}$

Therefore, $\ln\left(\frac{\sqrt{1+2x}}{2-3x}\right)$ is valid

when $-\frac{1}{2} \leqslant x < \frac{1}{2}$.

For both to be valid, you need the smaller interval.

EXERCISE 11D

1. Find the first three non-zero terms of the Maclaurin series for each expression.

   a i e^{-3x} ii e^{x^3} b i $\ln(1+3x)$ ii $\ln(1-2x)$

   c i $\sin\left(-\frac{x}{2}\right)$ ii $\sin(3x^2)$ d i $\cos\left(\frac{x^2}{3}\right)$ ii $\cos(-2x)$

   e i $(1-4x)^{\frac{1}{2}}$ ii $\left(1+\frac{x}{3}\right)^{-4}$

2. By first manipulating it into an appropriate form, find the first three non-zero terms of the Maclaurin series for each expression.

   a i $\ln(3+x)$ ii $\ln\left(\frac{1}{2}-x\right)$

   b i $(2-3x)^{-3}$ ii $\left(\frac{1}{4}+2x\right)^{-\frac{1}{2}}$

   c i $(8x-27)^{\frac{1}{3}}$ ii $(3x-4)^{-2}$

3. By combining Maclaurin series of different functions, find the series expansion as far as the term in x^4 for each expression.

   a i $\ln(1+x)\sin 2x$ ii $\ln(1-x)\cos 3x$

   b i $\frac{e^x}{1+x}$ ii $\frac{\sin x}{1-2x}$

   c i $\ln(1+\sin x)$ ii $\ln(\cos x)$

4. Find the Maclaurin series for $\ln(1+4x^2+4x)$ and state the interval in which the series is valid.

5. Find the Maclaurin series as far as the term in x^4 for $e^{3x}\sin 2x$.

6. Show that $\sqrt{1+x^2}e^{-x} = 1 - x + x^2 - \frac{2}{3}x^3 + \frac{1}{6}x^4 + \ldots$.

7. a Find the first two non-zero terms of the Maclaurin series for $\tan x$.

   b Hence find the Maclaurin series of $e^{\tan x}$ up to and including the term in x^4.

A 8. a By using the Maclaurin series for $\cos x$, find the series expansion for $\ln(\cos x)$ up to the term in x^4.

   b Hence find the first two non-zero terms of the expansion of $\ln(\sec x)$.

   c Use your result from **b** to find the first two non-zero terms of the series for $\tan x$.

9. a Find the first four terms of the Maclaurin series for $f(x) = \ln[(2+x)^3(1-3x)]$.

   b Find the equation of the tangent to $f(x)$ at $x = 0$.

10. a Find the Maclaurin series for $\ln\sqrt{\frac{1+x}{1-x}}$, stating the interval in which the series is valid.

    b Use the first three terms of this series to estimate the value of $\ln 2$, stating the value of x used.

Checklist of learning and understanding

This is the last value taken by r, where counting ends.

The value of r changes with each new term.

Greek capital sigma means 'add up'. $\sum_{r=1}^{r=n} f(r) = f(1) + f(2) + ... + f(n)$

This is the first value taken by r, where counting starts.

- The formulae for the sums of integers, squares and cubes:
  - $\sum_{r=1}^{n} r = \frac{1}{2}n(n+1)$
  - $\sum_{r=1}^{n} r^2 = \frac{1}{6}n(n+1)(2n+1)$
  - $\sum_{r=1}^{n} r^3 = \frac{1}{4}n^2(n+1)^2$
- You can manipulate series in these ways:
  - $\sum (u_r + v_r) = \sum u_r + \sum v_r$
  - $\sum cu_r = c\sum u_r$
  - $\sum_{r=1}^{n} c = nc$

  where c is a constant.

- **Method of differences**
  If the general term of a series, u_r, can be written in the form $u_r = f(r+1) - f(r)$, then:

  $$\sum_{r=1}^{n} u_r = f(n+1) - f(1)$$

- Maclaurin series for some common functions and the values of x for which they are valid include:
  - $e^x = 1 + x + \frac{x^2}{2!} + \frac{x^3}{3!} + \cdots$ all $x \in \mathbb{R}$
  - $\ln(1+x) = x - \frac{x^2}{2} + \frac{x^3}{3} - \frac{x^4}{4} + \cdots$ $-1 < x \leqslant 1$
  - $\sin x = x - \frac{x^3}{3!} + \frac{x^5}{5!} - \frac{x^7}{7!} + \cdots$ all $x \in \mathbb{R}$
  - $\cos x = 1 - \frac{x^2}{2!} + \frac{x^4}{4!} - \frac{x^6}{6!} + \cdots$ all $x \in \mathbb{R}$
  - $(1+x)^n = 1 + nx + \frac{n(n-1)x^2}{2!} + \frac{n(n-1)(n-2)x^3}{3!} + \cdots$ $|x| < 1$

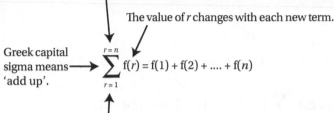

Mixed practice 11

1 Find an expression for $\sum_{r=n+1}^{2n}(2r+1)$.

Choose from these options.

A $3n^2+2n$ B $3n^2+3n-1$ C n^2+n+1 D n^2+2n

2 Use the formulae for $\sum_{r=1}^{n} r$ and $\sum_{r=1}^{n} r^2$ to show that $\sum_{r=1}^{n}(r+2)(r-1)=\frac{n}{3}(n+4)(n-1)$.

3 Use the formulae for $\sum_{r=1}^{n} r^2$ and $\sum_{r=1}^{n} r^3$ to find the value of $\sum_{r=5}^{40} r^2(2r-3)$.

4 a Show that $\sum_{r=1}^{n} 2r(2r^2-3r-1)=n(n+p)(n+q)^2$, where p and q are integers to be found.

b Hence find the value of $\sum_{r=11}^{20} 2r(2r^2-3r-1)$.

[©AQA 2013]

5 a Show that $\frac{1}{k+4}-\frac{1}{k+5}\equiv\frac{1}{k^2+9k+20}$.

b Hence show that $\sum_{k=1}^{n}\frac{1}{k^2+9k+20}=\frac{an}{b(n+5)}$, where a and b are integers to be found.

6 a Write down the expansion of e^{3x} in ascending powers of x up to and including the term in x^2.

b Hence, or otherwise, find the term in x^2 in the expansion, in ascending powers of x, of $e^{3x}(1+2x)^{-\frac{3}{2}}$.

[©AQA 2013]

7 Find the set of values of x for which the Maclaurin series of the function $f(x)=\sqrt{1+e^x}$ is valid.

Choose from these options.

A All $x\in\mathbb{R}$ B $|x|<1$ C $x<0$ D $x>0$

8 a Show that $\sum_{r=2}^{n} r(r-1)(r+1)=\frac{n}{4}(n^2-1)(n+2)$.

b Hence find the sum of $(11\times12\times13)+(12\times13\times14)+(13\times14\times15)+\cdots+(38\times39\times40)$.

9 a Show that $\sum_{r=1}^{n} r^3+\sum_{r=1}^{n} r$ can be expressed in the form $kn(n+1)(an^2+bn+c)$, where k is a rational number and a, b and c are integers.

b Show that there is exactly one positive integer n for which $\sum_{r=1}^{n} r^3+\sum_{r=1}^{n} r=8\sum_{r=1}^{n} r^2$.

[©AQA 2010]

10 a Show that $\frac{1}{r^2}-\frac{1}{(r+1)^2}\equiv\frac{2r+1}{r^2(r+1)^2}$.

b Hence show that $\sum_{r=1}^{n}\frac{2r+1}{r^2(r+1)^2}=\frac{n(n+2)}{(n+1)^2}$.

11 a Show that $\frac{1}{r!}-\frac{1}{(r+1)!}\equiv\frac{r}{(r+1)!}$.

b Find $\frac{1}{2!}+\frac{2}{3!}+\frac{3}{4!}+\ldots+\frac{n}{(n+1)!}$.

c Hence find $\sum_{r=1}^{\infty}\frac{r}{(r+1)!}$.

12 a Given that $f(r)=r^2(2r^2-1)$, show that $f(r)-f(r-1)=(2r-1)^3$.

b Use the method of differences to show that $\sum_{r=n+1}^{2n}(2r-1)^3=3n^2(10n^2-1)$.

[©AQA 2013]

13 a Find the first three terms in the expansion, in ascending powers of x, of $\ln(4+3x)$.

b Write down the first three terms in the expansion, in ascending powers of x, of $\ln(4-3x)$.

c Show that, for small values of x, $\ln\left(\frac{4+3x}{4-3x}\right)\approx\frac{3}{2}x$.

[©AQA 2010]

14 a Express $\frac{3}{(3r-1)(3r+2)}$ in partial fractions.

b Hence find an expression for $\sum_{r=1}^{n}\frac{3}{(3r-1)(3r+2)}$.

15 Use the method of differences to show that $\sum_{k=1}^{n}\frac{3k+4}{k(k+1)(k+2)}=\frac{n(an+b)}{c(n+1)(n+2)}$

where a, b and c are constants to be found.

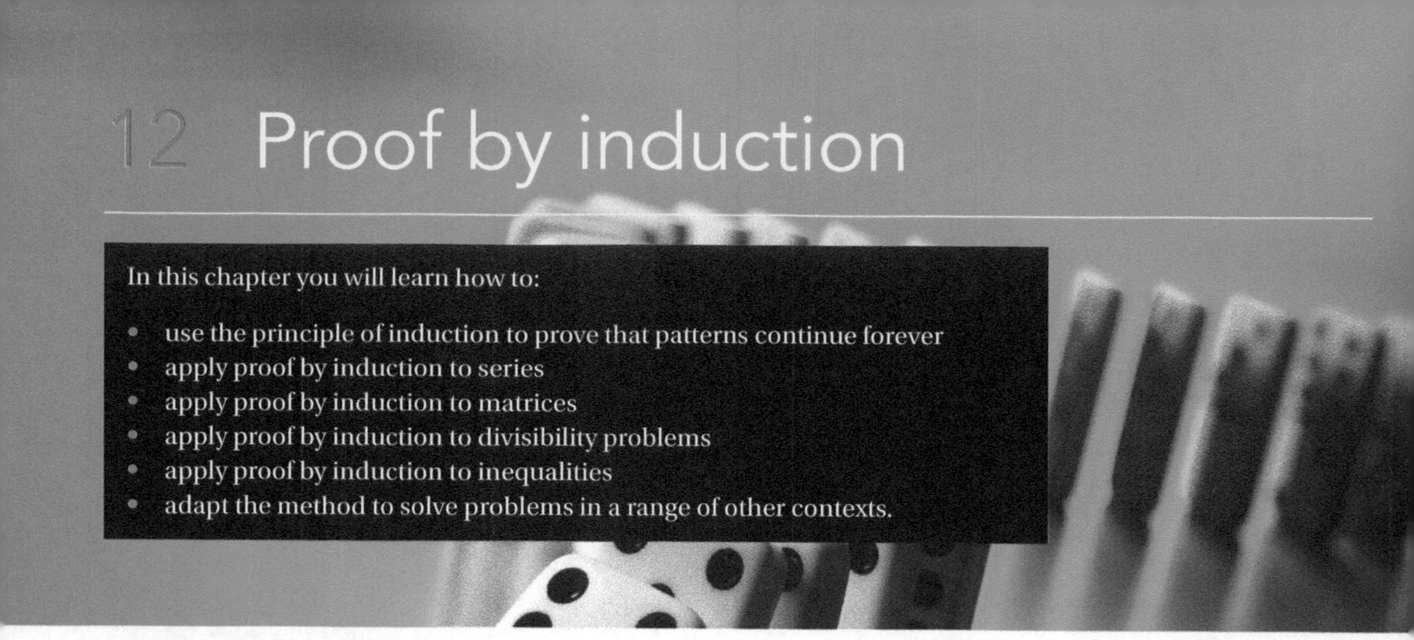

12 Proof by induction

In this chapter you will learn how to:

- use the principle of induction to prove that patterns continue forever
- apply proof by induction to series
- apply proof by induction to matrices
- apply proof by induction to divisibility problems
- apply proof by induction to inequalities
- adapt the method to solve problems in a range of other contexts.

Before you start …

GCSE	You should be able to use laws of indices.	1 a Given that $3^n - 7 = A$, express $3^{n+1} - 7$ in terms of A. b Simplify $5^{n+2} - 5^n$.
Chapter 7	You should be able to multiply matrices.	2 Given that $\mathbf{M} = \begin{pmatrix} 1 & a \\ 0 & 2 \end{pmatrix}$ find $\mathbf{M}^3$.
Chapter 11	You should be able to work with sigma notation for series.	3 Evaluate $\sum_{k=1}^{5}(2^k - 1)$.
A Level Mathematics Student Book 1, Chapter 9	You should be able to work with factorials.	4 Simplify $26! - 25!$

What is induction?

Mathematicians and scientists are very interested in patterns. In science, you can observe patterns and conjecture a general rule; this is called **inductive reasoning**. There is no way to prove that this rule is correct, and the pattern may not continue forever. In mathematics, you have so far used deductive reasoning, where you start from known facts and use logic to derive new results, but sometimes a potential new result is based on observations of a pattern. In that case, you need a way of proving that the pattern continues indefinitely. One of the most powerful ways of doing this is a method called **induction**.

Section 1: The principle of induction

Sequences often produce interesting patterns. Consider adding up consecutive odd numbers:

$$S_1 = 1 = 1^2$$

$$S_2 = 1 + 3 = 4 = 2^2$$

$$S_3 = 1 + 3 + 5 = 9 = 3^2$$

$$S_4 = 1 + 3 + 5 + 7 = 16 = 4^2$$

But you can't keep checking forever to be sure that this pattern continues.

Suppose that you have checked (by direct calculation) that the pattern continues up to the 15th odd number; so you know that the sum of the first 15 odd numbers is

$$S_{15} = 1 + 3 + 5 + \ldots + 29 = 225 = 15^2$$

To check that the pattern continues to the first 16 odd numbers, you do not have to add all the odd numbers from 1 to 31. You can use the result you already have, so

$$\begin{aligned} S_{16} &= 1 + 3 + 5 + \ldots + 29 + 31 \\ &= S_{15} + 31 \\ &= 225 + 31 \\ &= 256 \\ &= 16^2 \end{aligned}$$

Building upon the previous work you have done, rather than starting all over again, is called an **inductive step**. In this example it seems fairly straightforward but in other problems this can seem very difficult.

Now suppose that you have checked the pattern for the first k odd numbers. This means that

$$S_k = 1 + 3 + 5 + \ldots + (2k - 1) = k^2$$

To check that the pattern still holds for the first $k + 1$ odd numbers, you can use the result for the first k odd numbers (which you assume has already been checked):

$$\begin{aligned} S_{k+1} &= S_k + (2k + 1) \\ &= 1 + 3 + 5 + \ldots + (2k - 1) + (2k + 1) \\ &= k^2 + (2k + 1) \\ &= (k + 1)^2 \end{aligned}$$

Therefore the pattern still holds for the first $k + 1$ odd numbers.

In summary, for the pattern, 'The sum of the first n odd numbers equals n^2', you have found:

- the pattern holds for the first case (when $n = 1$)
- *if* you can check the pattern up to some whole number k, then it follows that it will also hold for $k + 1$.

Does this prove that the pattern continues forever? Yes!

Did you know?

This example is the first documented example of proof by induction, by the Italian mathematician Francesco Maurolico in 1575.

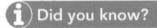

You know that it holds for $n=1$; but if it holds for any number k, then it holds for $k+1$, so it therefore holds for $n=2$. But if it holds for $n=2$ it therefore holds for $n=3$ and so on. You can continue this process to reach any number n, however large.

The reasoning described in the last paragraph is called the *Principle of Mathematical Induction*.

Key point 12.1

To prove a statement (or rule) about a positive integer n:

1. Prove that the initial case is true.
2. Assume that the kth case is true.
3. Show that if the proposition is true for k, it is also true for $k+1$.
4. Write a conclusion.

The hardest step is undoubtedly **3**. To do this you need to make a link between one proposition and the next – the **inductive step**. The exact way you do this depends upon the type of problem. In the following sections you will see how to apply proof by induction in various contexts.

Section 2: Induction and series

In Chapter 11 you used the method of differences to find the formula for the sum of the first n terms of certain sequences. To do this, however, you needed a way of splitting the general term of the sequence into a difference. In this section, you will prove results for a far wider class of series (including the formulae you assumed for $\sum r^2$ and $\sum r^3$) by induction.

Rewind

See Chapter 11, Section 1, for a reminder of sigma notation.

Tip

You must lay your proof out clearly, with the structure given in Key point 12.1, to show that you are following the correct logic.

Key point 12.2

When using induction to prove a result about series, use:

$$S_{k+1} = S_k + u_{k+1}$$

WORKED EXAMPLE 12.1

Prove by induction that

$$\sum_{r=1}^{n} r(r+2) = \frac{n(n+1)(2n+7)}{6} \text{ for all } n \in \mathbb{Z}^+.$$

For $n=1$:

LHS $= 1\times 3 = 3$

RHS $= \dfrac{1(1+1)(2\times 1+7)}{6} = \dfrac{1\times 2\times 9}{6} = 3$

So, the result is true for $n=1$.

Show that the statement is true for the starting value (in this case, $n=1$).

Continues on next page

Assume that the result is true for $n = k$:

$$\sum_{r=1}^{k} r(r+2) = \frac{k(k+1)(2k+7)}{6}$$

State the assumption for $n = k$.

Let $n = k + 1$:

$$\sum_{r=1}^{k+1} r(r+2) = \sum_{r=1}^{k} r(r+2) + (k+1)(k+3)$$

Consider S_{k+1} and relate it to S_k by using $S_{k+1} = S_k + u_{k+1}$.

$$= \frac{k(k+1)(2k+7)}{6} + (k+1)(k+3)$$

$$= (k+1)\left(\frac{2k^2+7k}{6} + \frac{6k+18}{6}\right)$$

Substitute in the result for $n = k$ (assumed to be true).

$$= \frac{(k+1)(2k^2+13k+18)}{6}$$

$$= \frac{(k+1)(k+2)(2k+9)}{6}$$

Combine this into one fraction and simplify. It is always a good idea to take out any common factors.

$$= \frac{(k+1)((k+1)+1)(2(k+1)+7)}{6}$$

So, the result is true for $n = k + 1$.

Show that this is in the required form by separating out $k + 1$ in each place it occurs.

The result is true for $n = 1$, and if it is true for $n = k$ it is also true for $n = k + 1$. Therefore, the result is true for all $n \in \mathbb{Z}^+$, by induction.

Make sure you write a conclusion.

Common error

In the conclusion, **do not** write: 'The statement is true for $n = k$ and true for $n = k+1$.' The logic must be correct: '**If** the statement is true for $n = k$, then it is true for $n = k+1$.'

EXERCISE 12A

1 Prove by induction that, for all $n \in Z^+$,

$$\sum_{r=1}^{n} 2 \times 3^{r-1} = 3^n - 1$$

2 Prove by induction that, for all integers $n \geqslant 1$,

$$\sum_{r=1}^{n} r^2 = \frac{n(n+1)(2n+1)}{6}$$

3 Using mathematical induction, prove that, for all positive integers,

$$\sum_{r=1}^{n} r^3 = \frac{n^2(n+1)^2}{4}$$

4 Prove by induction that, for all integers $n \geqslant 1$,

$$\sum_{r=1}^{n} \frac{1}{r(r+1)} = \frac{n}{n+1}$$

5 Use mathematical induction to show that, for all integers $n \geqslant 1$,

$$\sum_{r=1}^{n} r2^r = 2\left[(n-1)2^n + 1\right]$$

6 Prove by induction that, for all $n \in \mathbb{Z}^+$,

$$\frac{1}{1\times 3} + \frac{1}{3\times 5} + \frac{1}{5\times 7} + \ldots + \frac{1}{(2n-1)(2n+1)} = \frac{n}{2n+1}$$

7 Using mathematical induction, prove that, for all integers $n \geqslant 1$,

$$\sum_{r=1}^{n} r(r!) = (n+1)! - 1$$

8 Prove by induction that, for all positive integers,

$$1^2 - 2^2 + 3^2 - 4^2 + \ldots + (-1)^{n-1}n^2 = (-1)^{n-1}\frac{n(n+1)}{2}$$

9 Prove, using mathematical induction, that, for all $n \in \mathbb{Z}^+$,

$$(n+1) + (n+2) + (n+3) + \ldots + (2n) = \frac{1}{2}n(3n+1)$$

10 Prove by induction that, for all integers $n \geqslant 1$,

$$\sum_{k=1}^{n} k\,2^k = (n-1)2^{n+1} + 2$$

Section 3: Induction and matrices

With matrices, the inductive step involves relating one power to the next one.

Key point 12.3

When using induction to prove a result about powers of matrices, use:

$$\mathbf{A}^{k+1} = \mathbf{A}^k\mathbf{A}$$

Rewind

See Chapter 7, Section 2, for a reminder about matrix multiplication.

WORKED EXAMPLE 12.2

Use mathematical induction to prove that $\begin{pmatrix} 3 & -4 \\ 1 & -1 \end{pmatrix}^n = \begin{pmatrix} 1+2n & -4n \\ n & 1-2n \end{pmatrix}$ for all integers $n \geqslant 1$.

When $n = 1$:

$$\begin{pmatrix} 3 & -4 \\ 1 & -1 \end{pmatrix}^1 = \begin{pmatrix} 3 & -4 \\ 1 & -1 \end{pmatrix}$$

$$\begin{pmatrix} 1+2\times 1 & -4\times 1 \\ 1 & 1-2\times 1 \end{pmatrix} = \begin{pmatrix} 3 & -4 \\ 1 & -1 \end{pmatrix}$$

So the result is true for $n = 1$.

Show that the statement is true for the starting value (in this case, $n = 1$).

Assume that the result is true for $n = k$:

$$\begin{pmatrix} 3 & -4 \\ 1 & -1 \end{pmatrix}^k = \begin{pmatrix} 1+2k & -4k \\ k & 1-2k \end{pmatrix}$$

State the assumption for $n = k$.

Let $n = k + 1$:

$$\begin{pmatrix} 3 & -4 \\ 1 & -1 \end{pmatrix}^{k+1} = \begin{pmatrix} 3 & -4 \\ 1 & -1 \end{pmatrix}^k \begin{pmatrix} 3 & -4 \\ 1 & -1 \end{pmatrix}$$

Consider $\mathbf{A}^{k+1}$ and relate it to $\mathbf{A}^k$ using $\mathbf{A}^{k+1} = \mathbf{A}^k\mathbf{A}$.

$$= \begin{pmatrix} 1+2k & -4k \\ k & 1-2k \end{pmatrix}\begin{pmatrix} 3 & -4 \\ 1 & -1 \end{pmatrix}$$

Substitute in the result for $n = k$ (assumed to be true).

$$= \begin{pmatrix} 3+6k-4k & -4-8k+4k \\ 3k+1-2k & -4k-1+2k \end{pmatrix}$$

$$= \begin{pmatrix} 3+2k & -4k-4 \\ k+1 & -1-2k \end{pmatrix}$$

Multiply the two matrices together and simplify. You must show sufficient detail of your calculation.

$$= \begin{pmatrix} 1+2(k+1) & -4(k+1) \\ k+1 & 1-2(k+1) \end{pmatrix}$$

Hence the result is true for $n = k + 1$.

Show that this is clearly in the required form by separating out $k + 1$ in each place it occurs.

The result is true for $n = 1$, and if it is true for $n = k$ it is also true for $n = k + 1$. Therefore, the result is true for all $n \geqslant 1$, by induction.

Always write a conclusion.

EXERCISE 12B

1 Use mathematical induction to prove that, for all positive integers n,

$$\begin{pmatrix} 1 & 3 \\ 0 & 1 \end{pmatrix}^n = \begin{pmatrix} 1 & 3n \\ 0 & 1 \end{pmatrix}$$

2 Prove by induction that, for all $n \in \mathbb{N}$,

$$\begin{pmatrix} 1 & 0 \\ a & 1 \end{pmatrix}^n = \begin{pmatrix} 1 & 0 \\ na & 1 \end{pmatrix}$$

Tip

Remember that the symbol $\mathbb{Z}$ is used to denote the set of all integers (whole numbers) and $\mathbb{N}$ the set of positive integers.

3 Let $\mathbf{A} = \begin{pmatrix} 1 & 1 \\ 1 & 1 \end{pmatrix}$. Use induction to prove that, for all $n \geqslant 1$,

$$\mathbf{A}^n = \begin{pmatrix} 2^{n-1} & 2^{n-1} \\ 2^{n-1} & 2^{n-1} \end{pmatrix}$$

4 Let $\mathbf{M} = \begin{pmatrix} -2 & 9 \\ -1 & 4 \end{pmatrix}$. Prove by induction that, for all positive integer powers n,

$$\mathbf{M}^n = \begin{pmatrix} 1-3n & 9n \\ -n & 1+3n \end{pmatrix}$$

5 Given that $\mathbf{M} = \begin{pmatrix} 1 & a \\ 0 & 2 \end{pmatrix}$, prove by induction that, for all $n \geqslant 1, n \in \mathbb{Z}$,

$$\mathbf{M}^n = \begin{pmatrix} 1 & (2^n - 1)a \\ 0 & 2^n \end{pmatrix}$$

6 Using mathematical induction, show that, for all $n \geqslant 1, n \in \mathbb{Z}$,

$$\begin{pmatrix} 3 & a \\ 0 & 1 \end{pmatrix}^n = \begin{pmatrix} 3^n & \frac{3^n - 1}{2}a \\ 0 & 1 \end{pmatrix}$$

7 Prove by induction that, for all integers $n \geqslant 1$,

$$\begin{pmatrix} 1 & 1 & 1 \\ 1 & 1 & 1 \\ 1 & 1 & 1 \end{pmatrix}^n = \begin{pmatrix} 3^{n-1} & 3^{n-1} & 3^{n-1} \\ 3^{n-1} & 3^{n-1} & 3^{n-1} \\ 3^{n-1} & 3^{n-1} & 3^{n-1} \end{pmatrix}$$

8 Matrix A is given by $\mathbf{A} = \begin{pmatrix} 0 & 1 \\ 0 & 0 \end{pmatrix}$. Let p and q be real numbers.

a Find the matrix $p\mathbf{I} + q\mathbf{A}$, where $\mathbf{I}$ is the 2×2 identity matrix.

b Use mathematical induction to prove that

$$(p\mathbf{I} + q\mathbf{A})^n = \begin{pmatrix} p^n & np^{n-1}q \\ 0 & p^n \end{pmatrix}, \text{ for all integers } n \geqslant 1.$$

Section 4: Induction and divisibility

Number theory is an important area of pure mathematics: it is concerned with properties of natural numbers. One of the important tasks in number theory is studying divisibility.

Consider the expression $f(n) = 7^n - 1$ for $n = 0, 1, 2, \ldots$

Looking at the first few values of n:

$$f(0) = 7^0 - 1 = 0$$
$$f(1) = 7^1 - 1 = 6$$
$$f(2) = 7^2 - 1 = 48$$
$$f(3) = 7^3 - 1 = 342$$

It looks as if $f(n)$ is divisible by 6 for all values of n. You can use mathematical induction to prove this.

Tip

Note that, in this example, the first value of n is $n = 0$. Induction does not have to start from $n = 1$.

Key point 12.4

When using induction to prove that an expression is divisible by an integer d:

- write the assumption as $f(k) = dA$ for some integer A
- isolate any term in the expression for $f(k+1)$ that appears in $f(k)$ and substitute in from the assumption.

WORKED EXAMPLE 12.3

The expression $f(n)$is defined by $f(n) = 7^n - 1$ for all integers n.
Prove that $f(n)$ is divisible by 6 for all integers $n \geqslant 0$.

$f(0) = 7^0 - 1 = 0 = 0 \times 6$

So $f(0)$ is divisible by 6.

Show that the statement is true for the starting value (in this case, $n = 0$).

Assume that $f(k)$ is divisible by 6:

$7^k - 1 = 6A$ for some $A \in \mathbb{Z}$.

State the assumption for $n = k$.

Consider $f(k+1)$:

Consider $f(k+1)$ and relate it to $f(k)$ by isolating, say, 7^k.

$$\begin{aligned} 7^{k+1} - 1 &= 7(7^k) - 1 \\ &= 7(6A + 1) - 1 \\ &= 42A + 7 - 1 \\ &= 42A + 6 \\ &= 6(7A + 1) \end{aligned}$$

Substitute in the result for $n = k$ (assumed to be true).

Simplify, working towards taking out a factor of 6.

So $f(k+1)$ is divisible by 6.

$f(0)$ is divisible by 6, and if $f(k)$ is divisible by 6, then so is $f(k+1)$. Therefore $f(n)$ is divisible by 6 for all integers $n \geqslant 0$, by induction.

Write a conclusion.

Sometimes it can be more difficult to manipulate the expression for $n = k+1$ to find the factor you need.

WORKED EXAMPLE 12.4

Use mathematical induction to prove that, for all positive integers n, $3^{4n-2} + 2^{6n-3}$ is divisible by 17.

When $n = 1$:

$3^{4-2} + 2^{6-3} = 3^2 + 2^3 = 17$

Show that the statement is true for the starting value (in this case, $n = 1$).

which is divisible by 17 so the statement is true for $n = 1$.

Assume that the statement is true for $n = k$:

State the assumption for $n = k$.

$3^{4k-2} + 2^{6k-3} = 17A$ for some integer A.

Let $n = k+1$:

Consider $n = k+1$ and relate it to $n = k$ by isolating, say, 3^{4k-2}.

$$
\begin{aligned}
3^{4(k+1)-2} + 2^{6(k+1)-3} &= 3^{4k+2} + 2^{6n+3} \\
&= 3^4(3^{4k-2}) + 2^{6k+3} \\
&= 3^4(17A - 2^{6k-3}) + 2^{6k+3} \\
&= 81 \times 17A - 81(2^{6k-3}) + 2^{6k+3} \\
&= 81 \times 17A - 2^{6k-3}(81 - 2^6) \\
&= 81 \times 17A - 2^{6k-3}(17) \\
&= 17(81A - 2^{6k-3})
\end{aligned}
$$

Substitute in the result for $n = k$ (assumed to be true).

Work towards finding a common factor of 17. Factorise 2^{6k-3} from the second and third terms and see what is left.

which is divisible by 17 so the statement is true for $n = k+1$.

The statement is true for $n = 1$ and if it is true for $n = k$ then it is also true for $n = k+1$. Therefore the statement is true for all $n \geqslant 1$, by induction.

Write a conclusion.

EXERCISE 12C

1. Prove by induction that $5^n - 1$ is divisible by 4 for all $n \in \mathbb{N}$.
2. Prove by induction that $4^n - 1$ is divisible by 3 for all $n \in \mathbb{N}$.
3. Using mathematical induction, prove that $7^n - 3^n$ is divisible by 4 for all $n \in \mathbb{N}$.
4. Use induction to prove that $30^n - 6^n$ is divisible by 12 for all integers $n \geqslant 0$.
5. Show by using induction that $n^3 - n$ is divisible by 6 for all integers $n \geqslant 1$.
6. Using mathematical induction, prove that $n(n^2 + 5)$ is divisible by 6 for all integers $n \geqslant 1$.
7. Use induction to show that $7^n - 4^n - 3^n$ is divisible by 12 for all $n \in \mathbb{Z}^+$.

8 Prove, using mathematical induction, that $3^{2n+2} - 8n - 9$ is divisible by 64 for all positive integers n.

9 Prove by induction that $2^{4n+3} + 3^{3n+1}$ is divisible by 11 for all positive integers n.

10 Prove that the sum of the cubes of any three consecutive integers is divisible by 9.

Section 5: Induction and inequalities

One of the most surprising uses of mathematical induction is in proving inequalities. For example, consider the inequality $2^n > 6n+1$. Trying the first few integer values of n shows that this inequality is satisfied for $n \geqslant 5$. In fact, by looking at the graphs of $y = 2^n$ and $y = 6n+1$ you can see that the inequality is satisfied for all **real** numbers n greater than approximately 4.94. Note that the inequality cannot be solved algebraically. However, you can use induction to show that it holds for all **integers** $n \geqslant 5$.

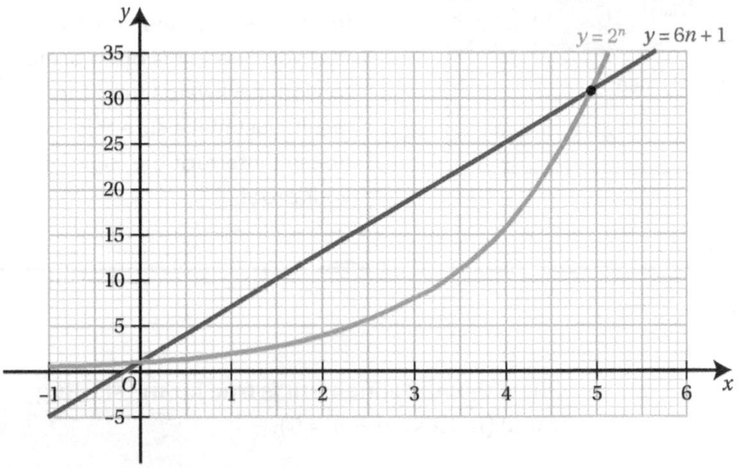

WORKED EXAMPLE 12.5

Use induction to prove that the inequality $2^n > 6n+1$ holds for all integers $n \geqslant 5$.

For $n = 5$:

LHS $= 2^5 = 32$

RHS $= 6 \times 5 + 1 = 31$

LHS > RHS, so the inequality holds for $n = 5$.

Show that the statement it true for the starting value (in this case, $n = 5$).

Assume that the inequality holds for $n = k$:

$2^k > 6k+1$

State the assumption for $n = k$.

Let $n = k+1$:

$2^{k+1} = 2(2^k)$

Consider $n = k+1$ and relate it to $n = k$.

Continues on next page

$> 2(6k+1)$

$= 12k+2$

$= 6k+(6k+2)$

$> 6k+7$

$= 6(k+1)+1$

Substitute in the result for $n = k$ (assumed to be true).

Working backwards, you need the RHS to be $6(k+1)+1 = 6k+7$, so separate the $6k$ term.

Use the fact that $k \geqslant 5$: then $6k+2$ is greater than 32, so it is definitely greater than 7.

LHS > RHS, so the inequality holds for $n = k+1$.

The inequality holds for $n = 5$ and if it holds for $n = k$, then it also holds for $n = k+1$. Therefore, it holds for all integers $n \geqslant 5$ by induction.

Write a conclusion.

Many interesting inequalities involve the factorial function. For the inductive step you often need to make a link between $n!$ and $(n+1)!$

Key point 12.5

When using induction to prove a result involving factorials, use:

$$(k+1)! = (k+1)\times k!$$

Tip

In order to get the expression for $n = k+1$ into the final form that you need, it is often useful to work backwards from the result. Just make sure this doesn't interrupt the logic of your proof.

WORKED EXAMPLE 12.6

Prove that $n! > 20n$ for all integers $n \geqslant 5$.

When $n = 5$:

LHS $= 5! = 120$

RHS $= 20\times 5 = 100$

LHS > RHS so the inequality holds for $n = 5$.

Show that the statement is true for the starting value (in this case, $n = 5$).

Assume the result is true for $n = k$:

$k! > 20k$

State the assumption for $n = k$.

Let $n = k+1$:

$(k+1)! = (k+1)\times k!$

Consider $n = k+1$ and relate it to $n = k$ using $(k+1)! = (k+1)\times k!$

$> (k+1)\times 20k$

$= 20k^2 + 20k$

$> 20k + 20$ (since $k > 1$)

$= 20(k+1)$

Substitute in the result for $n = k$ (assumed to be true).

Working backwards, you need the RHS to be $20(k+1) = 20k+20$, so divide through by $k > 1$.

Continues on next page

LHS > RHS so the inequality holds for $n = k + 1$. The statement is true for $n = 5$ and if it is true for $n = k$ then it is also true for $n = k + 1$. Therefore, it is true for all $n \geqslant 5$ by induction.

Write a conclusion.

EXERCISE 12D

1. Prove by induction that $2^n > 1 + n$ for all $n \geqslant 2$.
2. Prove by induction that $2^n > 2n$ for $n \geqslant 3$.
3. Use mathematical induction to show that $2^n > 11n$ for $n \geqslant 7$.
4. Prove, using induction, that $n! > 2^n$ for all $n \geqslant 4$.
5. Using mathematical induction, prove that $n! > 3^n$ for all $n \geqslant 7$.
6. Prove by induction that $2^n > n^2$ for all $n \geqslant 5$.
7. Use mathematical induction to prove that $3^n > n^3$ for all $n \geqslant 4$.
8. **a** Solve the inequality $x^2 - x - 1 > 0$.
   b Prove by induction that $n! > n^2$ for $n \geqslant 4$.
9. Use mathematical induction to prove that $n! > 3n^2 + 2n$ for all integers $n > 4$.

Checklist of learning and understanding

- In mathematics, no matter how many times a rule works this will never be enough to prove it. Mathematical induction provides a logically rigorous way of showing that a pattern will continue forever.
- The steps of a mathematical induction proof are:
  1. Prove that the initial case is true.
  2. Assume that the kth case is true.
  3. Show that if the proposition is true for k, it is also true for $k + 1$.
  4. Write a conclusion.
- The inductive step depends on the type of problem.

Problem type	**Inductive step**
Series	Use $S_{k+1} = S_k + u_{k+1}$
Powers of matrices	Use $\mathbf{A}^{k+1} = \mathbf{A}^k\mathbf{A}$
Divisibility	Substitute from $f(k) = kA$ into $f(k+1)$
Factorials	Use $(k+1)! = (k+1) \times k!$
Anything else	There will probably be a hint in the previous part of the question.

Mixed practice 12

> **Tip**
>
> Some questions in this exercise are different from all of the examples in this chapter. You need to be able to adapt familiar methods to new contexts!

1 Use mathematical induction to prove that, for all positive integers n,

$$\begin{pmatrix} 1 & 0 \\ 1 & 1 \end{pmatrix}^n = \begin{pmatrix} 1 & 0 \\ n & 1 \end{pmatrix}$$

2 Prove that $3^{2n} + 7$ is divisible by 8 for all $n \in \mathbb{N}$.

3 Prove by induction that $3^n > 5n + 2$ for all integers $n \geqslant 3$.

4 Use mathematical induction to prove that $12^n - 1$ is a multiple of 11 for all $n \in \mathbb{N}$.

5 Prove by induction that for positive integer n,

$$\sum_{r=1}^{n} r(r+1) = \frac{n}{3}(n+1)(n+2)$$

6 Prove by induction that, for all $n \in \mathbb{Z}^+$,

$$\begin{pmatrix} 5 & 2 \\ -8 & -3 \end{pmatrix}^n = \begin{pmatrix} 1+4n & 2n \\ -8n & 1-4n \end{pmatrix}$$

7 Use mathematical induction to show that $15^n - 2^n$ is a multiple of 13 for all $n \in \mathbb{N}$.

8 Prove by induction that $11^{n+2} + 12^{2n+1}$ is divisible by 133 for all integers $n \geqslant 0$.

9 Use induction to prove that, for all integers $n \geqslant 1$,

$$\frac{1}{2!} + \frac{2}{3!} + \frac{3}{4!} + \frac{4}{5!} + \ldots + \frac{n}{(n+1)!} = \frac{(n+1)! - 1}{(n+1)!}$$

10 Prove by induction that, for all $n \in \mathbb{N}$,

$$\sum_{r=1}^{r=n} \frac{r}{2^r} = 2 - \left(\frac{1}{2}\right)^n (n+2)$$

11 Prove that, for all $n \in \mathbb{N}$, $\left(1-\sqrt{5}\right)^n$ has the form $\left(a - b\sqrt{5}\right)$ where a, b are positive integers.

12 Prove by induction that, for all integers $n \geqslant 1$,

$$\sum_{r=1}^{n} (r^2+1)(r!) = n(n+1)!$$

[© AQA 2008]

13 **a** Given that $f(k) = 12^k + 2 \times 5^{k-1}$, show that

$$f(k+1) - 5\,f(k) = a \times 12^k$$

where a is an integer.

b Prove by induction that $12^n + 2 \times 5^{n-1}$ is divisible by 7 for all integers $n \geqslant 1$.

[© AQA 2011]

14 The expression $f(n)$ is given by $f(n) = 2^{4n+3} + 3^{3n+1}$.

a Show that $f(k+1) - 16\,f(k)$ can be expressed in the form $A \times 3^{3k}$, where A is an integer.

b Prove by induction that $f(n)$ is a multiple of 11 for all integers $n \geqslant 1$.

[© AQA 2015]

15 Prove by induction that, for any matrix M, $\det(\mathbf{M}^n) = (\det \mathbf{M})^n$ for all positive integers n.

(You may use the result that $\det(\mathbf{AB}) = \det(\mathbf{A})\det(\mathbf{B})$ for any two matrices **A** and **B**.)

16 Prove by induction that

$$\frac{2\times 1}{2\times 3} + \frac{2^2\times 2}{3\times 4} + \frac{2^3\times 3}{4\times 5} + \ldots + \frac{2^n \times n}{(n+1)(n+2)} = \frac{2^{n+1}}{n+2} - 1$$

for all integers $n \geqslant 1$.

[© AQA 2009]

17 **a** Show that, for any two complex numbers z and w, $(zw)^* = z^* w^*$.

b Prove by induction that $(z^n)^* = (z^*)^n$ for positive integer n.

18 De Moivre's theorem for complex numbers states that:

$$(\cos\theta + \mathrm{i}\sin\theta)^n = \cos n\theta + \mathrm{i}\sin n\theta$$

Use induction to prove de Moivre's theorem for all $n \in \mathbb{N}$.

Rewind

The result in Question 18 is an extension of the rule for multiplying complex numbers in modulus–argument form, which you learnt in Chapter 1, Section 5.

Fast forward

You will use de Moivre's theorem in Further Mathematics Student Book 2.

FOCUS ON ... PROOF 2

Proving properties of identity and inverse matrices

Throughout Chapters 7 and 8 you used various properties of identity and inverse matrices without really checking that they must be true. But in mathematics, all properties you use must be proved. This is particularly important when working with structures such as matrices, which have many properties similar to numbers but also some that are different. (For example, matrix multiplication is **not** commutative, and **not** all non-zero matrices have an inverse.) This means that our intuition about what 'works' may be wrong.

Key point 7.8 stated that the identity matrix $\mathbf{I}$ has the unique property that $\mathbf{IA} = \mathbf{AI} = \mathbf{A}$ for every matrix $\mathbf{A}$. It is reasonable to ask: *How can you be sure that there is only one matrix with this property? Would it be possible to have two different identity matrices?*

You are going to prove that the identity matrix is indeed unique.

PROOF 11

If $\mathbf{J}$ is a matrix such that $\mathbf{JA} = \mathbf{AJ} = \mathbf{A}$ for every matrix $\mathbf{A}$, then $\mathbf{J} = \mathbf{I}$.

Consider the product $\mathbf{IJ}$ — One way to prove that $\mathbf{J} = \mathbf{I}$ is to use the identity property directly.

By the definition of $\mathbf{I}$, $\mathbf{IJ} = \mathbf{J}$ — $\mathbf{I}$ is the identity matrix so $\mathbf{IA} = \mathbf{A}$ for any conformable matrix $\mathbf{A}$.

But by the proposed property of $\mathbf{J}$, $\mathbf{IJ} = \mathbf{I}$ — You are proposing that $\mathbf{AJ} = \mathbf{A}$ for any conformable matrix $\mathbf{A}$.

Then $\mathbf{J} = \mathbf{IJ} = \mathbf{I}$ so $\mathbf{J} = \mathbf{I}$ — State your conclusion.

The statement in Key point 7.8 requires that both $\mathbf{IA} = \mathbf{A}$ and $\mathbf{AI} = \mathbf{A}$. You know that matrix multiplication is generally not commutative ($\mathbf{AB} \neq \mathbf{BA}$ for some matrices). So it is sensible to ask: *Is there some other matrix* $\mathbf{J}$*, such that* $\mathbf{JA} = \mathbf{A}$ *but* $\mathbf{AJ} \neq \mathbf{A}$? (In advanced mathematics, you would say that $\mathbf{J}$ is a 'left identity' but not a 'right identity'.)

It turns out that this is not the case. You are going to prove this result.

PROOF 12

If **J** is a matrix such that **JA** = **A** for all **A**, then also **AJ** = **A**.

If **JA** = **A** for all **A**, then:	
$\mathbf{AJ} = (\mathbf{AJ})\mathbf{I}$ (by the definition of the identity matrix **I**)	Multiplication by **I**, by definition, does not change a matrix.
$= \mathbf{A}(\mathbf{JI})$	Group the matrices as you like, but do not change the order.
$= \mathbf{AI}$ (by the proposed property of **J**)	Use the given property of **J** in its product with **I**.
$= \mathbf{A}$	
Hence if **JA** = **A**, for all = **A** then also **AJ** = **A**.	State your conclusion.

QUESTIONS

1 Use arguments similar to those in proofs 11 and 12 to prove these properties of inverse matrices.

a Every non-singular matrix has a unique inverse.

b If **A** is a non-singular matrix with inverse $\mathbf{A}^{-1}$, and if **B** is another matrix such that **AB** = **I**, then $\mathbf{B} = \mathbf{A}^{-1}$.

c If **A** in a non-singular matrix, then its 'right inverse' and its 'left inverse' are the same, so if $\mathbf{AA}^{-1} = \mathbf{I}$ then also $\mathbf{A}^{-1}\mathbf{A} = \mathbf{I}$.

2 In proofs 11 and 12, it was stressed several times that you cannot simply 'cancel' a matrix from both sides of an equation. In other words, **AB** = **AC** does not necessarily imply that **B** = **C**.

a Find a counterexample to prove that $\mathbf{AB} = \mathbf{AC} \Rightarrow \mathbf{B} = \mathbf{C}$ is not true.

b Find an example of two non-zero matrices whose product is zero. (Hence it is not true that $\mathbf{AB} = \mathbf{O} \Rightarrow \mathbf{A} = \mathbf{O}$ or $\mathbf{B} = \mathbf{O}$.)

FOCUS ON ... PROBLEM SOLVING 2

Alternative approaches to calculating distances

In Worked example 9.19, you looked at this problem.

Line l has equation $r = \begin{pmatrix} 3 \\ -1 \\ 0 \end{pmatrix} + \lambda \begin{pmatrix} 1 \\ -1 \\ 1 \end{pmatrix}$ and point A has coordinates (3, 9, −2).

Find the shortest distance from A to l.

You used this strategy to solve the problem.

Strategy 1

- Write down the position vector of a general point P on the line (in terms of λ):

$$\mathbf{p} = \begin{pmatrix} 3+\lambda \\ -1-\lambda \\ \lambda \end{pmatrix}$$

- To get the shortest distance, AP must be perpendicular to the line. This means that $\overrightarrow{AP} \bullet \begin{pmatrix} 1 \\ -1 \\ 1 \end{pmatrix} = 0$, which gives an equation for λ:

$$\begin{pmatrix} (3+\lambda)-3 \\ (-1-\lambda)-9 \\ \lambda+2 \end{pmatrix} \bullet \begin{pmatrix} 1 \\ -1 \\ 1 \end{pmatrix} = 0$$

$$\Rightarrow 1(\lambda) - 1(-10-\lambda) + 1(\lambda+2) = 0$$

- Find λ and hence find the coordinates of P.
- Calculate the distance AP.

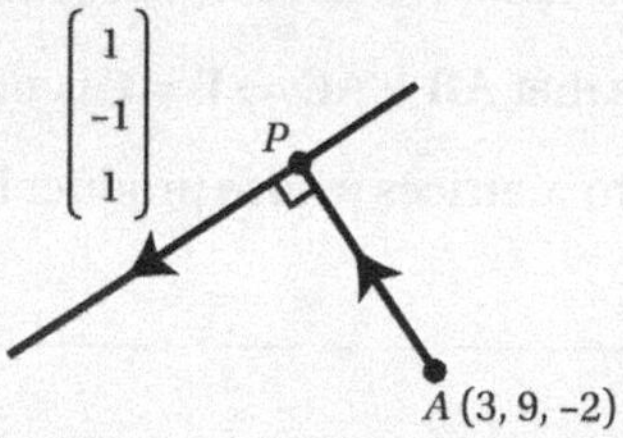

You can solve this problem in several different ways. Strategies 2 and 3 give two alternative solutions: decide for yourself which is the simpler.

Strategy 2

This also starts by expressing the vector $\overrightarrow{AP}$ in terms of λ, but you then use algebra or calculus to minimise this expression.

- Show that $(AP)^2 = 3\lambda^2 + 24\lambda + 104$.
- Either by completing the square, or by using differentiation, find the minimum value of $(AP)^2$.
- Hence show that the required minimum distance AP is $2\sqrt{14}$.

Strategy 3

This strategy uses a geometrical approach. You find the minimum distance directly, without finding the coordinates of P (or the value of λ).

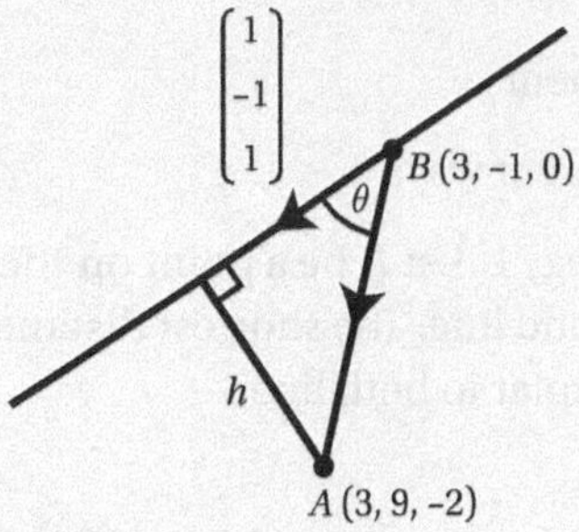

The point $B(3, -1, 0)$ lies on the line. The shortest distance, marked h, equals $|\overrightarrow{AB}|\sin\theta$.

- Find the exact value of $|\overrightarrow{AB}|$.
- Use the scalar product to show that $\cos\theta = -\frac{6}{\sqrt{78}}$ and find the exact value of $\sin\theta$.
- Hence show that $h = 2\sqrt{14}$.

QUESTIONS

1 Strategy 3 can be simplified slightly if you introduce the concept of *projection*. Consider a line with direction vector **d** through the point with position vector **b**, and a point off the line with position vector **a**.

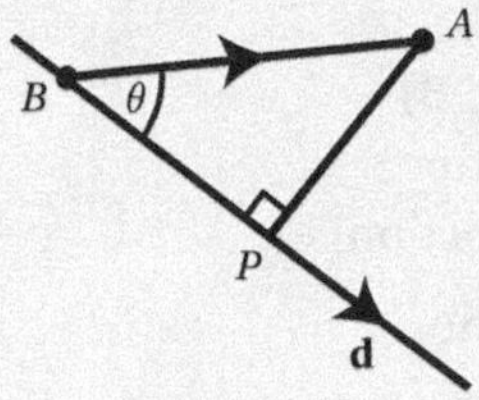

Let P be the foot of the perpendicular from A to the line. The length of BP is the projection of BA onto the line.

a Use the right-angled triangle ABP to express the length BP in terms of BA and θ.

b Use the scalar product to write $\cos\theta$ in terms of $(\mathbf{a} - \mathbf{b})$ and $\mathbf{d}$.

c Hence show that $BP = \frac{(\mathbf{a}-\mathbf{b})\cdot\mathbf{d}}{|\mathbf{d}|}$.

d Use this formula for the given example. Hence use $AP^2 = BA^2 - BP^2$ to find the shortest distance from A to the line.

2 The lines with equations $\mathbf{r} = \begin{pmatrix} -1 \\ 2 \\ 0 \end{pmatrix} + \lambda \begin{pmatrix} 3 \\ -1 \\ 1 \end{pmatrix}$ and $\mathbf{r} = \begin{pmatrix} -1 \\ 3 \\ 8 \end{pmatrix} + \mu \begin{pmatrix} -1 \\ 2 \\ 3 \end{pmatrix}$

are skew.

Find the shortest distance between them.

Try solving the problem in three different ways.

i Use the scalar product, as in Strategy 1. Let P be a point on the first line and Q a point on the second line. The shortest distance is achieved when PQ is perpendicular to both lines.

a Write $\overrightarrow{PQ}$ in terms of λ and μ.

b Use the scalar product to obtain two equations for λ and μ.

c Hence find the minimum distance PQ.

ii Starting from part **a**, show that $|\overrightarrow{PQ}|^2 = 14\mu^2 + 11\lambda^2 + 4\mu\lambda + 52\mu - 14\lambda + 65$.
How could you find the minimum value of this expression?

iii You can use the scalar product to find a vector perpendicular to the direction vectors of both lines. This vector will then be parallel to $\overrightarrow{PQ}$.

a Use the scalar product to form two equations and hence find a vector, $\mathbf{n} = \begin{pmatrix} n_1 \\ n_2 \\ n_3 \end{pmatrix}$.

b Point $A(-1, 2, 0)$ lies on the first line and point $B(-1, 3, 8)$ lies on the second line. Since $\mathbf{n}$ is perpendicular to both lines, PQ is the projection of AB onto the line with direction $\mathbf{n}$. Adapt the argument from Question 1 to show that the length of PQ is $\frac{(\mathbf{a} - \mathbf{b}) \cdot \mathbf{n}}{|\mathbf{n}|}$. Hence show that the shortest distance between the two lines is $\sqrt{6}$.

Fast forward

In Further Mathematics Student Book 2 you will see how you can use the vector product to find a vector perpendicular to two other vectors.

FOCUS ON ... MODELLING 2

Counting paths in networks: choosing the right representation

Describing the situation

There are many situations where you are interested in connections between places or people. Examples include road systems and social networks. For many purposes you can model such a situation as a *network* (or *graph*). This is a system of points (*nodes or vertices*) connected by lines (*edges*).

This model only shows which nodes are connected by an edge; it ignores things like shape of the road, the exact position of towns, or the nature of 'friendship' in a social network (close friend, family, etc.) The model can be refined in several ways: the edges can have numbers associated with them (representing, for example, length, time or cost); sometimes edges can be directed (for example, in a network of one-way streets). There can also be more than one edge connecting two nodes; for example, there may be both a motorway and an A-road connecting two cities, or two people may be friends on several social networking sites.

Did you know?

One of the most famous network models is the map of the London Underground (the Tube).

The network in this diagram has 3 nodes (labelled A, B, C) and 4 edges (labelled p, q, r, s).

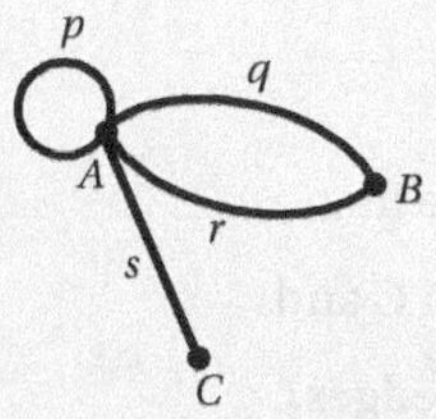

Selecting a mathematical representation

Although there are some problems that can be solved efficiently by using pictures (for example, in geometry and some probability), using equations is a far more common way in which to represent mathematical models. Furthermore, most network problems deal with very large networks and need to be solved using a computer; this means that you need to represent the information in a way that can be used in a computer program.

A useful way of representing a network is as a matrix. Each row and column corresponds to a node, and each entry is the number of edges connecting two nodes. This is called the **adjacency matrix** for the network. For example, the network shown can be represented by this matrix.

$$\begin{array}{c} \\ A \\ B \\ C \end{array} \begin{array}{c} \begin{array}{ccc} A & B & C \end{array} \\ \begin{bmatrix} 1 & \textcircled{2} & 1 \\ 2 & 0 & 0 \\ 1 & 0 & 0 \end{bmatrix} \end{array}$$

The circled number 2 shows that there are two edges between A and B. (Notice that since the edges are not directed the matrix is symmetric.)

You will consider networks in which:

- there can be more than one edge connecting two nodes
- a node can be connected to itself
- the edges are **not** directed; this means that if A is connected to B then B is also connected to A
- there are no numbers on the edges; you are only interested in the number of edges between two nodes, not their length or shape.

Using the model

A **path** between two nodes is a sequence of edges starting at the first node and ending at the second. For example, one path between C and B is $B \overset{r}{—} A \overset{p}{—} A \overset{s}{—} C$; this path has length 3 (it consists of three edges).

A path can use the same edge more than once; for example, $A \overset{s}{—} C \overset{s}{—} A \overset{q}{—} B \overset{r}{—} A \overset{q}{—} B$ is a path of length 5 between A and B.

You are going to solve this problem.

How can you find the number of paths of a given length between two given nodes?

For example, in the network shown, there are six paths of length 3 between A and C.

$$A \overset{p}{—} A \overset{p}{—} A \overset{s}{—} C$$
$$A \overset{s}{—} C \overset{s}{—} A \overset{s}{—} C$$
$$A \overset{q}{—} B \overset{q}{—} A \overset{s}{—} C$$
$$A \overset{q}{—} B \overset{r}{—} A \overset{s}{—} C$$
$$A \overset{r}{—} B \overset{q}{—} A \overset{s}{—} C$$
$$A \overset{r}{—} B \overset{r}{—} A \overset{s}{—} C$$

QUESTIONS A

To get a feel for the problem, answer these questions.

1 List all the paths of length 2 between:

a A and B b A and A.

2 List all the paths of length 3 between:

a B and C b B and B.

Even listing all the paths of length 3 seems reasonably complicated. How would you count all the paths of length 10? What would you do if your network had 1000 nodes?

It turns out that, with the matrix representation, it is possible to prove an extremely useful result.

The number of paths of length n between two nodes is given by the corresponding entry in the matrix $\mathbf{M}^n$.

For example, $\mathbf{M}^3 = \begin{pmatrix} 11 & 12 & 6 \\ 12 & 4 & 2 \\ 6 & 2 & 1 \end{pmatrix}$, which shows that there are 6 paths of length 3 between A and C (as you have seen).

> **Tip**
>
> You can actually prove this result by induction. To count the number of paths of length $k+1$ between A and B, think about using paths of length k to get from A to another node, and then taking the final step from that node to B. The way you add up the numbers of different paths is the same as the way you multiply matrices.

QUESTIONS B

3 By representing the network as an adjacency matrix, find the number of paths:

a of length 2 between A and B

b of length 5 between A and C.

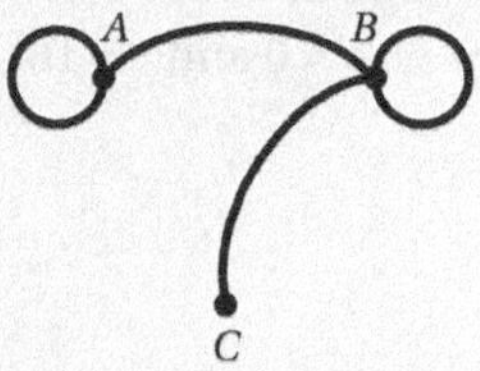

4 Find the number of paths:

a of length 3 between A and C

b of length 4 between A and A.

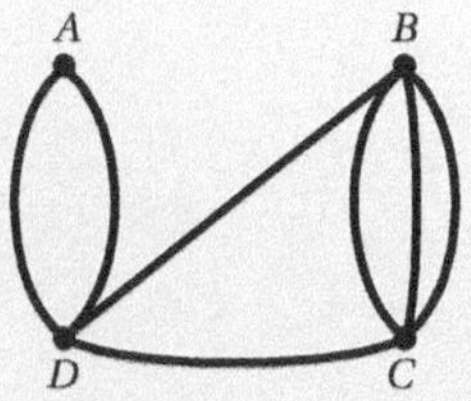

5 Can you adapt the model to answer the question about the number of paths in a directed network, such as this one? Find the number of paths of length 6:

a from B to A

b from A to B

c from C to C.

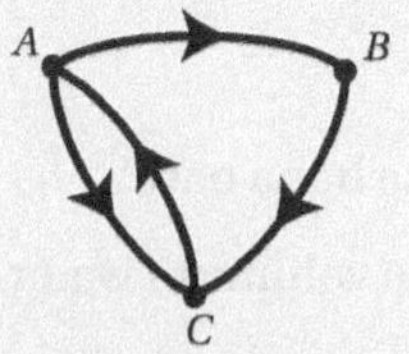

CROSS-TOPIC REVIEW EXERCISE 2

1 Find the determinant of $\begin{pmatrix} 2\text{i} & -1 \\ 3\text{i} & \text{i} \end{pmatrix}$.

2 Find the value of c for which the matrix $\begin{pmatrix} 3 & -c \\ 1 & c+5 \end{pmatrix}$ is singular.

3 Consider the matrix $\mathbf{A} = \begin{pmatrix} -3 & 1 \\ 1 & 2 \end{pmatrix}$.

a Find det(**A**).

S is a square with vertices (1, 0), (3, 0), (1, 2) and (3, 2). The transformation with matrix **A** transforms S into T.

b Find the area of T.

4 **A** and **B** are 2×2 non-singular matrices.

Simplify $\mathbf{B}(\mathbf{AB})^{-1}\mathbf{A}$.

5 **a** Find a vector equation of the line L_1 passing through the points $P(3, 12, -5)$ and $Q(-1, 1, 6)$.

b The line L_2 has equation $\mathbf{r} = (\mathbf{i} + 8\mathbf{j} + 2\mathbf{k}) + s(2\mathbf{i} + 6\mathbf{j} - 5\mathbf{k})$.

Find the point of intersection of L_1 and L_2.

6 The position vectors of three points A, B and C relative to an origin O are given respectively by:

$\overline{OA} = 7\mathbf{i} + 3\mathbf{j} - 3\mathbf{k}$,

$\overline{OB} = 4\mathbf{i} + 2\mathbf{j} - 6\mathbf{k}$ and

$\overline{OC} = 5\mathbf{i} - \mathbf{j} - 5\mathbf{k}$.

a Find the angle between AB and AC.

b Find the area of triangle ABC.

7 The shape of a vase can be modelled by rotating the curve with equation $16x^2 - (y-8)^2 = 32$ between $y = 0$ and $y = 16$ completely about the y-axis.

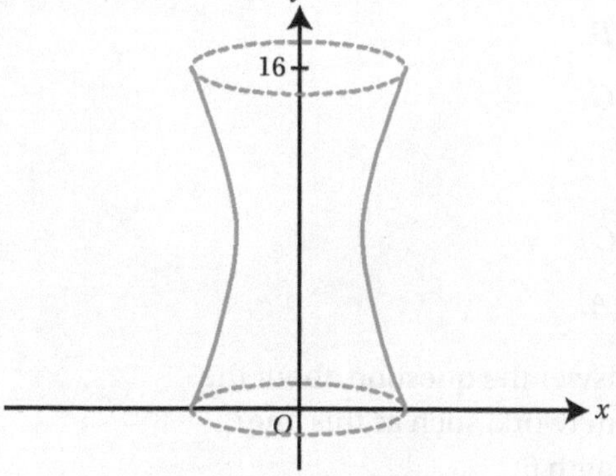

The vase has a base.

Find the volume of water needed to fill the vase, giving your answer as an exact value.

[© AQA 2013]

8 The shaded region in the diagram is bounded by the curve with equation $y = 4x^{-\frac{3}{2}} + 1$, the x-axis and the lines $x = 1$ and $x = 4$.

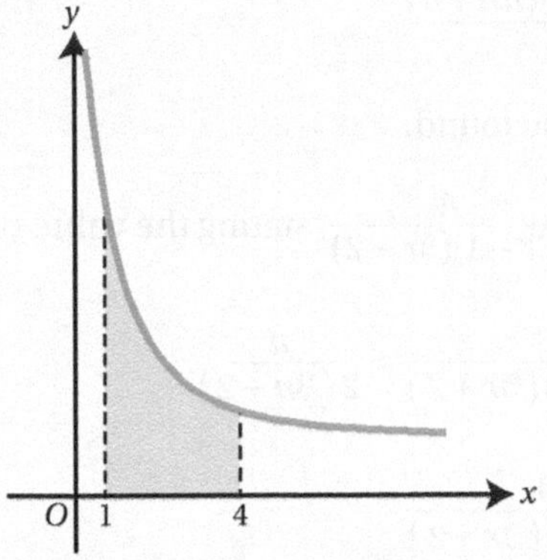

This region is rotated through 360° about the x-axis. Find the exact volume of the solid formed.

9 The 3×3 matrix **R** represents reflection in the plane $y = 0$. The 3×3 matrix **S** represents a 90° rotation about the x-axis.

a By considering the images of the unit vectors **i**, **j** and **k**, find:

i **R** **ii** **S**.

b Find the single matrix that represents reflection in the plane $y = 0$ followed by rotation about the x-axis.

c Hence find the image of the point (2, −3, 5) under this combination of transformations.

10 Line l_1 passes through points $M(3, -7, 1)$ and $N(5, -5, 5)$.

a Find the vector equation of l_1.

b Given that l_1 intersects the line with equation $\mathbf{r} = (\mathbf{i} + 6\mathbf{k}) + \mu(-2\mathbf{i} + \mathbf{j} + p\mathbf{k})$, find the value of p.

11 Determine whether the lines $\frac{x-3}{2} = \frac{y+3}{1} = \frac{z-1}{-4}$ and $\frac{x-3}{6} = \frac{y-4}{-4} = \frac{z-2}{1}$ intersect or are skew.

12 The coordinates of the points A and B are (3, −2, 1) and (5, 3, 0) respectively.

The line l has equation $\mathbf{r} = \begin{pmatrix} 5 \\ 3 \\ 0 \end{pmatrix} + \lambda \begin{pmatrix} 1 \\ 0 \\ -3 \end{pmatrix}$.

a Find the distance between A and B.

b Find the acute angle between the lines AB and l. Give your answer to the nearest degree.

c The points B and C lie on l such that the distance AC is equal to the distance AB. Find the coordinates of C.

[©AQA 2008]

13 Use the formulae for $\sum r$ and $\sum r^2$ to show that

$$\sum_{r=0}^{n}(r+1)^2 = \frac{(n+1)(n+2)(an+b)}{6}$$

where a and b are constants to be found.

14 a Show that $\frac{1}{3r-1} - \frac{1}{3r+2} \equiv \frac{A}{(3r-1)(3r+2)}$, stating the value of the constant A.

b Hence show that $\sum_{r=1}^{n}\frac{1}{(3r-1)(3r+2)} = \frac{n}{2(3n+2)}$.

c Find the value of $\sum_{r=1}^{\infty}\frac{1}{(3r-1)(3r+2)}$.

15 Prove by induction that $2^n + 6^n$ is divisible by 8 for all positive integers n.

16 a Show that $\frac{1}{(k+2)!} - \frac{k+1}{(k+3)!} \equiv \frac{2}{(k+3)!}$

b Prove by induction that, for all positive integers n,

$$\sum_{r=1}^{n}\frac{r\times 2^r}{(r+2)!} = 1 - \frac{2^{n+1}}{(n+2)!}$$

[©AQA 2010]

17 The transformation represented by the matrix $\mathbf{M} = \begin{pmatrix} -3 & 4 \\ 4 & 3 \end{pmatrix}$ has an invariant line $y = cx$ with $c > 0$.

a Find the value of c.

b Show that this is not a line of invariant points.

18 The equation of a straight line l is $\mathbf{r} = \begin{pmatrix} 1 \\ -1 \\ 2 \end{pmatrix} + \lambda\begin{pmatrix} -2 \\ 3 \\ 4 \end{pmatrix}$.

Find the shortest distance between l and the point with coordinates $(7, 5, -2)$. Give your answer to 2 decimal places.

19 The line l_1 has Cartesian equation $\frac{x-2}{3} = \frac{y+1}{4} = \frac{z-3}{5}$.

a Find a vector equation for l_1.

The line l_2 has vector equation $\mathbf{r} = \mathbf{i} - \mathbf{k} + t(\mathbf{i} + \mathbf{j} + \mathbf{k})$.

b Show that l_1 and l_2 do not intersect.

c i Find points P and Q on l_1 and l_2 respectively such that $\overline{PQ}$ is perpendicular to both lines.

ii Hence find the shortest distance between l_1 and l_2. Give your answer in exact form.

20 The mean value of the function $f(x) = 2 - \frac{1}{2\sqrt{x}}$ between 1 and k is $\frac{8}{5}$.

Find the value of k.

21 **a** Find, up to the term in x^3, the Maclaurin series for $\ln\left(\frac{2+x}{2-x}\right)$.

b Find the set of x values for which the expansion is valid.

c By evaluating the series in **a** at an appropriate value of x, find a rational approximation to ln 3.

22 **a** Express $(k+1)^2+5(k+1)+8$ in the form k^2+ak+b where a and b are constants.

b Prove by induction that, for all integers $n \geqslant 1$,

$$\sum_{r=1}^{n} r(r+1)\left(\frac{1}{2}\right)^{r-1} = 16-(n^2+5n+8)\left(\frac{1}{2}\right)^{n-1}$$

[©AQA 2014]

23 A linear transformation is represented by the matrix $\mathbf{M}=\begin{pmatrix} 3 & -1 \\ 0 & a \end{pmatrix}$.

Line l_1 has equation $\mathbf{r}=\begin{pmatrix} 1 \\ 0 \end{pmatrix}+t\begin{pmatrix} -3 \\ 1 \end{pmatrix}$. Line l_2 is the image of l_1 under the transformation **M**.

Find the value of a such that l_2 is perpendicular to l_1.

24 **a** The transformation **U** of three-dimensional space is represented by the matrix $\begin{pmatrix} 1 & 4 & -3 \\ 2 & -1 & 0 \\ 1 & 1 & -1 \end{pmatrix}$.

i Write down a vector equation for the line L with Cartesian equation $\frac{x-1}{2}=\frac{y-2}{3}=\frac{z-3}{6}$.

ii Find a vector equation for the image of L under **U**, and deduce that it is a line through the origin.

b The plane transformation **V** is represented by the matrix $\begin{pmatrix} 1 & 4 \\ 2 & -1 \end{pmatrix}$.

L_1 is the line with equation $y=\frac{1}{2}x+k$, and L_2 is the image of L_1 under **V**.

i Find, in the form $y=mx+c$, the Cartesian equation for L_2.

ii Deduce that L_2 is parallel to L_1 and find, in terms of k, the distance between these two lines.

[©AQA 2011]

25 The matrix **A** is given by $\mathbf{A}=\begin{pmatrix} 1 & 0 \\ 3 & 1 \end{pmatrix}$.

a Show that $\mathbf{A}^2=2\mathbf{A}-\mathbf{I}$.

b Prove by induction that for $n \geqslant 1$,

$\mathbf{A}^n = n\mathbf{A}-(n-1)\mathbf{I}$

c The matrix $\mathbf{A}^5$ represents a linear transformation. The image of the point P under this transformation has coordinates (−25, 12). Find the coordinates of P.

PRACTICE PAPER

1 hour 30 minutes, 80 marks

1 $z_1 = 6\left(\cos\frac{\pi}{4} + \text{i}\sin\frac{\pi}{4}\right)$ and $z_2 = 2\left(\cos\frac{\pi}{6} + \text{i}\sin\frac{\pi}{6}\right)$

Find $\arg\left(\frac{z_1}{z_2}\right)$.

Choose from these options.

A $\frac{3}{2}$ **B** 3 **C** $\frac{\pi}{12}$ **D** $\frac{7\pi}{6}$ [1 mark]

2 Find the equation of the asymptote of the curve $\frac{2x^2 - 3x + 5}{x^2 + 4}$

Choose from these options.

A $y = 2$ **B** $y = \frac{1}{2}$ **C** $x = 2$ **D** $x = -2$ [1 mark]

3 Two complex numbers are given by $z = 3 - 2\text{i}$ and $w = 2 + \text{i}$. Showing your method clearly, and giving your answer in the form $x + y\text{i}$, calculate:

a $w - z^*$ [2 marks]

b $\frac{z}{w}$. [3 marks]

4 Two vectors are given by $\mathbf{a} = 2\mathbf{i} + (p-1)\mathbf{j} - 2\mathbf{k}$ and $\mathbf{b} = p\mathbf{i} + 2\mathbf{j} + (2p-1)\mathbf{k}$, where $p \in \mathbb{R}$.

a Show that **a** and **b** are perpendicular for all values of p. [2 marks]

b Find the value of p for which $\mathbf{a} + \mathbf{b}$ is parallel to the vector $2\mathbf{j} + 14\mathbf{k}$. [3 marks]

5 The cubic equation $2x^3 - 5x + 6 = 0$ has roots α, β and γ. Find the value of $\frac{1}{\alpha} + \frac{1}{\beta} + \frac{1}{\gamma}$. [5 marks]

6 Use the formulae for $\sum_{r=1}^{n} r$ and $\sum_{r=1}^{n} r^2$ to show that $\sum_{r=1}^{n} (2r-1)^2 = \frac{n}{3}(2n-1)(2n+1)$. [6 marks]

7 The region bounded by the curve $y = \frac{1}{\sqrt[4]{x}} + 1$, the x-axis and the lines $x = 1$ and $x = 4$ is rotated through 360° about the x-axis.

Show that the volume generated is $\frac{(a + b\sqrt{2})\pi}{c}$, where a, b and c are integers to be found. [5 marks]

8 Show that the lines with Cartesian equations $\frac{x-2}{5} = \frac{y+1}{-1} = \frac{z-1}{1}$ and $\frac{x+1}{2} = \frac{y-1}{2} = \frac{z-2}{7}$ are skew. [5 marks]

9 Consider the matrix $\mathbf{A} = \begin{pmatrix} 3c & c-1 \\ 2 & -3 \end{pmatrix}$.

a Find the value of c for which **A** is a singular matrix. [3 marks]

b Given that c is such that **A** is non-singular, find $\mathbf{A}^{-1}$ in terms of c. [2 marks]

10 The complex number a is given by $a = -\sqrt{3} + \text{i}$.

a Find the modulus and the argument of a [2 marks]

b On an Argand diagram represent the locus of points satisfying the inequalities $|z-a| \leqslant |z-\text{i}|$ and $0 < \arg(z-a) < \frac{\pi}{4}$. [4 marks]

11 A curve C has polar equation $r(1-\cos\theta) = 3$.

a Find its Cartesian equation in the form $y^2 = \text{f}(x)$. [4 marks]

b The curve C intersects the line $3r = \frac{4}{\cos\theta}$.

Find the value of r at the points of intersection. [3 marks]

12 **a** Show that the Maclaurin series of the function $\ln(1+\sin x)$ up to the term in x^4 is $x - \frac{x^2}{2} + \frac{x^3}{6} - \frac{x^4}{12} + \cdots$ [6 marks]

b A student claims that this series is valid for all $x \in \mathbb{R}$. Show by means of a counterexample that he is wrong. [2 marks]

13 The hyperbola H has equation $3x^2 - 16y^2 = 48$.

a Find the equations of the asymptotes to H. [2 marks]

The line $y = mx + 1$ is a tangent to H.

b **i** Show that $(3-16m^2)x^2 - 32mx - 64 = 0$

ii Hence find the possible values of m. [5 marks]

14 A linear transformation is represented by the matrix $\mathbf{M} = \begin{pmatrix} 1 & -2 & 2 \\ 0 & 0 & 3 \\ 1 & 0 & -2 \end{pmatrix}$.

A combined transformation consists of the reflection in the x–y plane followed by the transformation represented by $\mathbf{M}$.

a Find the matrix representing the combined transformation. [3 marks]

b Find the image of the point $(-1, 2, 4)$ under the combined transformation. [1 mark]

c Show that the only invariant point under the combined transformation is $(0, 0, 0)$. [3 marks]

15 Prove by induction that $9^n + 7^n - 2^n$ is divisible by 14 for all $n \in \mathbb{Z}^+$. [7 marks]

FORMULAE

Further Pure Mathematics

Differentiation

f(x)	f′(x)
$\sin^{-1} x$	$\frac{1}{\sqrt{1-x^2}}$
$\cos^{-1} x$	$-\frac{1}{\sqrt{1-x^2}}$
$\tan^{-1} x$	$\frac{1}{1+x^2}$
$\tanh x$	$\text{sech}^2\, x$
$\sinh^{-1} x$	$\frac{1}{\sqrt{1+x^2}}$
$\cosh^{-1} x$	$\frac{1}{\sqrt{x^2-1}}$
$\tanh^{-1} x$	$\frac{1}{1-x^2}$

Integration

f(x)	$\int$ f(x) dx
$\tanh x$	$\ln \cosh x$
$\frac{1}{\sqrt{a^2-x^2}}$	$\sin^{-1}\left(\frac{x}{a}\right)$ $\quad(\lvert x\rvert < a)$
$\frac{1}{a^2+x^2}$	$\frac{1}{a}\tan^{-1}\left(\frac{x}{a}\right)$
$\frac{1}{\sqrt{x^2-a^2}}$	$\cosh^{-1}\left(\frac{x}{a}\right)$ or $\ln\left\{x+\sqrt{x^2-a^2}\right\}$ $\quad(x > a)$
$\frac{1}{\sqrt{a^2+x^2}}$	$\sinh^{-1}\left(\frac{x}{a}\right)$ or $\ln\left\{x+\sqrt{x^2+a^2}\right\}$
$\frac{1}{a^2-x^2}$	$\frac{1}{2a}\ln\left\lvert\frac{a+x}{a-x}\right\rvert = \frac{1}{a}\tanh^{-1}\frac{x}{a}$ $\quad(\lvert x\rvert < a)$
$\frac{1}{x^2-a^2}$	$\frac{1}{2a}\ln\left\lvert\frac{x-a}{x+a}\right\rvert$

Complex numbers

$[r(\cos\theta + \mathrm{i}\sin\theta)]^n = r^n(\cos n\theta + \mathrm{i}\sin n\theta)$

The roots of $z^n = 1$ are given by $z = \mathrm{e}^{\frac{2\pi k \mathrm{i}}{n}}$ for $k = 0, 1, 2, \ldots, n-1$

Matrix transformations

Anticlockwise rotation through θ about O: $\begin{bmatrix} \cos\theta & -\sin\theta \\ \sin\theta & \cos\theta \end{bmatrix}$

Reflection in the line $y = (\tan\theta)x$: $\begin{bmatrix} \cos 2\theta & \sin 2\theta \\ \sin 2\theta & -\cos 2\theta \end{bmatrix}$

The matrices for rotations (in three dimensions) through an angle θ about one of the axes are:

$\begin{bmatrix} 1 & 0 & 0 \\ 0 & \cos\theta & -\sin\theta \\ 0 & \sin\theta & \cos\theta \end{bmatrix}$ for the x-axis

$\begin{bmatrix} \cos\theta & 0 & \sin\theta \\ 0 & 1 & 0 \\ -\sin\theta & 0 & \cos\theta \end{bmatrix}$ for the y-axis

$\begin{bmatrix} \cos\theta & -\sin\theta & 0 \\ \sin\theta & \cos\theta & 0 \\ 0 & 0 & 1 \end{bmatrix}$ for the z-axis

Summations

$$\sum_{r=1}^{n} r^2 = \frac{1}{6}n(n+1)(2n+1)$$

$$\sum_{r=1}^{n} r^3 = \frac{1}{4}n^2(n+1)^2$$

Maclaurin's series

$\mathrm{f}(x) = \mathrm{f}(0) + x\,\mathrm{f}'(0) + \frac{x^2}{2!}\mathrm{f}''(0) + \ldots + \frac{x^r}{r!}\mathrm{f}^{(r)}(0) + \ldots$

$\mathrm{e}^x = \exp(x) = 1 + x + \frac{x^2}{2!} + \ldots + \frac{x^r}{r!} + \ldots$ for all x

$\ln(1+x) = x - \frac{x^2}{2} + \frac{x^3}{3} - \ldots + (-1)^{r+1}\frac{x^r}{r} + \ldots \quad (-1 < x_n \leqslant 1)$

$\sin x = x - \frac{x^3}{3!} + \frac{x^5}{5!} - \ldots + (-1)^r \frac{x^{2r+1}}{(2r+1)!} + \ldots$ for all x

$\cos x = 1 - \frac{x^2}{2!} + \frac{x^4}{4!} - \ldots + (-1)^r \frac{x^{2r}}{(2r)!} + \ldots$ for all x

Vectors

The resolved part of **a** in the direction of **b** is $\frac{\mathbf{a}.\mathbf{b}}{|\mathbf{b}|}$

The vector product $\mathbf{a}\times\mathbf{b} = |\mathbf{a}||\mathbf{b}|\sin\theta\,\hat{\mathbf{n}} = \begin{vmatrix} \mathbf{i} & a_1 & b_1 \\ \mathbf{j} & a_2 & b_2 \\ \mathbf{k} & a_3 & b_3 \end{vmatrix} = \begin{bmatrix} a_2b_3 - a_3b_2 \\ a_3b_1 - a_1b_3 \\ a_1b_2 - a_2b_1 \end{bmatrix}$

If A is the point with position vector $\mathbf{a} = a_1\mathbf{i} + a_2\mathbf{j} + a_3\mathbf{k}$, then

- the straight line through A with direction vector $\mathbf{b} = b_1\mathbf{i} + b_2\mathbf{j} + b_3\mathbf{k}$ has equation $\frac{x-a_1}{b_1} = \frac{y-a_2}{b_2} = \frac{z-a_3}{b_3} = \lambda$ (Cartesian form) or

  $(\mathbf{r}-\mathbf{a})\times\mathbf{b} = 0$ (vector product form)

- the plane through A and parallel to **b** and **c** has vector equation $\mathbf{r} = \mathbf{a} + s\mathbf{b} + t\mathbf{c}$

Area of a sector

$A = \frac{1}{2}\int r^2\,\mathrm{d}\theta$ (polar coordinates)

Hyperbolic functions

$\cosh^2 x - \sinh^2 x \equiv 1$

$\sinh 2x \equiv 2\sinh x \cosh x$

$\cosh 2x \equiv \cosh^2 x + \sinh^2 x$

$\cosh^{-1}x = \operatorname{arcosh} x = \ln\left\{x + \sqrt{x^2-1}\right\} \quad (x \geqslant 1)$

$\sinh^{-1}x = \operatorname{arsinh} x = \ln\left\{x + \sqrt{x^2+1}\right\}$

$\tanh^{-1}x = \operatorname{artanh} x = \frac{1}{2}\ln\left\{\frac{1+x}{1-x}\right\} \quad (|x| < 1)$

Conics

	Ellipse	Parabola	Hyperbola
Standard form	$\frac{x^2}{a^2} + \frac{y^2}{b^2} = 1$	$y^2 = 4ax$	$\frac{x^2}{a^2} - \frac{y^2}{b^2} = 1$
Parametric form	$x = a\cos\theta$ $y = b\sin\theta$	$x = at^2$ $y = 2at$	$x = a\sec\theta$ $y = b\tan\theta$
Asymptotes	none	none	$\frac{x}{a} = \pm\frac{y}{b}$

Further numerical integration

The mid-ordinate rule: $\int_a^b y \, \mathrm{d}x \approx h\left(y_{\frac{1}{2}} + y_{\frac{3}{2}} + \ldots + y_{n-\frac{3}{2}} + y_{n-\frac{1}{2}}\right)$

where $h = \dfrac{b-a}{n}$

Simpson's rule: $\int_a^b y \, \mathrm{d}x \approx \frac{1}{3}h\left\{\left(y_0 + y_n\right) + 4\left(y_1 + y_3 + \ldots + y_{n-1}\right) + 2\left(y_2 + y_4 + \ldots + y_{n-2}\right)\right\}$

where $h = \dfrac{b-a}{n}$ and n is even

Numerical solution of differential equations

For $\dfrac{\mathrm{d}y}{\mathrm{d}x} = \mathrm{f}(x)$ and small h, recurrence relations are:

Euler's method: $y_{n+1} = y_n + h\mathrm{f}(x_n)$, $\quad x_{n+1} = x_n + h$

For $\dfrac{\mathrm{d}y}{\mathrm{d}x} = \mathrm{f}(x, y)$:

Euler's method: $y_{r+1} = y_r + h\mathrm{f}(x_r, y_r)$, $\quad x_{r+1} = x_r + h$

Improved Euler method: $y_{r+1} = y_{r-1} = 2h\mathrm{f}(x_r, y_r)$, $\quad x_{r+1} = x_r + h$

Arc length

$s = \int \sqrt{1 + \left(\frac{\mathrm{d}y}{\mathrm{d}x}\right)^2} \, dx$ (Cartesian coordinates)

$s = \int \sqrt{\left(\frac{\mathrm{d}x}{\mathrm{d}t}\right)^2 + \left(\frac{\mathrm{d}y}{\mathrm{d}t}\right)^2} \, \mathrm{d}t$ (parametric form)

Surface area of revolution

$S_x = 2\pi \int y \sqrt{1 + \left(\frac{\mathrm{d}y}{\mathrm{d}x}\right)^2} \, \mathrm{d}x$ (Cartesian coordinates)

$S_x = 2\pi \int y \sqrt{\left(\frac{\mathrm{d}x}{\mathrm{d}t}\right)^2 + \left(\frac{\mathrm{d}y}{\mathrm{d}t}\right)^2} \, \mathrm{d}t$ (parametric form)

Pure mathematics

Binomial series

$$(a+b)^n = a^n + \binom{n}{1} a^{n-1}b + \binom{n}{2} a^{n-2}b^2 + \ldots + \binom{n}{r} a^{n-r}b^r + \ldots + b^n \quad (n \in \mathbb{Z}^+)$$

where $\binom{n}{r} = {}^nC_r = \dfrac{n!}{r!(n-r)!}$

$$(1+x)^n = 1 + nx + \frac{n(n-1)}{1.2}x^2 + \ldots + \frac{n(n-1)\ldots(n-r+1)}{1.2\ldots r}x^r + \ldots \quad (|x| < 1, n \in \mathbb{Q})$$

Arithmetic series

$$S_n = \frac{1}{2}n(a+l) = \frac{1}{2}n[2a+(n-1)d]$$

Geometic series

$$S_n = \frac{a(1-r^n)}{1-r}$$

$$S_\infty = \frac{a}{1-r} \quad \text{for} \quad |r|<1$$

Trigonometry: small angles

For small angle θ,

$\sin\theta \approx \theta$

$\cos\theta \approx 1 - \frac{\theta^2}{2}$

$\tan\theta \approx \theta$

Trigonometric identities

$\sin(A \pm B) \equiv \sin A\cos B \pm \cos A\sin B$

$\cos(A \pm B) \equiv \cos A\cos B \mp \sin A\sin B$

$\tan(A \pm B) \equiv \frac{\tan A \pm \tan B}{1 \mp \tan A\tan B} \quad \left(A \pm B \neq \left(k+\frac{1}{2}\right)\pi\right)$

$\sin A + \sin B \equiv 2\sin\frac{A+B}{2}\cos\frac{A-B}{2}$

$\sin A - \sin B \equiv 2\cos\frac{A+B}{2}\sin\frac{A-B}{2}$

$\cos A + \cos B \equiv 2\cos\frac{A+B}{2}\cos\frac{A-B}{2}$

$\cos A - \cos B \equiv 2\sin\frac{A+B}{2}\sin\frac{A-B}{2}$

Differentiation

$\mathbf{f(x)}$	$\mathbf{f'(x)}$
$\tan kx$	$k\sec^2 kx$
$\operatorname{cosec} x$	$-\operatorname{cosec} x\cot x$
$\sec x$	$\sec x\tan x$
$\cot x$	$-\operatorname{cosec}^2 x$
$\frac{f(x)}{g(x)}$	$\frac{f'(x)g(x)-f(x)g'(x)}{(g(x))^2}$

Differentiation from first principles

$$f'(x) = \lim_{h\to 0}\frac{f(x+h)-f(x)}{h}$$

Integration

$$\int u \frac{\mathrm{d}v}{\mathrm{d}x}\,\mathrm{d}x = uv - \int v \frac{\mathrm{d}u}{\mathrm{d}x}\,\mathrm{d}x$$

(+ constant; $a > 0$ where relevant)

f(x)	$\int$ **f(x) dx**
$\tan x$	$\ln\lvert\sec x\rvert$
$\cot x$	$\ln\lvert\sin x\rvert$
$\operatorname{cosec} x$	$-\ln\lvert\operatorname{cosec} x + \cot x\rvert = \ln\lvert\tan \frac{1}{2}x\rvert$
$\sec x$	$\ln\lvert\sec x + \tan x\rvert = \ln\lvert\tan(\frac{1}{2}x + \frac{1}{4}\pi)\rvert$
$\sec^2 kx$	$\frac{1}{k}\tan kx$

Numerical solution of equations

The Newton-Raphson iteration for solving $f(x) = 0$: $x_{n+1} = x_n - \dfrac{f(x_n)}{f'(x_n)}$

Numerical integration

The trapezium rule: $\int_a^b y\,\mathrm{d}x \approx \frac{1}{2}h\{(y_0 + y_n) + 2(y_1 + y_2 + \ldots + y_{n-1})\}$, where $h = \dfrac{b-a}{n}$

Answers to exercises

1 Complex numbers

BEFORE YOU START

1 $x=1\pm\frac{\sqrt{3}}{3}$

2

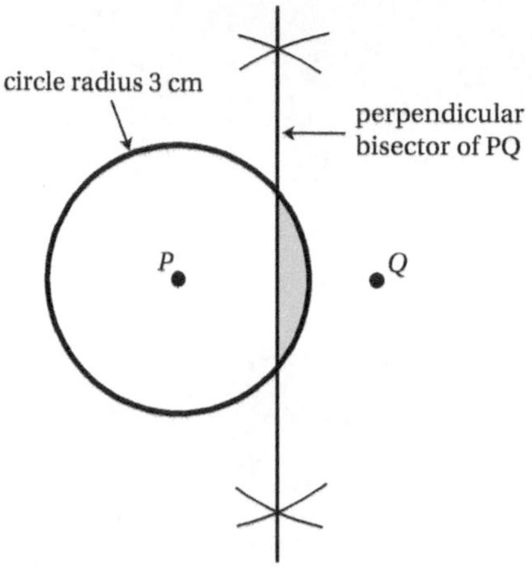

3 $x=-1,\ y=3$ or $x=3,\ y=1$

EXERCISE 1A

1 a i 5 ii -2
b i 1 ii -1
c i 2 ii 7
d i 0 ii -3
e i 0 ii 0
f i $a-1$ ii $-4-b$

2 a i 5i ii -8i
b i -5 ii 1
c i -9 ii -16
d i $-2\text{i}+5$ ii $-14+10\text{i}$

3 a i $-1+\text{i}$ ii $15+6\text{i}$
b i 5i ii $14+23\text{i}$
c i $8-\text{i}$ ii $16+2\text{i}$
d i $8+6\text{i}$ ii $7-24\text{i}$
e i 1 ii 13

4 a i $3+4\text{i}$ ii $3-\text{i}$
b i $\frac{1}{2}+\frac{1}{5}\text{i}$ ii $-\frac{1}{2}+\frac{1}{8}\text{i}$
c i $\frac{3}{2}+\frac{3}{2}\text{i}$ ii $-3+11\text{i}$

5 a i 2i ii 7i
b i $2\sqrt{2}\text{i}$ ii $5\sqrt{2}\text{i}$
c i $\frac{4}{3}-2\text{i}$ ii $-\frac{1}{3}+\frac{5}{3}\text{i}$
d i $\frac{1}{3}+\frac{1}{2}\text{i}$ ii $\frac{5}{4}-\frac{\sqrt{5}}{2}\text{i}$

6 a i $x=\pm3\text{i}$ ii $x=\pm6\text{i}$
b i $x=\pm\sqrt{10}\text{i}$ ii $x=\pm\sqrt{13}\text{i}$
c i $x=1\pm2\text{i}$ ii $x=\frac{1}{2}\pm\frac{\sqrt{39}}{2}\text{i}$
d i $x=1\pm\frac{\sqrt{51}}{3}\text{i}$ ii $x=-\frac{3}{5}\pm\frac{4}{5}\text{i}$

7 a i $-\text{i}$ ii 1
b i 16 ii 125i
c i -8 ii 8i
d i i ii i

8 a i $a=-\frac{7}{13},\ b=\frac{17}{13}$ ii $a=\frac{12}{37},\ b=-\frac{2}{37}$
b i $a=8,\ b=-18$ ii $a=1,\ b=0$
c i $a=1,\ b=0$ ii $a=-6,\ b=6$

9 a i $z=\pm\left(\sqrt{2}-\sqrt{2}\text{i}\right)$
ii $z=\pm\left(\frac{3\sqrt{2}}{2}+\frac{3\sqrt{2}}{2}\text{i}\right)$
b i $z=\pm\left(\sqrt{3}+\text{i}\right)$
ii $z=\pm\left(\sqrt{\frac{\sqrt{26}+5}{2}}+\text{i}\sqrt{\frac{\sqrt{26}-5}{2}}\right)$

10 $a=\pm\sqrt{3},\ b=\mp\frac{\sqrt{3}}{3}$

11 $a=8,\ b=1$ or $a=-1,\ b=10$

12 $z=-\frac{1}{2}+\frac{1}{2}\text{i}$

13 a $x=-\frac{1}{5},\ y=-\frac{2}{5}$
b $-\frac{3\text{i}}{2+\text{i}}=-\frac{3}{5}-\frac{6}{5}\text{i}$

14 $z=\pm(1-2\text{i})$

15 $z=\pm\frac{\sqrt{2}}{2}(1+\text{i})$

16 $z=\frac{5\sqrt{2}}{2}-\frac{\sqrt{2}}{2}\text{i}$ or $z=-\frac{5\sqrt{2}}{2}+\frac{\sqrt{2}}{2}\text{i}$

EXERCISE 1B

1 **a** **i** $2+3\text{i}$ **ii** $4-4\text{i}$

b **i** $-\text{i}-3$ **ii** $-3\text{i}+2$

c **i** -3i **ii** i

d **i** -45 **ii** 9

2 **a** **i** $-\frac{1}{5}-\frac{8}{5}\text{i}$ **ii** $-\frac{10}{17}+\frac{6}{17}\text{i}$

b **i** -4i **ii** i

c **i** $\frac{15}{17}+\frac{8}{17}\text{i}$ **ii** $\frac{3}{5}-\frac{4}{5}\text{i}$

d **i** $-1+\text{i}$ **ii** $\frac{2}{5}-\frac{11}{5}\text{i}$

3 **a** $z=\frac{7}{5}-\frac{3}{5}\text{i}$ **b** $z=6-\frac{1}{2}\text{i}$

4 **a** $z=\frac{9}{2}-\frac{9}{2}\text{i},\ w=-3-\frac{4}{3}\text{i}$

b $z=\frac{3}{10}+\frac{1}{10}\text{i},\ w=\frac{3}{5}+\frac{1}{5}\text{i}$

5 **a** $z=\frac{1}{2}-2\text{i}$ **b** $z=-\frac{2}{3}-3\text{i}$

6 **a** $z=\frac{2}{3}+7\text{i}$ **b** $z=-\frac{5}{3}+\frac{1}{3}\text{i}$

7 **a** **i** $z^*=3+x-\text{i}y$ **ii** $z^*=x-2-\text{i}y$

b **i** $z^*=x+2-\text{i}(3y-1)$

ii $z^*=3-x-\text{i}(y+3)$

c **i** $z^*=\frac{x(x^2+y^2+1)}{x^2+y^2}-\text{i}\frac{y(x^2+y^2-1)}{x^2+y^2}$

ii $z^*=\frac{x(x^2+y^2-1)}{x^2+y^2}-\text{i}\frac{y(x^2+y^2+1)}{x^2+y^2}$

d **i** $z^*=\text{i}\frac{2xy}{x^2+y^2}$ **ii** $z^*=\frac{2x^2}{x^2+y^2}$

8 $-\frac{13}{5}-\frac{1}{5}\text{i}$

9 **a** Re: $2x-3y$; Im: $3x-2y$

b $z=-4-4\text{i}$

10 $x=6,\ y=3$

11 Proof.

12 $z=-\frac{1}{2}\text{i}$

13 No solutions.

14 Re: $\frac{x(x+1)+y^2}{(x+1)^2+y^2}$; Im: $\frac{y}{(x+1)^2+y^2}$

EXERCISE 1C

1 **a** **i**

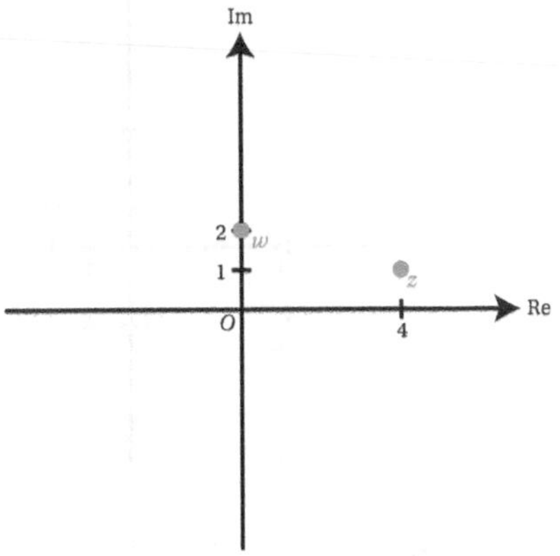

ii

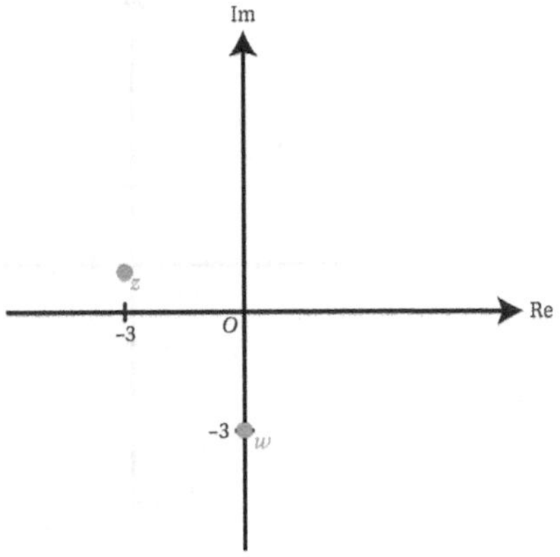

b **i**

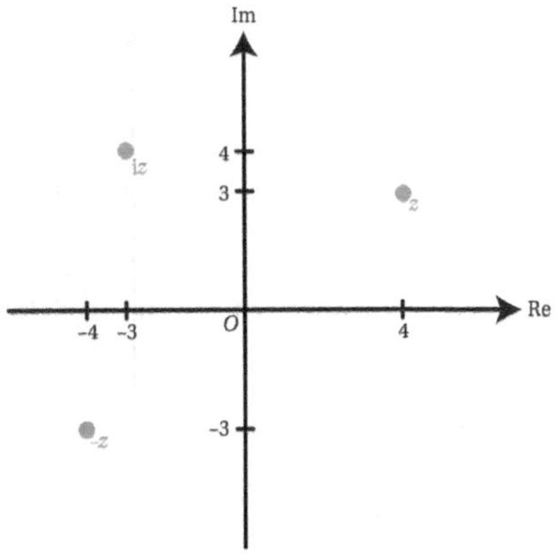

ii

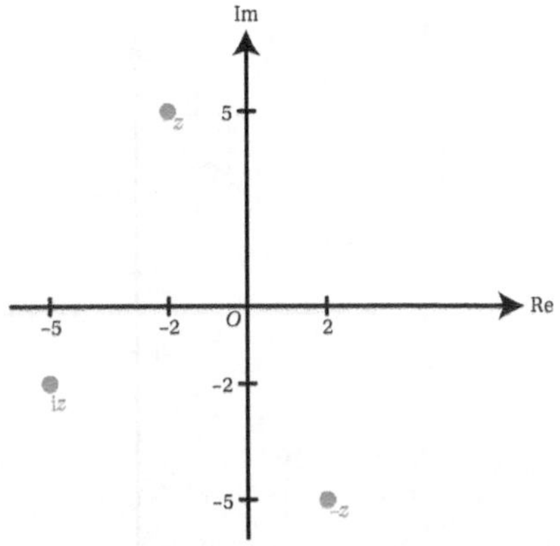

c **i**

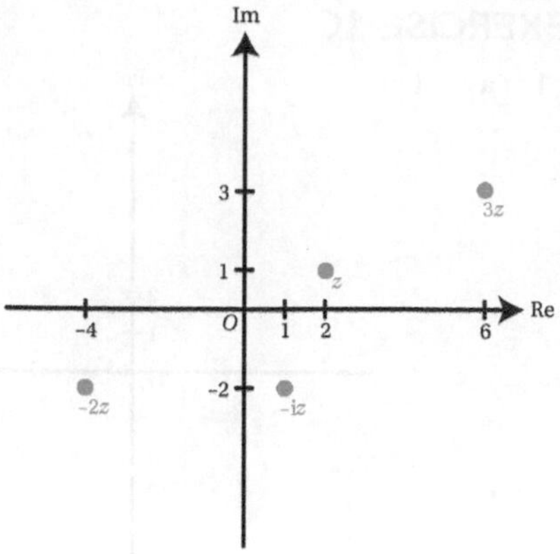

ii

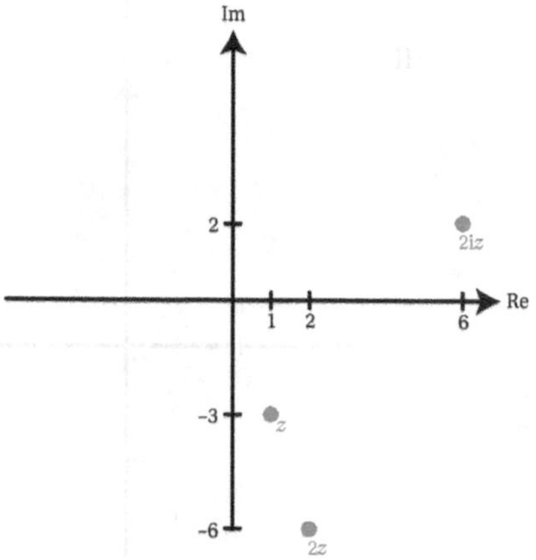

2 **a** **i**

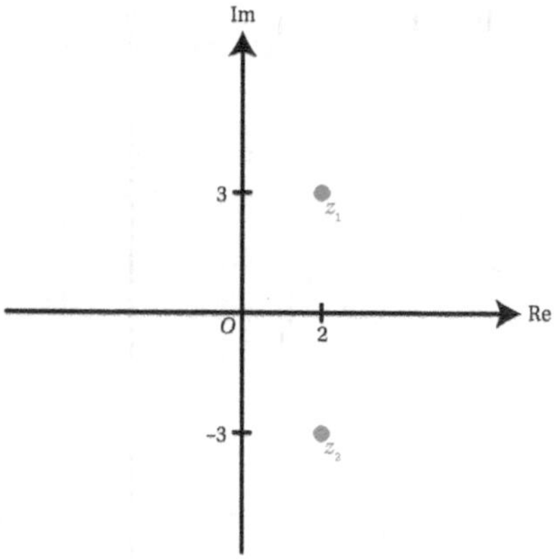

ii

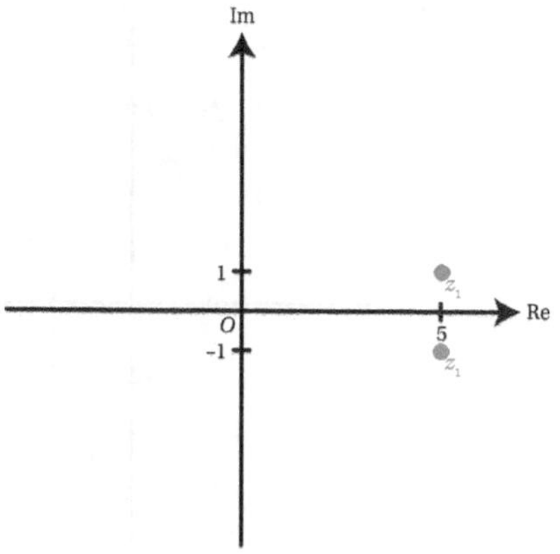

b **i**

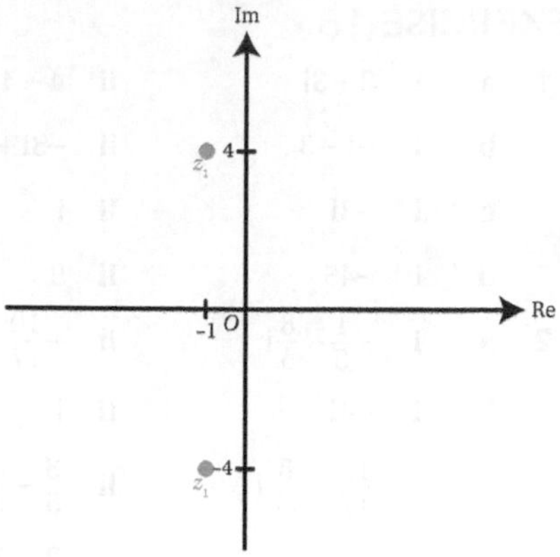

ii

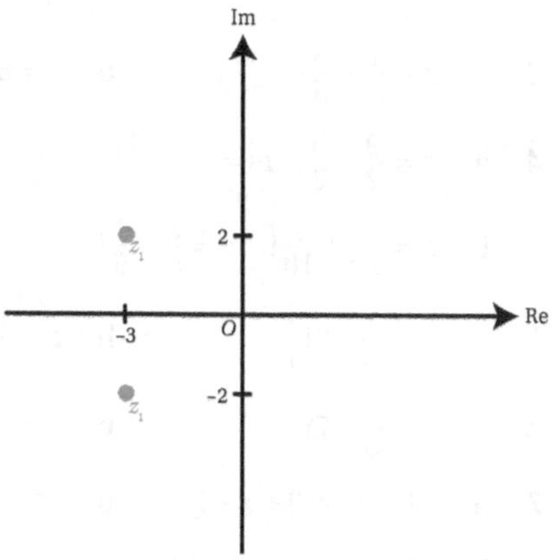

3 **a** **i**

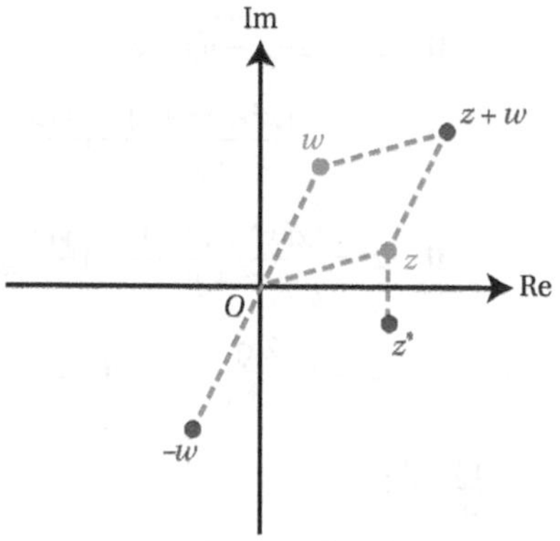

ii

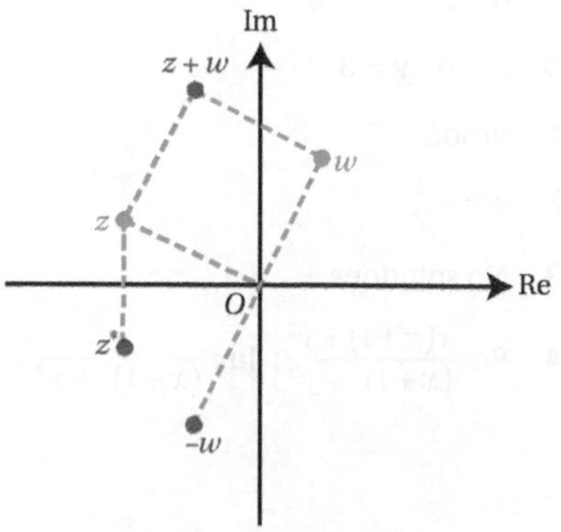

b **i**

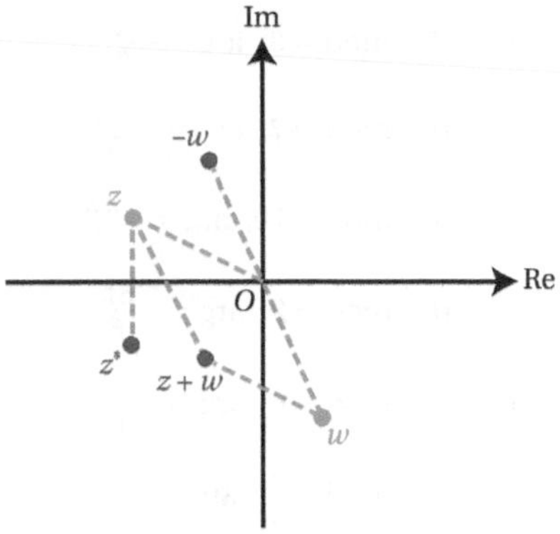

ii

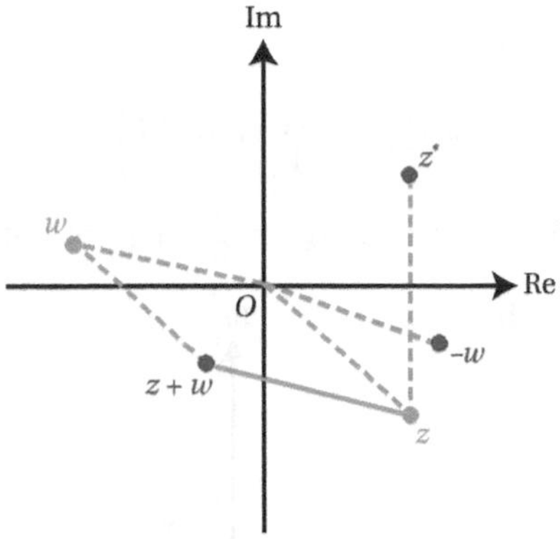

EXERCISE 1D

1	**a**	**i** $\frac{3\pi}{4}$		**ii** $\frac{\pi}{4}$	
	b	**i** $\frac{\pi}{2}$		**ii** $\frac{3\pi}{2}$	
	c	**i** $\frac{2\pi}{3}$		**ii** $\frac{5\pi}{6}$	
	d	**i** $\frac{5\pi}{18}$		**ii** $\frac{4\pi}{9}$	
2	**a**	**i** 5.585		**ii** 0.349	
	b	**i** 4.712		**ii** 1.1571	
	c	**i** 1.134		**ii** 2.531	
	d	**i** 1.745		**ii** 1.449	
3	**a**	**i** 60°		**ii** 45°	
	b	**i** 150°		**ii** 120°	
	c	**i** 270°		**ii** 300°	
	d	**i** 69.9°		**ii** 265°	

WORK IT OUT 1.1

Solution 1 is correct.

EXERCISE 1E

1 **a** **i** mod = 6; arg = 0

ii mod = 13; arg = 0

b **i** mod = 3; arg = π

ii mod = 1.6; arg = π

c **i** mod = 4; arg = $\frac{\pi}{2}$

ii mod = 0.5; arg = $\frac{\pi}{2}$

d **i** mod = 2; arg = $-\frac{\pi}{2}$

ii mod = 5; arg = $-\frac{\pi}{2}$

e **i** mod = $\sqrt{2}$; arg = $\frac{\pi}{4}$

ii mod = $\sqrt{7}$; arg = 0.714

f **i** mod = 2; arg = $-\frac{2\pi}{3}$

ii mod = $4\sqrt{2}$; arg = $-\frac{\pi}{4}$

2 **a** **i** mod = $2\sqrt{5}$; arg = 0.464

ii mod = 5; arg = 5.64

b **i** mod = 2; arg = $\frac{5\pi}{6}$

ii mod = $\sqrt{38}$; arg = 1.34

c **i** mod = $\sqrt{10}$; arg = 3.46

ii mod = $\sqrt{13}$; arg = 2.55

3 **a** **i** $2+2\sqrt{3}i$ **ii** $1+i$

b **i** $-\sqrt{2}+\sqrt{2}i$ **ii** $-1+\sqrt{3}i$

c **i** $-3i$ **ii** -4

4 **a** **i**

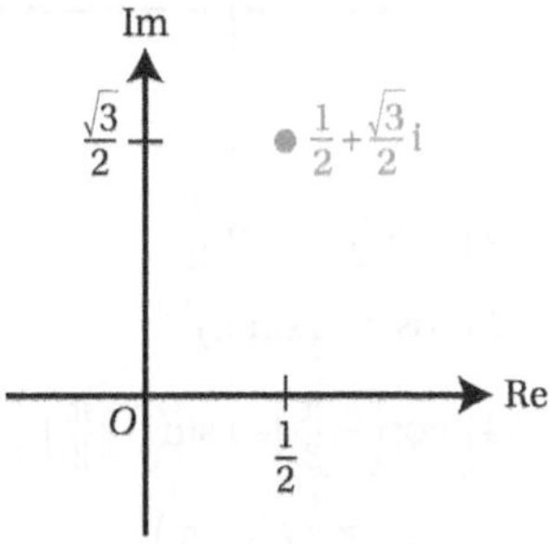

ii

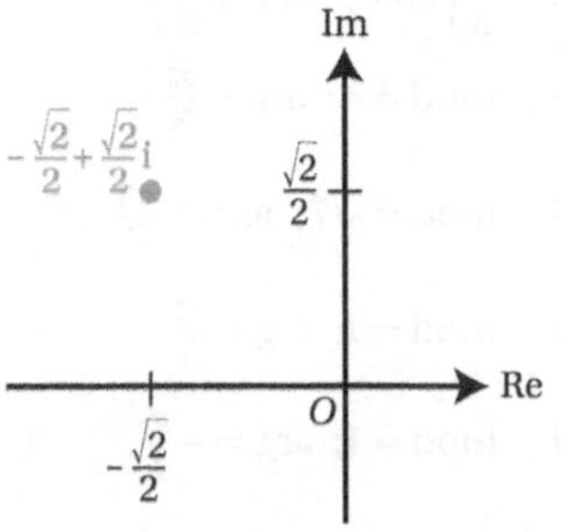

b **i**

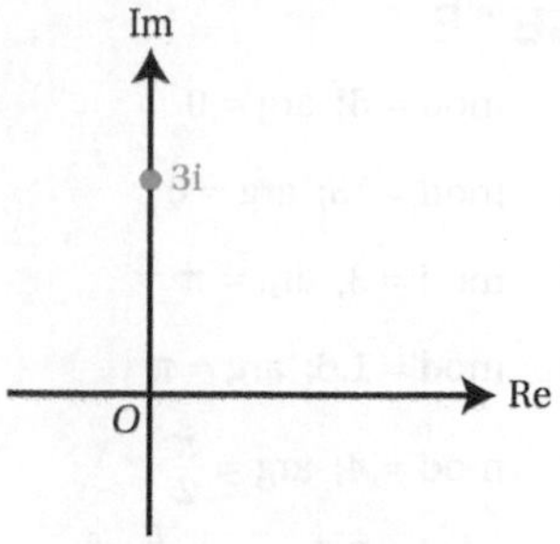

ii

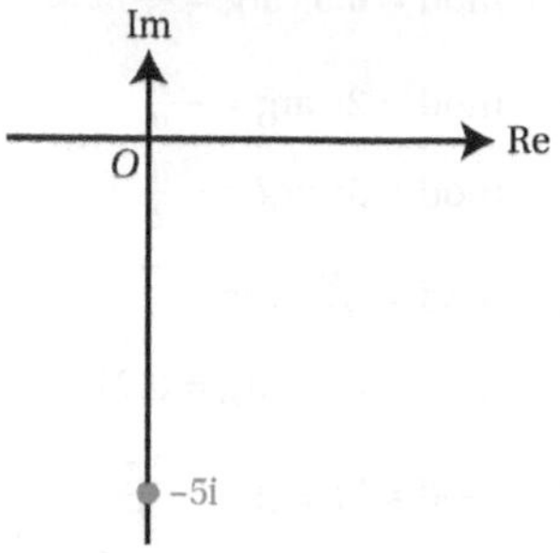

c **i**

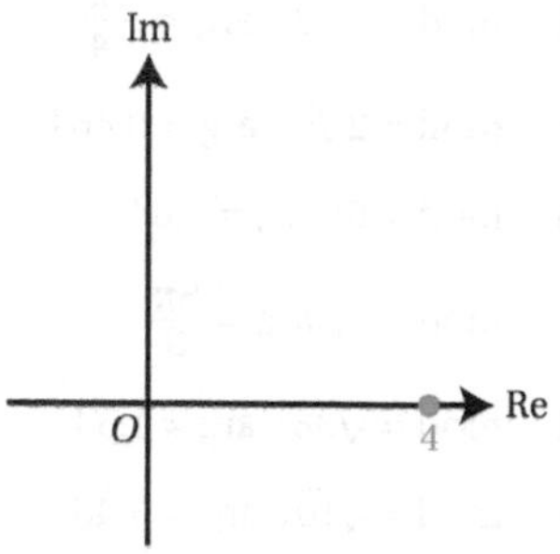

ii

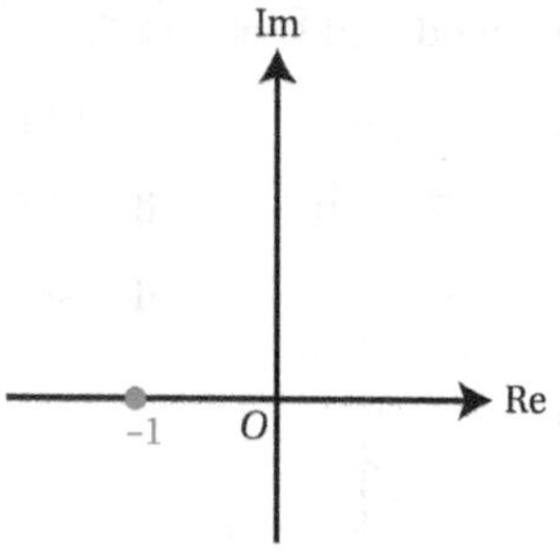

5 **a** **i** $4\left(\cos\frac{\pi}{2}+\text{i}\sin\frac{\pi}{2}\right)$

ii $5(\cos\pi+\text{i}\sin\pi)$

b **i** $4\left(\cos\left(-\frac{\pi}{3}\right)+\text{i}\sin\left(-\frac{\pi}{3}\right)\right)$

ii $\frac{2}{3}\left(\cos\frac{\pi}{6}+\text{i}\sin\frac{\pi}{6}\right)$

6 **a** **i** mod $=4$; arg $=\frac{\pi}{3}$

ii mod $=\sqrt{7}$; arg $=\frac{3\pi}{7}$

b **i** mod $=1$; arg $=\frac{\pi}{5}$

ii mod $=1$; arg $=-\frac{\pi}{4}$

c **i** mod $=3$; arg $=-\frac{\pi}{8}$

ii mod $=7$; arg $=-\frac{4\pi}{5}$

d **i** mod $=10$; arg $=-\frac{2\pi}{3}$

ii mod $=2$; arg $=-\frac{5\pi}{6}$

e **i** mod $=6$; arg $=-\frac{\pi}{10}$

ii mod $=\frac{1}{2}$; arg $=\frac{\pi}{3}$

7 **a** $\frac{3\sqrt{2}}{2}-\frac{3\sqrt{2}}{2}\text{i}$

b $4\sqrt{2}\left(\cos\frac{3\pi}{4}+\text{i}\sin\frac{3\pi}{4}\right)$

8 **a** $|z|=2$; $\arg z=\frac{\pi}{3}$ $|w|=6$; $\arg w=-\frac{\pi}{6}$

b

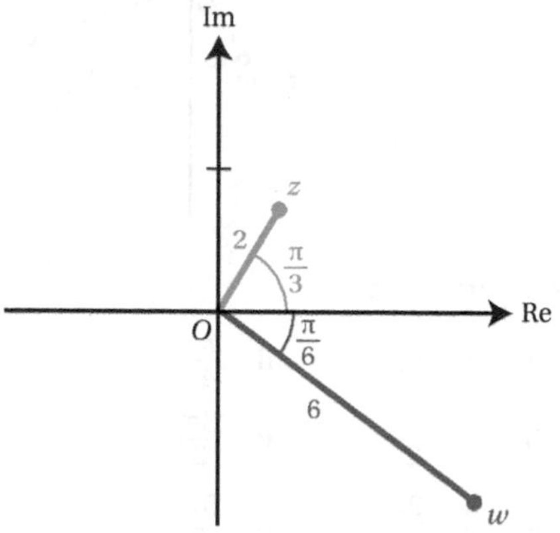

c $|zw|=12$; $\arg(zw)=\frac{\pi}{6}$

$|zw|=|z||w|$

$\arg(zw)=\arg(z)+\arg(w)$

9 **a** $2r\cos\theta$ **b** r^2 **c** $\cos 2\theta+\text{i}\sin 2\theta$

10 Re $=0$; Im $=\frac{1-\cos\theta}{\sin\theta}\left(\text{or, equivalently, }\frac{\sin\theta}{1+\cos\theta}\right)$

EXERCISE 1F

1 **a** **i**

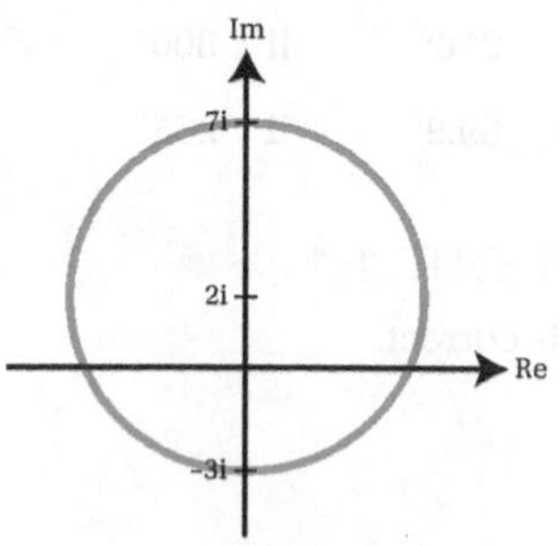

ii

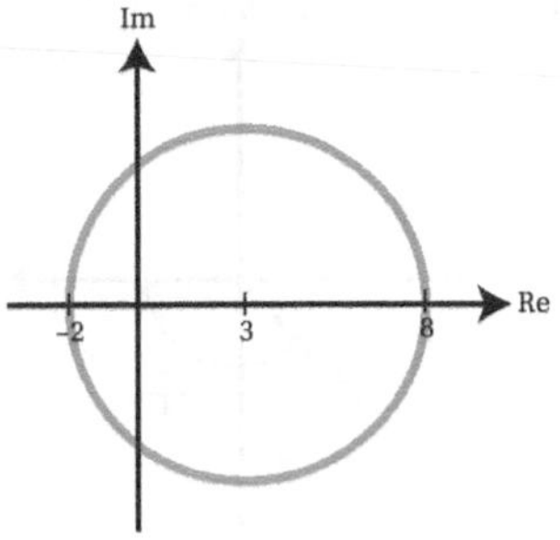

b i

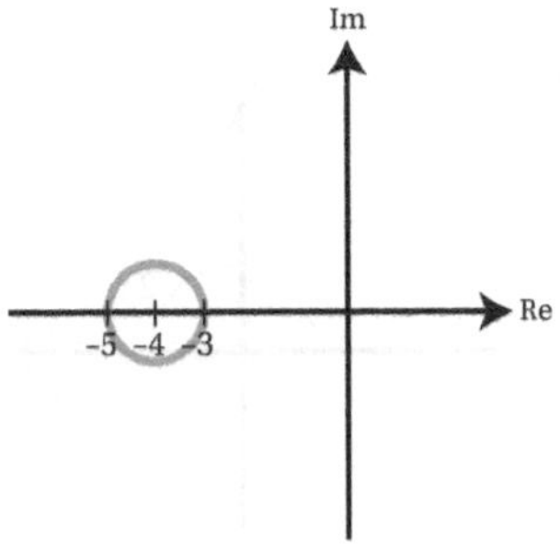

ii

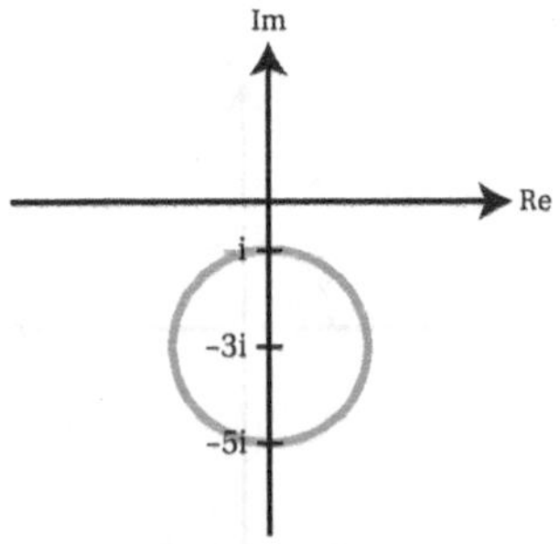

c i

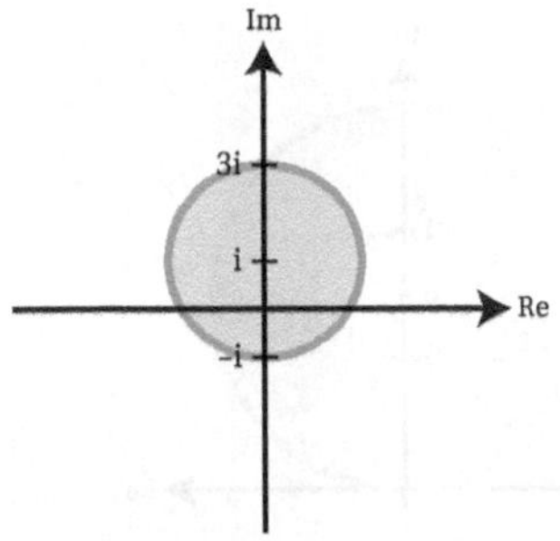

ii

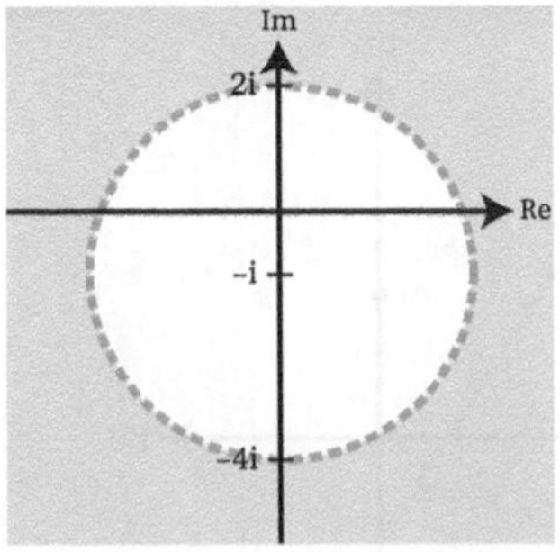

d i

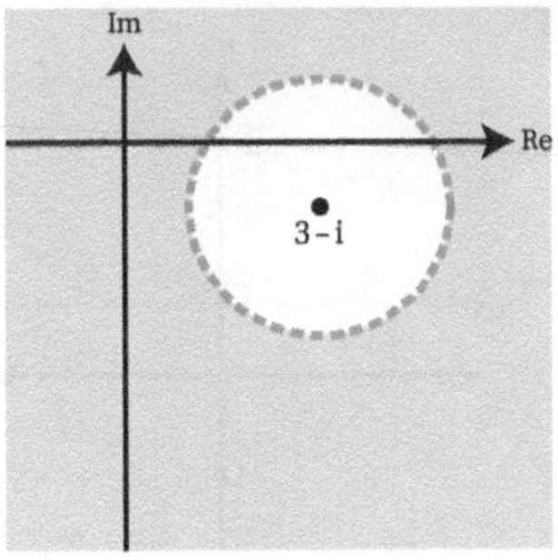

ii

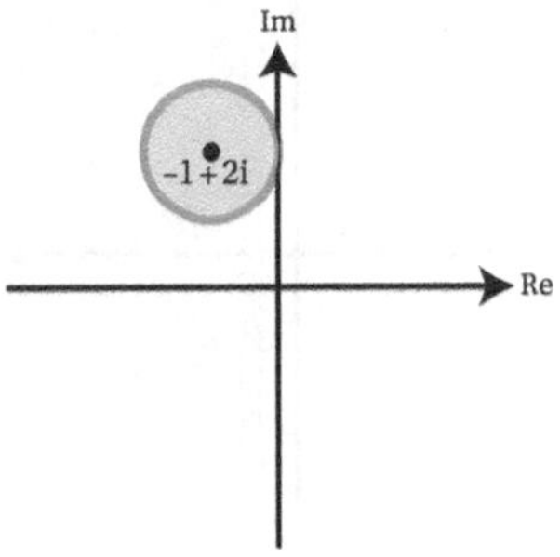

2 a ii $\{x + iy : (x-3)^2 + y^2 = 25\}$

b i $\{x + iy : (x+4)^2 + y^2 = 1\}$

ii $\{x + iy : x^2 + (y+3)^2 = 4\}$

c i $\{x + iy : x^2 + (y-1)^2 \leqslant 4\}$

ii $\{x + iy : x^2 + (y+1)^2 > 9\}$

d i $\{x + iy : (x-3)^2 + (y+1)^2 > 4\}$

ii $\{x + iy : (x+1)^2 + (y-2)^2 \leqslant 1\}$

3 a i

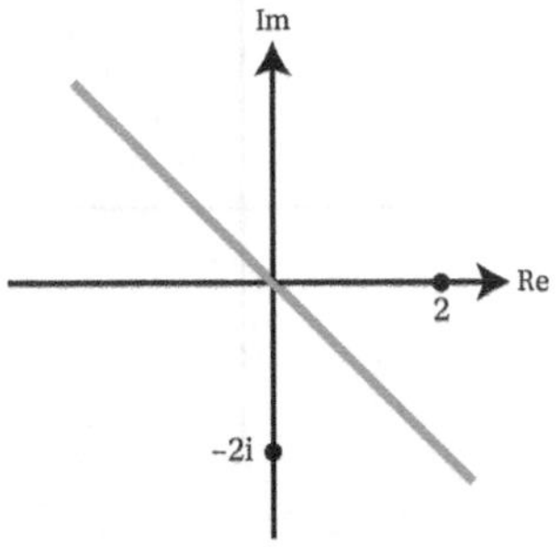

ii

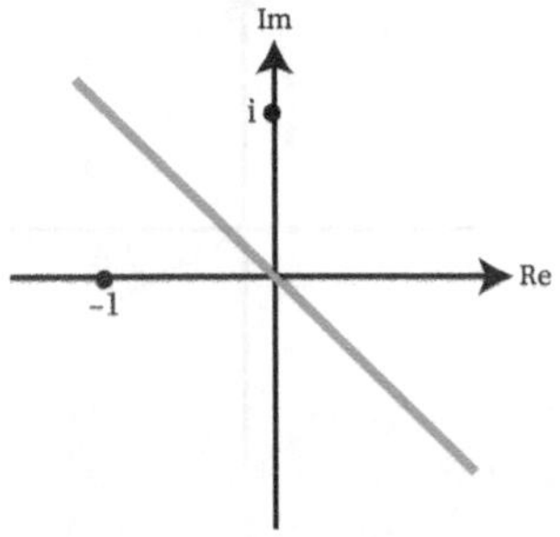

b **i**

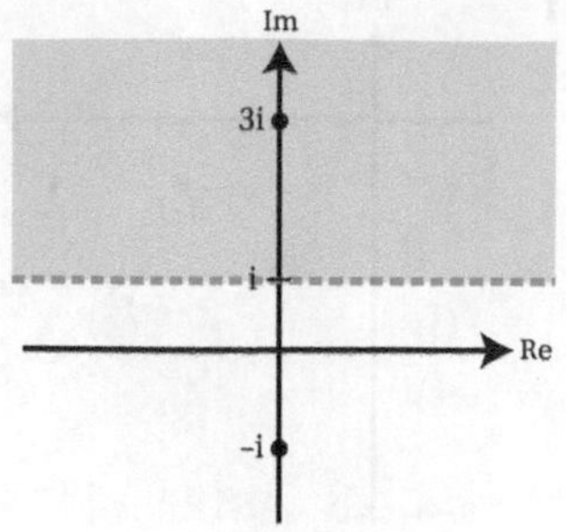

ii

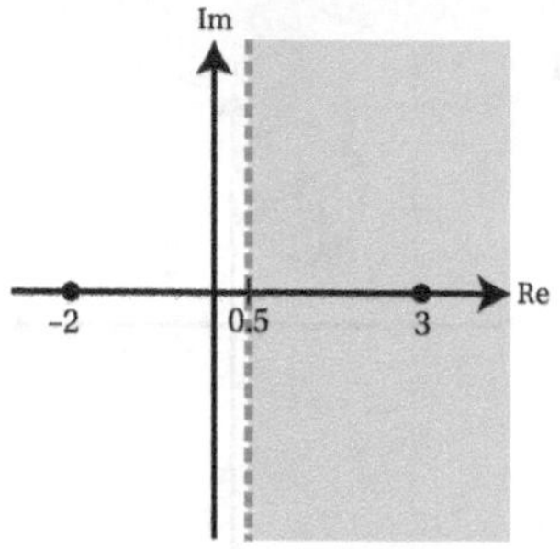

4 **a** **i**

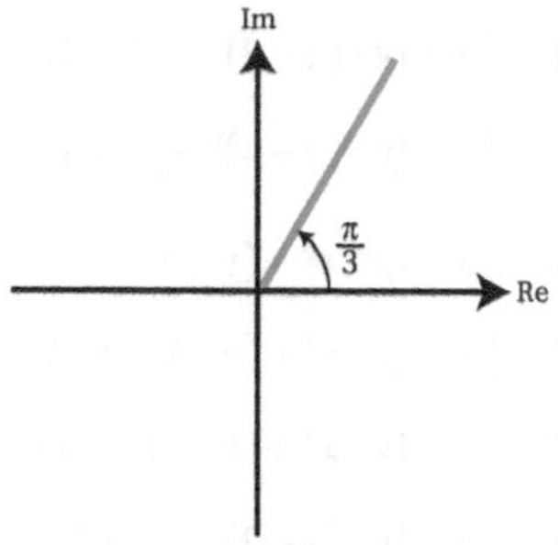

ii

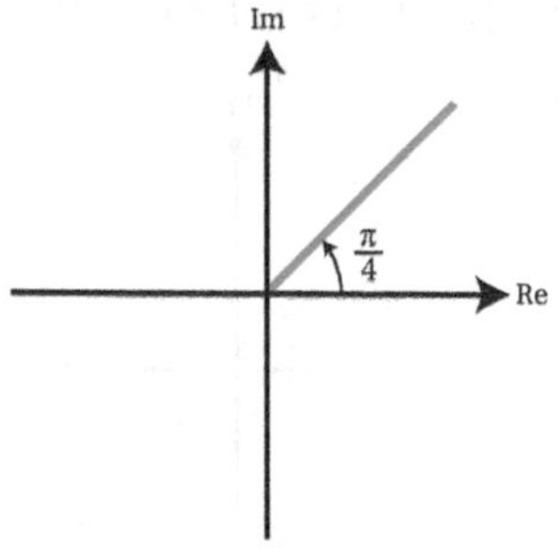

b **i**

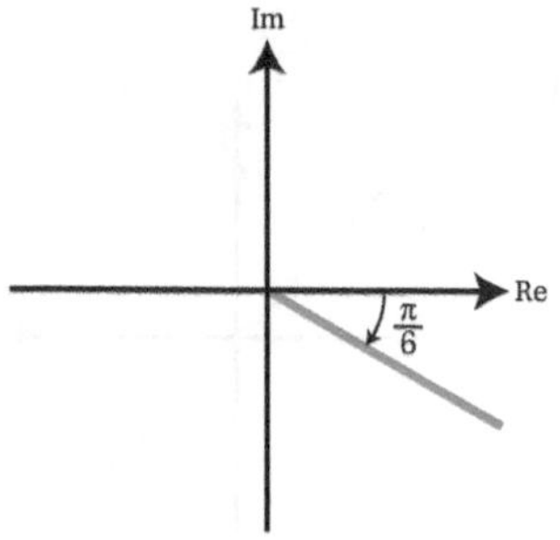

ii

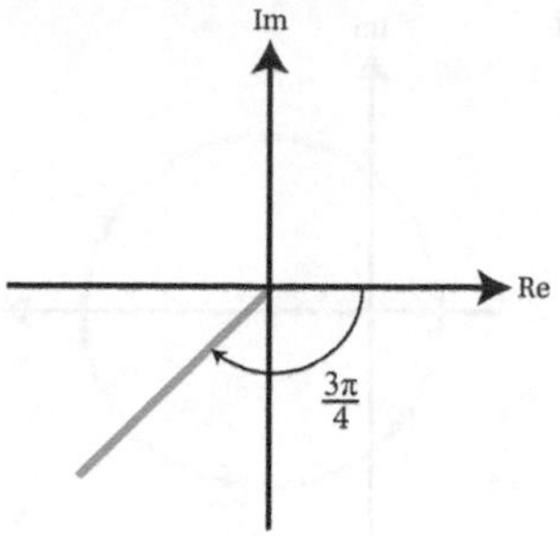

c **i**

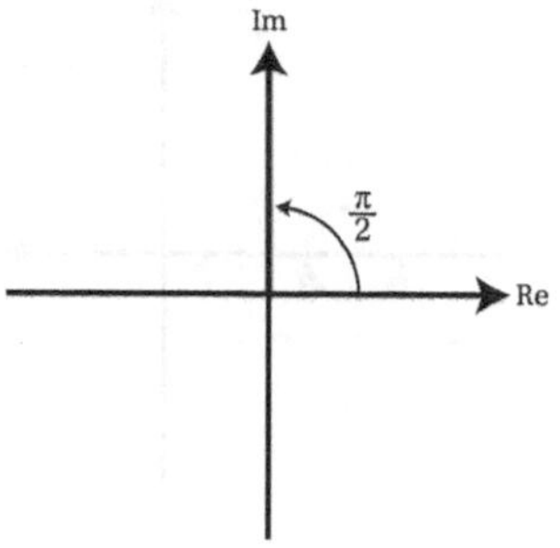

ii

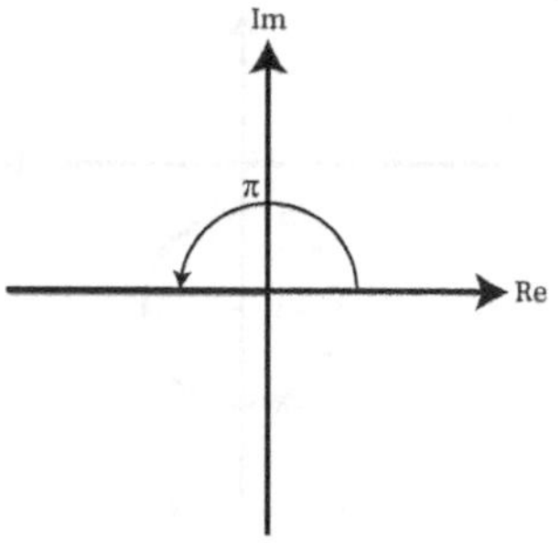

5

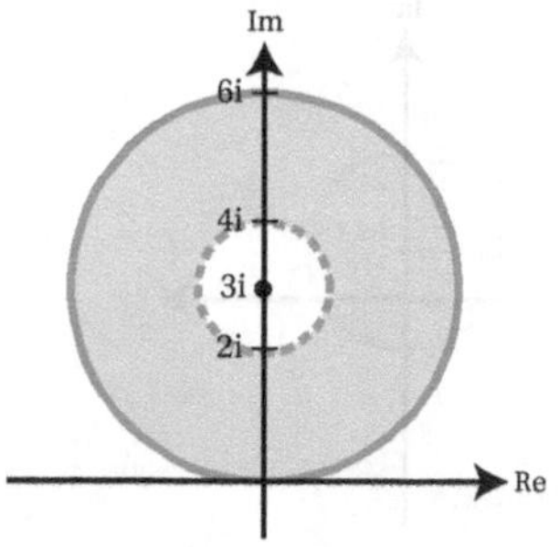

6

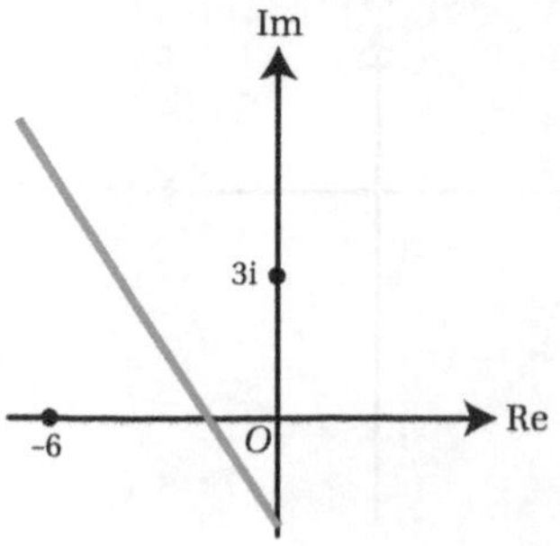

7

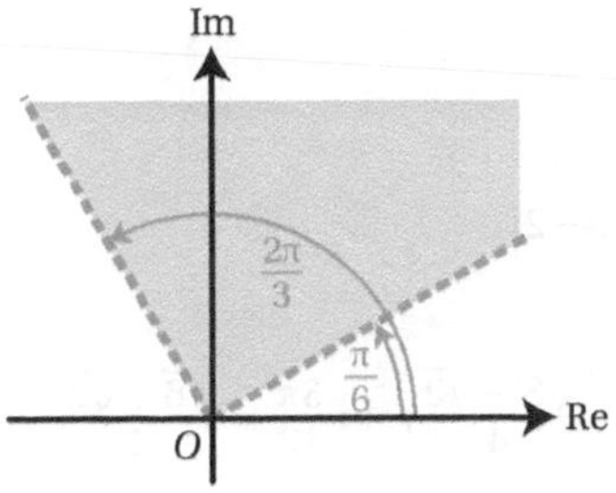

8 **a**

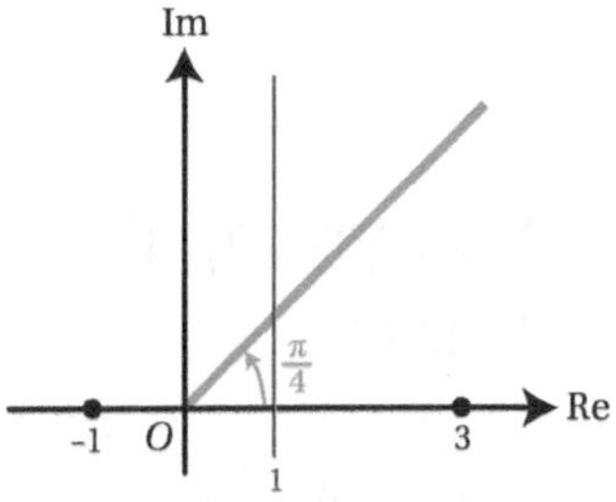

b $z = 1 + \mathrm{i}$

9

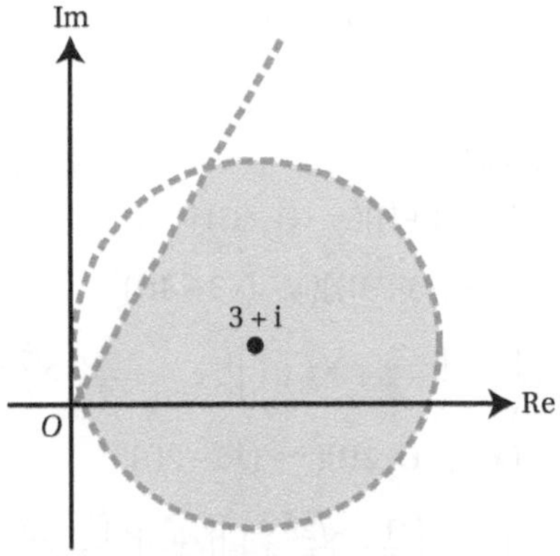

10

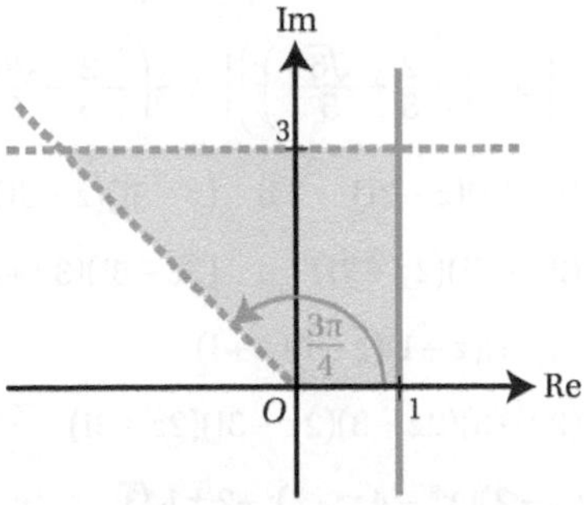

11 $z = 1 + \mathrm{i}$ **12** $\frac{\pi}{2}$

WORK IT OUT 1.2

Solution 2 is correct.

EXERCISE 1G

1 **a** **i** $21\left(\cos\frac{11\pi}{30} + \mathrm{i}\sin\frac{11\pi}{30}\right)$

ii $4\left(\cos\frac{2\pi}{9} + \mathrm{i}\sin\frac{2\pi}{9}\right)$

b **i** $4(\cos 4 + \mathrm{i}\sin 4)$

ii $3\left(\cos\left(-\frac{5\pi}{14}\right) + \mathrm{i}\sin\left(-\frac{5\pi}{14}\right)\right)$

2 **a** **i** $\cos\left(-\frac{11\pi}{12}\right) + \mathrm{i}\sin\left(-\frac{11\pi}{12}\right)$

ii $\cos\frac{13\pi}{20} + \mathrm{i}\sin\frac{13\pi}{20}$

b **i** $\cos\frac{\pi}{12} + \mathrm{i}\sin\frac{\pi}{12}$

ii $\cos\frac{4\pi}{15} + \mathrm{i}\sin\frac{4\pi}{15}$

c **i** $\cos\frac{7\pi}{20} + \mathrm{i}\sin\frac{7\pi}{20}$

ii $\cos\left(-\frac{5\pi}{12}\right) + \mathrm{i}\sin\left(-\frac{5\pi}{12}\right)$

d **i** $\cos\frac{9\pi}{20} + \mathrm{i}\sin\frac{9\pi}{20}$

ii $\cos\frac{13\pi}{20} + \mathrm{i}\sin\frac{13\pi}{20}$

3 $3\left(\cos\frac{5\pi}{12} + \mathrm{i}\sin\frac{5\pi}{12}\right)$ **4** $\sqrt{3} + \mathrm{i}$

5 **a** $\frac{\sqrt{2} - \sqrt{6}}{4} + \frac{\sqrt{2} + \sqrt{6}}{4}\mathrm{i}$ **b** $-2 - \sqrt{3}$

6 Proof.

MIXED PRACTICE 1

1 C **2** B

3 **a** $8 + 2\mathrm{i}$ **b** $-\frac{1}{17} - \frac{4}{17}\mathrm{i}$

4 $-\frac{\sqrt{3}}{2} + \frac{7}{2}\mathrm{i}$

5 **a** **i** $\sqrt{29}$ **ii** 1.95

b modulus = $\sqrt{29}$; argument = −1.95

6 **a** $z = -7 \pm 2\mathrm{i}$

b

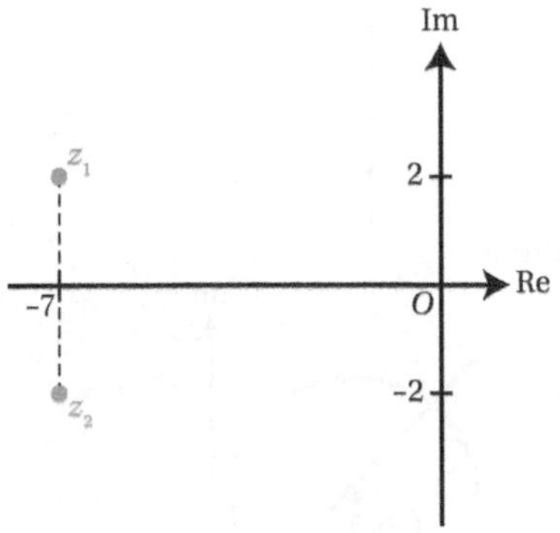

7 $-2 - \frac{3}{8}\mathrm{i}$

8

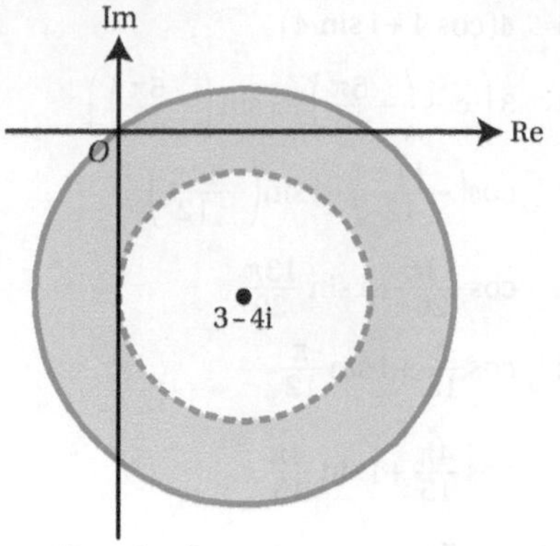

9 **a** $\text{Re} = 2y$; $\text{Im} = 2y - 2x$ **b** $5 + 10\text{i}$

10 C

11 Proof.

12 **a** $w = \text{i}z$

b Proof.

13

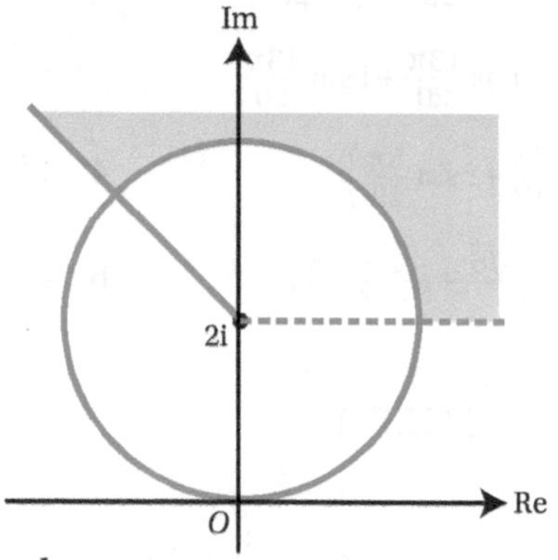

14 **a, b**

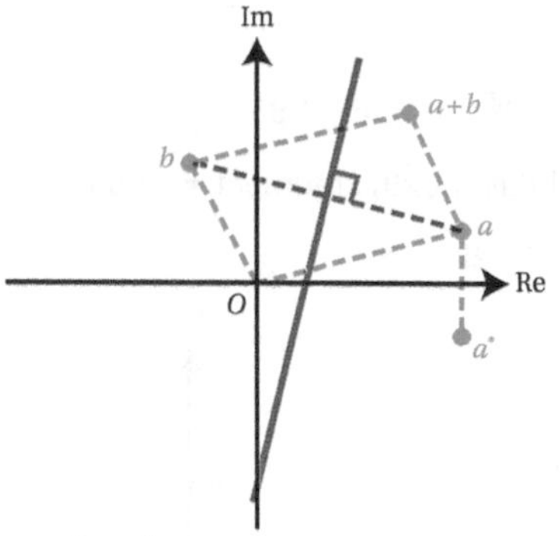

15 **a** Proof.

b

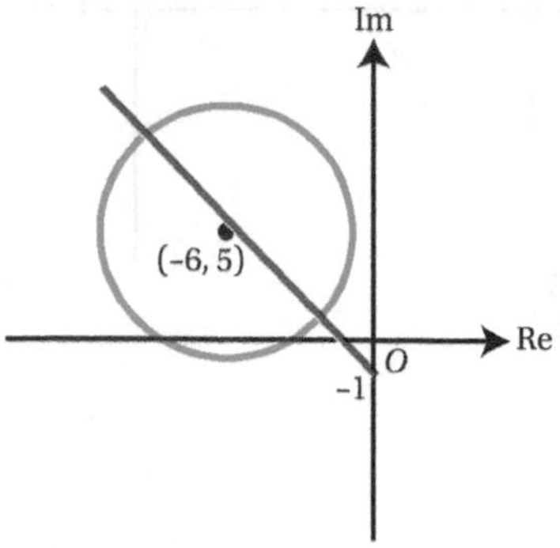

c $-10 + 9\text{i}$

16 $\frac{1}{2}$

17 Proof.

18 $w = 5\text{i}$, $z = 3 + 2\text{i}$

19 **a** Proof.

b $\cos\frac{5\pi}{12} = \frac{\sqrt{6} - \sqrt{2}}{4}$; $\sin\frac{5\pi}{12} = \frac{\sqrt{6} + \sqrt{2}}{4}$

20 Proof.

21 θ

2 Roots of polynomials

BEFORE YOU START

1 **a** Proof.

b $\text{f}(x) = (2x + 1)(x - 2)(x - 3)$

2 **a** $17 - 7\text{i}$ **b** $x = 1 + 2\text{i}, 1 - 2\text{i}$

3 **a** 6 **b** 34

EXERCISE 2A

1 **a** **i** $(x - (1 + \text{i}))(x - (1 - \text{i}))$

ii $(x - (-3 + 4\text{i}))(x - (-3 - 4\text{i}))$

b **i** $\left(x - \left(-\frac{3}{2} + \frac{\sqrt{7}}{2}\text{i}\right)\right)\left(x - \left(-\frac{3}{2} - \frac{\sqrt{7}}{2}\text{i}\right)\right)$

ii $(x - (-1 + 2\text{i}))(x - (-1 - 2\text{i}))$

c **i** $3\left(x - \left(\frac{1}{3} + \frac{\sqrt{29}}{3}\text{i}\right)\right)\left(x - \left(\frac{1}{3} - \frac{\sqrt{29}}{3}\text{i}\right)\right)$

ii $5\left(x - \left(-\frac{2}{5} + \frac{\sqrt{6}}{5}\text{i}\right)\right)\left(x - \left(-\frac{2}{5} - \frac{\sqrt{6}}{5}\text{i}\right)\right)$

2 **a** **i** $(z - 2\text{i})(z + 2\text{i})$ **ii** $(z - 5\text{i})(z + 5\text{i})$

b **i** $(2z - 7\text{i})(2z + 7\text{i})$ **ii** $(3z - 8\text{i})(3z + 8\text{i})$

c **i** $(z - 1)(z + 1)(z - \text{i})(z + \text{i})$

ii $(2z - 3)(2z + 3)(2z - 3\text{i})(2z + 3\text{i})$

3 **a** **i** $(x - 2)(x^2 + 4x + 7)$; $-2 \pm \text{i}\sqrt{3}$

ii $(x + 1)(x^2 + 2x + 5)$; $-1 \pm 2\text{i}$

b **i** $(x + 2)(2x^2 - 4x + 3)$; $1 \pm \frac{\sqrt{2}}{2}\text{i}$

ii $(x - 1)(3x^2 + 2x + 2)$; $-\frac{1}{3} \pm \frac{\sqrt{5}}{3}\text{i}$

4 Proof; $x = 3, 5 \pm \text{i}$

5 **a** Proof.

b $(2x + 1)(x^2 + 4x + 6)$
$= (2x + 1)(x + 2 + \sqrt{2}\text{i})(x + 2 - \sqrt{2}\text{i})$

6 a Proof.

b $f(x) = x(x-3)(2x+1)(2x+3)$; $x = 0, 3, -0.5$ or -1.5

7 a Proof.

b $(x+1)(x-2)(x^2+4x+5)$; $x = -1, 2, -2+\mathrm{i}$ or $-2-\mathrm{i}$

8 $x = 2, -2, 5\mathrm{i}$ or $-5\mathrm{i}$

WORK IT OUT 2.1

Solution 3 is correct.

EXERCISE 2B

1 a i $a = 0$; $b = 25$ ii $a = 0$; $b = 9$

b i $a = -6$; $b = 25$ ii $a = -2$; $b = 5$

2 a i $(x-5)(x^2-6x+13)$; $x = 5$, $x = 3 \pm 2\mathrm{i}$

ii $(x+3)(x^2-4x+5)$; $x = -3$, $x = 2 \pm \mathrm{i}$

b i $(x-1)(x^2-2x+5)$; $x = 1$, $x = 1 \pm 2\mathrm{i}$

ii $(x+4)(x^2-6x+10)$; $x = -4$, $x = 3 \pm \mathrm{i}$

3 a i $(x^2-2x+10)(x^2+4)$; $x = 1 \pm 3\mathrm{i}$, $x = \pm 2\mathrm{i}$

ii $(x^2-6x+10)(x^2+1)$; $x = 3 \pm \mathrm{i}$, $x = \pm \mathrm{i}$

b i $(x^2-4x+13)(x^2+2x+3)$; $x = 2 \pm 3\mathrm{i}$, $x = -1 \pm \mathrm{i}\sqrt{2}$

ii $(x^2-4x+29)(x^2+x+2)$; $x = 2 \pm 5\mathrm{i}$, $x = -\frac{1}{2} \pm \mathrm{i}\frac{\sqrt{7}}{2}$

c i $(x^2+4)(x^2+2x+6)$; $x = \pm 2\mathrm{i}$, $x = -1 \pm \mathrm{i}\sqrt{5}$

ii $(x^2+16)(x^2-4x+5)$; $x = \pm 4\mathrm{i}$, $x = 2 \pm \mathrm{i}$

4 $x = -2$, $x = 5 + \mathrm{i}$

5 a Proof.

b $x = -3$, $x = \pm 4\mathrm{i}$

6 a $2 - 5\mathrm{i}$ b $x = \pm \mathrm{i}$

7 $x = -2\mathrm{i}$, $x = 4 + \mathrm{i}$; $(x^2+4)(x^2-8x+17)$

8 a Proof.

b $(z^2+4)(z^2+z+1)$

c $z = -2\mathrm{i}$, $z = \frac{-1 \pm \sqrt{3}\mathrm{i}}{2}$

9 $x^4 - 6x^3 + 29x^2 - 96x + 208$

EXERCISE 2C

1 a k b $4k$ c $k^2 - 4k$

d $\frac{1}{2}$ e $k^3 - 6k^2$ f $\frac{k-4}{4k}$

2 a $-\frac{9}{a}$ b a^2

c $\frac{9}{a^2} + 2a$ d $\frac{9}{a^2} + 4a$

EXERCISE 2D

1 a -2 b 3

c $-\frac{8}{3}$ d -4

2 a $-\frac{1}{2}; -\frac{3}{2}$ b $0; -\frac{2}{3}$ c $0; -\frac{8}{5}$

3 $-\frac{1}{8}$

4 a $\frac{1}{5}$ b $\frac{2}{5}$

5 a 4 b $\frac{1}{3}$

6 a 1 b Proof.

7 $\frac{9}{4}$

8 Proof.

9 a $-2a^2$

b $-2a^2 < 0$; but if p, q, r were all real, the sum of their squares would be positive. (This is an example of proof by contradiction.)

EXERCISE 2E

1 a i $a = -6$; $b = -6$ ii $a = -8$; $c = -4$

b i $a = 3$; $b = 12$ ii $b = -5$; $d = -5$

2 a i $x^2 - 10x + 29$ ii $x^2 - 6x + 10$

b i $x^3 - x^2 + 9x - 9$ ii $x^3 - 5x^2 + x - 5$

c i $x^4 - 3x^3 + 5x^2 - x - 10$

ii $x^4 - 2x^3 - 6x^2 + 22x - 15$

d i $x^3 - 6x^2 + 9x + 50$

ii $x^3 + 3x^2 + 9x - 13$

3 a $4 + \mathrm{i}$ b $x^3 - 10x^2 + 33x - 34 = 0$

4 a $-3\mathrm{i}$, $3 + \mathrm{i}$ b $a = 6$; $d = 90$

5 $x^3 - 9x^2 + 36x + 27 = 0$

6 $4x^3 + 24x^2 + 45x + 31 = 0$

7 $2x^2 - 3x + 5 = 0$

8 $x^4 - 6x^3 + 8x + 32 = 0$

9 **a** $\frac{2}{5}; -\frac{11}{25}$ **b** $25x^2 + 11x + 4 = 0$

10 **a** $-\frac{3}{5}$ **b** $5x^3 - 3x + 1 = 0$

11 $16x^3 + 12x^2 - 49 = 0$

12 **a** $(\alpha\beta)^2 + (\beta\gamma)^2 + (\gamma\alpha)^2 + 2\alpha\beta\gamma(\alpha + \beta + \gamma)$

b $x^3 + 3x^2 + 16x - 36 = 0$

WORK IT OUT 2.2

Solution 2 is correct.

EXERCISE 2F

1 **a** **i** $8u^3 - 6u + 1 = 0$

ii $27u^3 + 18u^2 + 5 = 0$

b **i** $3u^3 - 18u^2 + 35u - 18 = 0$

ii $2u^3 + 7u^2 + 8u + 4 = 0$

c **i** $8u^3 + 2u + 17 = 0$

ii $9u^3 - 15u^2 + 2u + 6 = 0$

2 $3u^3 + 9u^2 + 5u + 1 = 0$

3 $32u^4 + 4u + 5 = 0$

4 **a** Proof.

b $x = -2, -2 \pm 3\mathrm{i}$

5 **a** $c = 3$ **b** $x = 3, 3, 6$

6 $\frac{2}{9}$ **7** $9u^2 - 75u + 1 = 0$

8 **a** $k = 1$ **b** $x = -1 \pm \mathrm{i}$ or $-1 \pm 2\mathrm{i}$

MIXED PRACTICE 2

1 D **2** C **3** $-3\mathrm{i}, 2$

4 $a = -3; b = 7; c = -5$

5 **a** $1 - \mathrm{i}\sqrt{2}; x^2 - 2x + 3$

b $x = -5, x = 1 \pm \mathrm{i}\sqrt{2}$

6 $x^4 - 4x^3 - x^2 - 16x - 20 = 0$

7 **a** $\alpha + \beta = \frac{7}{5}; \alpha\beta = \frac{1}{5}$

b Proof.

c $5x^2 - 42x + 65 = 0$

8 $z = 1 - 2\mathrm{i}, z = -3$

9 **a** $2 + 3\mathrm{i}$ **b** $b = -2; c = 5; d = 26$

10 $a = -9; b = 33$

11 **a** Proof. **b** $x^2 + 301x + 8 = 0$

12 $3u^3 + 23u^2 + 64u + 67 = 0$

13 $20u^3 - 9u + 2 = 0$

14 **a** **i** $\alpha + \beta + \gamma = 0$ **ii** $\alpha\beta\gamma = -q$

b Proof.

c **i** $\beta = 4 - 7\mathrm{i}; \gamma = -8$ **ii** $p = 1; q = 520$

d $520w^3 + w^2 + 1 = 0$

15 **a** **i** $-\frac{b}{a}, -\frac{d}{a}$ **ii** Proof.

b **i–ii** Proof.

16 **ai–ii** Proof.

b **i** $-\frac{b}{a}, -\frac{d}{a}$

ii Proof.

c **i** Proof.

ii $\frac{25}{4}; 1$

iii $4x^3 - 20x^2 + 25x - 4 = 0$

3 The ellipse, hyperbola and parabola

BEFORE YOU START

1 $k < \frac{1}{5}$

2 $\frac{11}{4}$

3 $x = 0, y = 0$

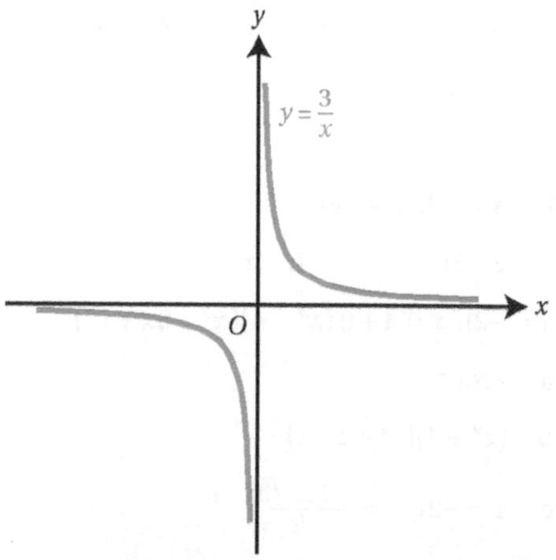

4 **a** $y = (x-3)^2 - 3(x-3) = x^2 - 9x + 18$

b $y = 3x^2 - 9x$

5 $(-2, 5); 6$

EXERCISE 3A

1 **a** **i**

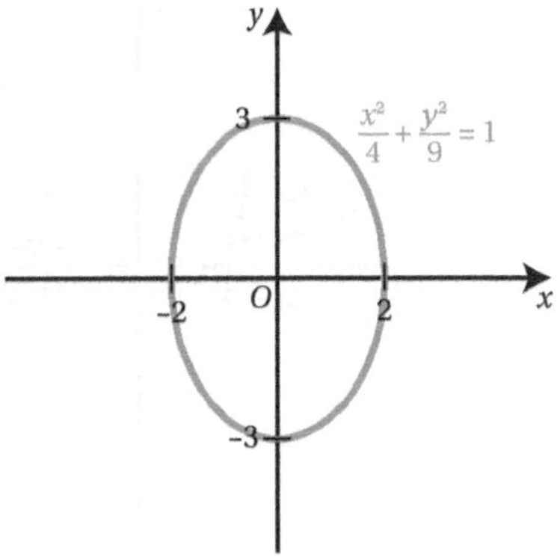

ii

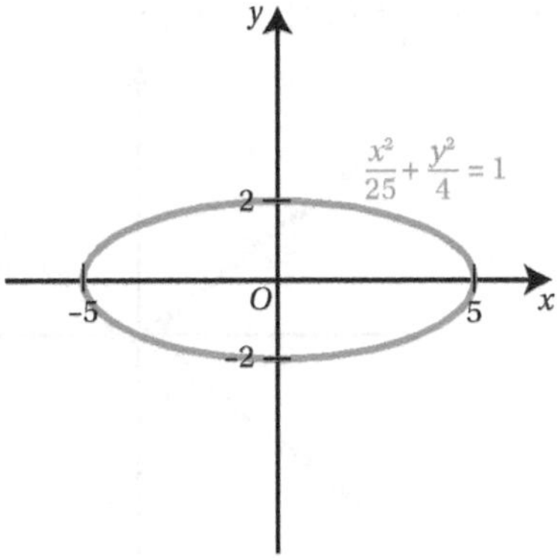

b **i**

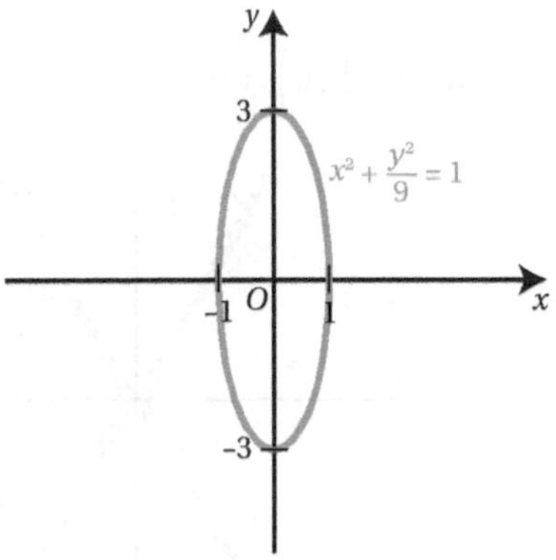

ii

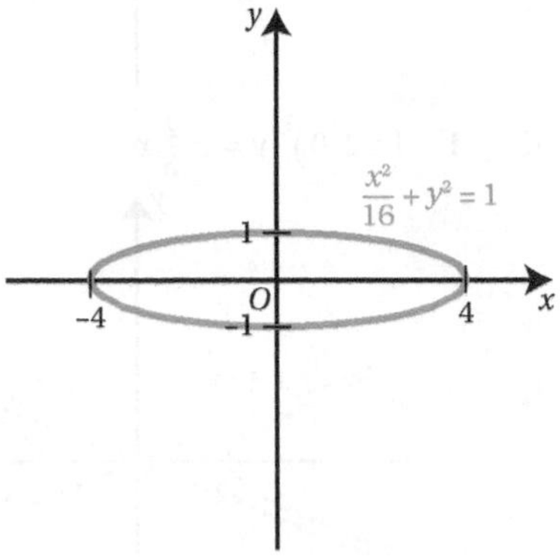

c **i**

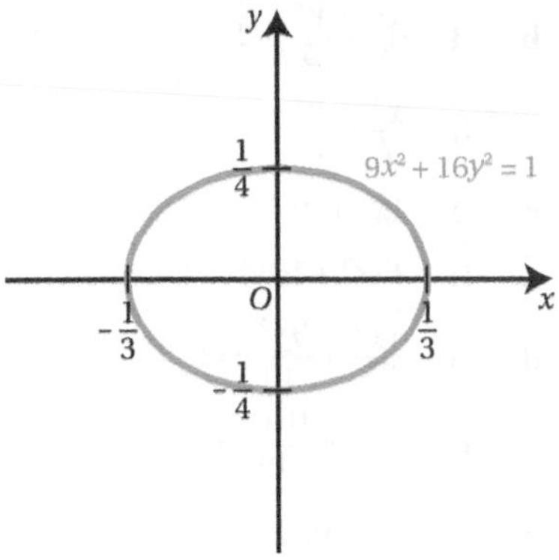

ii

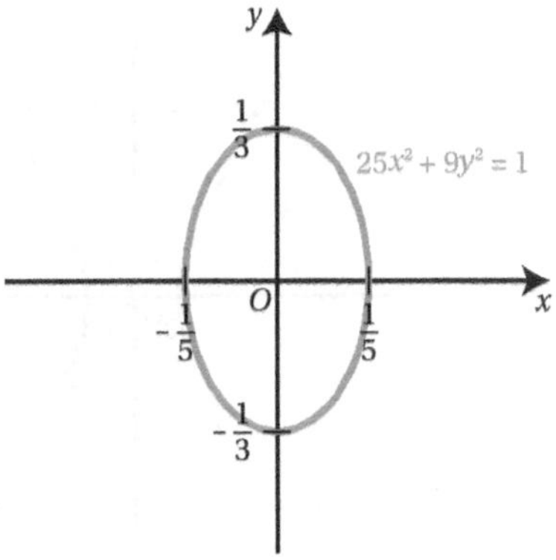

d **i**

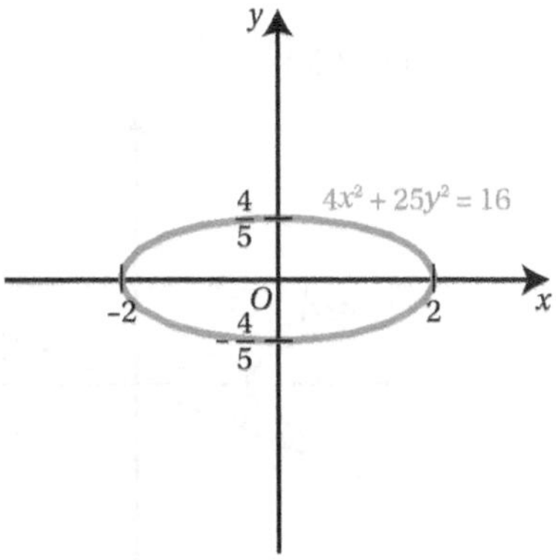

ii

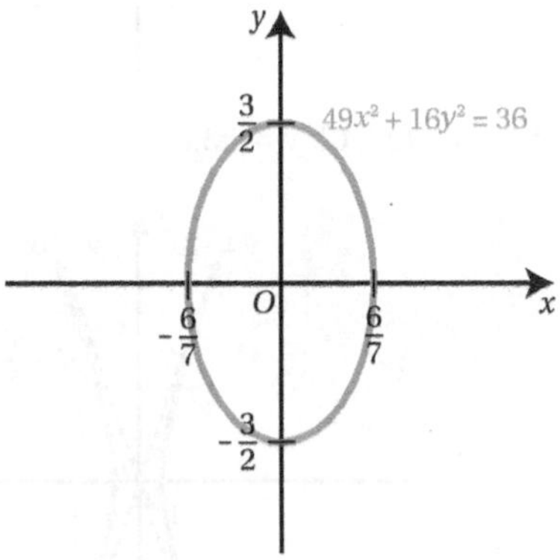

2 **a** **i** $\frac{x^2}{25}+\frac{y^2}{9}=1$

ii $\frac{x^2}{4}+\frac{y^2}{16}=1$

b **i** $\frac{x^2}{3}+\frac{y^2}{5}=1$

ii $\frac{x^2}{7}+\frac{y^2}{10}=1$

c **i** $4x^2+9y^2=1$

ii $25x^2+16y^2=1$

d **i** $\frac{4x^2}{25}+\frac{9y^2}{49}=1$

ii $\frac{9x^2}{4}+\frac{16y^2}{49}=1$

3 **a** **i** $(\pm 2, 0)$, $y=\pm\frac{3}{2}x$

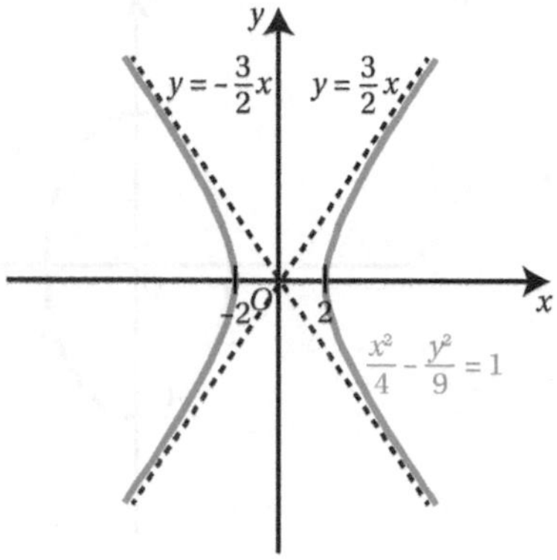

ii $(\pm 5, 0)$, $y=\pm\frac{2}{5}x$

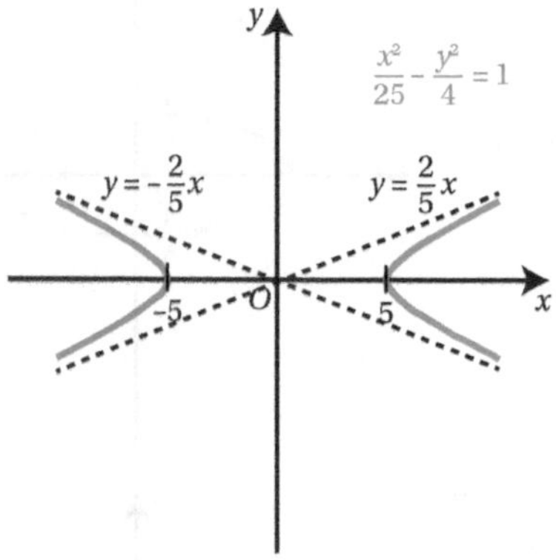

b **i** $(\pm 1, 0)$, $y=\pm 3x$

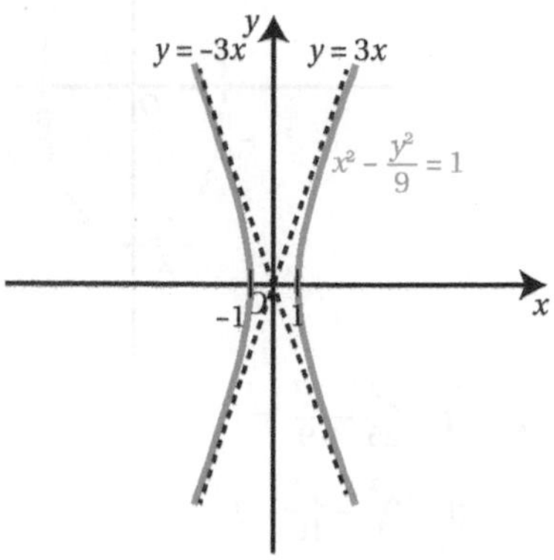

ii $(\pm 4, 0)$, $y=\pm\frac{1}{4}x$

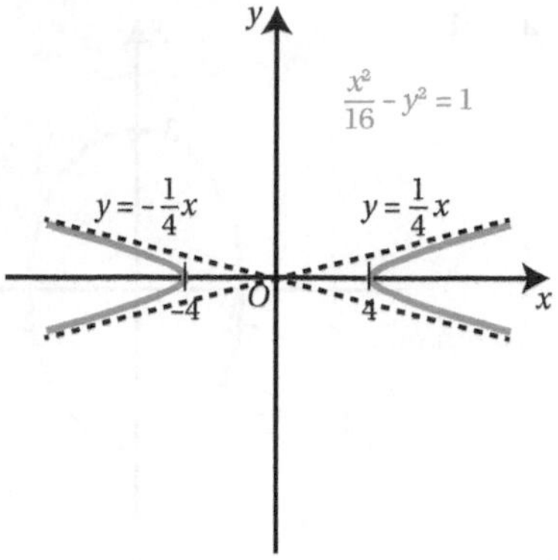

c **i** $\left(\pm\frac{1}{3}, 0\right)$, $y=\pm\frac{3}{4}x$

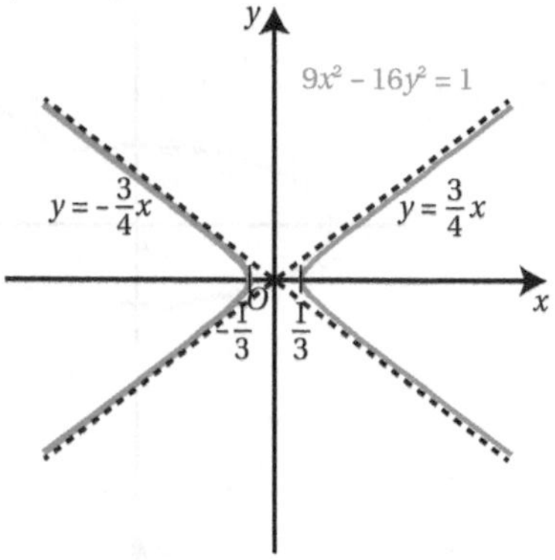

ii $\left(\pm\frac{1}{5}, 0\right)$, $y=\pm\frac{5}{3}x$

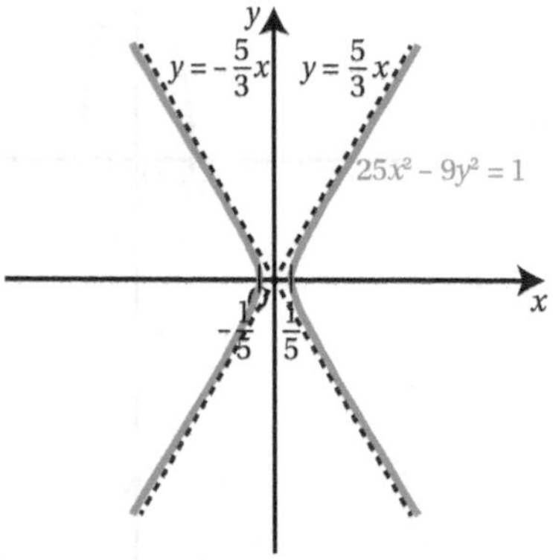

d **i** $(\pm 2, 0)$, $y=\pm\frac{2}{5}x$

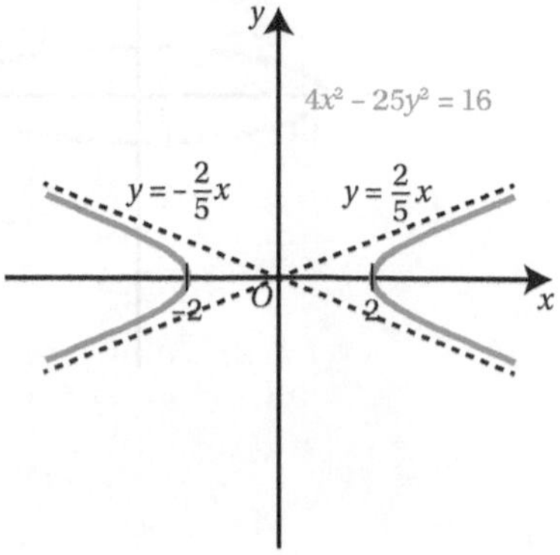

ii $\left(\pm\frac{6}{7}, 0\right)$, $y = \pm\frac{7}{4}x$

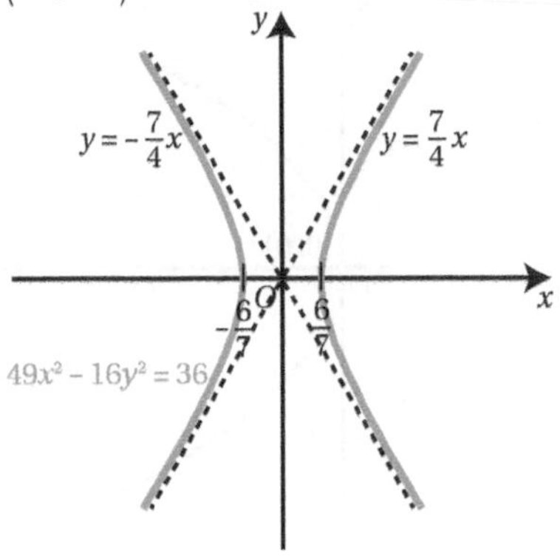

4 **a** **i** (5, 5), (−5, −5)

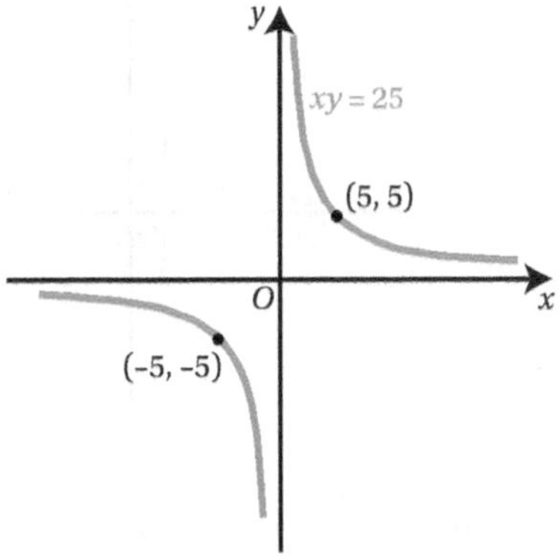

ii (4, 4), (−4, −4)

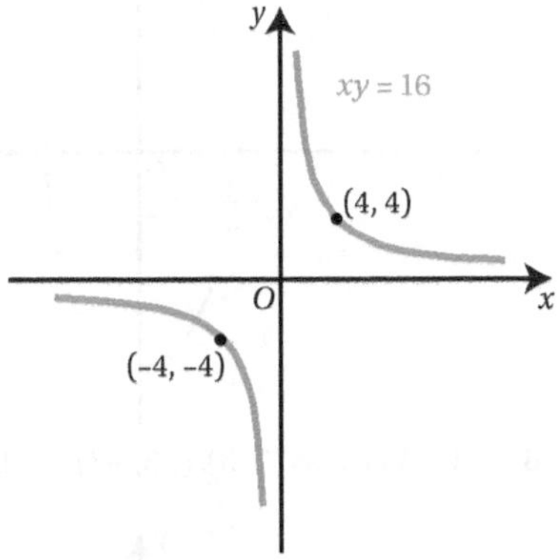

b **i** $\left(\frac{1}{2}, \frac{1}{2}\right), \left(-\frac{1}{2}, -\frac{1}{2}\right)$

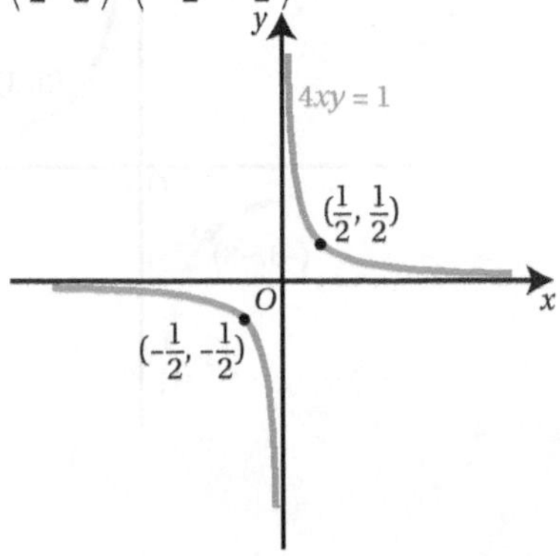

ii $\left(\frac{1}{7}, \frac{1}{7}\right), \left(-\frac{1}{7}, -\frac{1}{7}\right)$

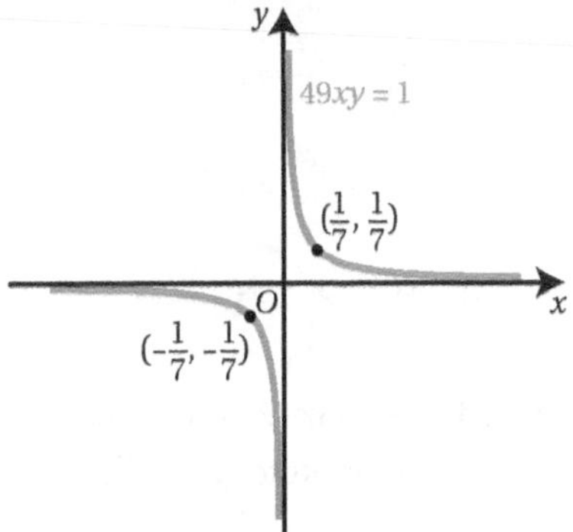

c **i** $\left(\frac{\sqrt{5}}{2}, \frac{\sqrt{5}}{2}\right), \left(-\frac{\sqrt{5}}{2}, -\frac{\sqrt{5}}{2}\right)$

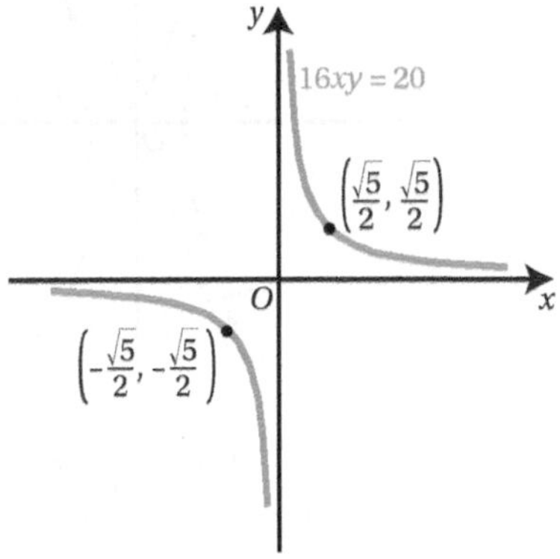

ii $\left(\frac{5\sqrt{3}}{3}, \frac{5\sqrt{3}}{3}\right), \left(-\frac{5\sqrt{3}}{3}, -\frac{5\sqrt{3}}{3}\right)$

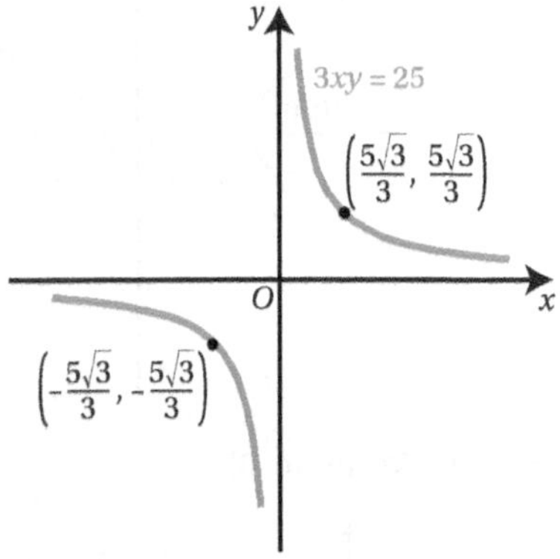

5 **a** **i** $\frac{x^2}{25} - \frac{y^2}{9} = 1$

ii $\frac{x^2}{9} - \frac{y^2}{4} = 1$

b **i** $\frac{x^2}{5} - \frac{y^2}{2} = 1$

ii $\frac{x^2}{3} - \frac{y^2}{7} = 1$

c **i** $\frac{x^2}{\frac{1}{4}} - \frac{y^2}{\frac{1}{3}} = 1$

ii $\frac{x^2}{\frac{1}{9}} - \frac{y^2}{\frac{1}{5}} = 1$

d **i** $\frac{x^2}{\frac{4}{9}} - \frac{y^2}{\frac{25}{9}} = 1$

ii $\frac{x^2}{\frac{9}{16}} - \frac{y^2}{\frac{1}{4}} = 1$

6 **a** **i** Vertices $(2, 0), (-2, 0)$; asymptotes $y = \pm\frac{5}{2}x$

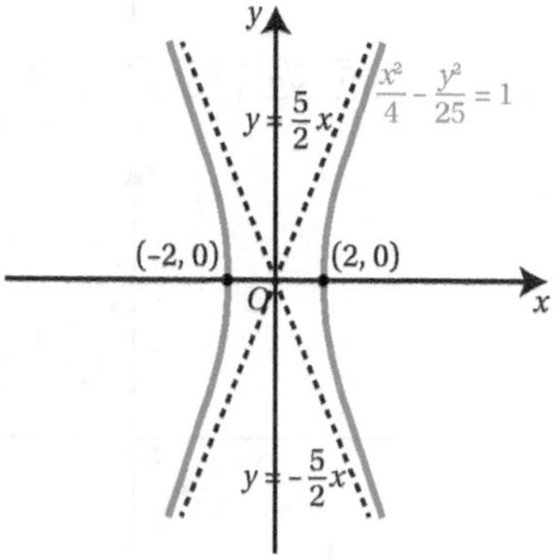

ii Vertices $(10, 0), (-10, 0)$; asymptotes $y = \pm\frac{3}{5}x$

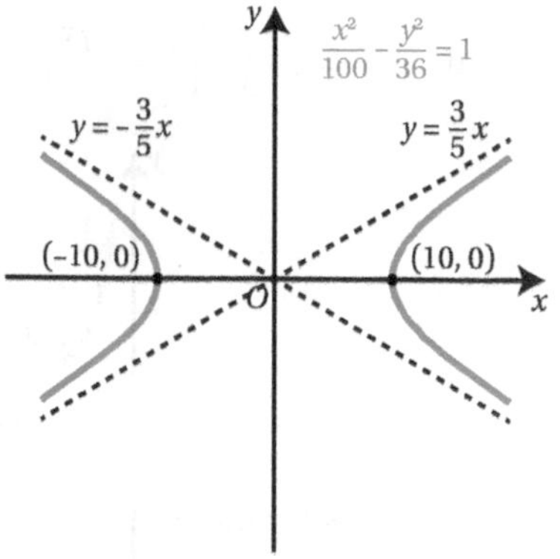

b **i** Vertex $(0, 0)$

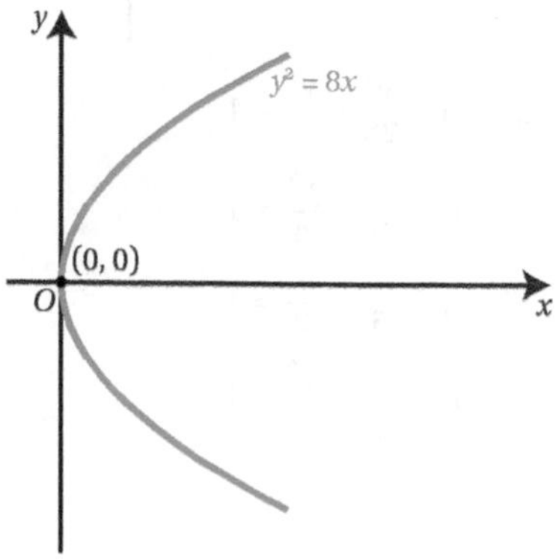

ii Vertex $(0, 0)$

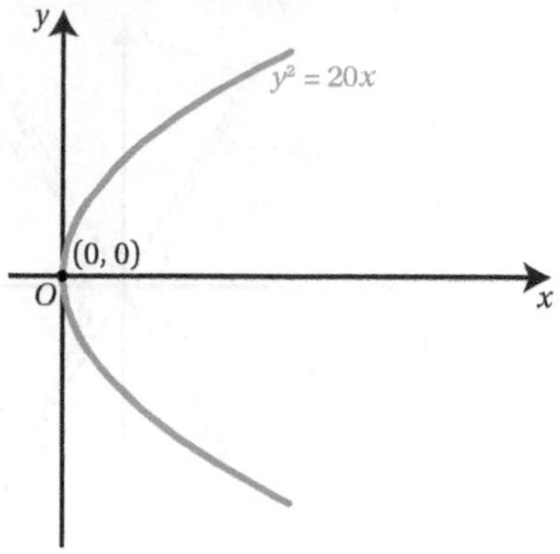

c **i**

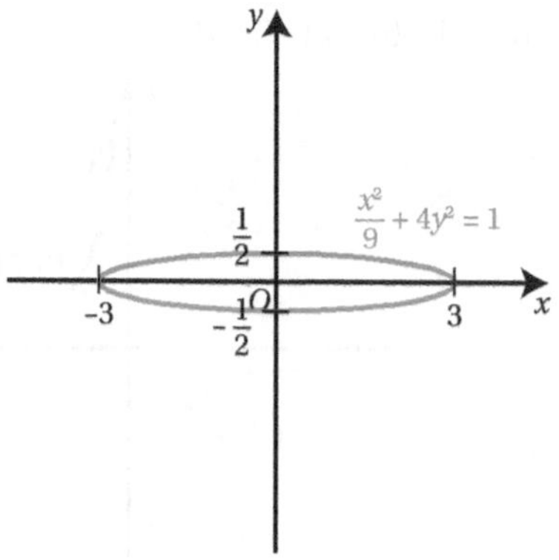

ii

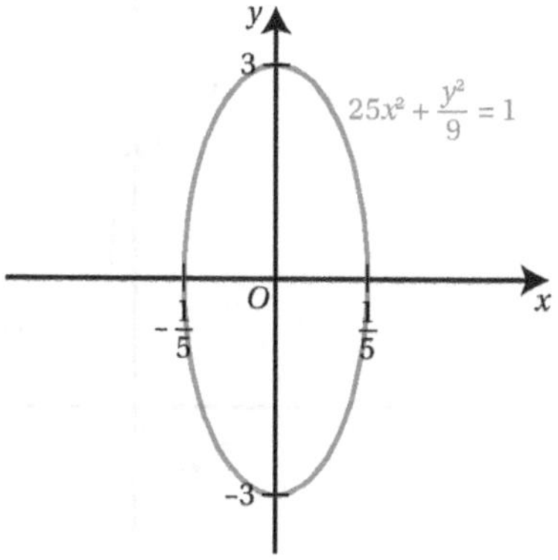

d **i** Vertices $(3, 3), (-3, -3)$; $x = 0, y = 0$

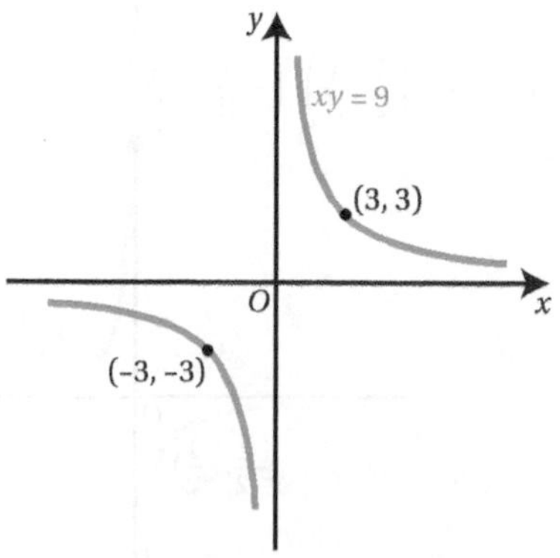

ii Vertices $(12, 12), (-12, -12)$; $x = 0, y = 0$

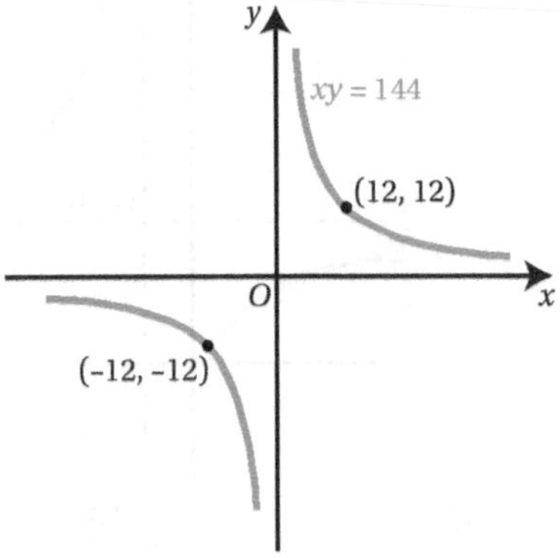

7 C

8 C

9 Proof.

EXERCISE 3B

1 $\frac{5}{8}$

2 a $y = \pm\frac{5x}{4}$

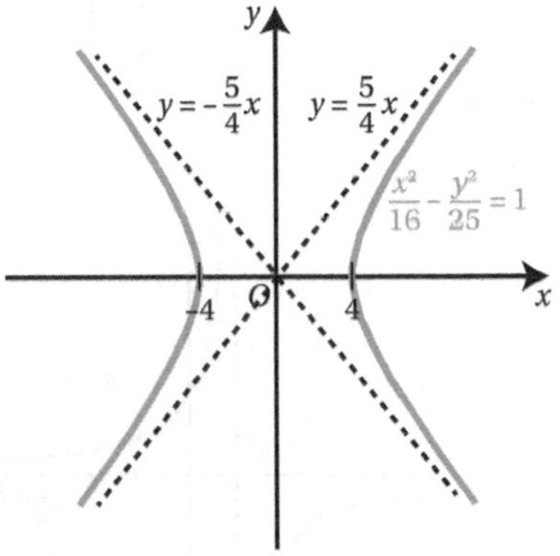

b $\pm\sqrt{39}$

3 $\pm\sqrt{3}$

4 a Proof; $(1.5, 3)$ b $(13.5, -9)$

5 a $(-1.2, 1.4)$

b $(1.38, 0.538)$

6 a

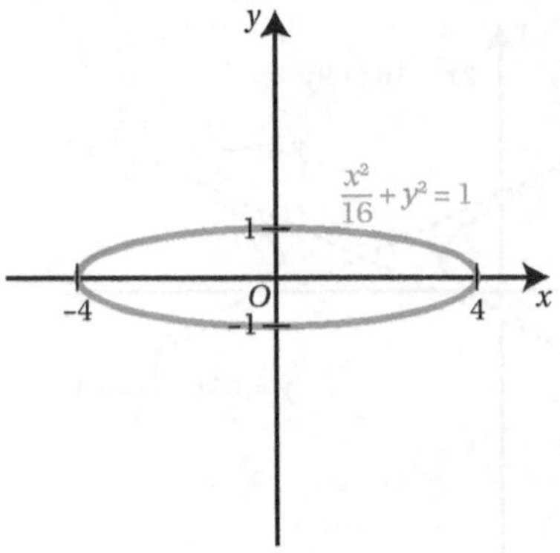

b $m < -\frac{\sqrt{3}}{4}$ or $m > \frac{\sqrt{3}}{4}$

7 a $y = 2x$

b Proof.

8 a $\pm\sqrt{2}$

b Proof.

EXERCISE 3C

1 a i $x^2 - 6x + 3y^2 + 12y + 15 = 0$

ii $10x^2 + 20x + 3y^2 - 24y + 28 = 0$

b i $xy - 3x + 5y - 31 = 0$

ii $xy - 10x - 10y = 0$

c i $y^2 = 162x$

ii $y^2 = 6x$

d i $3y^2 - 2x^2 = 18$

ii $y^2 - 2x^2 = 10$

e i $y^2 = -4x$

ii $y^2 = 4x$

2 a i

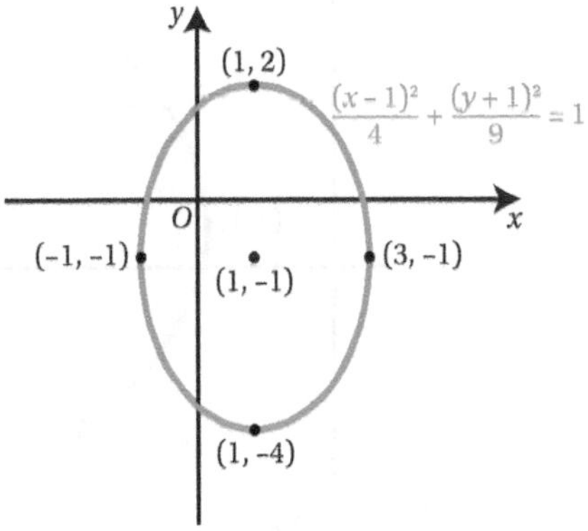

ii

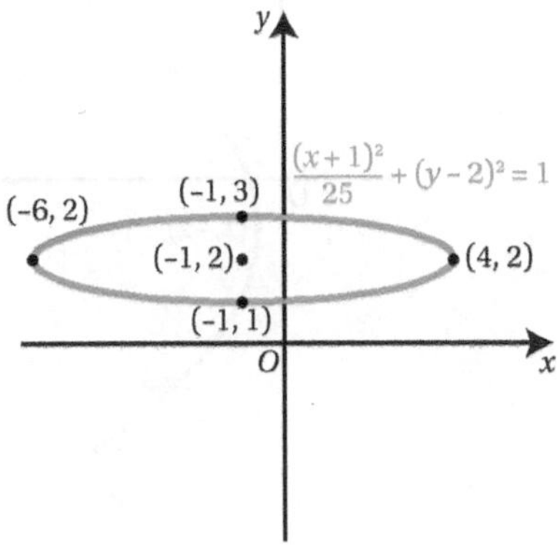

b **i**

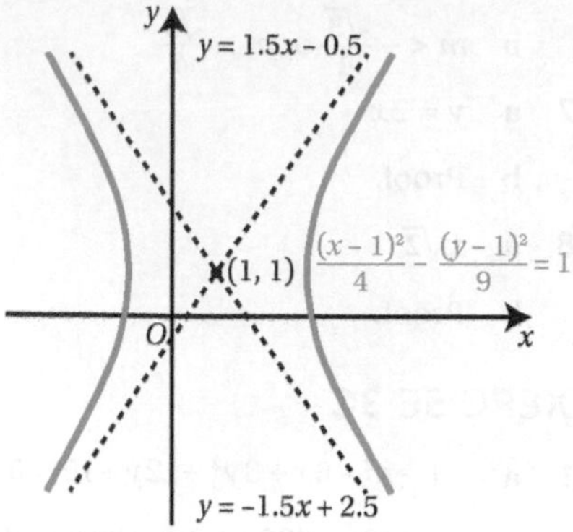

ii

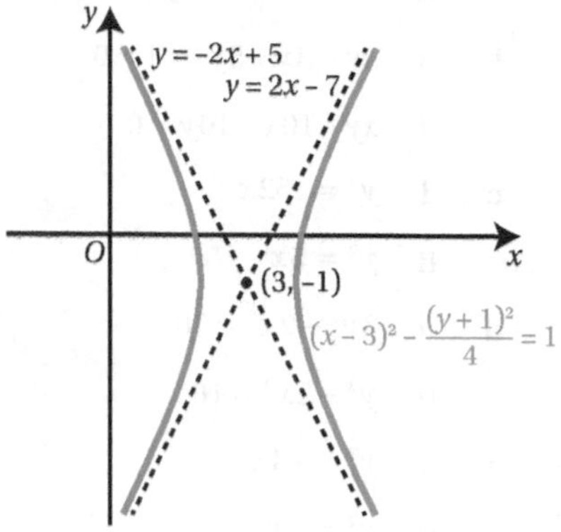

c **i**

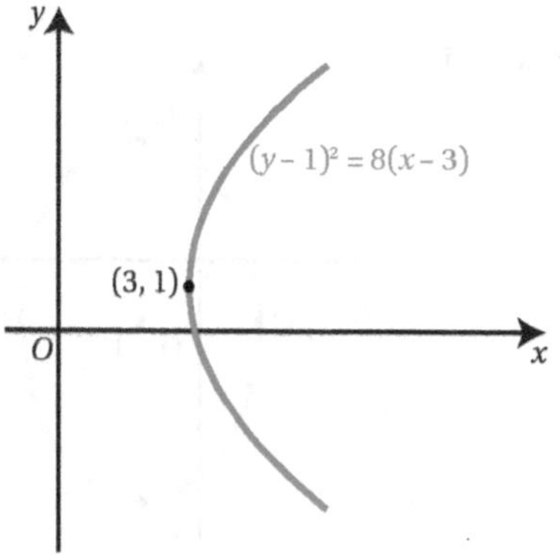

ii

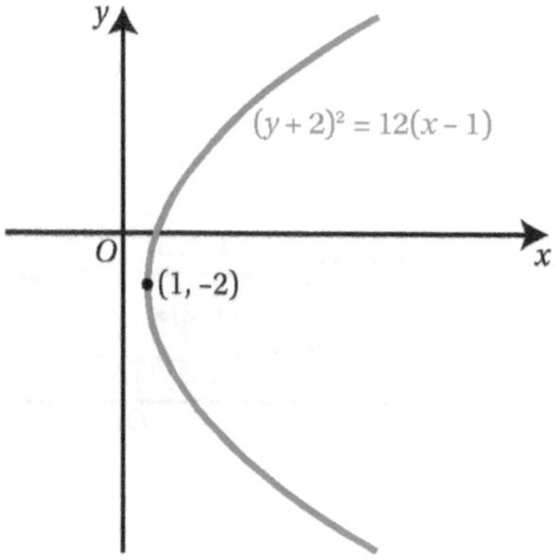

d **i**

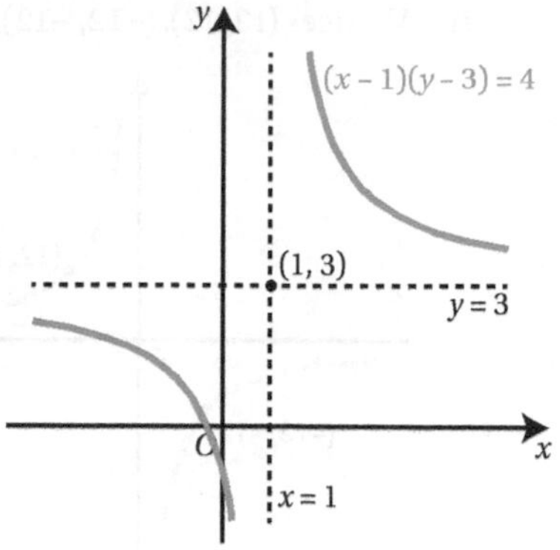

ii

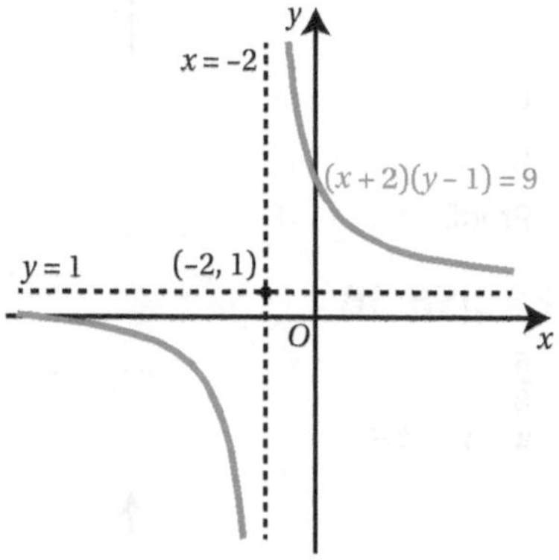

3 $p = 3; q = -1$

4 $\frac{\sqrt{2}}{4}$

5 **a**

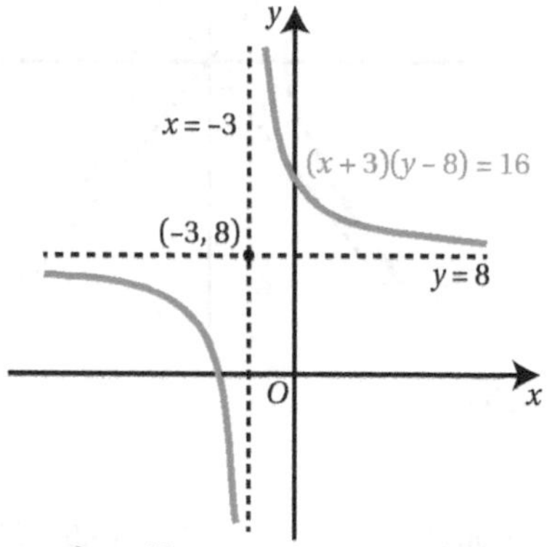

b $y = \frac{8x+40}{x+3}$

6 20

7 $(4a, 4a)$

8

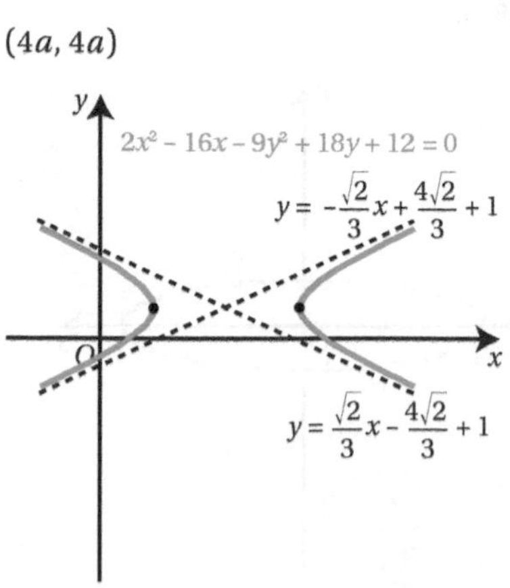

Vertices $\left(4\pm\sqrt{\frac{11}{2}}, 1\right)$; asymptotes

$y=-\frac{\sqrt{2}}{3}x+\frac{4\sqrt{2}}{3}+1$ and $y=\frac{\sqrt{2}}{3}x-\frac{4\sqrt{2}}{3}+1$

9 **a** Vertices $\left(\pm\sqrt{3}, 0\right)$; asymptotes $y=\pm\sqrt{\frac{2}{3}}x$

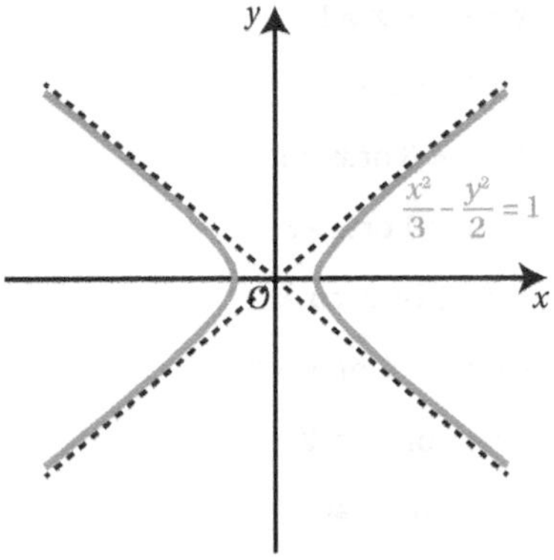

b 2.5

10 $a=3; b=2$

11 **a** $\pm\sqrt{\frac{5}{14}}$

b $y=\pm\sqrt{\frac{5}{14}}(x+4)$

12 $y=-2x-1$

13 **a** $\frac{2\sqrt{3}}{3}$

b $(3\sqrt{3}, 3)$

c $y=-\sqrt{3}x+12$

d $y=\frac{2}{3}x-1$

14 **a** $\frac{5}{4}$

b $y=x+3$ and $y=-x-3$

c $\frac{10\sqrt{41}}{3}$

MIXED PRACTICE 3

1 D

2 A

3 D

4 $c=\pm5\sqrt{2}$

5 $m=\pm\sqrt{9.5}$

6 **a** $\frac{3}{2}$; $\left(\frac{4}{3}, 4\right)$

b $y^2=-12x$

c $y=-\frac{3}{2}x+2$

7 **a** Proof.

b 2

c $3x+8y=50$

8 B

9 **a** 1

b (2, 4)

c (0, −1), (0, 7), $\left(-\frac{7}{8}, 0\right)$

d

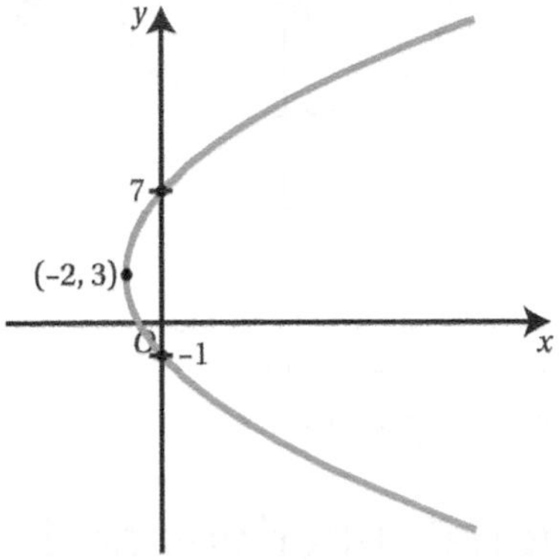

e $y=x+7$

10 **a** $x^2=5y$

b (5, 5)

c $y=2x-5$

d $y=\frac{1}{2}x+\frac{5}{2}$

11 5

12 **a**

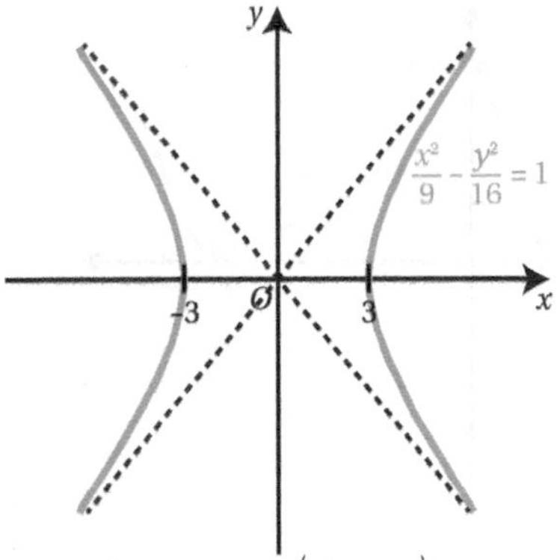

b $y=\pm\frac{4}{3}(x+3)=\pm\left(\frac{4}{3}x+4\right)$

13 **a** A is at $x=2$; B is at $x=6$.

b **i** Proof.

ii $m=\frac{\sqrt{12}}{12}=\frac{\sqrt{3}}{6}$

iii $\left(3, \frac{\sqrt{3}}{2}\right)$

14 B

15 $b>1$

16 **a** Proof.

b $y=2x+11$, $y=2x-11$, $y=-2x+11$ and $y=-2x-11$

17 a

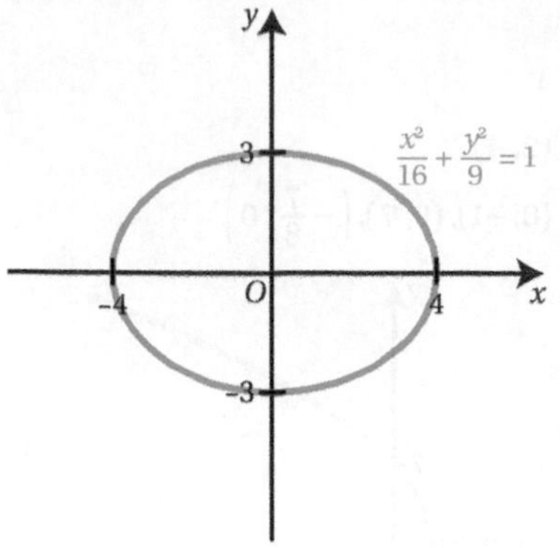

b Proof.

c $a=-1; b=2; c=71$

d $y=x+8$ and $y=x-2$

4 Rational functions and inequalities

BEFORE YOU START

1 $x<1$

2 $x<2$ or $x>3$

3 $k>1$

4

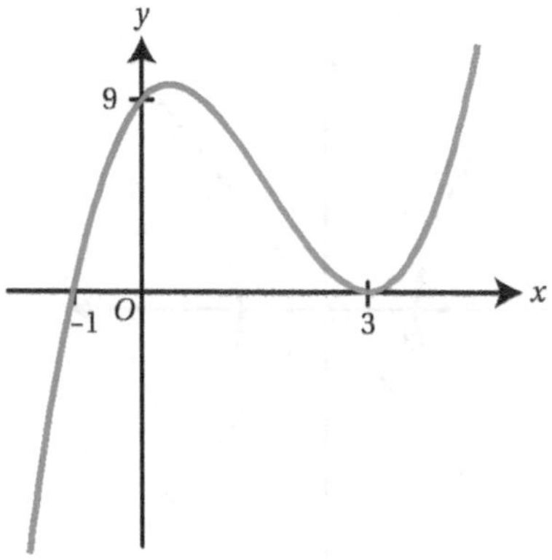

5 $(x-1)(x-2)^2$

6 Translation 3 units to the right.

7 $y=0$ and $x=0$

8 $1+\frac{3}{x^2+1}$

EXERCISE 4A

1 a i $0<x<1$ or $x>2$

ii $-1<x<0$ or $x>4$

b i $x\leqslant -3$ or $2\leqslant x\leqslant 5$

ii $x\leqslant 1$ or $3\leqslant x\leqslant 4$

c i $1\leqslant x\leqslant 2$ or $3\leqslant x\leqslant 4$

ii $-2\leqslant x\leqslant -1$ or $3\leqslant x\leqslant 7$

d i $x\geqslant -1$

ii $x=-3$

e i $x<-3$ or $x>1$

ii $x<2$ or $x>4$

f i $2<x<3$ or $x>5$

ii $1<x<4$ or $x>8$

2 a i $x<-2$ or $0<x<2$

ii $-3<x<0$ or $x>3$

b i $x\leqslant 1$ or $x\geqslant 2$

ii $x\leqslant 3$ or $x\geqslant 4$

c i $x<-1$ or $1<x<2$

ii $x<-2$ or $1<x<2$

d i $-2-\sqrt{5}<x<-2+\sqrt{5}$ or $x>1$

ii $x<-2$ or $-\frac{1}{2}-\frac{\sqrt{5}}{2}<x<-\frac{1}{2}+\frac{\sqrt{5}}{2}$

3 $x<-1.5$ or $0<x<5$

4 $b<x<c$ or $x>d$

5 a Proof.

b $-\frac{5}{2}-\frac{\sqrt{29}}{2}<x<-\frac{5}{2}+\frac{\sqrt{29}}{2}$, $x>2$

6 $x\leqslant -a$ or $x\geqslant a$

7 $a=-24; b=33; c=62$

8 $a=-5; b=7; c=-1$

9 $x>1$

10 a For example, $x^4>-1$

b For example, $(x^2-1)^2\geqslant 0$

EXERCISE 4B

1 a i $x=-4$, $y=2$; $(0, -0.25)$, $(0.5, 0)$

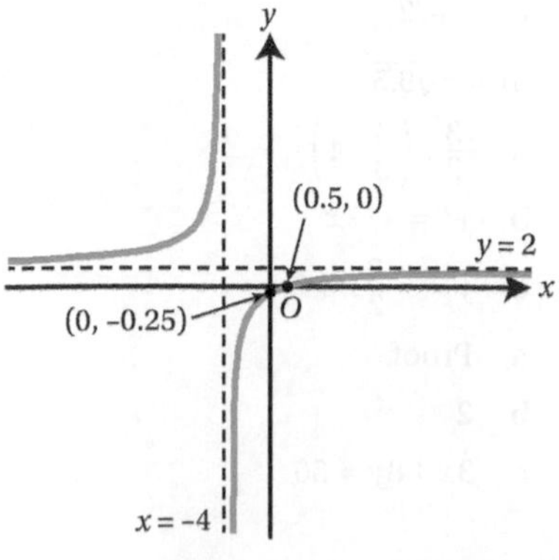

ii $x = 1,\ y = 3;\ (0, 2), \left(\frac{2}{3}, 0\right)$

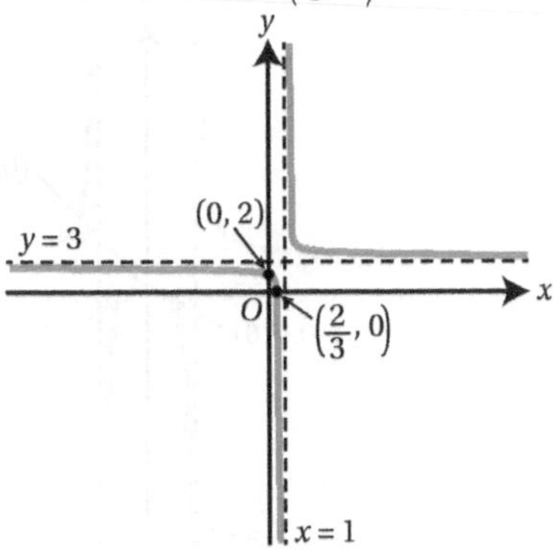

b **i** $x = -4,\ y = 2;\ (0, -2), (4, 0)$

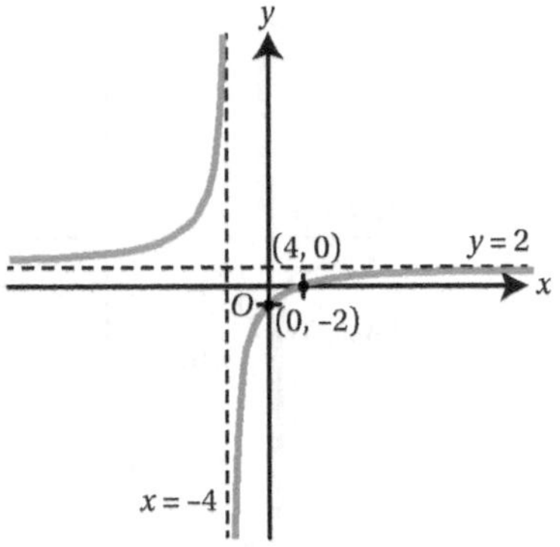

ii $x = 3,\ y = 3;\ (0, 0)$

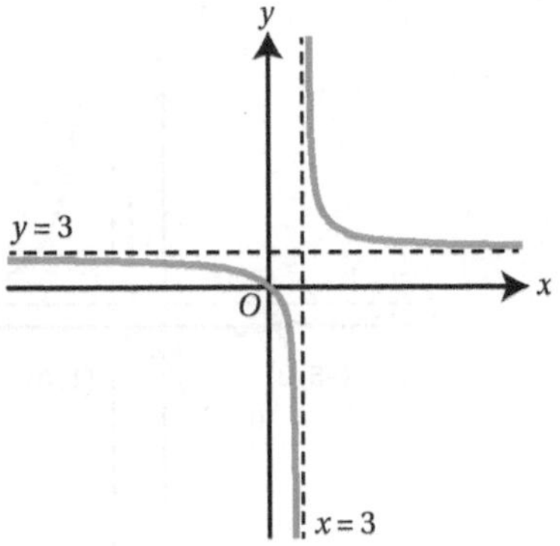

2 **a** **i** $\frac{3x-6}{x-1}$ **ii** $\frac{x+4}{2x+2}$

b **i** $\frac{x-4}{x+4}$ **ii** $\frac{3x-6}{2x-6}$

3 **a** **i** $x < -2$ **ii** $-2 < x < -1.5$

b **i** $x \leqslant -16$ or $x > -5$

ii $x < -1$ or $x \geqslant 6$

4 **a** **i** $1-\sqrt{6} < x < 1+\sqrt{6}$ or $x < -2$

ii $x < \frac{1-\sqrt{21}}{2}$ or $-\frac{3}{2} < x < \frac{1+\sqrt{21}}{2}$

b **i** $x > -1$

ii $x \leqslant -2$ or $-1 < x \leqslant 2$

5 **a** $x = -1.5$

b **i**

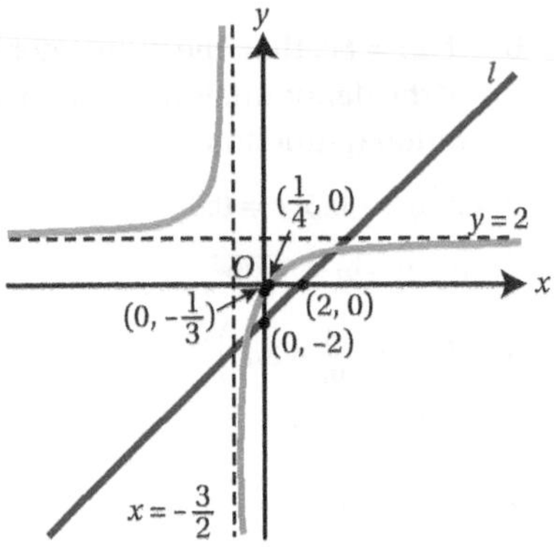

ii $x < -1.5$ or $\frac{5-\sqrt{65}}{4} < x < \frac{5+\sqrt{65}}{4}$

6 **a**

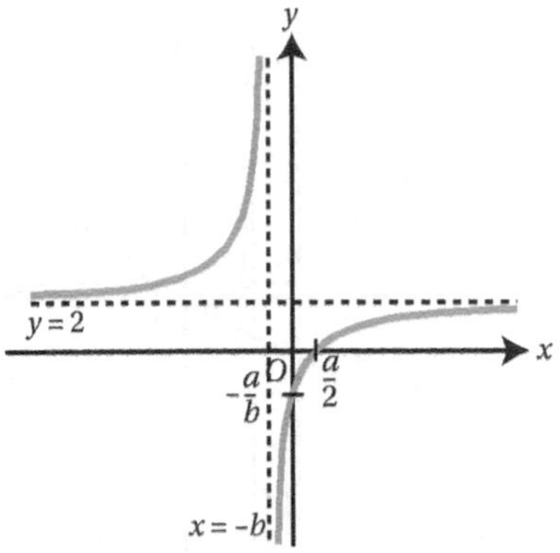

b $-a - 3b < x < -b$

7 **a** $m = \frac{3 \pm \sqrt{5}}{2}$

b **i**

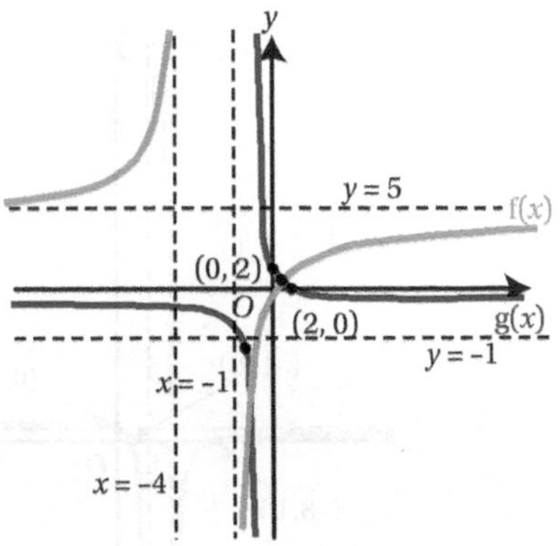

ii $x < -4$ or $\frac{-7-\sqrt{241}}{12} < x < -1$ or $x > \frac{-7+\sqrt{241}}{12}$

8 **a** **i** $k = -\dfrac{d}{c}$ **ii** $p = \dfrac{a}{c}$

b If $ad = bc$, then the numerator is a multiple of the denominator, and $f(x)$ is a degenerate rational function.

9 $p = 2$; $q = -0.5$; $r = 0.5$

10 **a** $c = 2 - m \pm \sqrt{12m}$

b $d^2 = 4\left(\dfrac{3}{m} + 3m\right)$

c $2\sqrt{6}$

EXERCISE 4C

1 **a** **i** $\left(0, -\dfrac{2}{15}\right)$, $(1, 0)$, $(-2, 0)$;
$x = -3$, $x = -5$, $y = 1$

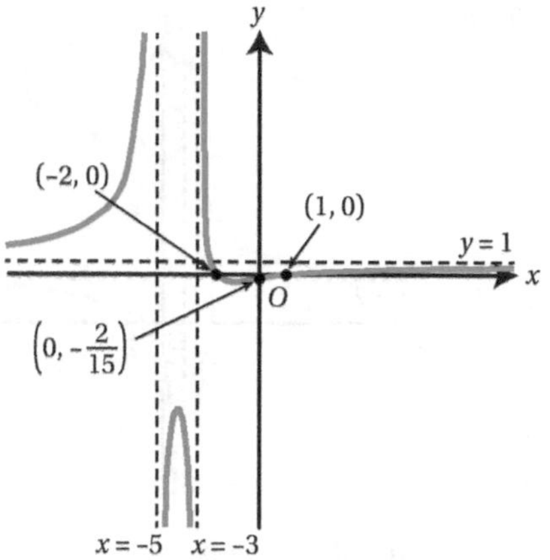

ii $(0, 1)$, $(-1, 0)$, $(-6, 0)$;
$x = -3$, $x = -2$, $y = 1$

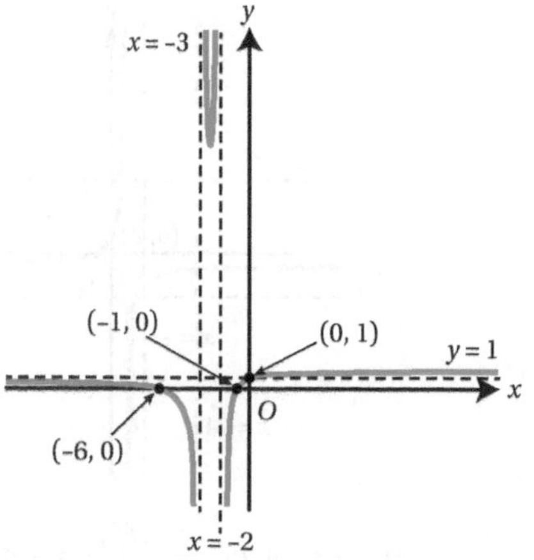

b **i** $(0, 0.5)$, $(-0.8, 0)$, $(1, 0)$; $x = \pm 2$, $y = 2.5$

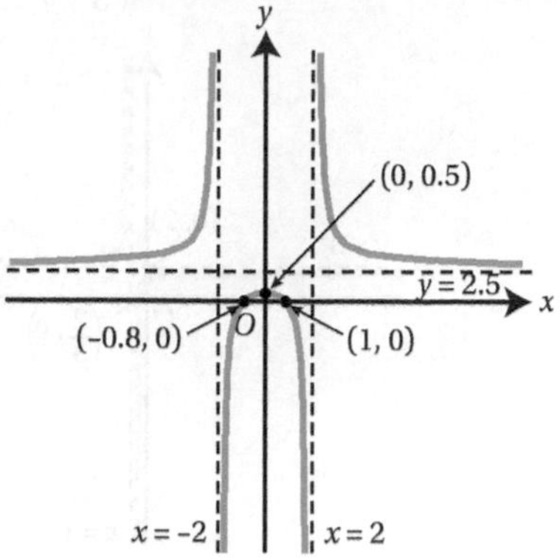

ii $(-0.5, 0)$, $(3, 0)$; $x = 0$, $x = 7$, $y = 2$

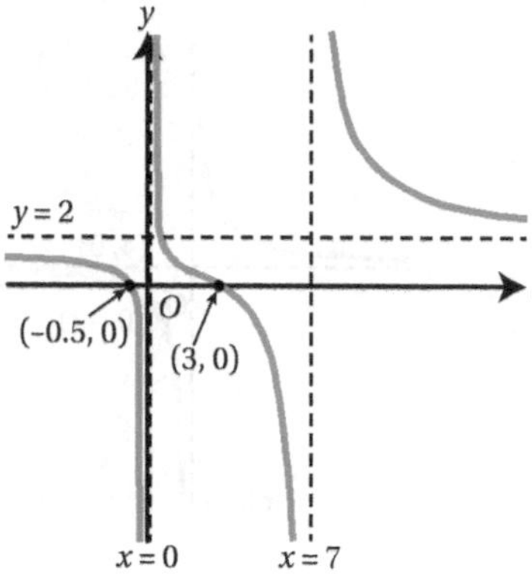

c **i** $\left(0, -\dfrac{5}{4}\right)$, $(-5, 0)$, $(1, 0)$; $x = 2$, $y = 1$

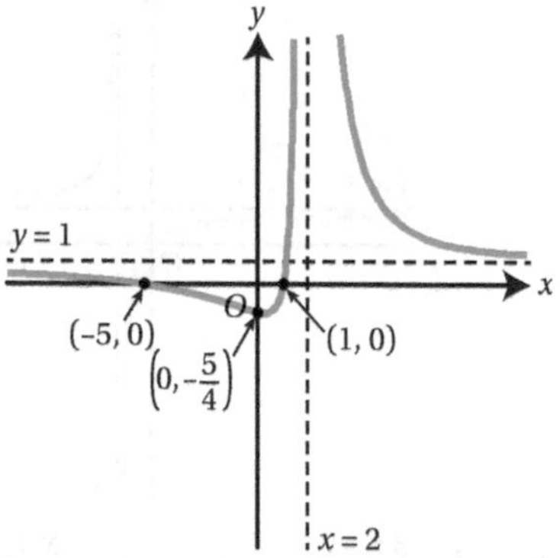

ii $(0, -3)$, $(-0.5, 0)$, $(-3, 0)$; $x = \pm 1$, $y = 2$

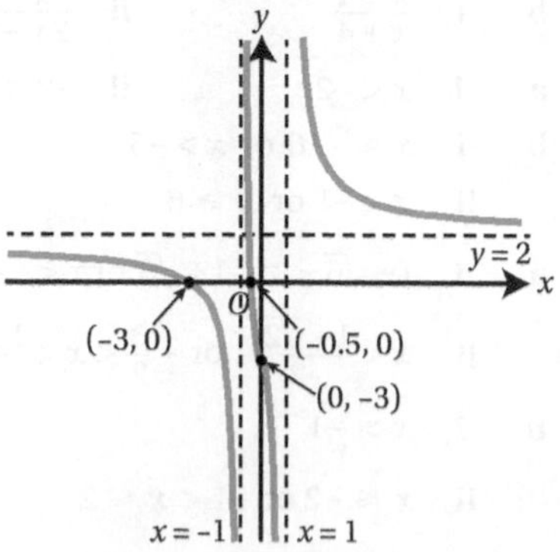

d **i** (0, 4); $y = 1$

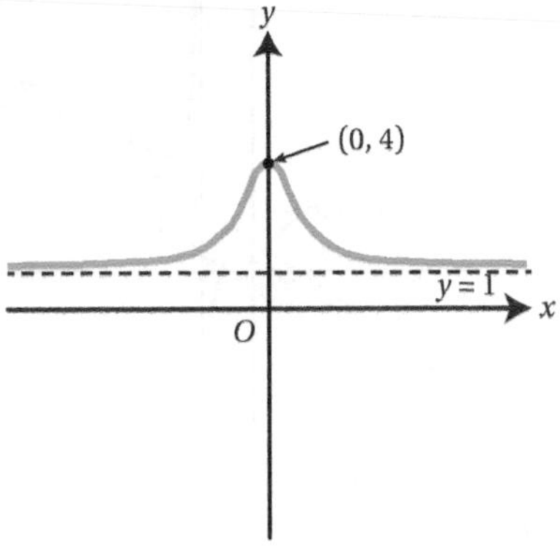

ii (0, −4), (−4, 0), (4, 0); $y = 1$

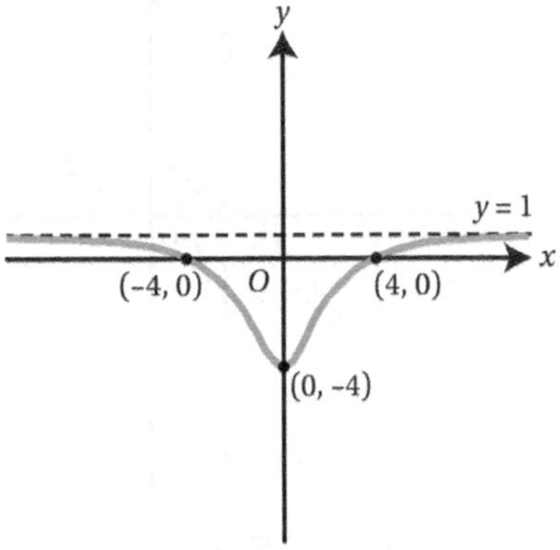

2 **a** **i** 4 **ii** 3

b **i** No stationary points.

ii No stationary points.

c **i** $2 \pm \sqrt{5}$ **ii** $\frac{3}{7}, \frac{27}{7}$

3 **a**

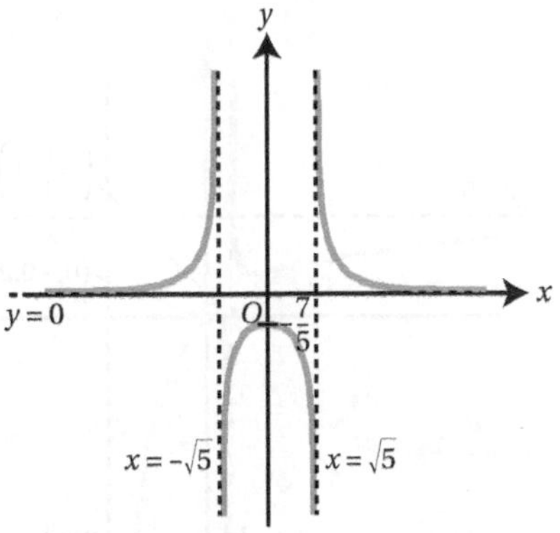

b $-\frac{3\sqrt{3}}{2} < x < -\sqrt{5}$ or $\sqrt{5} < x < \frac{3\sqrt{3}}{2}$

4 $k = 4$; $y = 1$, $x = 1$

5 $x < -5$, $-2 < x < 1$, $x > 1.5$

6 **a** **i** $y = 0$

ii $(3, 0), \left(0, -\frac{3}{16}\right)$

b **i** Proof.

ii $\left(8, \frac{1}{16}\right), \left(-2, -\frac{1}{4}\right)$

c

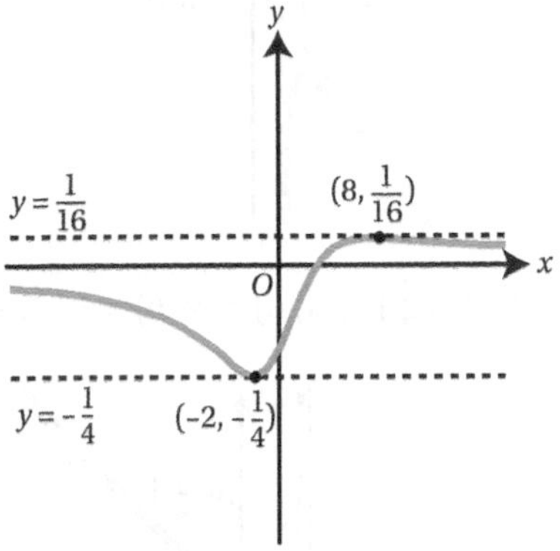

7

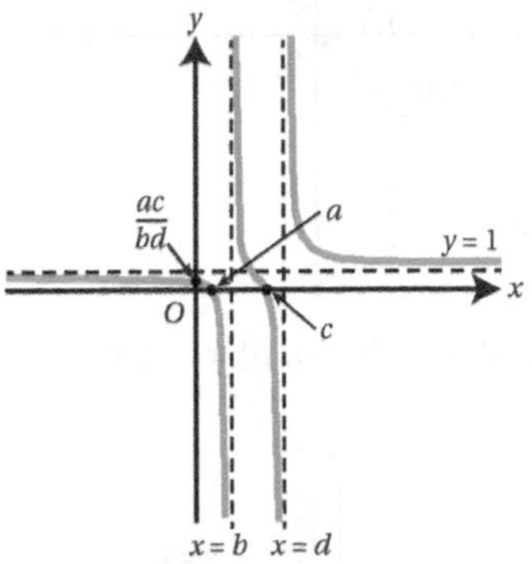

8 $a = b = 2$

9 **a** Proof.

b $a = 1$; $b = 2$; $c = 3$; $d = 0$; $e = 2$

10 **a** $p(t) = \frac{t^2 - 10t}{t^2 - 20t - 1}$

b 0.450 (3 s.f.)

A EXERCISE 4D

1 **a** **i** (0, 2.5), (−5, 0), (1, 0); $x = 2$, $y = x + 6$

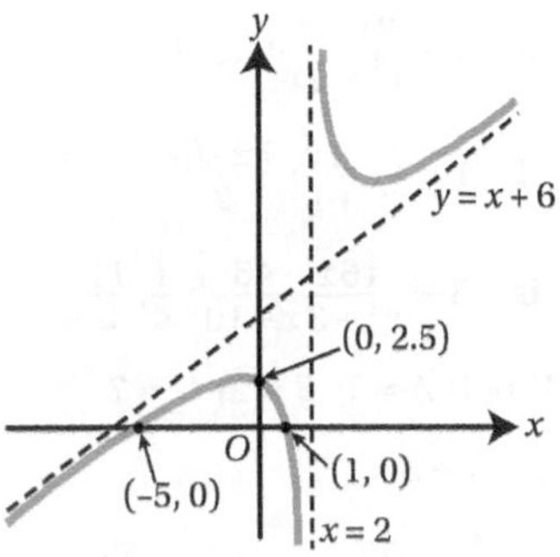

ii (0, 0), (−0.25, 0); $x = -1$, $y = 4x - 3$

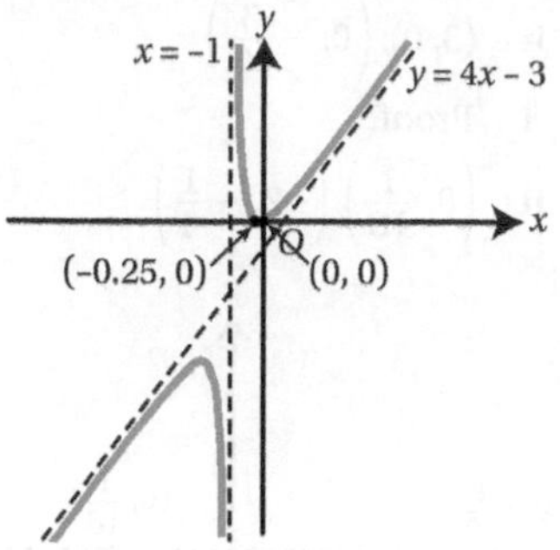

b **i** $\left(0, \frac{2}{3}\right)$, (−1, 0), (−2, 0); $x = -3$, $y = x$

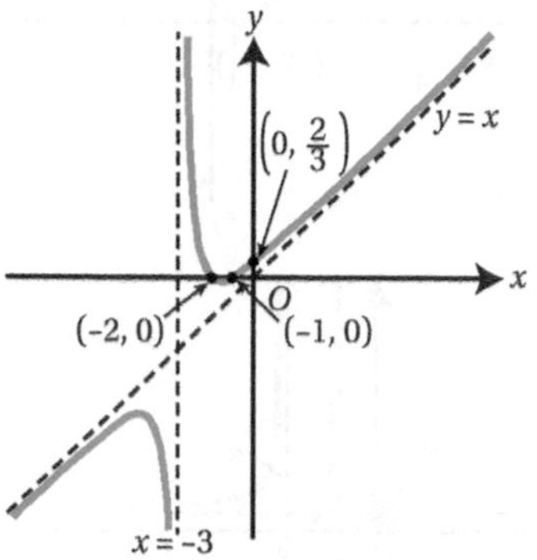

ii (0, 3), (−1, 0), (6, 0); $x = 2$, $y = x - 3$

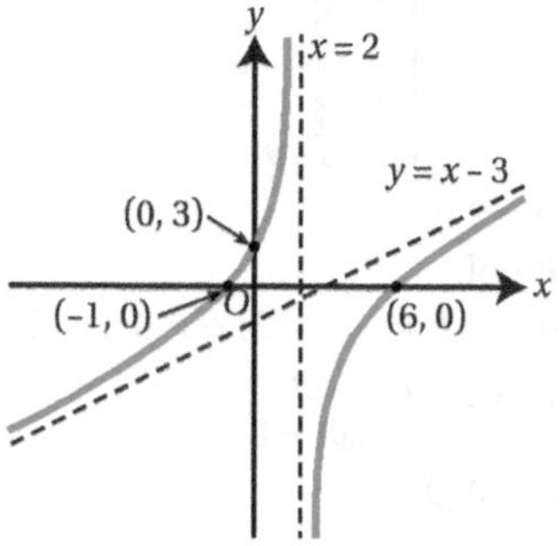

2 **a** **i** $1 - \frac{3}{x^2 - 1}$; 0

ii $1 - \frac{3x+1}{x^2 + 3x}$; none

b **i** $1 + \frac{2-4x}{x^2+1}$; $\frac{1 \pm \sqrt{5}}{2}$

ii $5 - \frac{16x - 46}{x^2 + 3x - 10}$; $\frac{1}{4}$, $\frac{11}{2}$

3 **a** Proof; $A = 1$; $B = 2$; $C = 7$

b $y = x + 2$

c $k = \frac{9}{2} \pm \sqrt{14}$; $y = 4.5 \pm \sqrt{14}$

d $x = \frac{5}{2}$; $\left(0, \frac{3}{5}\right)$, $\left(\frac{3}{2}, 0\right)$, (−1, 0)

e

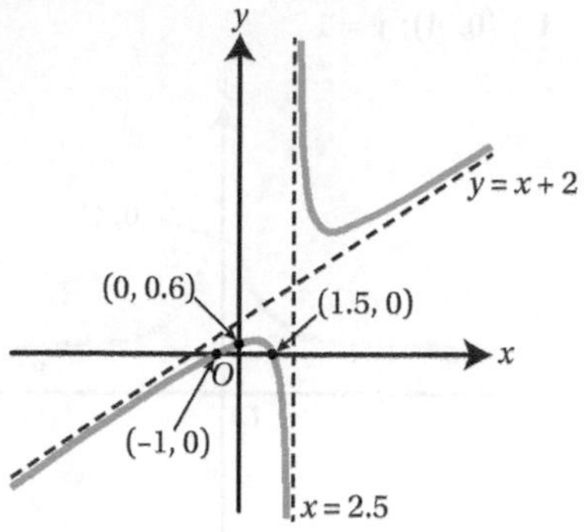

4 **a** $y = x + a$

b Proof.

c

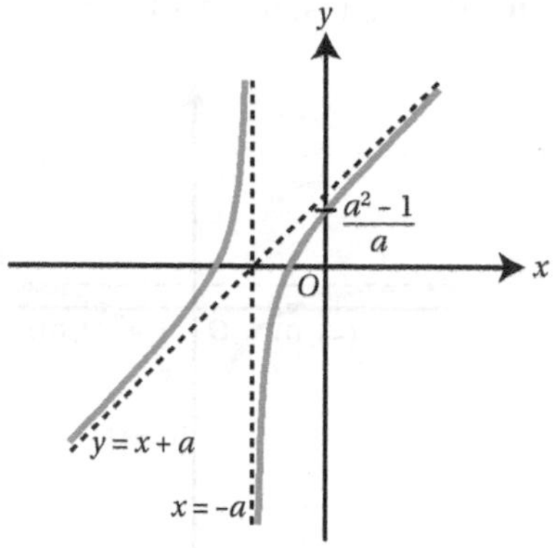

MIXED PRACTICE 4

1 A

2 A

3 $x < 0$, $0 < x < 2$ and $x > 5$

4 **a** (0, −0.25), (−0.5, 0), $\left(-\frac{1}{3}, 0\right)$; $y = 6$, $x = -1$, $x = 4$

b

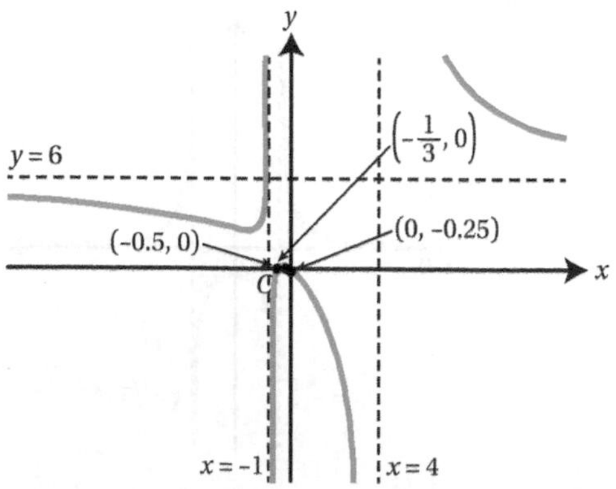

5 **a** **i** $x = 3$, $y = 0$

ii, iii

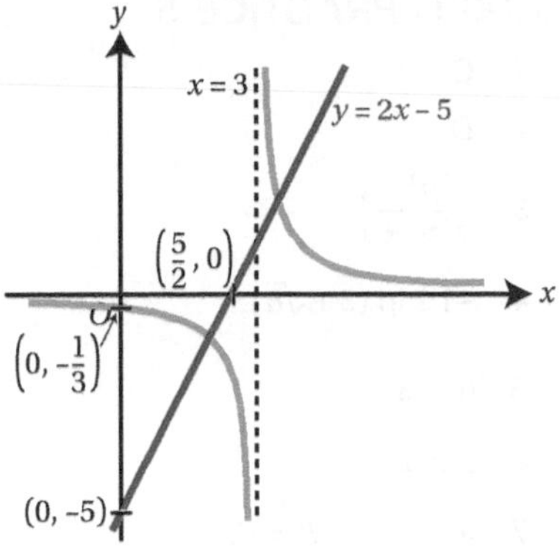

b **i** $x = 2, 3.5$

ii $2 < x < 3$ or $x > 3.5$

6 **a** Proof.

b $-4 < b < 4$

7 **a** $x = \pm a,\ y = 0$

b

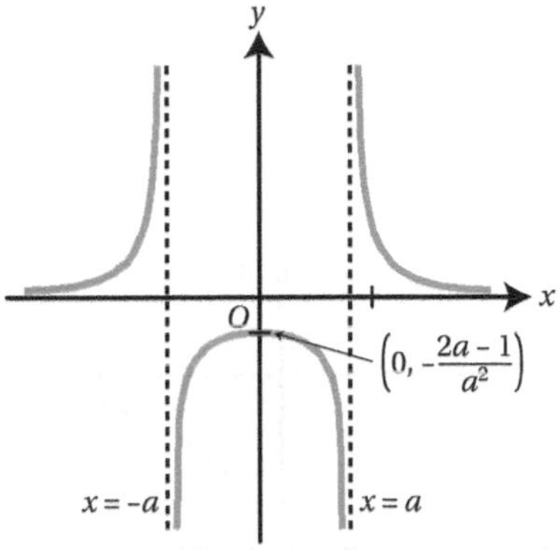

c $-a < x < 1 - a$ or $a - 1 < x < a$

8 **a** $(0, 1), \left(\frac{5 \pm \sqrt{17}}{2}, 0\right); y = 3,\ x = 2,\ x = 3$

b

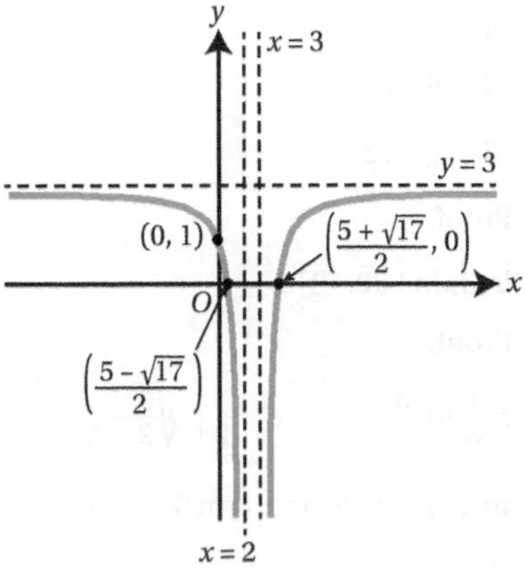

9 $k < 0$ or $k > 0.75,\ k \neq 1$

10 **a** $a = k;\ b = -k^2;\ c = -2k^2;\ d = -1;\ e = k - k^2$

b $k = 3$

11 $-2 < c < 0$

12 **a** $(3, 0)$ **b** $x = 1,\ y = 1$

c $(-3, 1.125);\ k = 1.125$

13 **a** $x = 0,\ x = -2,\ y = 0$

b **i** -1

ii

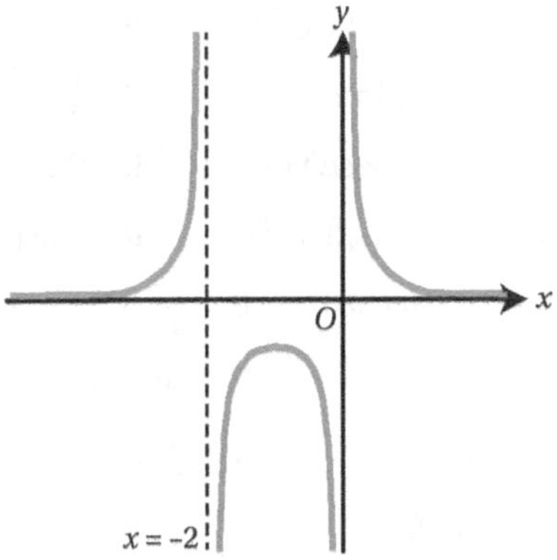

c $x \leqslant -4,\ -2 < x < 0,\ x \geqslant 2$

14 **a** $y = 1$ **b** Proof.

c $(-3, 1.5)$ and $(1, -0.5)$

A **15** $x = \frac{-1 \pm \sqrt{19}}{2}$

5 Hyperbolic functions

BEFORE YOU START

1 $x = \frac{-3 \pm \sqrt{13}}{2}$

2 Vertical stretch factor three and horizontal translation one to the left.

3 $\frac{\ln 5}{\ln 2}$

4 $8 + 12x + 6x^2 + x^3$

5 $-a$

6 3

EXERCISE 5A

1 **a** **i** 1.54 **ii** 10.0

b **i** 0.58 **ii** 101

c **i** 1.60 **ii** 10.3

d **i** 0.481 **ii** 1.32

e **i** Not possible.

ii Not possible.

2 a i -1.44 ii 0.0998

b i ± 1.57 ii ± 2.06

c i 0.973 ii 0.424

d i No solution. ii No solution.

e i 74.2 ii 27.3

3 a i $\ln\left(1+\sqrt{2}\right)$ ii $\ln\left(2+\sqrt{5}\right)$

b i $\ln\left(2+\sqrt{3}\right)$ ii $\ln\left(3+\sqrt{8}\right)$

c i $\frac{1}{2}\ln 2$ ii $\frac{1}{2}\ln\left(\frac{5}{3}\right)$

d i Doesn't exist. ii Doesn't exist.

e i $\ln\left(\sqrt{2}+\sqrt{3}\right)$ ii $\frac{1}{2}\ln\left(2+\sqrt{3}\right)$

4 2.10 or -0.0986

5 $\frac{5}{4}$

6 $\frac{4}{5}$

7 ± 0.515

8 0.549

9 $\sqrt{1+x^2}$

10–12 Proof.

EXERCISE 5B

1 Proof.

2 $\cosh x$

3–8 Proof.

9 a Proof.

b $4\cosh^3 x - 3\cosh x$

10 Proof.

WORK IT OUT 5.1

Solution 3 is correct.

EXERCISE 5C

1 $\ln 5$

2 $\ln\frac{1}{2}$

3 $0.481, 2.06, -2.06$

4 $x=\ln\frac{1}{2}$

5 $\ln 3$

6 $\ln 4$

7 $\ln\left(\frac{\sqrt{5}-1}{2}\right)$ or $\ln\left(3+\sqrt{10}\right)$

8 $\ln(2\pm\sqrt{3})$

9 $\ln(1+\sqrt{2})$

10 $\frac{1}{2}\ln(2+\sqrt{5})$

11 a Proof.

b $x=\ln 2, y=\ln 4$ or $x=\ln 4, y=\ln 2$.

12 $r^4 \geqslant p^4-q^4$

MIXED PRACTICE 5

1 C

2 D

3 $\frac{e^2p^2-1}{e^2p^2+1}$

4 $-1\pm\ln(3+\sqrt{8})$

5 0.128

6 $\pm\ln 2$

7 a $A=3, B=-2$

b $\ln\frac{1}{3}$

8 0 or $\ln\frac{1}{3}$

9 0 or $\frac{1}{3}\ln\left(\sqrt{5}\pm 2\right)$

10 $\ln\left(4\pm\sqrt{15}\right)$

11 Proof.

12 $\frac{x}{\sqrt{1+x^2}}$

13 Proof.

14 a

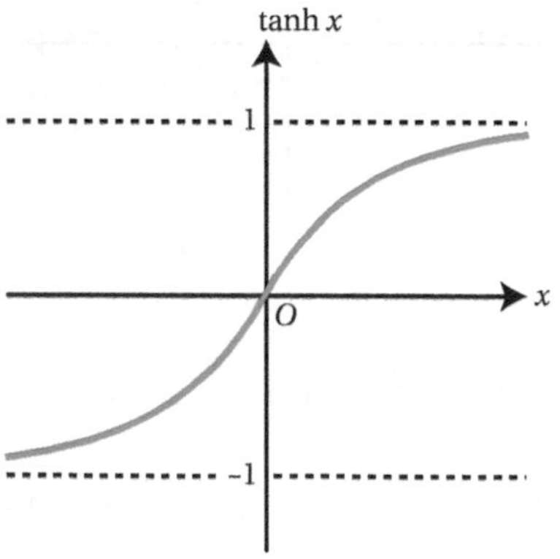

b Proof.

c i Proof.

ii $\frac{1}{2}\ln 2$

15 a Proof.

b $\ln 3$, $\ln(\sqrt{2}-1)$

16 a Proof.

b $\pm\frac{1}{3}\ln\frac{3}{2}$

c $\frac{1}{2}\left(\sqrt[3]{\frac{3}{2}}+\sqrt[3]{\frac{2}{3}}\right)$

17 $x=\ln 2, y=\ln 5$ or $x=\ln 5, y=\ln 2$.

18 Proof.

19 Proof.

20 a Proof.

b $\frac{1}{3}\ln 2$

c $p=-\frac{2}{3}, q=-\frac{4}{3}$

6 Polar coordinates

BEFORE YOU START

1 **a** $210°$ **b** $-\frac{\sqrt{3}}{2}$

2 **a** $\theta \in \left(\frac{\pi}{2}, \frac{3\pi}{2}\right)$ **b** 7

EXERCISE 6A

1 **a–d**

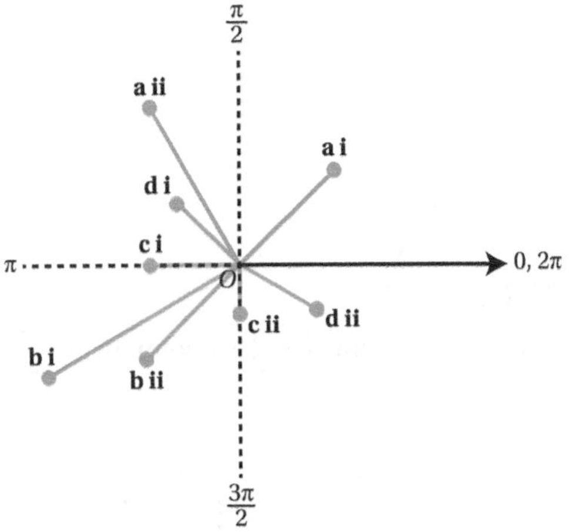

2 **a** **i** Distance $AB = 2.53$; area $AOB = 4.53$

ii Distance $AB = 3.42$; area $AOB = 2.5$

b **i** Distance $AB = 17.8$; area $AOB = 10.4$

ii Distance $AB = 8.92$; area $AOB = 2.59$

c **i** Distance $AB = 2.91$; area $AOB = 0.5$

ii Distance $AB = 7.21$; area $AOB = 12$

3 **a** **i**

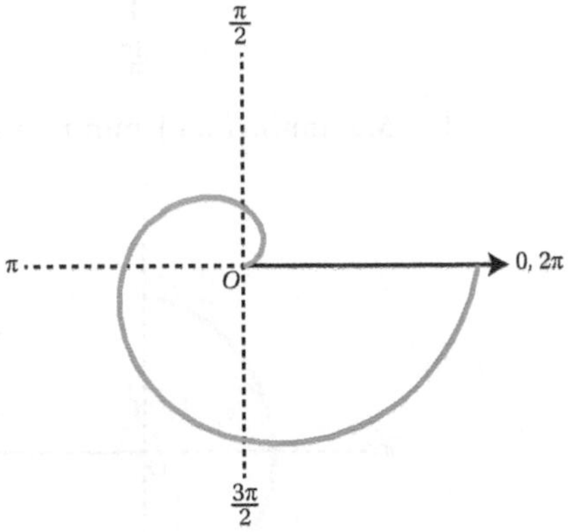

ii

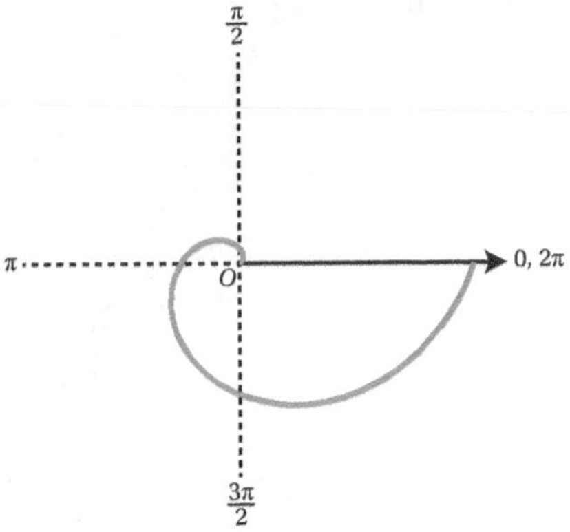

b **i**

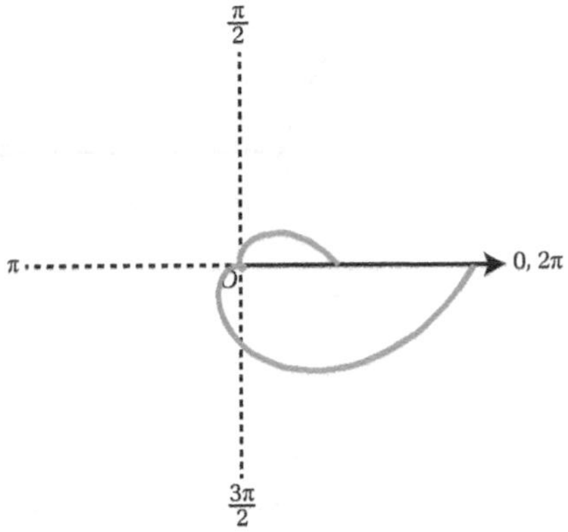

ii

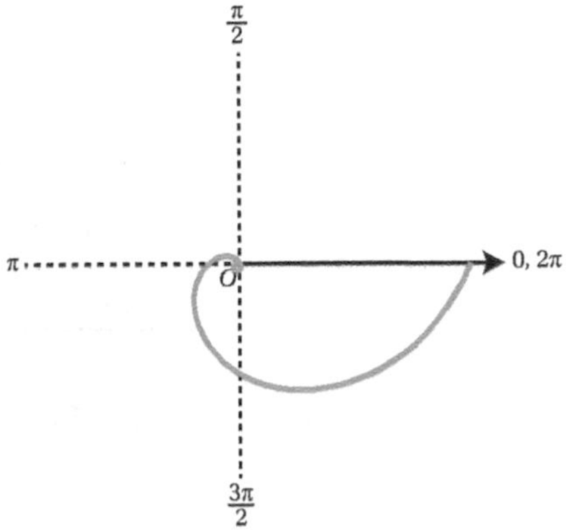

c **i**

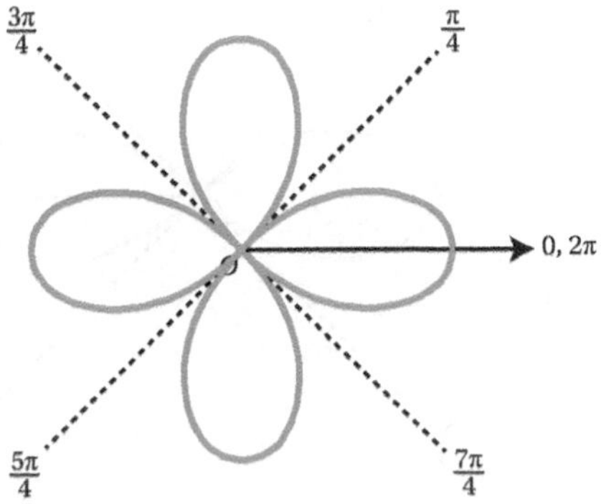

ii

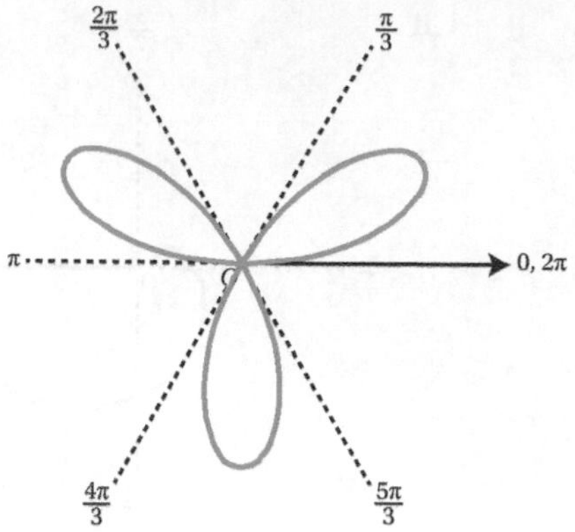

4 **a**

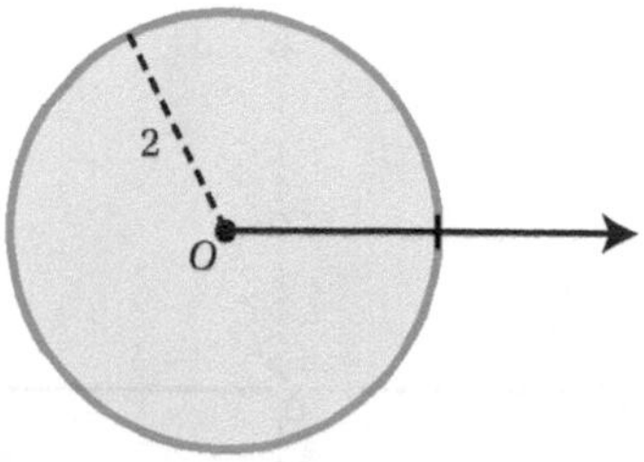

b

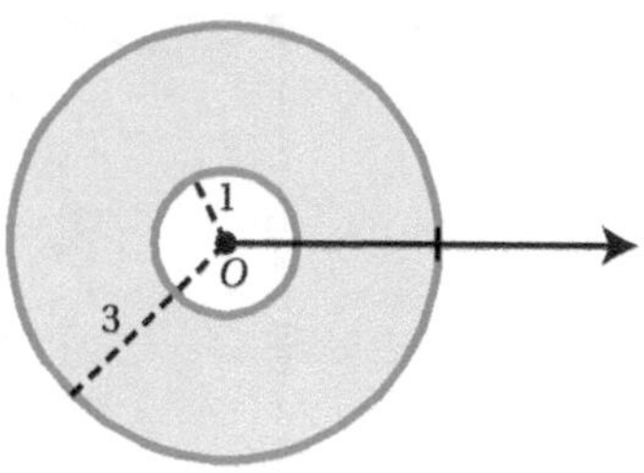

c

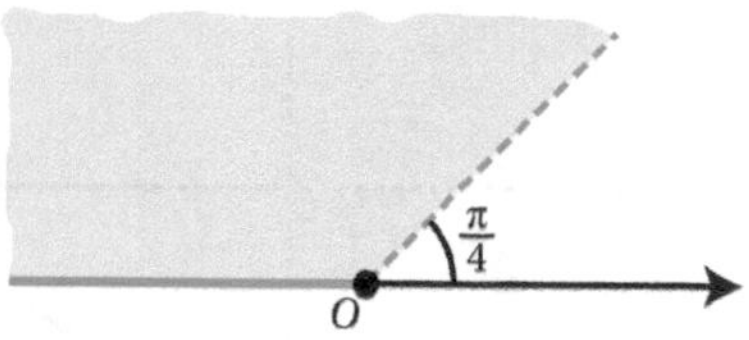

5 **a** $\frac{\pi}{4} < \theta < \frac{3\pi}{4}$ and $\frac{5\pi}{4} < \theta < \frac{7\pi}{4}$

b

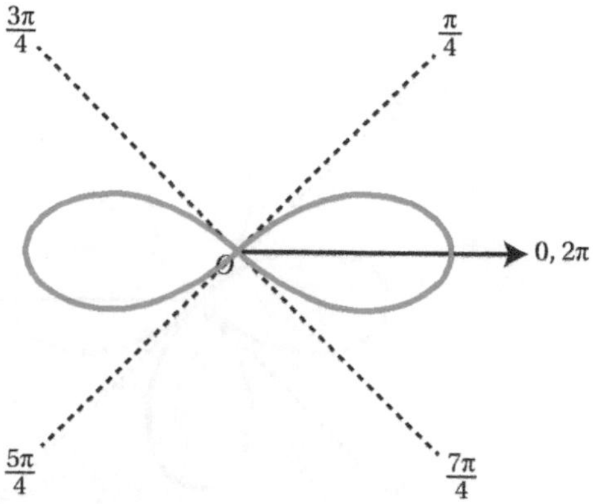

6 **a** Proof.

b

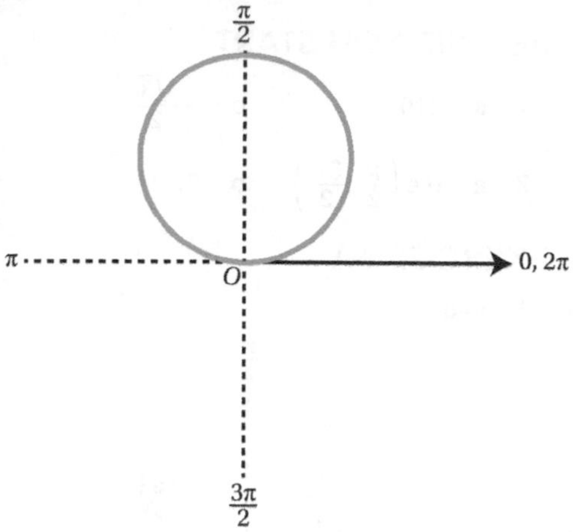

c $2\sqrt{3}$

EXERCISE 6B

1 **a** **i** Maximum $\left(5, \frac{\pi}{2}\right)$; minimum $\left(1, \frac{3\pi}{2}\right)$

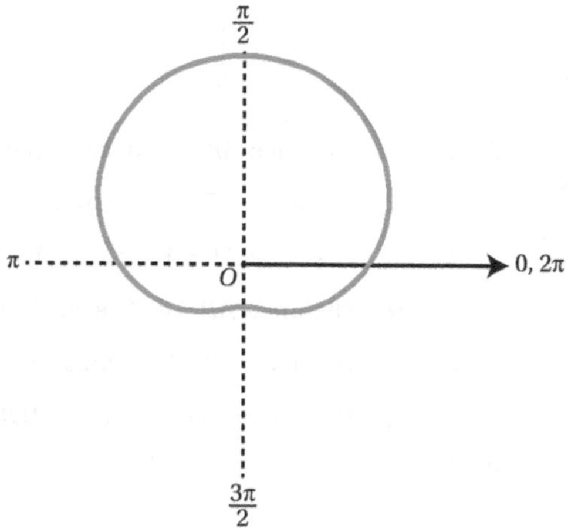

ii Maximum (6, 0); minimum (4, π)

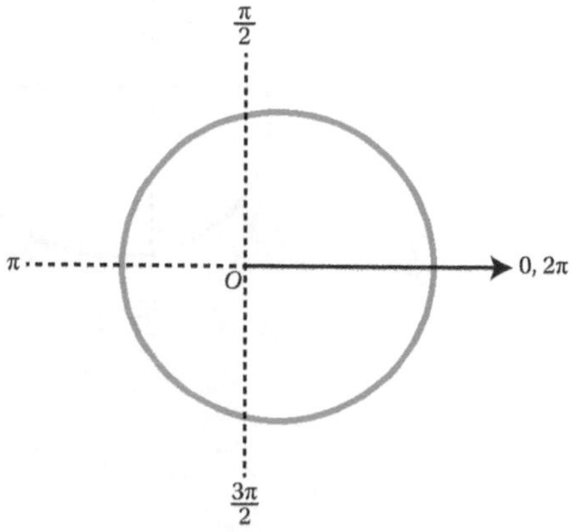

b **i** Maxima $\left(10, \frac{\pi}{2}\right)$ and $\left(10, \frac{3\pi}{2}\right)$; minima $(4, 0)$ and $(4, \pi)$

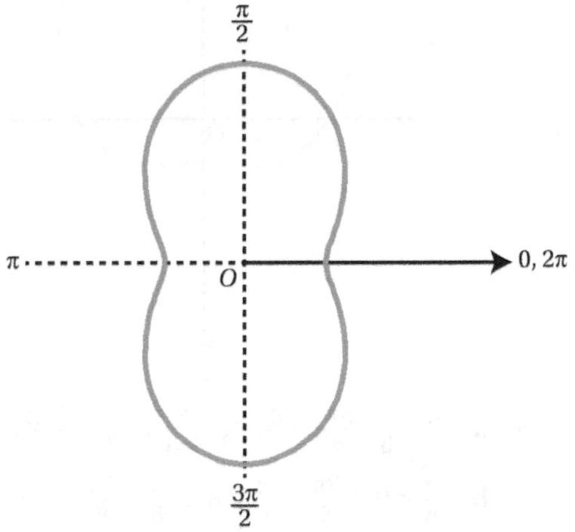

ii Maxima $\left(7, \frac{3\pi}{4}\right)$ and $\left(7, \frac{7\pi}{4}\right)$; minima $\left(3, \frac{\pi}{4}\right)$ and $\left(3, \frac{5\pi}{4}\right)$

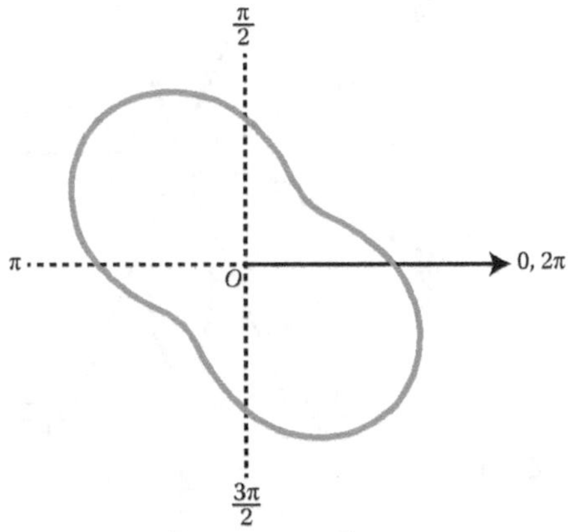

2 **a** **i** $\theta = 0, \frac{\pi}{3}, \frac{2\pi}{3}, \pi, \frac{4\pi}{3}, \frac{5\pi}{3}$

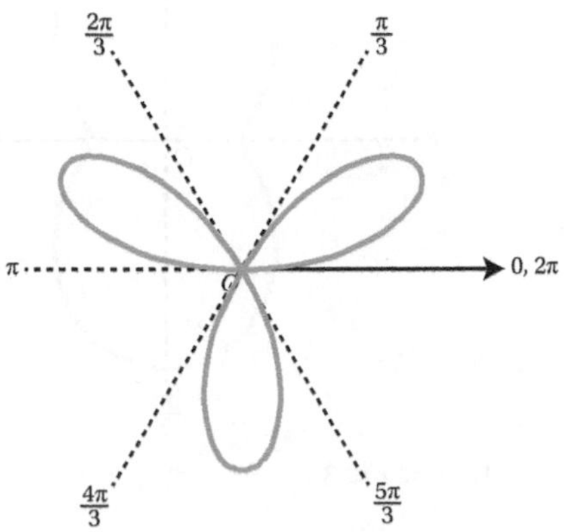

ii $\theta = \frac{\pi}{4}, \frac{3\pi}{4}, \frac{5\pi}{4}, \frac{7\pi}{4}$

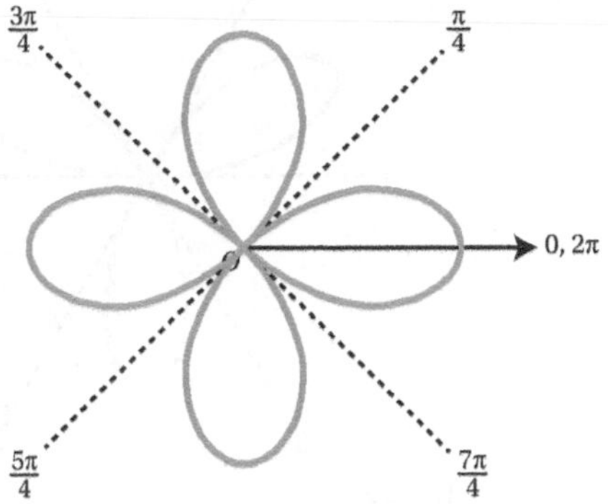

b **i** $\theta = \frac{2\pi}{9}, \frac{4\pi}{9}, \frac{8\pi}{9}, \frac{10\pi}{9}, \frac{14\pi}{9}, \frac{16\pi}{9}$

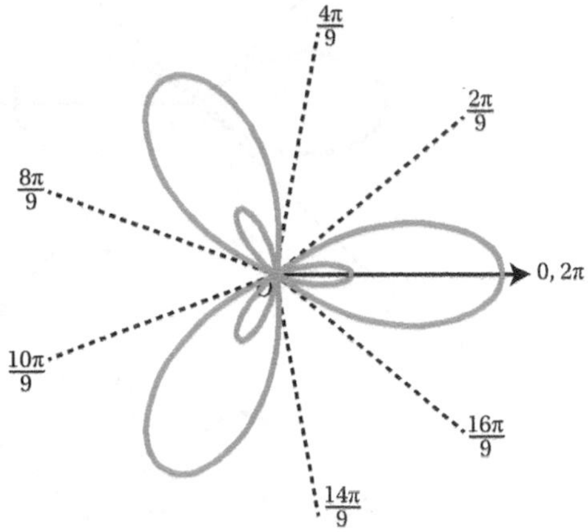

ii $\theta = \frac{\pi}{16}, \frac{3\pi}{16}, \frac{9\pi}{16}, \frac{11\pi}{16}, \frac{17\pi}{16}, \frac{19\pi}{16}, \frac{25\pi}{16}, \frac{27\pi}{16}$

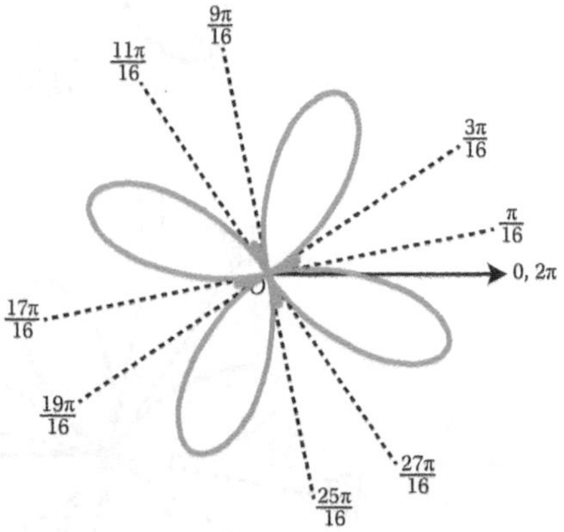

3 a i

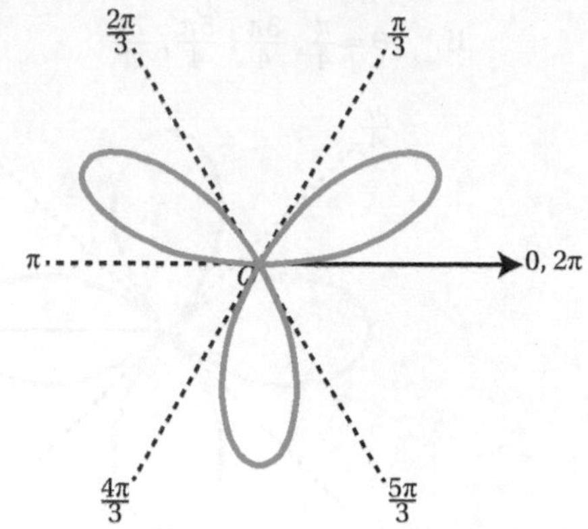

ii

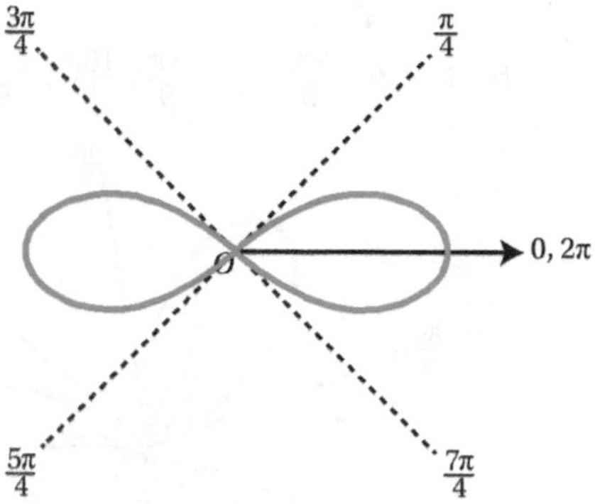

b i

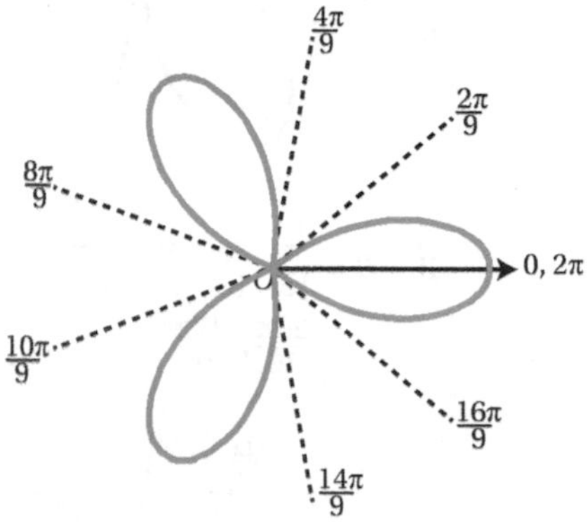

ii

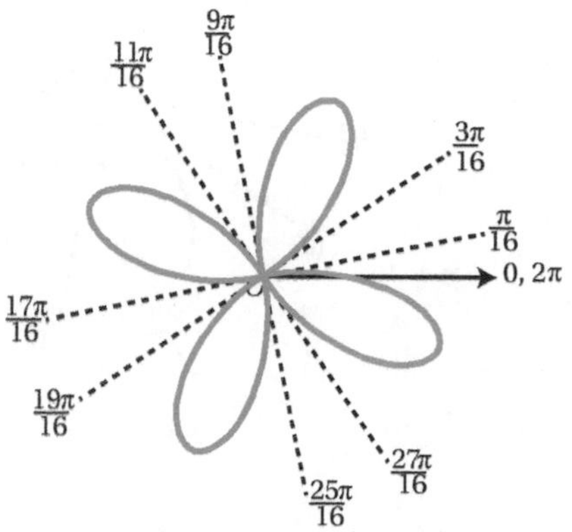

4 a Maximum $\left(5, \frac{\pi}{2}\right)$; minimum $\left(1, \frac{3\pi}{2}\right)$

b

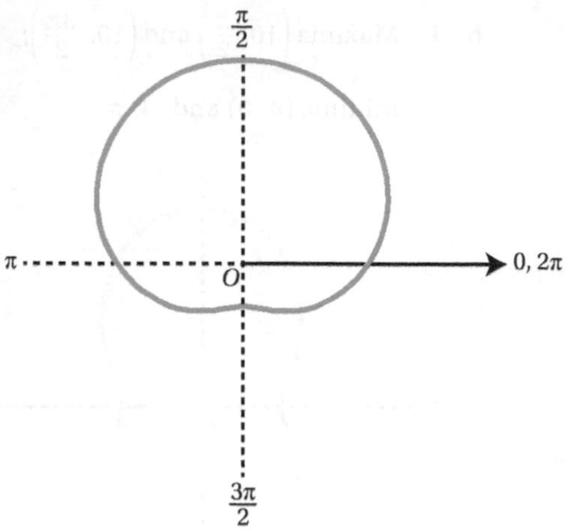

5 a $\theta = \frac{\pi}{8}, \frac{3\pi}{8}, \frac{5\pi}{8}, \frac{7\pi}{8}, \frac{9\pi}{8}, \frac{11\pi}{8}, \frac{13\pi}{8}, \frac{15\pi}{8}$

b $\frac{\pi}{8} < \theta < \frac{3\pi}{8}$, $\frac{5\pi}{8} < \theta < \frac{7\pi}{8}$, $\frac{9\pi}{8} < \theta < \frac{11\pi}{8}$ and $\frac{13\pi}{8} < \theta < \frac{15\pi}{8}$

c

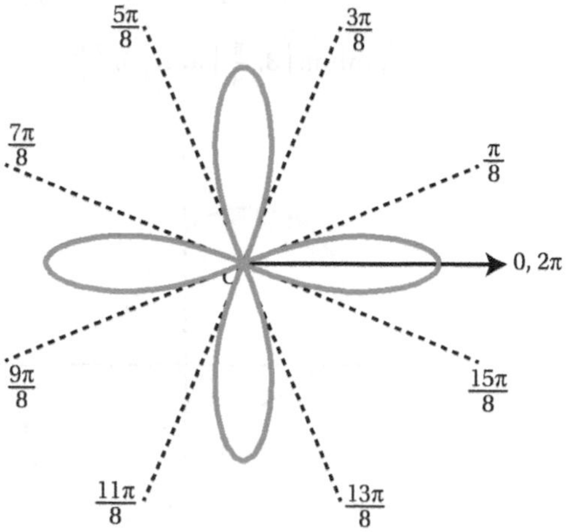

6 a

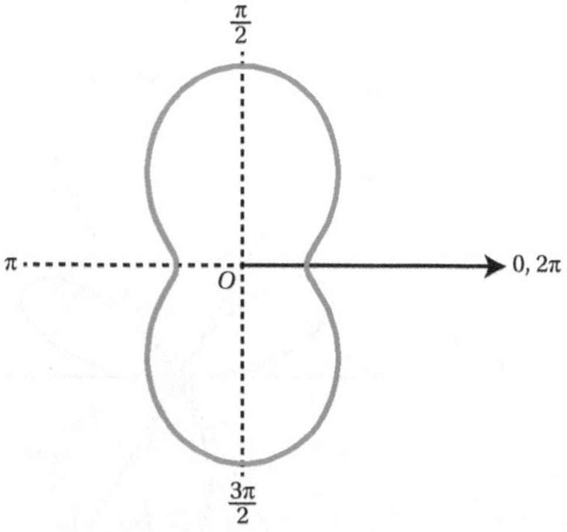

b $1 \leqslant r \leqslant 3$

7 **a** $\theta = \frac{7\pi}{12}, \frac{11\pi}{12}, \frac{19\pi}{12}, \frac{23\pi}{12}$

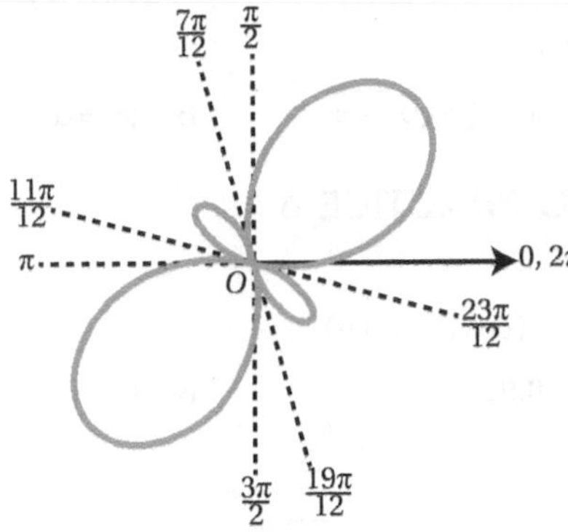

b Maximum $|r|$ at $\left(3, \frac{\pi}{4}\right)$ and $\left(3, \frac{5\pi}{4}\right)$

8 **a** Largest $r = 3$ when $\theta = \frac{\pi}{2}$; smallest $r = 1$ when $\theta = \frac{3\pi}{2}$.

b

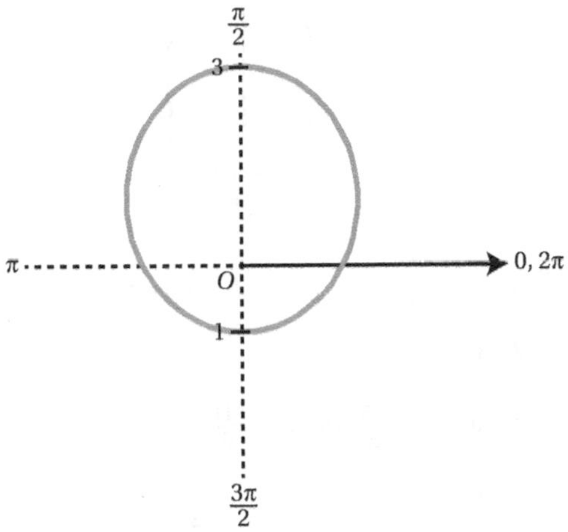

9 **a**

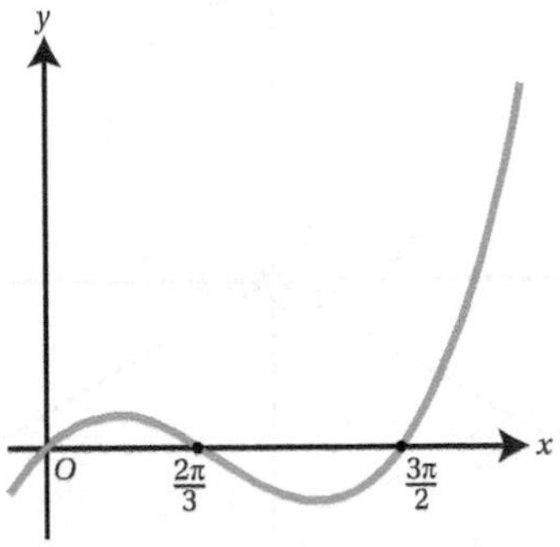

b

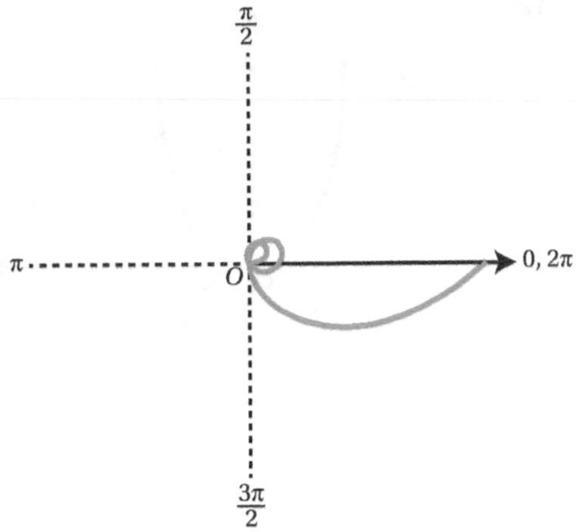

10 **a** Maximum y is $4\pi^2$; minimum y is π^2

b

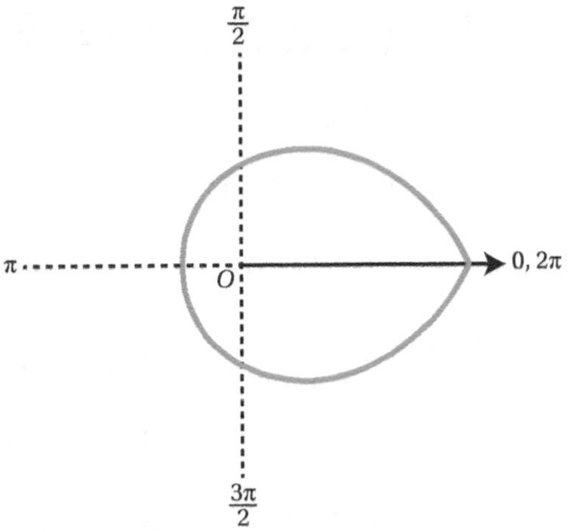

11

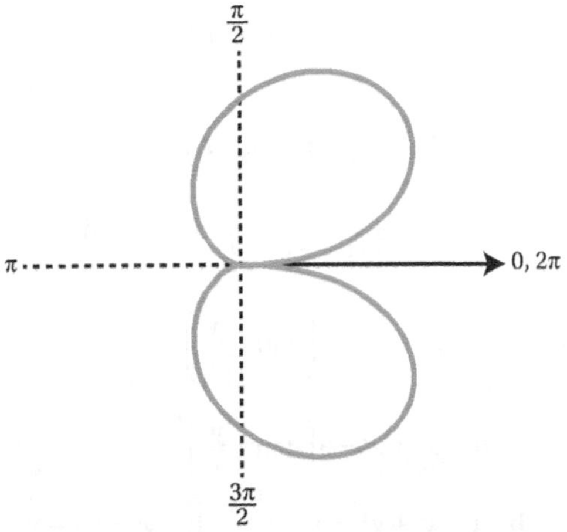

12

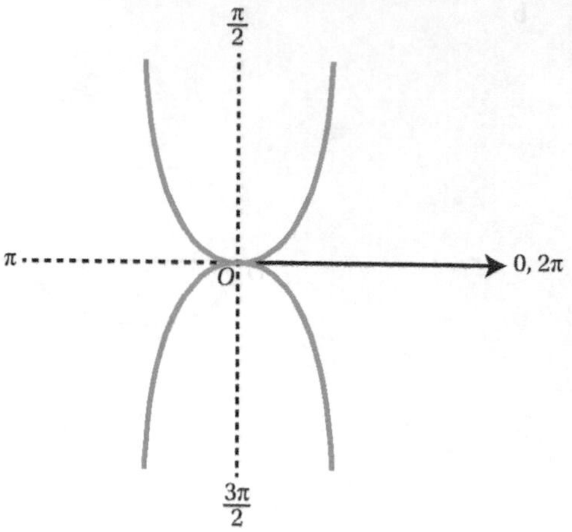

EXERCISE 6C

1 a i $\left(\frac{5}{\sqrt{2}}, \frac{5}{\sqrt{2}}\right)$ ii $\left(\frac{3}{2}, \frac{3\sqrt{3}}{2}\right)$

b i $(-1, 1)$ ii $\left(-\frac{3}{2}, \frac{\sqrt{3}}{2}\right)$

c i $(-3.45, -4.91)$ ii $(2.50, -1.65)$

2 a i $(\sqrt{29}, 0.381)$ ii $(5, 0.927)$

b i $\left(2, \frac{\pi}{2}\right)$ ii $\left(3, \frac{3\pi}{2}\right)$

c i $(\sqrt{26}, 4.51)$ ii $(\sqrt{17}, 6.04)$

3 a i $\sin 2\theta = \frac{2}{3}$ ii $r = 2\cos^2\theta \sin\theta$

b i $r = 5(\sec\theta + \operatorname{cosec}\theta)$

ii $r = \left(\frac{3}{\sin^3\theta + \cos^3\theta}\right)^{\frac{1}{3}}$

c i $r = \frac{1}{\sin\theta - 3\cos\theta}$ ii $r = \sqrt{6}$

4 a i $x^2 + y^2 = 4\left(\tan^{-1}\frac{y}{x}\right)^2$

ii $x^2 + y^2 = 9\left(\tan^{-1}\frac{y}{x}\right)^4$

b i $x^2 + y^2 = 4y$ ii $y^2 = 2x - x^2$

c i $x^4 + x^2y^2 - 4y^2 = 0$ ii $x^2 + y^2 = \frac{y}{x}$

5 a $\alpha = 0.841$ b $\left(\frac{4}{3}, \frac{2\sqrt{5}}{3}\right)$

c $x^2 + y^2 = 3x$

6 $r = 2(\cos\theta + \sin\theta)$

7 $x^4 + x^2y^2 - 9y^2 = 0$

8 Proof.

9 a $y = \sqrt{x\sqrt{x} - x^2}$ b $y^2 = x^3 - x^2$

MIXED PRACTICE 6

1 D

2 $r = a(\cos\theta - \sin\theta)$

3 a 8.92 b 13.5

4

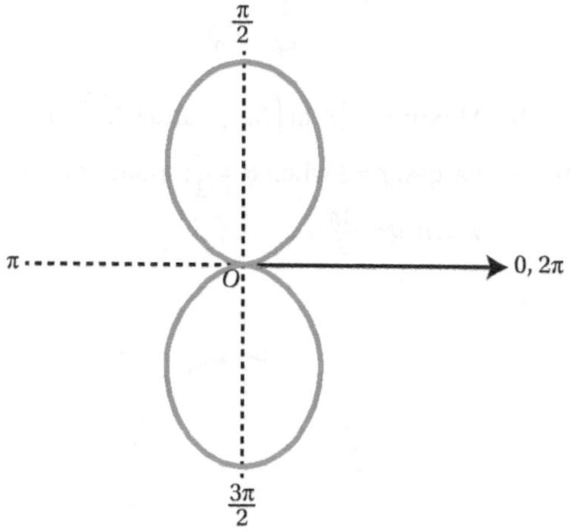

5 $y = \frac{x^2}{8} - 2$

6 $y^2 = 1 + \frac{24x - 7x^2}{16}$

7 C

8 a $\theta = \frac{7\pi}{6}$ and $\theta = \frac{11\pi}{6}$

b

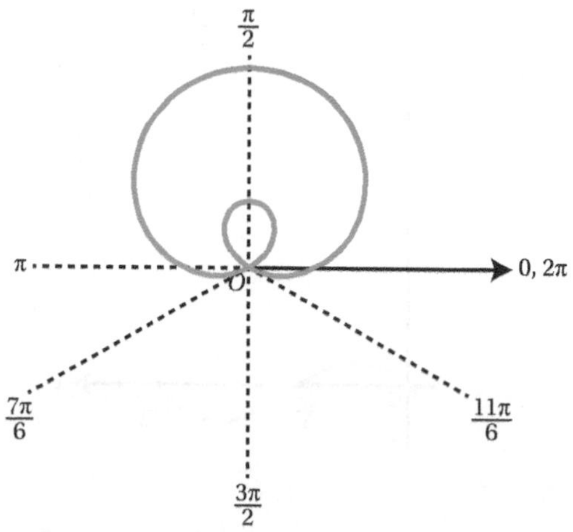

c $(x^2 + y^2 - 4y)^2 = 4(x^2 + y^2)$

9 **a**

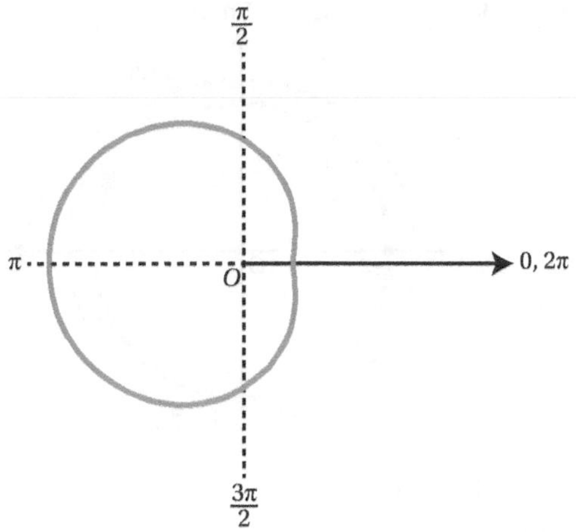

b $(-8, 0)$

10 **a** $\theta = \frac{\pi}{12}, \frac{5\pi}{12}, \frac{13\pi}{12}, \frac{17\pi}{12}$

b $\left(6, \frac{3\pi}{4}\right)$ and $\left(6, \frac{7\pi}{4}\right)$

c

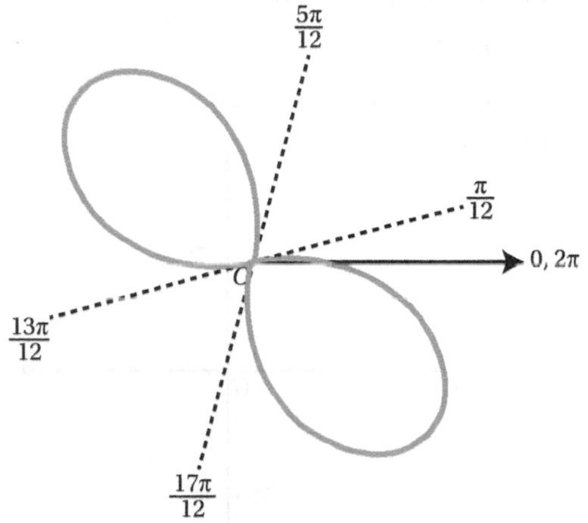

11 **a**

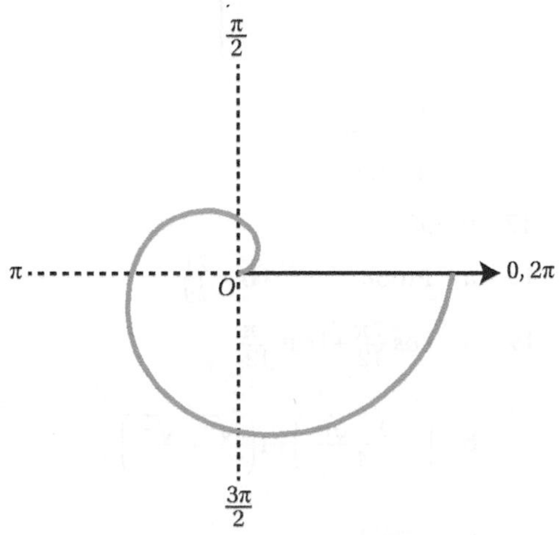

b $(-\pi, \pi\sqrt{3})$

12

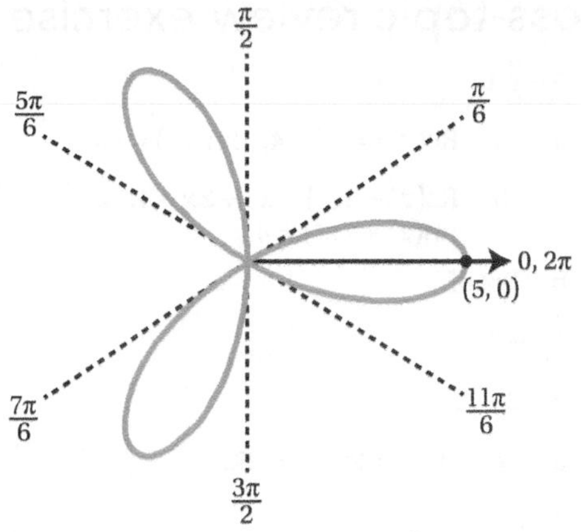

13 **a** Proof. **b** $r = \frac{1}{1+\sin\theta}$

14 $2\sqrt{3}-1$

15 **a** Proof.

b **i** $M\left(1, \frac{2\pi}{3}\right); N\left(1, \frac{4\pi}{3}\right)$

ii $\frac{3\sqrt{3}}{4}$

c $9y^2 = 4 - 8x - 5x^2$

Focus on ... Proof 1

1–4 Proof.

Focus on ... Problem solving 1

1 $x = 2, -1+3\text{i}, -1-3\text{i}$

2 $u = \text{i}, \frac{\sqrt{3}}{2} - \frac{1}{2}\text{i}, -\frac{\sqrt{3}}{2} - \frac{1}{2}\text{i}$

3 Discussion.

Focus on ... Modelling 1

1 2.64×10^{-13}

2 **a** $\frac{V_{in}}{R}$

b The resonant current would grow without limit.

c Even the wires will have some resistance.

3 The width of the resonant peak is smaller, so there will be less interference from stations with slightly different frequencies.

4 **a** $\arctan\left(\frac{\omega L - \frac{1}{\omega C}}{R}\right)$ **b** 0

Cross-topic review exercise 1

1 $5-\frac{2}{5}\mathrm{i}$

2 **a** **i** $\mathrm{Re}(z^2)=x^2-4$; $\mathrm{Im}(z^2)=4x$

ii $\mathrm{Re}(z^2+2z^*)=x^2+2x-4$;
$\mathrm{Im}(z^2+2z^*)=4x-4$

b Proof.

3 $\frac{k^2}{4}-5$

4 **a** $2-3\mathrm{i}$

b $a=-8$, $b=29$, $c=-52$

5 $\sqrt[3]{10}-2$

6 **a** $y^2=\frac{4ax}{3}$ **b** $k=\frac{1}{\sqrt{3}}$

7 **a** $x=-2$; $y=3$

b

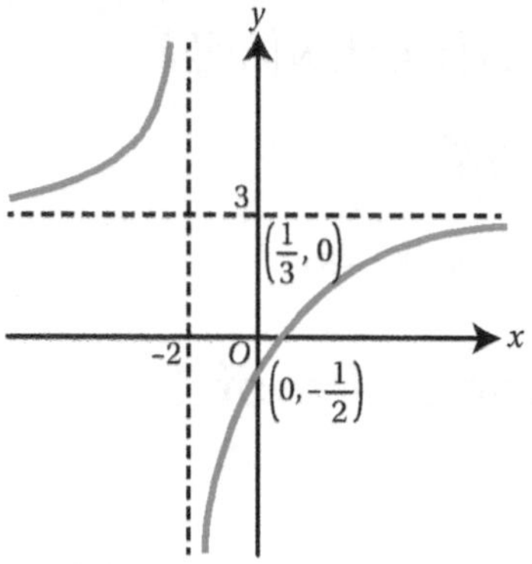

c $x>\frac{1}{3}$

8 $\ln\sqrt{5}$

9 $\ln 3$ or $\ln(\sqrt{5}-2)$

10 $r=14\sin\theta-8\cos\theta$

11 **a** $\sqrt{x^2+(y-1)^2}$

b

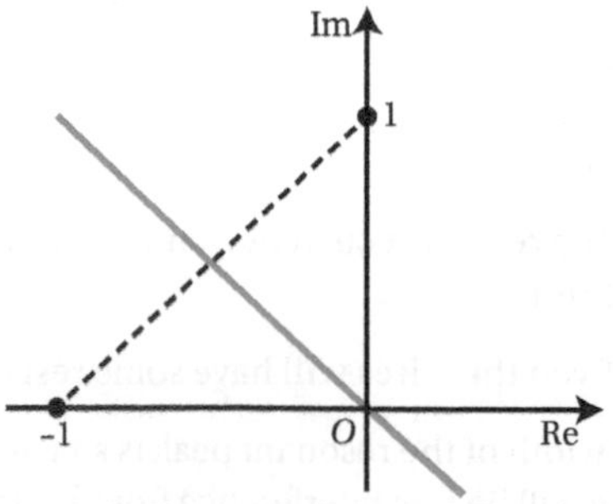

12 **a** $z=\pm\left(\sqrt{\frac{\sqrt{2}-1}{2}}+\mathrm{i}\sqrt{\frac{\sqrt{2}+1}{2}}\right)$

b $w=-\mathrm{i}\pm\left(\sqrt{\frac{\sqrt{2}-1}{2}}+\mathrm{i}\sqrt{\frac{\sqrt{2}+1}{2}}\right)$

13 **a** -2 **b** $-2, 1\pm 2\mathrm{i}$

c

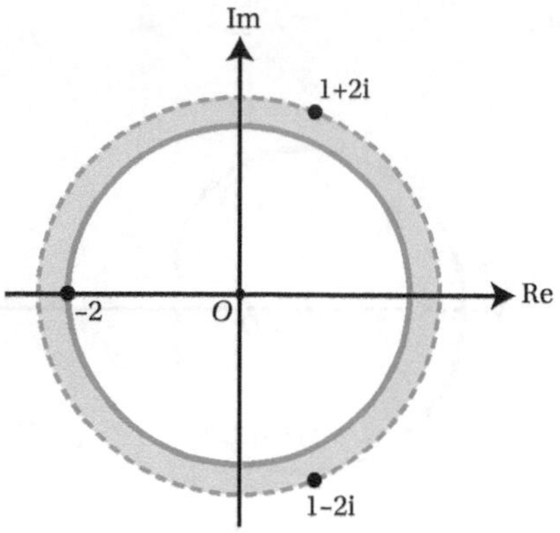

d 2.21

14 **a** Proof.

b **i** -5

ii Proof.

iii $2+\mathrm{i}$ and $2-\mathrm{i}$

15 **a** $\alpha+\beta=-\frac{3}{2}$; $\alpha\beta=-3$

b Proof.

c $24x^2+81x-146=0$

16 **a**

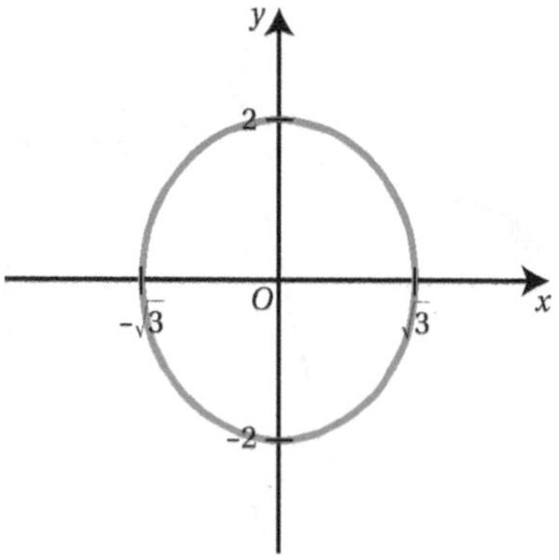

b $\frac{x^2}{3}+\frac{y^2}{16}=1$

c $a=1$, $b=-1$

17 Proof.

18 **a** Proof. **b** $\frac{21}{29}$

19 **a** $\cos\frac{7\pi}{12}+\mathrm{i}\sin\frac{7\pi}{12}$

b $\left(\frac{\sqrt{2}-\sqrt{6}}{4}\right)+\mathrm{i}\left(\frac{\sqrt{2}+\sqrt{6}}{4}\right)$

c $2-\sqrt{3}$

20 **a** $|z+1+\mathrm{i}|=5$

b

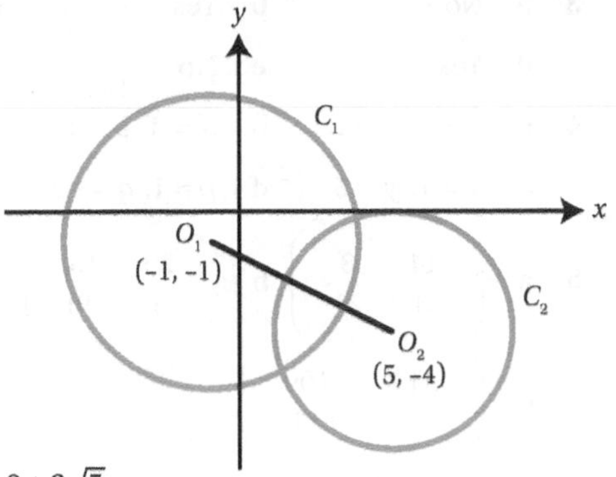

c $9+3\sqrt{5}$

21 a i (8, 8)

ii $y^2 = 8x$

iii Proof.

b i Proof.

ii $(-4, 2)$; $(2, -4)$

22 a $y=1$; $x=-1$; $x=3$

b i–ii Proof.

iii Proof; (1, 0).

c

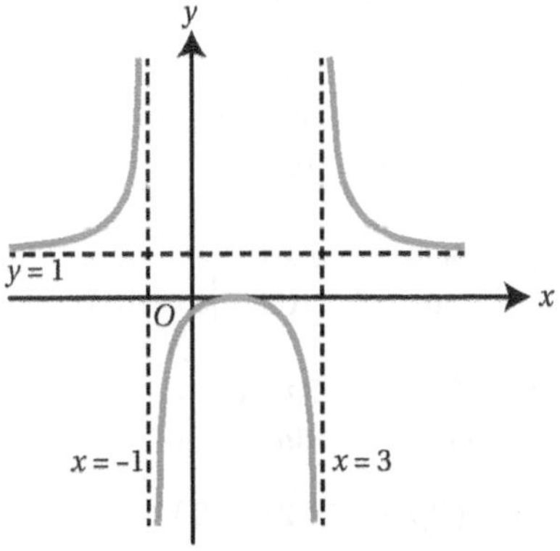

23 $k \geqslant \sqrt{5}$

24 a $P=\left(2, \frac{7\pi}{6}\right)$; $Q=\left(2, \frac{11\pi}{6}\right)$

b i $A=\left(4, \frac{\pi}{6}\right)$ ii $AQ=\sqrt{12}=2\sqrt{3}$

iii Proof.

7 Matrices

BEFORE YOU START

1 $\begin{pmatrix} 0 \\ 11 \end{pmatrix}$

2 $x=1$, $y=-1$

EXERCISE 7A

1 a i 2×2 ii 3×2 iii 2×3

iv 4×3

b i $\begin{pmatrix} 1 & 1 \\ 2 & 3 \end{pmatrix}$ ii $\begin{pmatrix} 1 & 2 & 1 \\ 5 & 3 & -3 \end{pmatrix}$

iii $\begin{pmatrix} 2 & 4 \\ 6 & 1 \\ 0 & 0 \end{pmatrix}$

iv $\begin{pmatrix} 1 & 8 & -1 & -5 \\ 2 & -3 & 7 & -2 \\ -4 & 3 & 22 & 0 \end{pmatrix}$

2 a i $\begin{pmatrix} 5 & 4 \\ 7 & -5 \end{pmatrix}$ ii $\begin{pmatrix} 5 & 4 \\ 3 & -6 \end{pmatrix}$

iii $\begin{pmatrix} -3 & 5 \\ 4 & 4 \\ 4 & -2 \end{pmatrix}$

b i $\begin{pmatrix} 1 & -1 \\ 6 & -6 \end{pmatrix}$ ii Not possible.

iii $\begin{pmatrix} -2 & -1 \\ 1 & 5 \end{pmatrix}$

c i $\begin{pmatrix} 2 & 4 \\ 2 & 6 \end{pmatrix}$ ii $\begin{pmatrix} -9 & -9 \\ 0 & 6 \end{pmatrix}$

iii $\begin{pmatrix} 4 & 20 \\ 8 & 12 \\ 4 & -12 \end{pmatrix}$

d i $\begin{pmatrix} -7 & -5 \\ 2 & 12 \end{pmatrix}$ ii $\begin{pmatrix} 0 & 0 \\ 0 & 0 \end{pmatrix}$

iii $\begin{pmatrix} 7 & 15 \\ 4 & 8 \\ 0 & -10 \end{pmatrix}$

3 a $\begin{pmatrix} 1 & 4 \\ 5 & 6 \end{pmatrix}$ b Not possible.

c Not possible. d $\begin{pmatrix} 5 & 0 \\ 1 & 7 \end{pmatrix}$

e Not possible. f $\begin{pmatrix} 10 & -2 & 4 \\ 0 & 2 & 1 \\ 8 & 1 & 7 \end{pmatrix}$

4 a $\begin{pmatrix} 3 & -1 \\ 7 & -2 \end{pmatrix}$ b $\begin{pmatrix} 1 & -3 \\ 9 & 4 \end{pmatrix}$

c $\begin{pmatrix} 10 & -3 \\ -2 & 19 \end{pmatrix}$ d $\begin{pmatrix} 6 & -3 \\ 1 & 14 \end{pmatrix}$

5 **a** $x = 3, y = 8$ **b** $x = 2, y = 5$

c $x = 3, y = -1$ **d** No solution.

6 **a** $\begin{pmatrix} -5 & -3 \\ -5 & 1 \end{pmatrix}$ **b** $\begin{pmatrix} \frac{19}{3} & -3 \\ \frac{13}{3} & \frac{19}{3} \end{pmatrix}$

c $\begin{pmatrix} 23 & -5 \\ 15 & 18 \end{pmatrix}$ **d** $\begin{pmatrix} -\frac{1}{3} & \frac{5}{3} \\ -\frac{1}{3} & -\frac{5}{3} \end{pmatrix}$

7 $a = 3, b = -\frac{3}{2}, s = 2, t = 7$

8 $(x, y) = (-1, -4)$ or $(-3, 0)$

9 $a = -\frac{2}{7}, b = -\frac{1}{7}, c = \frac{3}{11}, d = \frac{2}{11}$

10 Switching rows for columns before or after addition makes no difference; the same elements are added and the sum ends in the same position either way.

EXERCISE 7B

1 **a** Yes; 2×2 **b** No **c** No

d Yes; 1×2 **e** Yes; 2×3 **f** Yes; 1×1

g Yes; 2×2 **h** Yes; 2×2 **i** Yes; 3×3

j Yes; 3×3 **k** Yes; 3×3

2 **a** $\begin{pmatrix} 12 & 6 \\ 24 & 13 \end{pmatrix}$ **b** Not possible.

c Not possible. **d** $(0\ 0)$

e $\begin{pmatrix} 3 & -1 & 1 \\ 2 & 1 & 3 \end{pmatrix}$ **f** (10)

g $\begin{pmatrix} 8 & -5 \\ 6 & -3 \end{pmatrix}$

h $\begin{pmatrix} 0 & 4a \\ 2-2a & 4+4a \end{pmatrix}$

i $\begin{pmatrix} 7 & 2 & 9 \\ 8 & -2 & 11 \\ 3 & 0 & 21 \end{pmatrix}$

j $\begin{pmatrix} 2-a & 1+a^2 & 1+5a \\ 2+2a & 2+2a & 12+a^2 \\ 12 & 1 & 1+5a \end{pmatrix}$

k $\begin{pmatrix} -2 & 2 & 6 \\ -1 & 1 & 3 \\ -3 & 3 & 9 \end{pmatrix}$

3 **a** No **b** Yes **c** Yes

d Yes **e** No **f** No

4 **a** $x = 3, y = 2$ **b** $a = 4, b = 1$

c $x = 1, y = 5$ **d** $p = 1, q = -2$

5 **a** $\begin{pmatrix} 14 & 8 \\ 3 & 21 \end{pmatrix}$ **b** $\begin{pmatrix} 20 & 15 \\ 5 & 30 \end{pmatrix}$

c $\begin{pmatrix} 343 & -109 \\ 0 & 125 \end{pmatrix}$

6 **a** $(1\ \ 4)$ **b** (-13)

c $\begin{pmatrix} -1 & 7 \\ -5 & 35 \end{pmatrix}$

7 **a** $\begin{pmatrix} 4 & -3 & 15 \\ 0 & 1 & 12 \\ -12 & 0 & 4 \end{pmatrix}$

b $\begin{pmatrix} -1 & 13 & -7 \\ -2 & 14 & 5 \\ -14 & 6 & 4 \end{pmatrix}$

c $\begin{pmatrix} 10 & -3 & -3 \\ 5 & 19 & 8 \\ 3 & 6 & 16 \end{pmatrix}$

8 **a** $\begin{pmatrix} -6a+2 & a+6 \\ 0 & -10 \end{pmatrix}$

b $\begin{pmatrix} a^2+4a & 3a-4 \\ 6a^2-a & -6a-3 \end{pmatrix}$

9 **a** **i** $(2p-2\quad 2p+10)$

ii $\begin{pmatrix} 3p^2-4 \\ -p \\ -3p-4 \end{pmatrix}$

b Dimension mismatch.

10 $b = -5$

11 Proof.

12 $d = \frac{3}{2}c$

13 **a** $\mathbf{X} = \mathbf{AB} = \begin{pmatrix} a_{11}b_{11}+a_{12}b_{21} & a_{11}b_{12}+a_{12}b_{22} \\ a_{21}b_{11}+a_{22}b_{21} & a_{21}b_{12}+a_{22}b_{22} \end{pmatrix}$

$\mathbf{Y} = \mathbf{BC} = \begin{pmatrix} b_{11}c_{11}+b_{12}c_{21} & b_{11}c_{12}+b_{12}c_{22} \\ b_{21}c_{11}+b_{22}c_{21} & b_{21}c_{12}+b_{22}c_{22} \end{pmatrix}$

b Proof.

14 Proof; the result always applies for matrices **A** and **B** where the product **AB** exists.

EXERCISE 7C

1 a i 19 ii 5

b i 13 ii 11

c i $-5a$ ii $7a$

d i $5a^2+6$ ii $2a^2+3$

2 13, 11, 143; Proof.

3 $a=-1.5$ or 4

4 $x=0, 3$ or -4

EXERCISE 7D

1 a i $\frac{1}{19}\begin{pmatrix}4 & 1\\ -7 & 3\end{pmatrix}$ ii $\frac{1}{5}\begin{pmatrix}2 & -1\\ 3 & 1\end{pmatrix}$

b i $\frac{1}{13}\begin{pmatrix}5 & 3\\ -1 & 2\end{pmatrix}$ ii $\frac{1}{11}\begin{pmatrix}3 & 2\\ -1 & 3\end{pmatrix}$

c i $-\frac{1}{5a}\begin{pmatrix}-1 & -a\\ -3 & 2a\end{pmatrix}, a\neq 0$

ii $\frac{1}{7a}\begin{pmatrix}2a & a\\ -5 & 1\end{pmatrix}, a\neq 0$

d i $\frac{1}{5a^2+6}\begin{pmatrix}5a & 3\\ -2 & a\end{pmatrix}$

ii $\frac{1}{2a^2+3}\begin{pmatrix}3 & -2a\\ a & 1\end{pmatrix}$

2 a i -1.5 ii 0.2

b i 0 ii 0

c i ± 0.5 ii ± 2

d i 0, -2 ii 0, 0.2

3 a $\frac{1}{4}\begin{pmatrix}1 & 4\\ -5 & 0\end{pmatrix}$ b $\frac{1}{4}\begin{pmatrix}2 & -2\\ 11 & -1\end{pmatrix}$

c $\frac{1}{5}\begin{pmatrix}-8 & 2\\ 7 & 7\end{pmatrix}$ d $\begin{pmatrix}-1 & 0\\ -1 & 1\end{pmatrix}$

e $\begin{pmatrix}1.25\\ -0.25\end{pmatrix}$ f $\begin{pmatrix}2.2 & -0.4\end{pmatrix}$

4 $\begin{pmatrix}-0.6 & 0.4\\ -1.4 & 3.6\end{pmatrix}$

5 a $\det \mathbf{A}=-25$ b $\begin{pmatrix}-0.6 & -0.12\\ 0.2 & 0.64\end{pmatrix}$

6 $\begin{pmatrix}3\\ -8\end{pmatrix}$ **7** $k=\pm\sqrt{3}$

8 a $|\mathbf{A}|\geqslant 5$ b $\frac{1}{5+3c^2}\begin{pmatrix}1 & -3c\\ c & 5\end{pmatrix}$

9 $\frac{1}{4}\begin{pmatrix}7 & 9\\ 15 & 1\end{pmatrix}$

10 a Proof. b $\frac{1}{3}\mathbf{B}^{-1}$

11 $\mathbf{A}^{-1}\mathbf{BA}$

12 a $\mathbf{Q}^{-1}$ b $\frac{2}{3}\mathbf{I}$

13 Proof.

14 a $\frac{1}{\det \mathbf{A}}$ b $\det \mathbf{A}$ c $(\det \mathbf{A})^n$

EXERCISE 7E

1 a i $x=-1, y=2$ ii $x=2.5, y=1.5$

b i $x=0, y=0.5$ ii $x=\frac{-3}{4}, y=\frac{5}{3}$

c i $x=-4, y=-3$ ii $x=-18, y=-13$

2 $k=3, -4$

3 a $k=-4$ b $x=\frac{5}{k+4}, y=\frac{2-2k}{k+4}$

4 a $a\neq -3, 5$

b $x=\frac{2a-7}{a^2-2a-15}, y=\frac{a-10}{a^2-2a-15}$

MIXED PRACTICE 7

1 B **2** C

3 $a=\frac{4}{3}, b=6, c=\frac{7}{3}, k=-3$

4 a $k=-2$

b Proof; for example:

$\mathbf{C}=\begin{pmatrix}1 & 0\\ 0 & 0\end{pmatrix}, \mathbf{D}=\begin{pmatrix}0 & 0\\ 1 & 0\end{pmatrix}$

5 $\mathbf{B}=\frac{1}{4}\begin{pmatrix}5 & 19\\ -2 & -10\end{pmatrix}$

6 a $a=-1$

b $\mathbf{A}^{-1}=\frac{1}{4(a+1)}\begin{pmatrix}4 & 2\\ -2 & a\end{pmatrix}$

7 a $\begin{pmatrix}0 & -4\\ 4 & 0\end{pmatrix}$

b Proof; $k=4$

c Proof.

8 a i $k=\frac{35}{3}$

ii $\mathbf{M}^{-1}=\frac{1}{3k-35}\begin{pmatrix}k & -5\\ -7 & 3\end{pmatrix}$

b $x=53, y=-30$

9 **a** Proof.

b $\frac{1}{6}\begin{pmatrix} -27 & -3 \\ 20 & 2 \end{pmatrix}$

10 **a** Proof; $k = 12$ **b** $p = -9$

11 **a** $\begin{pmatrix} x^2+9 & 7x-3 \\ 7x-3 & 50 \end{pmatrix}$

b Proof.

c $x = -1$

12 **a** Proof.

b For example:

$\mathbf{C} = \begin{pmatrix} 1 & 2 \\ 2 & 4 \end{pmatrix}, \mathbf{D} = \begin{pmatrix} 2 & 6 \\ -1 & -3 \end{pmatrix}$

Importantly, the second column of **C** must be a multiple k of the first, and the upper row of **D** must be a multiple $-k$ of the lower.

13 $x = -0.5,\ y = 0$

14 **a** Proof.

b For example: $\mathbf{A} = \begin{pmatrix} 1 & 2 \\ 3 & 4 \end{pmatrix}$; $\mathbf{B} = \begin{pmatrix} 2 & 0 \\ 1 & 1 \end{pmatrix}$

8 Matrix transformations

BEFORE YOU START

1 **a** $P'(-2, 1),\ Q'(-4, 1),\ R'(-2, 2)$

b $P'(-2, -1),\ Q'(-4, -1),\ R'(-2, -2)$

c $P'(2, 4),\ Q'(2, 8),\ R'(4, 4)$

2 $\mathbf{a} = 5\mathbf{i} + 3\mathbf{j} - 2\mathbf{k},\ \mathbf{b} = -\mathbf{j} + \mathbf{k}$

3 $\theta = 150°$

4 $\begin{pmatrix} 8 & 9 \\ -12 & 3 \end{pmatrix}$

5 **a** 14 **b** $\frac{1}{14}\begin{pmatrix} -3 & -5 \\ 4 & 2 \end{pmatrix}$

6 10

EXERCISE 8A

1 **a**

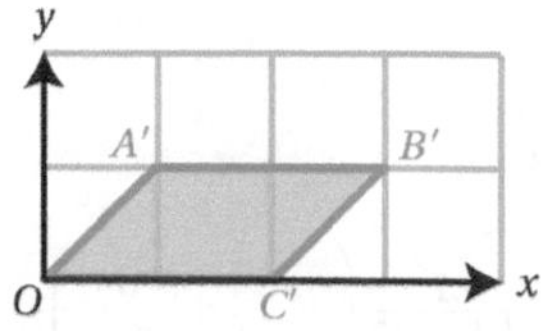

b

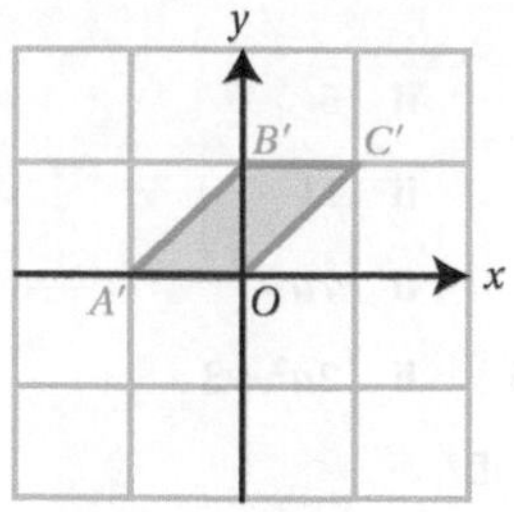

c

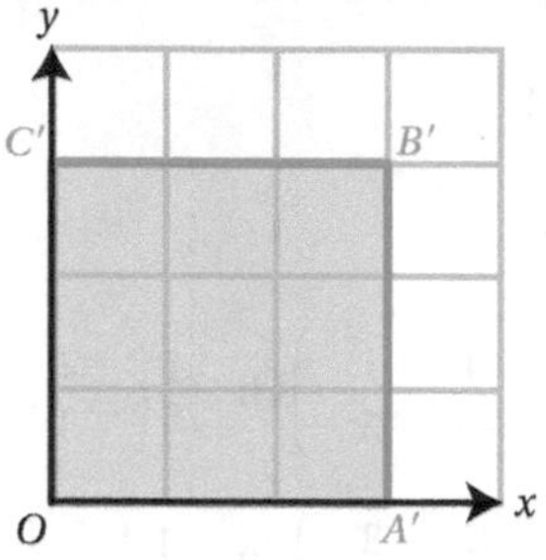

d

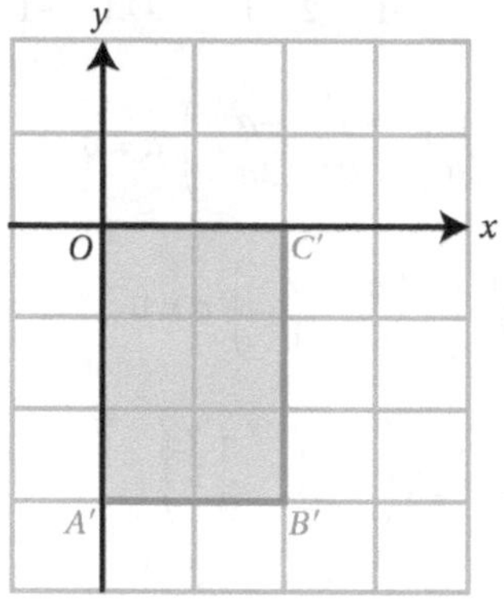

2 **a** $\begin{pmatrix} 2 & 1 \\ 0 & 2 \end{pmatrix}$ **b** $\begin{pmatrix} 2 & 3 \\ 1 & -1 \end{pmatrix}$

c $\begin{pmatrix} 1 & -2 \\ -4 & 1 \end{pmatrix}$ **d** $\begin{pmatrix} 2 & 1 \\ -1 & 1 \end{pmatrix}$

3 **a** Reflection; $y = -x$.

b Reflection; $y = x$.

c Neither.

d Reflection; $x = 0$.

e Rotation; 180°.

f Neither.

4 **a** **i** $(5, 3)$

ii $(-3, 5)$

b **i** $(3, 1)$ **ii** $(1, -3)$

c **i** $(21, 12)$ **ii** $(-28, 16)$

d **i** $\left(-2, -\frac{1}{2}\right)$ **ii** $\left(\frac{4}{3}, -\frac{1}{3}\right)$

5 **a** **i** $(6, -3)$ **ii** $(-6, 3)$

b **i** $(7, -2)$ **ii** $(-2, -7)$

c **i** $\left(\frac{2}{3}, -\frac{5}{3}\right)$ **ii** $\left(1, \frac{2}{5}\right)$

d **i** $(-2, -6)$ **ii** $(-2, -6)$

6 **a** $\begin{pmatrix} 0 & 1 \\ -1 & 0 \end{pmatrix}$; 270° rotation about the origin.

b A rotation through θ_1 followed by a rotation through θ_2 is equivalent to a rotation through $(\theta_1 + \theta_2)$.

c $\begin{pmatrix} 0 & -1 \\ 1 & 0 \end{pmatrix}$; 90° rotation about the origin.

d Two successive reflections are always equal to a rotation, of double the angle between the reflection lines, in the same direction as from the first line to the second. (In part **c**, $y = x$ to the y-axis is a 45° rotation, so the combination is the same as a 90° rotation).

7 **a** Reflection in the same line.

b Rotation 90° clockwise about the origin.

8 $P'(-1, 1)$, $Q'(-1, 6)$, $R'(2, 4)$

9 $A'(-3, 1)$, $B'(-4, -2)$, $C'(-3, -2)$.

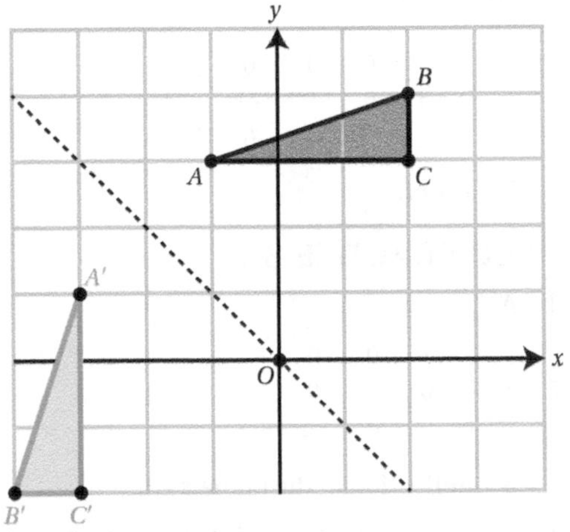

10 **a** $\mathbf{T} = \begin{pmatrix} 0 & 3 \\ 3 & 0 \end{pmatrix}$ **b** $P = (-3, 1)$

c 90 square units

EXERCISE 8B

1 **a** 60° **b** 53.1°

c 135° **d** 233°

2 **a** $y = \frac{\sqrt{3}}{3}x$ **b** $y = x$

c $y = 0.380x$

d $y = -0.268x$ or $y = -\tan(15°)x$

3 **a** Stretch with scale factor 6, parallel to x-axis.

b Stretch with scale factor 2, parallel to y-axis.

c Stretch with scale factor −3, parallel to x-axis.

d Stretch with scale factor −1.5, parallel to x-axis.

4 **a** $\begin{pmatrix} \frac{1}{2} & -\frac{\sqrt{3}}{2} \\ \frac{\sqrt{3}}{2} & \frac{1}{2} \end{pmatrix}$ **b** $\begin{pmatrix} 3 & 0 \\ 0 & 1 \end{pmatrix}$

c $\begin{pmatrix} -1 & 0 \\ 0 & 1 \end{pmatrix}$ **d** $\begin{pmatrix} 1 & 0 \\ 0 & \frac{1}{2} \end{pmatrix}$

e $\begin{pmatrix} -0.6 & -0.8 \\ -0.8 & 0.6 \end{pmatrix}$

f $\begin{pmatrix} -\frac{\sqrt{3}}{2} & \frac{1}{2} \\ -\frac{1}{2} & -\frac{\sqrt{3}}{2} \end{pmatrix}$ **g** $\begin{pmatrix} 5 & 0 \\ 0 & 5 \end{pmatrix}$

5 **a** $(1, -3)$ **b** $(\sqrt{2}, -2\sqrt{2})$

c $(1, 3)$ **d** $(3, -4)$

e $\left(\frac{\sqrt{3}-3}{2}, \frac{3\sqrt{3}+1}{2}\right)$ **f** $\left(1, -\frac{1}{3}\right)$

g $(1, -1)$

6 **a** $(2, -1)$ **b** $(2, 1)$

c $\left(\frac{1}{2} - \sqrt{3}, 1 + \frac{\sqrt{3}}{2}\right)$ **d** $(-0.5, 1)$

e $(-1, 1)$ **f** $\left(\sqrt{3} - \frac{1}{2}, 1 + \frac{\sqrt{3}}{2}\right)$

g $(-2, 2)$

7 $\left(\frac{\sqrt{2}}{2}, \frac{5\sqrt{2}}{2}\right)$

8 **a** $P(-6, 1)$, $Q(9, 1)$, $R(9, -2)$, $S(-6, -2)$

b 3

9 **a** $\begin{pmatrix} -0.8 & 0.6 \\ 0.6 & 0.8 \end{pmatrix}$ **b** $(-0.4, 2.8)$

10 **a** $\begin{pmatrix} 1 & 0 \\ 0 & 1.5 \end{pmatrix}$

b $(0, 0)$, $(1, 2)$, $\left(-3, \frac{2}{3}\right)$

EXERCISE 8C

1 **a** Line of invariant points $y = 0$; invariant line $y = 3x$.

b Invariant lines $y = 0$ and $y = 0.4x$.

c Invariant lines $y = 0.5x$ and $y = -x$

d Line of invariant points $y = x$; invariant line $y = 2x$.

2 **a** Line of invariant points $y = -3x$; invariant line $3y = x$.

b None.

c Line of invariant points $y = -0.5x$; invariant line $y = 2x$.

d Invariant lines $y = \frac{3 \pm \sqrt{21}}{6}x$.

e No invariant lines.

f Line of invariant points $x = 0$; invariant line $y = 0$.

3 **a** $a = -2$ **b** No; **A** is singular.

4 **a** $y = -x$ **b** $y = -x$ and $y = -3x$

5 **a** $y = -x$, $y = -4.5x$; Neither is a line of invariant points.

b $y = -x$, $y = -4.5x$

6 $\theta = \tan^{-1}\left(-\frac{1}{3}\right)$ or $\theta = \tan^{-1} 3$

7 **a** Line of invariant points $y = -2x$; invariant line $y = 0$.

b $b = 1$

8 **a** $p = -1$, $q = 2$

b **i** $|\mathbf{T}| = 0$ so $\mathbf{T}^{-1}$ does not exist.

ii Any point $(0, y, -y)$ will have image at the origin.

9 **a** $a = -2$, $b = -4$ **b** $y = -x$

EXERCISE 8D

1 **a** $\begin{pmatrix} 1 & 0 & 0 \\ 0 & 0 & -1 \\ 0 & 1 & 0 \end{pmatrix}$

b $\begin{pmatrix} \frac{\sqrt{3}}{2} & -\frac{1}{2} & 0 \\ \frac{1}{2} & \frac{\sqrt{3}}{2} & 0 \\ 0 & 0 & 1 \end{pmatrix}$

c $\begin{pmatrix} -1 & 0 & 0 \\ 0 & 1 & 0 \\ 0 & 0 & 1 \end{pmatrix}$ **d** $\begin{pmatrix} 1 & 0 & 0 \\ 0 & -1 & 0 \\ 0 & 0 & 1 \end{pmatrix}$

2 **a** $\left(1+\frac{\sqrt{3}}{2}, -1, \frac{1}{2}-\sqrt{3}\right)$ **b** $(-2, 1, 1)$

c $(2, -1, -1)$ **d** $(2, 1, 1)$

3 $(-2, -5, 1)$

4 **a** $\begin{pmatrix} 1 & 0 & 0 \\ 0 & -1 & 0 \\ 0 & 0 & 1 \end{pmatrix}$; $\begin{pmatrix} 1 & 0 & 0 \\ 0 & -1 & 0 \\ 0 & 0 & -1 \end{pmatrix}$

b $\begin{pmatrix} 1 & 0 & 0 \\ 0 & 1 & 0 \\ 0 & 0 & -1 \end{pmatrix}$; Reflection in $z = 0$.

5 $A'(1, 1, -1)$, $B'(1, 1, -3)$, $C'(1, 4, -1)$.

6 $(-\sqrt{2}, -3\sqrt{2}, -1)$

7 $\left(\frac{-3\sqrt{3}-1}{2}, 1, \frac{\sqrt{3}-3}{2}\right)$

8 $\begin{pmatrix} 0 & 1 & 0 \\ 1 & 0 & 0 \\ 0 & 0 & 1 \end{pmatrix}$

9 $\begin{pmatrix} 0 & -1 & 0 \\ 0 & 0 & -1 \\ 1 & 0 & 0 \end{pmatrix}$

10 $(1, -3, 4)$

11 **a** $\mathbf{T} = \begin{pmatrix} 0 & 1 & 0 \\ 0 & 0 & -1 \\ -1 & 0 & 0 \end{pmatrix}$

b Proof.

MIXED PRACTICE 8

1 A

2 **a** $\begin{pmatrix} 1 & 0 & 0 \\ 0 & 1 & 0 \\ 0 & 0 & -1 \end{pmatrix}$

b Reflection in the plane $z = 0$.

3 **a** Line of invariant points $y = -3x$; invariant line $y = -9x$.

b Invariant line $y = -x$.

4 **a** **i** Enlargement, scale factor -0.5.

ii Stretch parallel to the y-axis, scale factor 4.

b $\frac{1}{2}\begin{pmatrix} 2 & 0 & 0 \\ 0 & 1 & -\sqrt{3} \\ 0 & \sqrt{3} & 1 \end{pmatrix}$

5 **a** x-axis **b** $127°$

6 D

7 a $\mathbf{A} = \begin{pmatrix} 0 & 1 \\ -1 & 0 \end{pmatrix}$, $\mathbf{B} = \begin{pmatrix} 1 & 0 \\ 0 & 3 \end{pmatrix}$

b $\det \mathbf{C} = 0$

8 a Rotation 45° about O.

b Reflection in $y = x \tan 22.5°$.

c Rotation 90° about O.

d Identity transformation.

e Reflection in $y = x$.

9 $a = 1$ or -0.5, $b = 2$

10 a $\mathbf{T} = \begin{pmatrix} 3 & 0 \\ 0 & 2 \end{pmatrix}$

b i $k = 2$

ii Stretch parallel to the x-axis, scale factor 1.5.

11 a i $\begin{pmatrix} 0 & -1 \\ -1 & 0 \end{pmatrix}$ ii $\begin{pmatrix} 1 & 0 \\ 0 & 7 \end{pmatrix}$

b $\begin{pmatrix} 0 & -1 \\ -7 & 0 \end{pmatrix}$

c i Proof; $k = 12$

ii Scale factor $= \sqrt{12} = 2\sqrt{3}$; line of reflection $y = (\tan 105°)x$.

12 a i (0, 5)

ii $P'(2, 1)$, $Q'(30, -25)$, $R'(28, -26)$

b i $y = \frac{1}{2}x$ ii $y = -x$

13 a $y = 3x$; $y = (4 - 3k)x$

b i $k = 1, 2$

ii $y = 3x$

iii For $k = 1$: $y = x$; for $k = 2$: $y = -2x$

14 a $a = -1, b = 0$

b i $\mathrm{T} = \begin{pmatrix} 1 & -3 \\ 0 & 1 \end{pmatrix}$; the x-axis is invariant.

ii Enlargement with scale factor 2.

15 a $s = 3, t = 4$

b i Proof; $k = -16$

ii Enlargement scale factor 2 and rotation 45° about origin.

iii $\mathbf{B}^{15} = 16\,384 \begin{pmatrix} \sqrt{2} & \sqrt{2} \\ -\sqrt{2} & \sqrt{2} \end{pmatrix}$

9 Further applications of vectors

BEFORE YOU START

1 a $3\mathbf{i} - 5\mathbf{j}$ b $\begin{pmatrix} 3 \\ 0 \\ -2 \end{pmatrix}$

2 $a = -1, b = -2$ 3 $p = -\frac{3}{5}$, $q = -10$

4 113.6° and 246.4°

WORK IT OUT 9.1

Solution 3 is correct.

EXERCISE 9A

1 a i $\mathbf{r} = \begin{pmatrix} 4 \\ -1 \end{pmatrix} + \lambda \begin{pmatrix} 1 \\ 4 \end{pmatrix}$

ii $\mathbf{r} = \begin{pmatrix} 4 \\ 1 \end{pmatrix} + \lambda \begin{pmatrix} 2 \\ -3 \end{pmatrix}$

b i $\mathbf{r} = \begin{pmatrix} 1 \\ 0 \\ 5 \end{pmatrix} + \lambda \begin{pmatrix} 1 \\ 3 \\ -3 \end{pmatrix}$

ii $\mathbf{r} = \begin{pmatrix} -1 \\ 1 \\ 5 \end{pmatrix} + \lambda \begin{pmatrix} 3 \\ -2 \\ 2 \end{pmatrix}$

c i $\mathbf{r} = \begin{pmatrix} 4 \\ 0 \end{pmatrix} + \lambda \begin{pmatrix} 2 \\ 3 \end{pmatrix}$

ii $\mathbf{r} = \begin{pmatrix} 0 \\ 2 \end{pmatrix} + \lambda \begin{pmatrix} 1 \\ -3 \end{pmatrix}$

d i $\mathbf{r} = \begin{pmatrix} 0 \\ 2 \\ 3 \end{pmatrix} + \lambda \begin{pmatrix} 1 \\ 0 \\ -3 \end{pmatrix}$

ii $\mathbf{r} = \begin{pmatrix} 4 \\ -3 \\ 0 \end{pmatrix} + \lambda \begin{pmatrix} 2 \\ 3 \\ -1 \end{pmatrix}$

2 a i $\mathbf{r} = \begin{pmatrix} 4 \\ 1 \end{pmatrix} + \lambda \begin{pmatrix} -3 \\ 1 \end{pmatrix}$

ii $\mathbf{r} = \begin{pmatrix} 2 \\ 7 \end{pmatrix} + \lambda \begin{pmatrix} 2 \\ -9 \end{pmatrix}$

b i $\mathbf{r} = \begin{pmatrix} -5 \\ -2 \\ 3 \end{pmatrix} + \lambda \begin{pmatrix} 9 \\ 0 \\ 0 \end{pmatrix}$

ii $\mathbf{r} = \begin{pmatrix} 1 \\ 1 \\ 3 \end{pmatrix} + \lambda \begin{pmatrix} 9 \\ -6 \\ -3 \end{pmatrix}$

3 a i Yes ii Yes

b i Yes ii No

4 a No b No

c Yes d No

5 a $\mathbf{r} = (3\mathbf{i} - \mathbf{j} + 4\mathbf{k}) + \lambda(5\mathbf{i} - \mathbf{j} + 2\mathbf{k})$

b Proof.

c $2\sqrt{30}$

6 a $\mathbf{r} = \begin{pmatrix} 4 \\ -1 \\ 5 \end{pmatrix} + \lambda \begin{pmatrix} 3 \\ 8 \\ -3 \end{pmatrix}$

b No

7 $\mathbf{r} = \begin{pmatrix} 4 \\ 1 \\ 7 \end{pmatrix} + \lambda \begin{pmatrix} 1 \\ -6 \\ 2 \end{pmatrix}$

8 $\mathbf{r} = (4 + 2\lambda)\mathbf{i} + (-1 - \lambda)\mathbf{j} + (2 + 3\lambda)\mathbf{k}$

9 a Proof.

b (0, 3, 0)

10 a $\mathbf{r} = \begin{pmatrix} 7 \\ 1 \\ 2 \end{pmatrix} + \lambda \begin{pmatrix} -4 \\ -2 \\ 3 \end{pmatrix}$

b (−5, −5, 11) or (19, 7, −7)

11 a $\mathbf{r} = \begin{pmatrix} 2 \\ 1 \\ 4 \end{pmatrix} + \lambda \begin{pmatrix} 2 \\ -3 \\ 6 \end{pmatrix}$

b 7

c (−8, 16, −26) or (12, −14, 34)

EXERCISE 9B

1 a i $4x + 7y = 5$ ii $3x - 2y + 13 = 0$

b i $\frac{x-4}{2} = \frac{y+1}{-1} = \frac{z-5}{7}$

ii $\frac{x-1}{-1} = \frac{y-7}{1} = \frac{z-2}{2}$

c i $x = -1, \frac{y-5}{-2} = \frac{z}{2}$

ii $\frac{x-3}{7} = y, z = 6$

2 a i $\mathbf{r} = \begin{pmatrix} 0 \\ 2 \end{pmatrix} + \lambda \begin{pmatrix} 5 \\ 3 \end{pmatrix}$

ii $\mathbf{r} = \begin{pmatrix} 0 \\ -1 \end{pmatrix} + \lambda \begin{pmatrix} 3 \\ -4 \end{pmatrix}$

b i $\mathbf{r} = \begin{pmatrix} 0 \\ -\frac{17}{5} \end{pmatrix} + \lambda \begin{pmatrix} 5 \\ 3 \end{pmatrix}$

ii $\mathbf{r} = \begin{pmatrix} 0 \\ -\frac{4}{3} \end{pmatrix} + \lambda \begin{pmatrix} 3 \\ -2 \end{pmatrix}$

3 a i $\mathbf{r} = \begin{pmatrix} 2 \\ 2 \\ -1 \end{pmatrix} + \lambda \begin{pmatrix} 5 \\ 3 \\ 7 \end{pmatrix}$

ii $\mathbf{r} = \begin{pmatrix} -1 \\ 6 \\ 5 \end{pmatrix} + \lambda \begin{pmatrix} 4 \\ -1 \\ 3 \end{pmatrix}$

b i $\mathbf{r} = \begin{pmatrix} -1 \\ 0 \\ 1 \end{pmatrix} + \lambda \begin{pmatrix} 3 \\ -7 \\ -5 \end{pmatrix}$

ii $\mathbf{r} = \begin{pmatrix} 3 \\ -1 \\ 0 \end{pmatrix} + \lambda \begin{pmatrix} 2 \\ -4 \\ 5 \end{pmatrix}$

c i $\mathbf{r} = \begin{pmatrix} 11 \\ -1 \\ -2 \end{pmatrix} + \lambda \begin{pmatrix} 3 \\ 6 \\ 0 \end{pmatrix}$

ii $\mathbf{r} = \begin{pmatrix} -1 \\ 1 \\ 3 \end{pmatrix} + \lambda \begin{pmatrix} 5 \\ 0 \\ -2 \end{pmatrix}$

4 a Neither. b Parallel.

c Neither. d Same line.

5 $\mathbf{r} = \begin{pmatrix} 6 \\ 0 \end{pmatrix} + \lambda \begin{pmatrix} -3 \\ 5 \end{pmatrix}$

6 No

7 a $\frac{x-1}{3} = \frac{4-y}{2} = \frac{z+1}{3}$ b $\frac{1}{\sqrt{22}} \begin{pmatrix} 3 \\ -2 \\ 3 \end{pmatrix}$

8 a $p = -4.5, q = -2$

b (4, −0.5, −2) or (−8, −8.5, −2)

EXERCISE 9C

1 a i (10, −7, −2) ii (4.5, 0, 0)

b i No intersection. ii No intersection.

c i (3, 2, −5) ii No intersection.

2 Proof.

3 Proof; (1, −5, 2) **4** No **5** 4.5

6 Proof; area is 4.36 square units.

WORK IT OUT 9.2

Solution 2 is correct.

EXERCISE 9D

1 a i 9 ii 2
 b i −3 ii 3
 c i 5 ii −1

2 a i 1.12 ii 1.17
 b i 1.88 ii 1.13
 c i 1.23 ii 1.77

3 a i $-\frac{5}{2\sqrt{21}}$ ii $-\frac{20}{\sqrt{570}}$
 b i $-\frac{2}{\sqrt{102}}$ ii $\frac{1}{\sqrt{35}}$
 c i 0 ii 0

4 a i No ii Yes
 b i Yes ii No

5 a i 44.5° ii 56.5°
 b i 26.6° ii 82.1°

6 a 61.0°, 74.5°, 44.5°
 b 94.3°, 54.2°, 31.5°

7 87.7°

8 40.0°

9 a 1.3 radians (76.9°)
 b Proof.

10 a Proof.
 b 106.8°, 73.2°
 c $\frac{5}{4}$

11 a Proof. b 41.8°, 48.2° c $6\sqrt{5}$

12 $\left(\frac{64}{9}, \frac{4}{9}, \frac{19}{9}\right)$

13 a $\mathbf{r} = \begin{pmatrix} 2 \\ -4 \\ 3 \end{pmatrix} + \lambda \begin{pmatrix} 2 \\ -1 \\ 4 \end{pmatrix}$
 b $5\sqrt{2}$ c $\frac{3\sqrt{42}}{35}$ d 5.88

14 a $\left(\frac{5}{6}, \frac{19}{6}, \frac{9}{2}\right)$ b 48.5°
 c Proof. d $\frac{11\sqrt{11}}{6}$ (= 6.08)
 e 4.55

15 3

16 $\frac{\sqrt{230}}{5}$

17 a (9, −5, 8)
 b Proof.
 c (3, 4, −1)

18 a Proof.
 b $11\mu - 6\lambda = 21$
 c $\sqrt{30}$

MIXED PRACTICE 9

1 B

2 $\mathbf{r} = \begin{pmatrix} 7 \\ 0 \\ 3 \end{pmatrix} + \lambda \begin{pmatrix} 4 \\ -1 \\ -1 \end{pmatrix}$

3 a (−1, −3, 1)
 b 60.5°

4 Proof.

5 $x = 1.5$ or 2

6 53.6°

7 a $\mathbf{r} = \begin{pmatrix} \frac{1}{2} \\ -2 \\ \frac{4}{3} \end{pmatrix} + \lambda \begin{pmatrix} 2 \\ 3 \\ -2 \end{pmatrix}$
 b Yes, at $\left(\frac{11}{6}, 0, 0\right)$
 c 61.0°

8 a (8, 7, 1)
 b Proof.

9 No intersection (skew lines).

10 a $p = 19$
 b $\left(14, 13, \frac{8}{3}\right)$

11 a Proof.
 b Proof.
 c $\left(\frac{184}{11}, -\frac{32}{11}, -\frac{1}{11}\right)$ d 6.99

12 a Proof.
 b $\mathbf{r} = \begin{pmatrix} 3 \\ -2 \\ 4 \end{pmatrix} + \mu \begin{pmatrix} -2 \\ -3 \\ 2 \end{pmatrix}$
 c (5, 1, 2)
 d (9, 7, −2) or (−3, −11, 10)

13 a i $\begin{pmatrix} 1 \\ 0 \\ -1 \end{pmatrix}$
 ii Proof.
 b $\mathbf{r} = \begin{pmatrix} 6 \\ 2 \\ -4 \end{pmatrix} + \lambda \begin{pmatrix} 1 \\ 0 \\ -1 \end{pmatrix}$
 c (1, 2, 1)

14 $\frac{\sqrt{66}}{11}$

15 **a** $\mathbf{r}=\begin{pmatrix}-2\\3\\4\end{pmatrix}+\lambda\begin{pmatrix}3\\2\\-7\end{pmatrix}$

b $\frac{6\sqrt{341}}{31}$

16 **a** Proof.

b $(2+\sqrt{6}, -1-2\sqrt{6}, 2\sqrt{6})$ or $(2-\sqrt{6}, 2\sqrt{6}-1, -2\sqrt{6})$

17 **a** (4, 1, −2) **b** Proof.

c (1, 1, 2) **d** $\frac{5\sqrt{26}}{2}$

18 **a** Proof.

b **i** Proof. **ii** $t-3s+7=0$

iii $\sqrt{14}$

19 **a** **i** $\begin{pmatrix}3\\2\\-1\end{pmatrix}$

ii 85.9°

b **i** $\mathbf{r}=\begin{pmatrix}7\\-4\\10\end{pmatrix}+\mu\begin{pmatrix}3\\2\\-1\end{pmatrix}$

ii $\left(\frac{64}{7}, -\frac{18}{7}, \frac{65}{7}\right)$

10 Further calculus

BEFORE YOU START

1 $160\pi\ \text{cm}^3$ **2** 52.4 **3** 7500 m

EXERCISE 10A

1 **a** **i** 0.4π **ii** $\frac{128\pi}{7}$

b **i** 304.8π **ii** $\frac{18\pi}{7}$

c **i** $\frac{\pi}{2}$ **ii** $\frac{21\pi}{64}$

A **2** **a** **i** $\frac{\pi}{2}(e^2-1)$ **ii** $\frac{\pi}{2}(1-e^{-6})$

b **i** $\pi\left(\frac{e^4}{4}+e^2-\frac{1}{4}\right)$

ii $\pi\left(\frac{25}{2}-4e^{-2}-\frac{e^{-4}}{2}\right)$

c **i** 2π **ii** π

3 **a** **i** 32π **ii** $\frac{85}{2}\pi$

b **i** $\frac{96}{5}\pi$ **ii** $\frac{28\sqrt{2}}{3}\pi$

c **i** 3π **ii** $\frac{35}{3}\pi$

A **4** **a** **i** $\frac{\pi}{2}(e^4-1)$ **ii** $\frac{\pi}{8}(e^8+4e^4+3)$

b **i** $\pi\ln 2$ **ii** $\pi\ln 3$

c **i** $\frac{\pi^2}{2}$ **ii** $(\pi-2)\frac{\pi}{4}$

5 **a** (4, 0) **b** $\frac{11\pi}{6}$

6 75.4

A **7** π

8 2

9 $\frac{4\pi a^5}{15}$

A **10** $\frac{4}{3}$

11 **a** (0, 3), (4, 19)

b 630 (3 s.f.)

12 184 (3 s.f.)

13 **a** (1, 4), (9, 12) **b** $\frac{736\pi}{15}$

14 Proof.

15 Proof. Use $y=\frac{rx}{h}$.

A **16** $\pi\left(\frac{1}{2}e^2-2e+\frac{5}{2}\right)$

EXERCISE 10B

1 **a** **i** $\frac{1}{3}$ **ii** $\frac{13}{3}$

b **i** $\frac{4}{3}$ **ii** $\frac{1}{5}$

c **i** 17 **ii** 1995

A **2** **a** **i** $\frac{2}{\pi}$ **ii** 0

b **i** $e-1$ **ii** $\frac{1}{e-1}$

c **i** $\frac{38}{15}$ **ii** $\frac{1}{\sqrt{\pi}}$

3 $20\sqrt{T}$

4 1.5

5 **a** $\frac{a^2}{3}$ **b** $\frac{a}{\sqrt{3}}$

6 Proof.

A **7** 1 : 2

8 Proof.

9 a

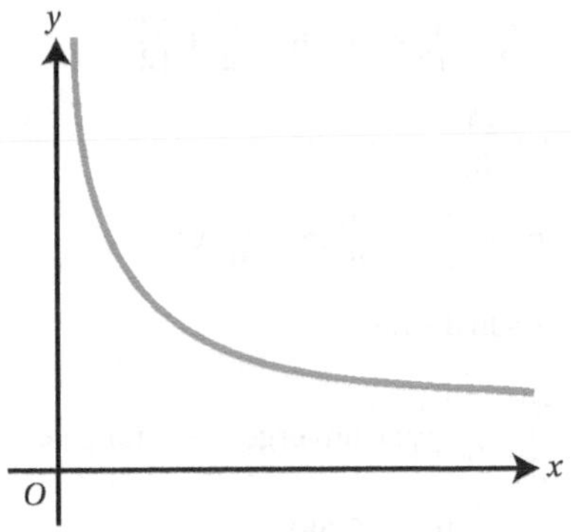

b Curve is concave up.

c Proof.

10 Not true. For example: $f(x) = x^2 - 1$ between -1 and 1.1.

MIXED PRACTICE 10

1 B **2** A

3 6 **4** $\frac{3}{2}$

5 4 **6** 57.8

7 $\frac{500\pi}{3}$ Ⓐ **8** $\frac{21\pi}{2}$

9 a $\frac{1}{a}$ **b** $\sqrt{a}$

10 $\frac{\pi a^5}{30}$

11 Proof.

Ⓐ **12 a** Proof **b** $\pi\left(\frac{5e^2}{6} - \frac{1}{2}\right)$

13 $\frac{\pi}{60}$

14 $3 + 5\sqrt{2}$

15 a (0, 12) and (4.5, −3.75)

b $\pi\int_0^{4.5} (14x^3 - 111x^2 + 216x)\,dx$

c 787

16 a

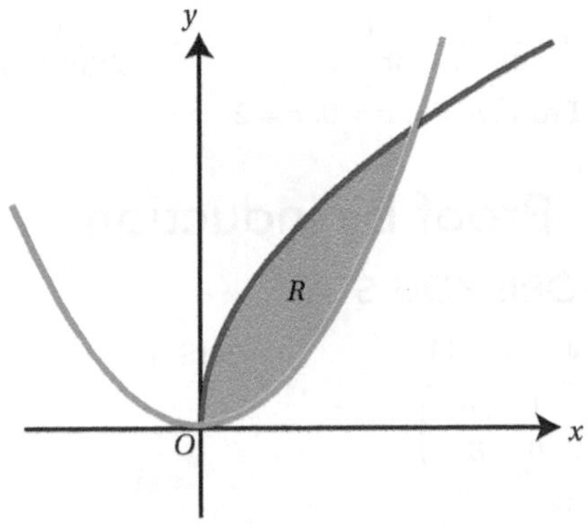

b $\frac{3\pi}{10}$ **c** $\frac{3\pi}{10}$

11 Series

BEFORE YOU START

1 $u_1 = 3,\ u_2 = 9,\ u_3 = 17$ **2** $3n^2(n+1)$

3 $16 + 96x + 216x^2 + 216x^3 + 81x^4$

EXERCISE 11A

1 a i 27 **ii** 39

b i 116 **ii** $\frac{195}{16}$

c i $21b$ **ii** $54p$

2 a i $\sum_{2}^{43} r$ **ii** $\sum_{3}^{30} 2r$

b i $\sum_{2}^{7} \frac{1}{2^r}$ **ii** $\sum_{0}^{5} \frac{2}{3^r}$

c i $\sum_{r=2}^{10} 7ra$ **ii** $\sum_{r=0}^{19} r^b$

WORK IT OUT 11.1

Solution 2 is correct.

EXERCISE 11B

1 a i 9455 **ii** 44 100

b i 1 379 609 **ii** 4750

2 a i $2n(4n+1)$ **ii** $\frac{n}{2}(3n+1)(6n+1)$

b i $\frac{n}{6}(n-1)(2n-1)$ **ii** $\frac{(n+1)^2(n+2)^2}{4}$

3–4 Proof.

5 a $3n^2 + 10n$ **b** 27

6 a Proof. **b** 25 225

7 Proof; $k = 3$

8 a Proof.

b $n(2n+1)(n-1)(2n+3)$

9 a Proof. **b** $\ln 3^{2660}$

10 Proof; $a = 4$

EXERCISE 11C

1–2 Proof.

3 a–b i Proof. **ii** $\frac{7}{78}$

Ⓐ **4 a** $\frac{1}{2r-1} - \frac{1}{2r+1}$

b Proof. **c** 1

5 Proof.

6 a Proof. b $\frac{3}{2}$

7 a Proof. b $\frac{2n}{3(2n+3)}$

c Proof.

8 $\frac{n^2+5n}{12(n+2)(n+3)}$

9 a Proof.

b $a=2, b=3, c=4$

c $\frac{5}{1848}$

10 a $\ln(n+1)$ b Proof.

EXERCISE 11D

1 a i $1-3x+\frac{9x^2}{2}$ ii $1+x^3+\frac{x^6}{2}$

b i $3x-\frac{9x^2}{2}+9x^3$ ii $-2x-2x^2-\frac{8}{3}x^3$

c i $-\frac{x}{2}+\frac{x^3}{48}-\frac{x^5}{3840}$ ii $3x^2-\frac{9x^6}{2}+\frac{81x^{10}}{40}$

d i $1-\frac{x^4}{18}+\frac{x^8}{1944}$ ii $1-2x^2+\frac{2x^4}{3}$

e i $1-2x-2x^2$ ii $1-\frac{4x}{3}+\frac{10x^2}{9}$

2 a i $\ln 3+\frac{x}{3}-\frac{x^2}{18}$ ii $-\ln 2-2x-2x^2$

b i $\frac{1}{8}+\frac{9x}{16}+\frac{27x^2}{16}$ ii $2-8x+48x^2$

c i $-3+\frac{8x}{27}+\frac{64x^2}{2187}$ ii $\frac{1}{16}+\frac{3x}{32}+\frac{27x^2}{256}$

3 a i $2x^2-x^3-\frac{2}{3}x^4+\cdots$

ii $-x-\frac{1}{2}x^2+\frac{25}{6}x^3+2x^4+\cdots$

b i $1+\frac{1}{2}x^2-\frac{1}{3}x^3+\frac{3}{8}x^4+\cdots$

ii $x+2x^2+\frac{23}{6}x^3+\frac{23}{3}x^4+\cdots$

c i $x-\frac{x^2}{2}+\frac{x^3}{6}-\frac{x^4}{12}+\cdots$

ii $-\frac{x^2}{2}-\frac{x^4}{12}+\cdots$

4 $\sum_{k=1}^{\infty}(-2)^{k+1}\frac{x^k}{k}; -\frac{1}{2}<x\leqslant\frac{1}{2}$

5 $2x+6x^2+\frac{23}{3}x^3+5x^4+\cdots$

6 Proof.

7 a $x+\frac{x^3}{3}+\cdots$

b $1+x+\frac{x^2}{2}+\frac{x^3}{2}+\frac{3}{8}x^4+\cdots$

8 a $-\frac{x^2}{2}-\frac{x^4}{12}+\cdots$ b $\frac{x^2}{2}+\frac{x^4}{12}+\cdots$

c $x+\frac{x^3}{3}+\cdots$

9 a $\ln 8-\frac{3}{2}x-\frac{39}{8}x^2-\frac{71}{8}x^3+\cdots$

b $y=\ln 8-1.5x$

10 a $\sum_{k=1}^{\infty}\frac{x^{2k-1}}{2k-1}$; Convergence is for $|x|<1$.

b $x=\frac{3}{5}$; $\ln 2\approx 0.688$.

MIXED PRACTICE 11

1 A

2 Proof.

3 1 278 270

4 a Proof; $p=-2$, $q=1$

b 149 080

5 a Proof. b Proof; $a=1$, $b=5$

6 a $1+3x+\frac{9}{2}x^2+\cdots$ b $3x^2$

7 C

8 a Proof. b 603 330

9 a Proof; $k=\frac{1}{4}$, $a=1$, $b=1$, $c=2$

b $n=10$

10 a, b Proof.

11 a Proof. b $1-\frac{1}{(n+1)!}$ c 1

12 a, b Proof.

13 a $\ln 4+\frac{3}{4}x-\frac{9}{32}x^2+\cdots$

b $\ln 4-\frac{3}{4}x-\frac{9}{32}x^2+\cdots$

c Proof.

14 a $\frac{1}{3r-1}-\frac{1}{3r+2}$ b $\frac{3n}{2(3n+2)}$

15 Proof; $a=5$, $b=9$, $c=2$

12 Proof by induction

BEFORE YOU START

1 a $3A+14$ b $24(5^n)$

2 $\begin{pmatrix}1 & 7a\\ 0 & 8\end{pmatrix}$

3 57

4 25(25!)

EXERCISE 12A

1–10 Proof.

EXERCISE 12B

1–7 Proof.

8 a $\begin{pmatrix} p & q \\ 0 & p \end{pmatrix}$

b Proof.

EXERCISE 12C

1–10 Proof.

EXERCISE 12D

1–7 Proof.

8 a $x < \frac{1}{2} - \frac{\sqrt{5}}{2}$ or $x > \frac{1}{2} + \frac{\sqrt{5}}{2}$

b Proof.

9 Proof.

MIXED PRACTICE 12

1–12 Proof.

13 a Proof; $a = 7$ b Proof.

14 a Proof; $A = 33$ b Proof.

15–18 Proof.

Focus on ... Proof 2

1, 2 Proof.

Focus on ... Problem solving 2

1 a $BP = BA\cos\theta$

b $\cos\theta = \frac{(\mathbf{a}-\mathbf{b})\bullet\mathbf{d}}{|\mathbf{a}-\mathbf{b}||\mathbf{d}|}$ where $\mathbf{d}$ is in the direction from B to P.

c Proof.

d $AP = 2\sqrt{14}$

2 i a $\overrightarrow{PQ} = \begin{pmatrix} -1-\mu \\ 3+2\mu \\ 8+3\mu \end{pmatrix} - \begin{pmatrix} -1+3\lambda \\ 2-\lambda \\ \lambda \end{pmatrix} = \begin{pmatrix} -\mu-3\lambda \\ 1+2\mu+\lambda \\ 8+3\mu-\lambda \end{pmatrix}$

b $\begin{pmatrix} -\mu-3\lambda \\ 1+2\mu+\lambda \\ 8+3\mu-\lambda \end{pmatrix} \cdot \begin{pmatrix} 3 \\ -1 \\ 1 \end{pmatrix} = 0 = 7-2\mu-11\lambda \quad (1)$

$\begin{pmatrix} -\mu-3\lambda \\ 1+2\mu+\lambda \\ 8+3\mu-\lambda \end{pmatrix} \cdot \begin{pmatrix} -1 \\ 2 \\ 3 \end{pmatrix} = 0 = 26+14\mu+2\lambda \quad (2)$

c $PQ = \sqrt{1^2+2^2+1^2} = \sqrt{6}$

ii Proof.

iii a $\mathbf{n} = \begin{pmatrix} 3 \\ -1 \\ 1 \end{pmatrix} \times \begin{pmatrix} -1 \\ 2 \\ 3 \end{pmatrix} = \begin{pmatrix} -5 \\ -10 \\ 5 \end{pmatrix}$

b Proof.

Focus on ... Modelling 2

1 a A – (*p*) – A – (*q*) – B; A – (*p*) – A – (*r*) – B

b A – (*p*) – A – (*p*) – A; A – (q) – B – (q) – A;
A – (q) – B – (r) – A; A – (r) – B – (q) – A;
A – (r) – B – (r) – A; A – (s) – C – (s) – A

2 a B – (q) – A – (p) – A – (s) – C;
B – (r) – A – (p) – A – (s) – C

b B – (q) – A – (p) – A – (q) – B;
B – (q) – A – (p) – A – (r) – B;
B – (r) – A – (p) – A – (q) – B;
B – (r) – A – (p) – A – (r) – B

3 a 2 b 11

4 a 6 b 24

5 a 1 b 2 c 2

Cross-topic review exercise 2

1 $-2+3i$

2 $-\frac{15}{4}$

3 a -7 b 28

4 I

5 a $\mathbf{r} = (3\mathbf{i}+12\mathbf{j}-5\mathbf{k}) + t(-4\mathbf{i}-11\mathbf{j}+11\mathbf{k})$

b $(-5, -10, 17)$

6 a 41.5° b 7.07

7 $\frac{160\pi}{3}$

8 $\frac{37\pi}{2}$

9 **a** **i** $\mathbf{R} = \begin{pmatrix} 1 & 0 & 0 \\ 0 & -1 & 0 \\ 0 & 0 & 1 \end{pmatrix}$

ii $\mathbf{S} = \begin{pmatrix} 1 & 0 & 0 \\ 0 & 0 & -1 \\ 0 & 1 & 0 \end{pmatrix}$

b $\begin{pmatrix} 1 & 0 & 0 \\ 0 & 0 & -1 \\ 0 & -1 & 0 \end{pmatrix}$

c (2, −5, 3)

10 **a** $\mathbf{r} = (3\mathbf{i} - 7\mathbf{j} + \mathbf{k}) + \lambda(\mathbf{i} + \mathbf{j} + 2\mathbf{k})$

b −1

11 Skew.

12 **a** $\sqrt{30}$ **b** 73° **c** (4, 3, 3)

13 Proof; $a = 2$, $b = 3$

14 **a** Proof; $A = 3$ **b** Proof. **c** $\frac{1}{6}$

15 Proof.

16 **a, b** Proof.

17 **a** 2

b Proof.

18 9.20

19 **a** $\mathbf{r} = \begin{pmatrix} 2 \\ -1 \\ 3 \end{pmatrix} + s\begin{pmatrix} 3 \\ 4 \\ 5 \end{pmatrix}$

b Proof.

c **i** $P\left(-\frac{5}{2}, -7, -\frac{9}{2}\right)$; $Q\left(-\frac{11}{3}, -\frac{14}{3}, -\frac{17}{3}\right)$

ii $\frac{7\sqrt{6}}{6}$

20 $\frac{9}{4}$

21 **a** $x + \frac{x^3}{12} + \cdots$ **b** $-2 < x < 2$ **c** $\frac{13}{12}$

22 **a** $k^2 + 7k + 14$

b Proof.

23 −30

24 **a** **i** $\mathbf{r} = \begin{pmatrix} 1 \\ 2 \\ 3 \end{pmatrix} + \lambda\begin{pmatrix} 2 \\ 3 \\ 6 \end{pmatrix}$

ii $\mathbf{r} = \begin{pmatrix} -4 \\ 1 \\ -1 \end{pmatrix}$

b **i** $y = \frac{1}{2}x - 3k$ **ii** Proof; $d = \frac{8\sqrt{5}k}{5}$

25 **a, b** Proof.

c (−25, 387)

Practice paper

1 C

2 A

3 **a** $-1 - i$ **b** $\frac{4}{5} - \frac{7i}{5}$

4 **a** Proof. **b** −2

5 $\frac{5}{6}$

6 Proof.

7 Proof; $a = 7$, $b = 16$, $c = 3$

8 Proof.

9 **a** $\frac{2}{11}$ **b** $\frac{1}{2 - 11c}\begin{pmatrix} -3 & 1-c \\ -2 & 3c \end{pmatrix}$

10 **a** 2; $\frac{5\pi}{6}$

b

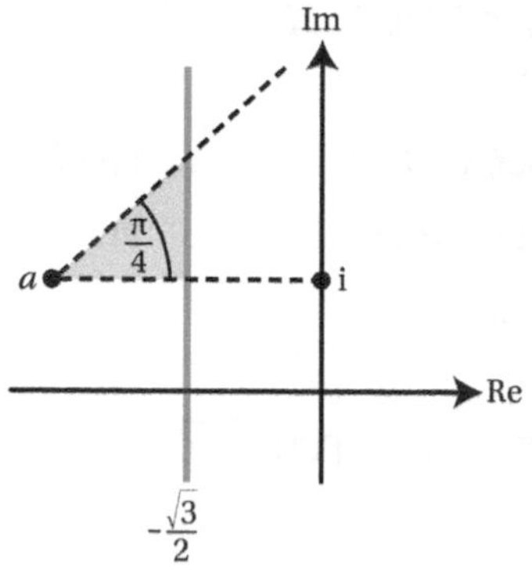

11 **a** $y^2 = 6x + 9$ **b** $r = \frac{13}{3}$

12 **a** Proof.

b For example, $x = \frac{3\pi}{2}$

13 **a** $y = \pm\frac{\sqrt{3}}{4}x$

b **i** Proof.

ii $m = \pm\frac{1}{2}$

14 **a** $\begin{pmatrix} 1 & -2 & -2 \\ 0 & 0 & -3 \\ 1 & 0 & 2 \end{pmatrix}$

b $(-13, -12, 7)$

c Proof.

15 Proof.

Glossary

Argand diagram: A diagram used to represent complex numbers geometrically.

Argument: The angle relative to the real axis of a complex number on an Argand diagram.

Associative: Matrix multiplication being associative means that $(\mathbf{AB})\mathbf{C} = \mathbf{A}(\mathbf{BC})$.

Axis of rotation: A line about which a rotation takes place in three dimensions.

Complex number: A number that can be written in the form $x + iy$, where x and y are real.

Complex conjugate pair: If $z = x + iy$, then the complex conjugate of z, $z^* = x - iy$.

Commutative: Matrices $\mathbf{A}$ and $\mathbf{B}$ are commutative (under multiplication) if $\mathbf{AB} = \mathbf{BA}$.

Conformable: Matrices are conformable if they have appropriate dimensions to allow an operation to take place. For example, two matrices are conformable for addition if they have the same dimensions, but for multiplication two matrices are conformable if the number of columns in the left matrix is the same as the number of rows in the right matrix.

Determinant: The determinant of a 2×2 matrix is the product of the lead diagonal elements minus the product of the reverse diagonal elements.

Discriminant: The part of the quadratic formula given by $b^2 - 4ac$.

Domain: The set of input values to a function.

Element: An entry in a rectangular matrix array.

Identity: A number that, for a given operation, produces no changes.

Identity matrix I: A square matrix with 1 on each element of the lead diagonal (upper left to lower right) and a zero everywhere else.

Image: The new shape/point that results after a transformation has been applied.

Invariant point: Any point that is unaffected by a transformation. The origin O is an invariant point because every linear transformation maps the origin to itself.

Inverse matrix: The inverse of a square matrix $\mathbf{A}$ has the property that $\mathbf{AA}^{-1} = \mathbf{A}^{-1}\mathbf{A} = \mathbf{I}$.

Line of invariant points: An invariant line l is a line for which the image of any point on l is also on l.

Modulus: The distance from the origin of a complex number on an Argand diagram.

Null matrix: *See* Zero matrix.

Object: The original shape before a transformation is applied.

Plane of reflection: A plane in which a shape is reflected in three-dimensional space.

Polar coordinates: A way of describing the position of a point by means of its distance from the origin and the angle relative to a fixed line.

Radian: Radians are the most commonly used unit of angle in advanced mathematics; 1 radian is the angle subtended at the centre of a circle of radius r by an arc of length r. 1 radian is about 57°.

Range: The set of outputs of a function.

Real polynomial: A polynomial all of whose coefficients are real numbers.

Series: The result of adding all the terms of a sequence.

Singular: A matrix $\mathbf{A}$ with det $\mathbf{A} = 0$ is called singular and has no inverse.

Skew lines: Lines that are not parallel but do not intersect.

Square matrix: A matrix that has the same number of rows and columns.

Transpose: The transpose of matrix $\mathbf{A}$ is denoted by $\mathbf{A}^{\mathrm{T}}$ and is such that the rows of $\mathbf{A}^{\mathrm{T}}$ are the columns of $\mathbf{A}$.

Transposing: Flipping a matrix around its lead diagonal, exchanging rows for columns.

Unit square: The square with vertices at $(0, 0)$, $(1, 0)$, $(1, 1)$ and $(0, 1)$.

Volume of revolution: A 3D shape formed by rotating a curve around the x- or y-axis.

Zero matrix or null matrix, Z: A matrix in which every element is equal to zero.

Index

Acknowledgements

The authors and publishers acknowledge the following sources of copyright material and are grateful for the permissions granted. While every effort has been made, it has not always been possible to identify the sources of all the material used, or to trace all copyright holders. If any omissions are brought to our notice, we will be happy to include the appropriate acknowledgements on reprinting.

Thanks to the following for permission to reproduce images:

Cover image: Peter Medlicott Sola/Getty Images
Back cover: Fabian Oefner www.fabianoefner.com

ARUNAS KLUPSAS/Getty Images; Dovapi/Getty Images; Jens Karlsson/Getty Images; PhotoFritz/Getty Images; Daniel Grill/Getty Images; Arran/photomuso/Getty Images; NonChanon/Getty Images; Saul Gravy/Getty Images; boygovideo/Getty Images; Earth Imaging/Getty Images; Philippe Lejeanvre/Getty Images; Richard McManus/Getty Images; Bryan Hollar/Getty Images; wragg/Getty Images; Paul Fearn/Alamy Stock Photo.

AQA material is reproduced by permission of AQA.